XXXIII Siberian Thermophysical Seminar (STS-33)

MATEC Web of Conferences Volume 115 (2017)

Novosibirsk, Russia
6 - 8 June 2017

Editors:

S.V. Alekseenko

ISBN: 978-1-5108-4432-2

Printed from e-media with permission by:

Curran Associates, Inc.
57 Morehouse Lane
Red Hook, NY 12571

Some format issues inherent in the e-media version may also appear in this print version.

Printed by Curran Associates, Inc. (2017)

For additional information, please contact EDP Sciences – Web of Conferences at the address below.

EDP Sciences – Web of Conferences
17, Avenue du Hoggar
Parc d'Activité de Courtabœuf
BP 112
F-91944 Les Ulis Cedex A
France

Phone: +33 (0) 1 69 18 75 75
Fax: +33 (0) 1 69 28 84 91

contact-edps@webofconferences.org

Additional copies of this publication are available from:

Curran Associates, Inc.
57 Morehouse Lane
Red Hook, NY 12571 USA
Phone: 845-758-0400
Fax: 845-758-2633
Email: curran@proceedings.com
Web: www.proceedings.com

TABLE OF CONTENTS

PREFACE

Proceedings of the STS-33 XXXIII Siberian Thermophysical Seminar

The Siberian Thermophysical Seminar has been traditionally held at Kutateladze Institute of Thermophysics SB RAS in Novosibirsk Akademgorodok since 1960. In 2017, the Seminar will be devoted to several anniversaries at once. The Kutateladze Institute of Thermophysics SB RAS is celebrating its 60 Year Anniversary. The conference took place in the city of Novosibirsk.

The conference will also host sections dedicated to the 100 Year Anniversary since the birth of Academician Ivan I. Novikov, the first Director of the Institute of Thermophysics, and the 80 Year Anniversary of Academician Eduard P. Volchkov. Works of these outstanding scientists are inseparably connected with the development of Russian thermal science and our Institute. The organizing committee sincerely believes that the XXXIII Siberian Thermophysical Seminar will attract young scientists to solving urgent scientific problems.

The present Conference covers the following topics: heat transfer and hydrodynamics in single phase and multiphase flows, phase transitions, reacting flows, detonation processes, experimental and numerical techniques in thermophysics and physical hydrodynamics, heat transfer and hydrodynamics on micro- and nanoscale and in industrial processes. The proceedings contain 108 papers grouped by a few topics.

The scientific committee appreciates the enormous work of the editorial board and reviewers in the preparation of this volume. We would like to express our sincere thanks to all authors for their research contributions, and also to organizers of the conference for their valuable spadework.

S.V. Alekseenko, V.E. Nakoryakov.

edp sciences

MATEC Web of Conferences

Statement of Peer review

In submitting conference proceedings to MATEC Web of Conferences, I certify to the Publisher that I adhere to the **Policy on Publishing Integrity** of the journal in order to safeguard good scientific practice in publishing.

1. All articles have been subjected to peer review administered by the proceedings editors.
2. Reviews have been conducted by expert referees, who have been requested to provide unbiased and constructive comments aimed, whenever possible, at improving the work.
3. Proceedings editors have taken all reasonable steps to ensure the quality of the materials they publish and their decision to accept or reject a paper for publication has been based only on the merits of the work and the relevance to the journal.

Title, date and place of the conference

STS-33, June 6th – June 8th, 2017, Novosibirsk, Russia

Proceedings editor(s): Professor S.V. Alekseenko

Date and editor's signature June 9th, 2017

17, avenue du Hoggar - PA de Courtabœuf – BP 112 - 91944 Les Ulis Cedex A (France)
Tél. : 33 (0)1 69 18 75 75 - Fax : 33(0)1 69 07 45 17 – www.edpsciences.org

Unsteady heat and mass transfer in gels, used as media for immobilizing micro bio-objects

Boris Pokusaev[1,*], *Andrey* Vyazmin[1,2], *Sergey* Karlov[1], *Dmitry* Nekrasov[1] and *Dmitry* Skladnev[3]

[1]Moscow Polytechnical University, Department of Chemical Engineering, 105066, Moscow, Russia
[2]Scientific Research Institute of Rubber Industry, 141312, Sergiev Posad, Russia
[3]Research Centre of Biotechnology RAS, 117312, Moscow, Russia

Abstract. The idea that gels can be used as a carrier during the formation of ordered bio-structured elements for medical additive technology seems quite promising. For experimental study we used agarose based gels. Such gels are appropriate for additive technologies used for the creation of matrix of bioreactors by 3D printing. We investigated the formation kinetics of different density gels, and studyid the diffusion in the gels using optical method. The paper presents new data on the thermophysical properties of the formed agarose based gels.

1 Introduction

The idea of growing tissues and organs in vitro using stem cells is not new. However, in order to implement this idea, it is necessary to create special bioreactors which would be capable of maintaining the required temperature, pH level, osmotic pressure, supplying cells with nutrients and oxygen, and removing their metabolic products [1]. The use of the additive manufacturing method to create tissues from stem cells [2] seems to be very promising. Such works are already carried out in the area of 3D-bioprinting [3].

Currently gels are considered as high-potential structuring material for formation of bioreactors with immobilized cells. Gel is a dispersive system with liquid dispersing medium, and the dispersion phase makes up a spatial structured mesh due to intermolecular interaction. The properties of gels allow forming bioreactors of complex configurations based on layer-by-layer application of gels of various concentrations and compositions. Agarose and other gels are used in microbiology to grow microorganisms [4].

The properties of gels are known to depend on the composition of a gel-forming medium and preparation method. There are many publications relating to the studies on the methods of gel synthesis and investigation of their properties (see, for instance, [5]). However, many properties of gels are common in terms of the application technology. For instance, the most significant peculiarities are manifested in the mass transfer processes in gels due to their instability and anisotropy [6].

Agarose gel is a typical gel which is formed from the solution when the temperature is lowered. We have studied the following aspects: structure of links formed in the disperse

* Corresponding author: pokusaev2005@yandex.ru

 DOI: 10.1051/matecconf/201711501001

phase of gel, physical-chemical properties of gels, as well as technology of gelation for various uses [7]. Nevertheless, many properties of gels, defining the possibility of their use for the creation of bioreactors based on additive technology, require further study. Here the aim is to carry out investigations of the kinetic of agarose gel formation depending on the concentration, determination of diffusion, and thermophysical properties of gels.

2 Additive method to form bioreactor

Consider the creation of technologically simple method of producing artificial biomaterials with application of 3D bioprinting. The problem is that the cultures of living cells are sensitive to external and internal influences. Deviation of the growing conditions from the optimum parameters leads to an inhibition of metabolic activity. To ensure minimal cells damage during 3D bioprinting a specific approach is proposed in order to avoid changes in the concentrations of the growth factors, and to protect immobilized cells from excess contact with oxygen. The approach is named "gel in gel". It is based on technology, where the basis for the layering of portions of the gel containing immobilized cells is in a thicker protective very dilute agarose gel, prepared with the addition of optimal concentration of essential growth factors. The scheme of bioreactor formation is shown in Fig. 1.

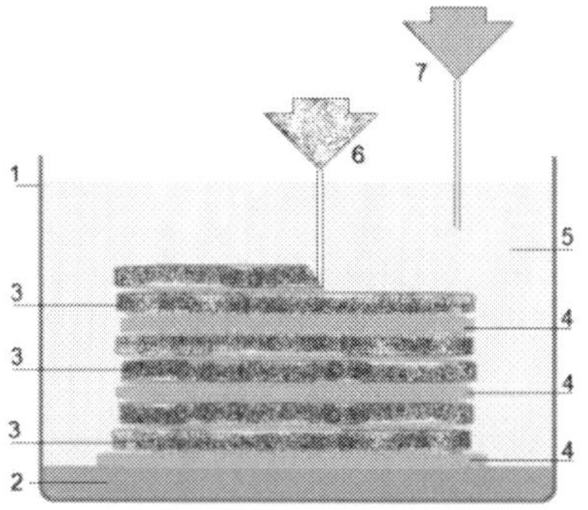

Fig. 1. The scheme of bioreactor created by using bioprinting method: 1 – the thermostatic cell, 2 – underlying layer of sterile silica gel, 3 - layers of agarose gel with immobilized living cells, 4 – drain layers of sterile agarose gel, 5 – layer protection of low concentration agarose gel, 6 – device for the 3D layer formation with target cells, 7 - device for 3D-forming layers of sterile agarose gel.

It is proposed to use horizontal and vertical layers of gels of different density and composition. Gel layers containing immobilized cells alternate with the drainage layers, containing a gel with a low concentration of agarose. Drainage layers should be considered as an analogue of the natural microvascular networks in tissue.

Technological process of creating bioreactor is the following: gel base of sterile silica gel is applied to the bottom of the thermostatic cell by 3D bio-printing at a temperature of 37°C. The working volume of thermostatic cell is filled with protective layer of agarose gel with low density (0.3% mass concentration of agsrose). Needle is injected into the cell for gel delivery for the formation of artificial biostructures. Agarose gel comprising a suspension of cells serves as a gel base at a temperature of 42°C. Agarose gel comprising cells alternates with layers of sterile agarose gel of low concentration.

3 Experimental equipments

In order to study the kinetics of gel formation and diffusion mass transfer in the gels of different concentrations, we used a spectroscopy method. The illumination system is made of 35W Xenon lamp emitting light waves within the range of 185–2000 nm. Optical fibers

0.2 mm in diameter are used. The basic part of the experimental setup consists of spectrophotometer USB2000+VIS-NIR including Sony ILX511 2048-element detector. An optical cuvette of 10×10 mm and 45 mm high contains the sample of gel and is installed in an air thermostat. A gel formation from the solution occurs when decreasing the temperature in the thermostat. The gel temperature range is chosen from 50 to 20°C.

In order to measure the thermophysical properties of agarose gels, the experimental setup is equipped with a special heat flow sensor. Using a gradient heat flow sensor based on the Seebeck thermoelectric effect allows directing measurement of the amount of heat flow. The sensor is mounted on the side surface of a glass container of cylindrical shape with a diameter of 50 mm and 95 mm high. Thermocouples are fixed to the diameter cap to control the distribution of the temperature fields in the volume during gel formation. The working area is placed in a thermostat, where a constant temperature of 20°C is maintained.

In experiments we used «Chemapol» agarose. The gels are obtained by mixing agarose with distilled water. Agarose gels with 0.6–1.5% of agarose are used in the experiments.

4 Results and discussion

Figure 2 shows the experimental data on the change of wavelength, which coincides with the maximum of the light transmission spectrum, depending on the temperatures and the concentrations of agarose equal to 0.8% and 1.5%. Along with the decrease of the temperature, the wavelengths suddenly increase at high temperatures, and then again take new constant values at low temperatures. This is caused by the increase of dispersion and light absorption of the selected wavelength when non-uniformities are less than a wavelength that is caused due to the change of the structure of the transmission medium.

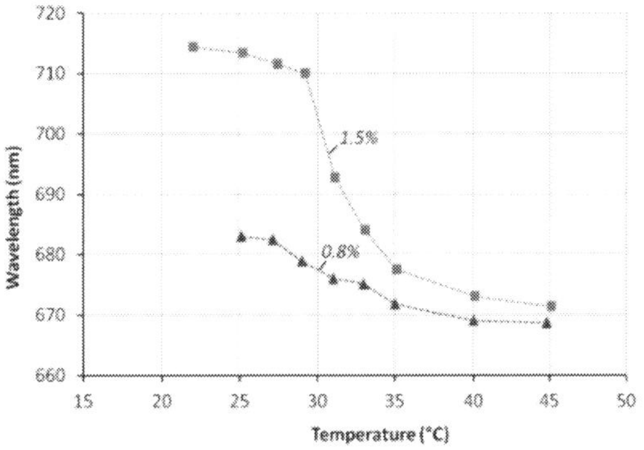

Fig. 2. The change in wavelength corresponding to the maximum of the spectrum of the transmitted light, depending on the temperature in the process of agarose gels formation

The dependencies shown in Fig. 2 are not equidistant for various agarose concentrations. This means that at high temperatures the solution shows weak dependency on the concentration. In the gel state, the structure of microdisperse medium significantly depends on the concentration.

Based on experimental data obtained by solving the inverse heat conduction problem, the average thermophysical coefficients for agarose gels of different concentrations within the temperature range of their formation are calculated. A mathematical model of non-stationary cooling of 1D cylindrical sample infinite in length with constant thermophysical properties (heat capacity and thermal conductivity) is considered for calculations.

To solve the inverse heat conduction problem, we used the extreme formulation and the principle of iterative regularization. The direct heat conduction problem is solved using an implicit difference scheme. The minimization problem is solved by the method of coordinate-wise descent.

Thermal conductivity coefficients and bulk heat capacities averaged over the gel formation time for gels with different concentrations are obtained. Their values are given in Table 1. The results show that within the experimental accuracy the heat capacity of gels of different concentrations is constant. Under the same conditions, the average thermal conductivity of gels decreases with increasing concentration of agarose.

Table 1. Averaged thermophysical properties of gels for different concentrations of agarose

Mass concentration of agarose, %	Average thermal conductivity of the gel, W/(m·K)	Average heat capacity of the gel, kJ(m^3·K)
0.6	0.538	2129.7
1.0	0.501	2121.8

The diffusion coefficients were determined experimentally by the method of moving boundaries in agarose gels with mass concentrations from 0.6 to 1.0% depending on the concentration of Peptone based LB bullion. The main results are shown on Fig. 3. Mass concentrations of the bullion ranged from 2.5 to 10%. This is bullion with agarose gel standard medium for cultivation of microorganisms. It is established that the diffusion coefficients have an order of magnitude less than that for diffusion in liquids.

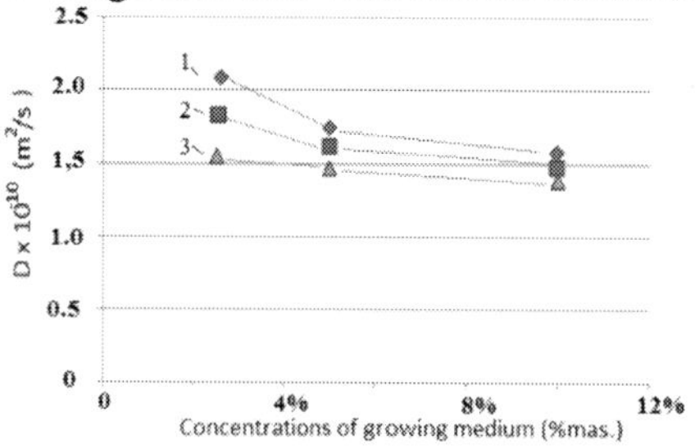

Fig. 3. The dependence of the diffusion coefficient of gel on concentration for nutrient bouillon. Notation: 1 –agarose gel concentration 0.6%; 2 – 0.8%; 3 – 1.0%.

The obtained data allow evaluating just qualitatively the dependence of diffusion coefficients on the concentration of agarose gel and nourishing medium. With the growth of both concentrations, the diffusion coefficients slightly decrease.

This work was financially supported by the Russian Science Foundation (project No. 15-19-00177).

References

1. C.A.V. Rodrigues, T.G. Fernandes, M.M. Diogo, C.L. da Silva, J.M.S. Cabral, Biotech. Adv. **29**, 815 (2011)
2. K. Hölzll, Sh. Lin, L. Tytgat, S. Van Vlierberghe, L. Gu, A. Ovsianikov, Biofabr. **8**, 032002 (2016)
3. M.Y. Wang, J.K. He, Y.X. Liu, M. Li, D. Li, Zh. Jin, Int. J. Bioprint. **1**, 15 (2015)
4. A.P. Hitchens, M.C. Leikind, J. Bacter. **37**, 485 (1939)
5. R.G. Weiss, P. Terech, *Molecular Gels: Materials with self-assembled fibrillar networks.* (Springer Science & Business Media, 2006)
6. B. Amsden, Macromolec. **31**, 8382 (1998)
7. G.A. Santos, *A manual for the processing of agar from Gracilaria* (ASEAN Regional Small-Scale Coastal Fisheries Development Project, Manila, 1990)

Multiscale processes of the flow and non-equilibrium interphase heat and mass transfer self-organization in multiphase systems

Vladimir V. Kuznetsov*[1,2]

[1]Kutateladze Institute of Thermophysics SB RAS, 630090 Novosibirsk, Russia
[2]Novosibirsk State University, Novosibirsk, 630090 Russia

Abstract. Multiscale processes of self-organization of the flow and non-equilibrium heat and mass transfer during the separation of a binary mixture in a multiphase system formed by a structural packing and a counter-flow vapor-liquid flow are considered. The conditions for the appearance of large-scale convective flow during separation of the binary mixture of refrigerants R21-R114 in a distillation column with Koch 1Y structure packing are obtained. The physical properties of this mixture are similar to the properties of air component at cryogenic temperatures.

1 Introduction

One of the main problems hindering the development of high-efficiency multiphase heat and mass transfer equipment with unique characteristics on the specific surface and the intensity of heat and mass transfer processes is the multiscale flow, caused by the interaction of processes at the micro- and macrolevel. For such equipment, the self-organization of the gas-liquid flow determines the intensity of heat and mass transfer processes, both at the microscale and the macro scale, due to blocking of the heat and mass transfer surface during the development of large-scale flow unevenness.

An example of multiphase heat and mass transfer equipment, when a high level of the specific surface and the intensity of heat and mass transfer processes are essential for increasing the productivity, are the distillation columns and two-phase compact heat exchangers [1]. Distillation is the most effective way for multicomponent mixture separation into pure components differing in boiling points, and it is carried out during counter-current contact of vapor and liquid [2]. Recently, a problem of increasing of the productivity of distillation columns has necessitated the use of structural packing to create a direct contact between the vapor and liquid. These packings provide less pressure loss and increase the phase contact surface compared to irregular packing or trays. For structure packing application, the physical processes in complex channel systems are studied, as well as methods for calculating the efficiency of binary mixture separation for the packing of various types, developed in [3].

* Corresponding author: vladkuz@itp.nsc.ru

The process of vapor enrichment with volatile distillate fraction and their depletion of liquid with the release of separation products in the upper and lower parts of the column is nonequilibrium. It occurs due to the difference in a concentration of the components in the liquid and vapor. Therefore, the occurrence of large-scale unevenness in the flow and fraction concentration for the liquid and gas causes the deterioration in the separation efficiency [4, 5]. In this paper, the processes of self-organization of the flow and nonequilibrium heat and mass transfer during separation of the binary mixture in complex channels systems of the structure packing and their influence on the process of separation of the R21-R114 binary mixture in a column filled with the Koch 1Y structure packing are considered. The physical properties of this mixture are similar to the properties of air components at cryogenic temperature.

2 Multiscale processes of counter-current gas-liquid flow self-organization in structure packing

Let us consider the counter-current flow of liquid and vapor in the distillation column filled with structure packing during binary gas mixture separation. In such a column, the liquid flows down under the action of gravity in the form of a thin film over the surface of structure packing, and the vapor rises upward under the conditions of nonequilibrium heat and mass transfer at the interphase. Structure packing represents the corrugated perforated sheets, tightly pressed together. It forms a complex system of channels for the vapor flow. The presence of a microstructure on the surface of the packing increases the wettability and uniformity of the flow as far as the heat and mass transfer coefficient for the liquid phase. The use of structure packing allows obtaining the specific surface area up to 750 m^2/m^3 with porosity of 0.98, and intense heat and mass transfer at a micro level that is achieved by a small thickness of the liquid film and spontaneous twist of the vapor flow.

The system under consideration is multiphase with fixed solid phase and counter-current vapor-liquid flow in the packing channels. The main cause of the spontaneous self-organization of the counter-current multicomponent flow with the formation of large-scale structures in complex channel system of the structure packing during the separation process is the convective flows arising in the column with a negative vapor density gradient along the gravity vector. The reduction of gas density is conditioned by the transition of the volatile distillate fraction from the liquid to vapor due to the mass transfer process.

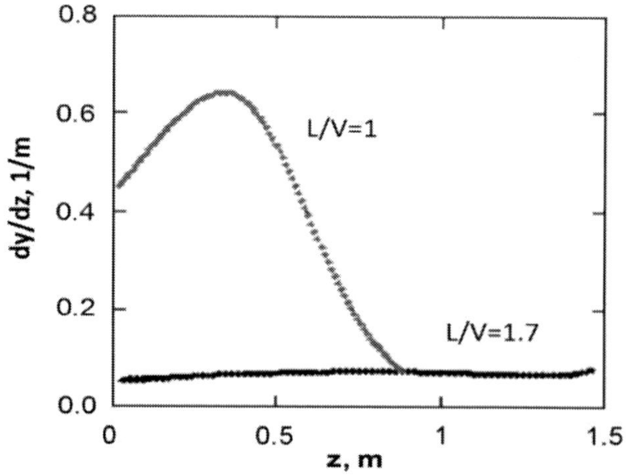

Fig. 1. Variation of the concentration gradient of volatile distillate fraction along the gravity vector during separation of R21-R114.

Numerical calculations based on the one-dimensional non-equilibrium separation model [6], used for the separation of R21-R114 mixture in a column filled with Koch 1Y structure packing for the conditions corresponding to experiments [4], showed that the highest concentration gradient of the volatile distillate fraction and vapor density are located in the top of the column. The concentration gradient, shown in Fig. 1, increases significantly with an increase in the ratio of molar flow rates of liquid and vapor from 1 to 1.7 that may be the cause of the convective flow development in the gravity field. Here the coordinate z is measured from the top of the column.

Let us consider the effect of negative gas density gradient on the stability of the flow in a column. During binary mixture separation, the gas density is changed along the column due to heat and mass transfer between liquid and gas. It is typical for separation that the gas mixture density depends mainly on the concentration of volatile distillate fraction because the temperature is varied slightly. Assuming small variation in the density of gas mixture along the column, we obtain:

$$\rho_G = \rho_{G0} + \rho'_G,\ \ \rho_G = \rho_G(Y_G, T),\ \ \rho_G = \rho_{G0} + (\partial\rho_G/\partial Y_G)_T Y'_G, \tag{1}$$

where $Y_G = Y_{G0} + Y'_G$ is the concentration of volatile distillate fraction. Introducing parameter $\beta_Y = -(\partial\rho_G/\partial Y_G)_T \rho_{G0}^{-1}$, we obtain the averaged momentum equation for the gas as follows:

$$\langle(\mathbf{u}_G\nabla)\mathbf{u}_G\rangle = -\frac{\nabla p_G}{\rho_{G0}} + \langle\nu_G\Delta\mathbf{u}_G\rangle - \beta_Y Y'_G\mathbf{g}, \tag{2}$$

where $\mathbf{u}_G$ is the velocity vector. Scaling equation (2) by the characteristic size l, the characteristic gas velocity U_G, the characteristic concentration $Y_0 = l\cdot dY/dz$ we obtain:

$$\langle(\tilde{\mathbf{u}}_G\nabla)\tilde{\mathbf{u}}_G\rangle = -\nabla\tilde{p}_G - N_{conv}\tilde{Y}_G\frac{\mathbf{g}}{g} + \frac{1}{\mathrm{Re}}\langle\nabla^2\tilde{\mathbf{u}}_G\rangle. \tag{3}$$

Here the dimensionless number that determines the ratio of gravitational and inertial forces in a convective flow and called as convection number is introduced as follows:

$$N_{conv} = \frac{l^2 g\beta_Y\, dY/dz}{U_G^2} = \frac{Ra\cdot \mathrm{Pr}}{Pe^2} \tag{4}$$

The convection number N_{conv} determines the ratio of gravitational and inertial forces during ascending vapor flow. Therefore, unstable vapor flow regimes correspond to the convection number, which is much higher than unity. The calculations of the separation of binary refrigerants mixture show that the convection number N_{conv} can significantly vary along column length. Figure 2 shows the variation in the convection number N_{conv} along the gravity vector for the molar liquid to vapor ratios $L/V = 1$ and $L/V = 1.7$. The perturbations with characteristic size l equal to the diameter of column D are considered in calculations.

As can be seen, the convection number considerably exceeds unity for the columns with a diameter exceeding 0.2 meters if $L/V = 1$. For this case, one can expect the appearance of convective flows in the column and significant heterogeneity of the components concentration in the column cross-section, which was observed experimentally in [4, 5].

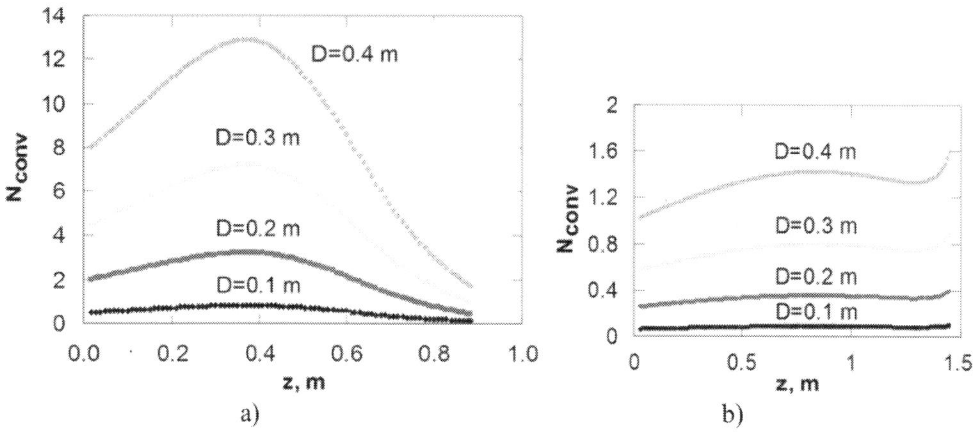

Fig. 2. Variation of the convection number along the column: a) $L/V=1$; b) $L/V=1.7$.

3 Conclusions

The obtained results allow characterizing the conditions for the development of large-scale convective structures during counter-current vapor-liquid flow in structure packing, as an example of the multiphase system with a fixed solid phase. It is necessary for the development of highly efficient heat and mass transfer devices. It is established that the ratio of the liquid to vapor molar flows in a distillation column with a structural packing closed to unity is the most dangerous and can lead to the development of convective flow in the large-diameter column with a negative gradient of the vapor density along the direction of the gravity vector. In this case, the separation efficiency will decrease because of the occurrence of large-scale heterogeneity in the concentration of mixture components.

This work was performed in Kutateladze Institute of Thermophysics SB RAS by a grant from the Russian Science Foundation (project No. 14-49-00010).

References

1. V.V. Kuznetsov, A.S. Shamirzaev, J. Eng. Thermophys. **24**, 357 (2015)
2. M.P. Malkov, I.B. Danilov, A.G. Zel'dovich, A.B. Fradkov, *Handbook of Physicotechnical Principles of Cryogenics* (Energoatomizdat, Moscow, 1985)
3. B. Hanley, C.C. Chen, AIChE J. **58**, 132 (2012)
4. A.N. Pavlenko, N.I. Pecherkin, V.Yu. Chekhovich, V.E. Zhukov, S. Sunder, P. Houghton, A.F. Serov, A.D. Nazarov, Theor. Found Chem. Eng. **40**, 329 (2006)
5. A.N. Pavlenko, X. Li, V.E. Zhukov, N.I. Pecherkin, O.A. Volodin, A.S. Surtaev, X. Gao, L. Zhang, H. Sui, H. Li, J. Eng. Thermophys. **24**, 210 (2015)
6. R. Taylor, R. Krishna, *Multicomponent Mass Transfer* (Wiley - IEEE, 1993)

MATEC Web of Conferences **115**, 02001 (2017) DOI: 10.1051/matecconf/201711502001
STS-33

An experimental study of the influence of pressure pulsations in the water turbine flow path on vibration parameters of the hydraulic unit

Dmitry Dekterev[1,2*], *Dmitry* Platonov[1,2], *Andrey* Minakov[1,2] , *Alexandra* Maslennikova[1] and *Artem* Abramov[1]

[1]Siberian Federal University, 79 Svobodny Ave., 660041 Krasnoyarsk, Russia
[2]Institute of Thermophysics, SB RAS, 1 Lavrentyev Ave., 630090, Novosibirsk, Russia

Abstract. The paper presents an experimental study of the relationship between the vibration of the water turbine unit and the pressure pulsations in the flow path. It is shown that the air injection into the flow path leads to a decrease in the hydraulic turbine vibration associated with the formation of the precessing vortex core.

1 Introduction

Hydropower is a significant part of the overall energy system. The reliability of the hydropower equipment operation is largely determined by its vibrational parameters. Vibrational signal is the sum of signals from all elements of the mechanism and hydromechanical processes that occur during the operation of a water turbine. It is generally believed that the most significant contribution of hydrodynamic processes is in the so-called transient modes of operation of the hydraulic unit (partial or high load). In such regimes a stable vortex structure with a pronounced precession frequency formed downstream the water turbine runner. The precessing vortex core (PVC) leads to generation of high amplitude low-frequency pressure pulsations [1, 2], which are the most dangerous in the case of resonance with the elements of the hydraulic turbine. In this case arising vibrations leads to a faster wear of the equipment and destruction of some unit elements. This work was carried out to determine the effect of pressure pulsations in the water turbine flow path on the vibration of its construction.

2 Experimental techniques

An experimental study was carried out at the research set-up with a model of a water turbine unit at a scale of 1/14.7. The main units of the set-up: the water storage tank, two regulated pumps, pressure tank, model of the hydraulic unit (spiral case, stator columns,

* Corresponding author: dekterev_da@mail.ru

guide vanes, runner (D = 0.3 m), transparent conical diffuser and draft tube), drain tank and set of flowmeters.

The opening of the guide vanes regulates by manual and electrical drives and is divided into 15 steps. The optimum operating mode by integral characteristics of the hydraulic unit and the pressure pulsation parameters in the flow path is the 9/15 opening. In this paper the modes with constant water head 3.5m for the guide vanes opening 1/15 - 8/15 was studied. The water turbine works at "free runner" mode. The frequency of the runner was determined using the optical tachometer (340 – 510 rpm depending on the guide vanes opening) and the frequency of the PVC was determined with the help of piezoelectric pressure sensors installed in wall of conical diffuser. A stable vortex core was visually observed at operating conditions corresponding to the openings 3/15- 5/15 (Fig. 1a), at the opening of 6/15 a large unstable spiral vortex was observed (Fig. 1b), in the mode of 7/15 a thin unstable vortex with the reconnection effect was observed (Fig. 1c), and at the 8/15 mode no vortex structures were observed.

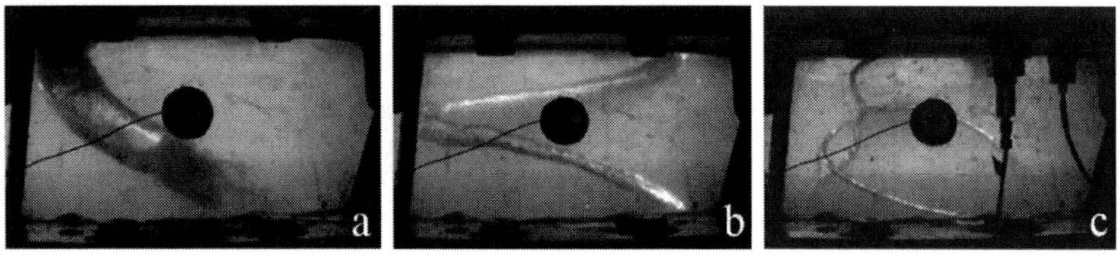

Fig. 1. Varieties of vortices downstream the runner

3 Experimental results

Investigation of the vibration parameters of the hydraulic turbine unit was carried out using a piezoelectric vibration acceleration sensor VD06A which was installed on conical diffuser flange. The signal was transmitted to a PC via ADC and recorded in mV. Spectra with evident frequency of the precession of the vortex core and the rotation frequency of the runner are obtained using a fast Fourier transform. Figure 2 shows the vibration spectrum for the mode with the opening of the guide vanes 4/15.

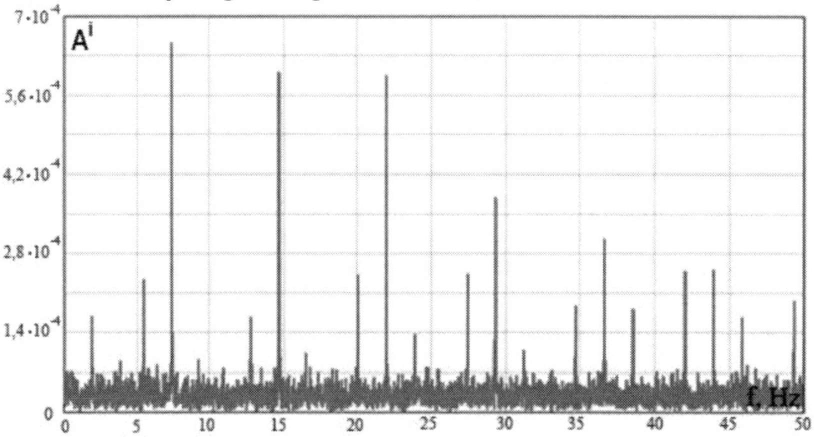

Fig. 2. Vibration spectrum

The first peak at a frequency of 1.9 Hz corresponds to the PVC frequency, the peak at a frequency of 7.3 Hz corresponds to the rotation frequency of the runner (438 rpm). Harmonics of the runner rotation frequency are also visible; in addition, noticeable peaks are equidistant from the rotation frequency (including harmonics) to a positive and negative value equal to the PVC frequency are observed. This means that pressure pulsations make a serious contribution to the appearance of low-frequency vibrations. A similar dependence is observed for the 1/15 - 5/15 openings. When the opening is 6/15 and above, only the double rotational frequency of the runner is observed.

The plot of the dependence of the vibration signal amplitude (peak to peak) on the opening of the guide vanes shows that the maximum of vibrations is observed in the case of openings 2/15 - 3/15, however, approaching the optimal operating modes of the turbine the signal intensity is significantly reduced.

Recently, various works have been conducted to study methods of decreasing of pressure pulsations in a flow with a precessing vortex core, including injection of air into the turbine flow path [3, 4]. In this work, air is also injected to the set-up, and its effect on the vibration level of the hydraulic unit is evaluated. Figure 3 shows the comparison of peak-to-peak vibration signal level with and without air injection.

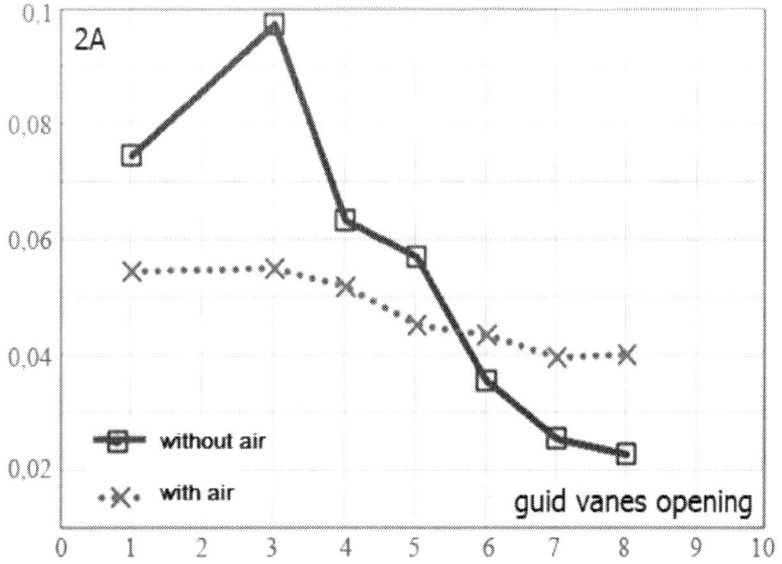

Fig. 3. Effect of air injection to vibration signal level

In this study, the airflow rate was constant and amounted to 6.3 m^3/h; it corresponds to 3.6% of the water flow rate when opening of guide vanes is 1/15 and 1.4% when opening is 8/15. It may be noted that the air injection reduces the level of vibration signal at the non-optimal operating modes of the hydraulic unit up to 45% for 3/15 opening. At the same time air injection leads to increase of signal in the cases close to the optimal mode, however, the increased signal does not exceed the signal level at non-optimal modes. Hence, one can draw conclusions about the advisability of using air to reduce vibrations only in non-optimal operating modes of hydraulic units.

Conclusions

It is showed that the pressure pulsations generated by the precessing vortex core make an appreciable contribution to the emergence of low-frequency vibration of the hydraulic unit.

It has been established that the injection of air into the flow path of a hydraulic unit in order to reduce the amplitude of vibration is appropriate only in cases of operating of water turbine in non-optimal modes. The question of determining the minimum sufficient volume of injected air remains rather interesting.

The work was performed at partial support of the projects funded by the Russian Foundation for Basic Research and Krasnoyarsk Regional Fund for Support of Scientific and Scientific-Technical Activities (16-48-243076).

References

1. A.K. Gupta, D.G. Lilley, N. Syred, *Swirl flows* (Abacus Press, 1984)
2. S. Alekseenko, P. Kuibin, V. Okulov, *Theory of concentrated vortices. An introduction*, (Springer, 2007)
3. G. Semenov, A. Smirnova, A. Dekterev, A. Minakov, A. Sentyabov, Proceedings of the Hydro 2012, Bilbao, Spain, 9 (2012)
4. S.V. Alekseenko, P.A. Kuibin, V.L. Okulov, S.I. Shtork, Vortex precession in a gas-liquid flow, Heat Transfer Research, **41** 465 (2010)

Aerodynamic and thermal interference of turbulent separated flows over building models

Aleksey Gnyria[1], *Sergey* Korobkov[1], *Anton* Koshin[1,*], and *Victor* Terekhov[2]

[1]Tomsk State University of Architecture and Building, Russia
[2] Kutateladze Institute of Thermal Physics SB RAS, Russia

Abstract. The paper presents experimental results of a physical simulation and computations of aerodynamic and thermal interference of turbulent flows over two building models. The experimental setup, conditions, and results are described, and the evaluation of obtained results is given in this paper. Conclusions are drawn on the basis of this research.

1 Introduction

Aerodynamic and thermal inteference of separated flows plays an important role in studying wind pressure and heat loss of buildings during their operation. The relevant literature shows a variety of approaches to experimental [1-3] and numerical [4] investigations of these phenomena. These works are mainly devoted to the aerodynamic behavior of airflows when facing a tandem of two square or rectangular prisms.

Although the literature in this field illustrates a wide range of information, the problem of aerodynamic and thermal inteference is still relevant due to mixed results, a complicated structure of separated flows and, as consequence, the impossibility of developing a correct model of turbulent transfer. Presently, the interference process and its effect on the airflow can be predicted only by virtue of systematic experimental research and construction of generalized curves under certain conditions.

2 Experimental setup

The experimental setup for aerodynamic tests includes a wind channel with a process chamber of 0.42×0.42 m size and 1.5 m length, experimental building models, and measuring devices. The schematic view of the arrangement of building models and the airflow visualization are presented in Fig. 1.

On the basis of the previous study [5], this work continues the investigations of aerodynamics and heat exchange between two square prisms arranged at different distances from each other (L/b = 0÷6) and the lateral shift (D/b = 1÷3), where b is the prism side equaling 50 mm (Fig. 1).

* Corresponding author: dawghood@mail.ru

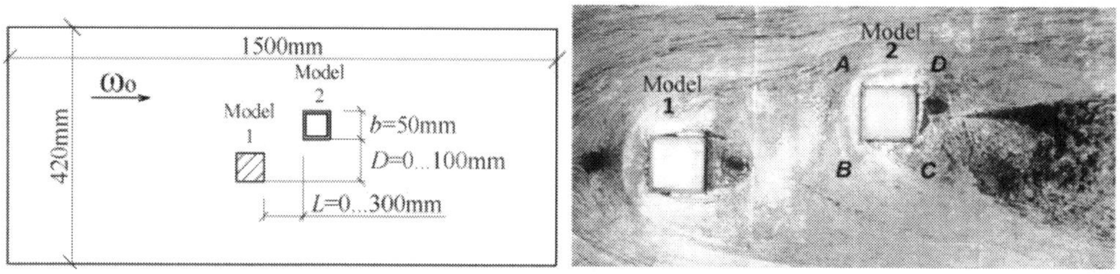

Fig. 1. Schematic view of bulding models arrangement and oil flow visualization.

During the experiment, the distribution of pressure and heat transfer coefficients is measured on each side of the square prism. Oil flow visualization is performed on the bottom of the wind channel as shown in Fig. 1. A detailed description of the experimental technique is presented in [5]. The Reynolds number in this experiment varies between $2{\cdot}10^4 - 4{\cdot}10^4$.

3 Results and discussions

As can be seen from Fig. 1, the vortex structure is observed in the aerodynamic trail. In front of Model 1, the airflow separates and forms a horseshoe vortex. Then it flows over Model 2 and forms another horseshoe vortex, becoming asymmetric at the same time.

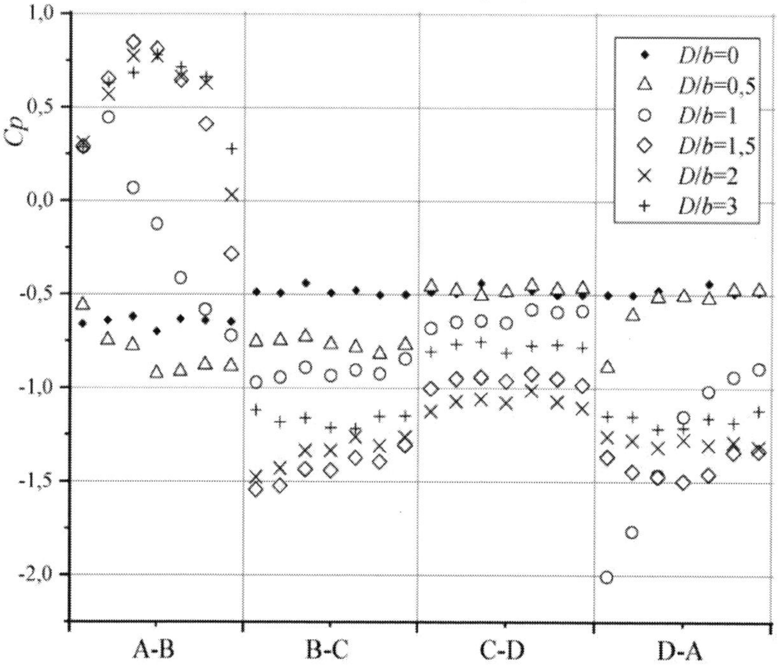

Fig 2. Distribution of pressure coefficients over sides of Model 2.

Measurement results on the pressure distribution demonstrate substantial differences in the wind pressure in Models 1 and 2. As Model 2 is positioned in the aerodynamic trail, it is subjected to a lesser pressure than Model 1. Moreover, the differential pressure occurs on *B–C* and *D–A* sides resulting in the lateral pressure on Model 2. The similar phenomena are observed during changes in heat transfer. Along *D–A* side, the intensity of heat transfer from the surface of Model 2 is higher than that along *B–C* side.

The coefficients of aerodynamic and thermal interference *IF* can be found in the following way [1–3]. The coefficient of aerodynamic interference IF_{Cp} can be obtained from the ratio between the mean and the local values of the pressure coefficient *Cp*:

$$IF_{Cp} = \frac{Cp_{mean}}{Cp_{mean(o)}} \tag{1}$$

The coefficient of thermal interference IF_{Nu} can be obtained from the ratio between the mean and the local value of the Nusselt number [1-3]:

$$IF_{Nu} = \frac{\overline{Nu}_{mean}}{\overline{Nu}_{mean(o)}} \tag{2}$$

The dependence between the parameter and the lateral shift *D/b* is presented in Fig. 3. The value of *IF* parameter is obtained *via* the minimum wind pressure which depends on *D/b* ratio at different distances *L/b* between square prisms (Fig. 1).

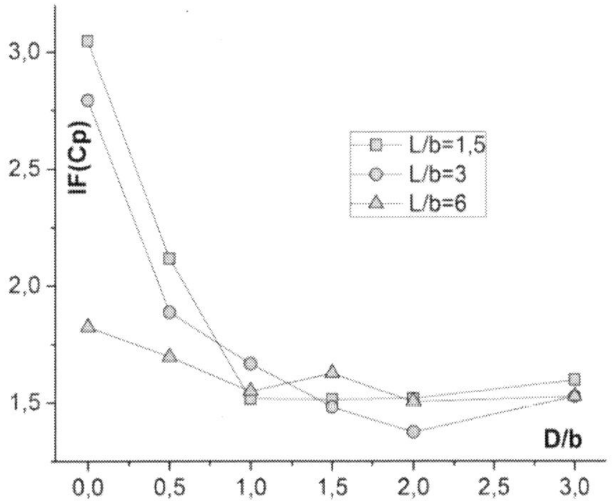

Fig. 3. Dependences between *IF* parameter and lateral shift *D/b*.

According to Fig. 3, the aerodynamic interference IF_{Cp} achieves larger values at lower lateral shift of square prisms. This indicates the growth in air rarefaction on the leeward prism, affected by the separated flow. At the same time, the heat transfer changes not so intensively and IF_{Nu} coefficient is close to unity in each case.

From the outcome of our investigation it is possible to conclude that the aerodynamic and thermal interference is subjected to the following impacts. First, *A–D* side when entering the airflow, enables the formation of differential pressure on the lateral sides of Model 2, thereby increasing the heat transfer on *D–A* side. Second, the separated flow behind Model 1, forms accelerated airstreams which scatter as early as at L/b = 1.5 and have a lower effect on the pressure coefficients and the Nusselt number at longer distances. Thermal interference turns to be more stable relative to the model arrangement.

This work was financially supported by Grant N 8780.2016.8 from the President of the Russian Federation.

References

1. X.F. Yu, Z.N. Xie, J.B. Zhu, J. Wind Eng. Indust. Aerod. **142**, 188 (2015)
2. Y. Hui, A.Yoshida, Y. Tamura, J. Fluids Structures **37** 120 (2013)
3. Y. Hui, A.Yoshida, Y. Tamura, J. Wind Eng. Indust. Aerod. **104-106**, 98 (2012)
4. A. Lankadasu, S. Vengadesan, Int. J. Numerical Methods Fluids **57** 1005 (2008)
5. A.A. Koshin, V.I. Terekhov, Conf. Energy Efficiency of Residential Buildings, 71 (2015)

Numerical algorithm improvements for the CFD-code ANES for turbulent flow simulations via the LES method

Valery Artemov[1], *Maksim* Makarov[1*], *Georgiy* Yankov[1] and *Konstantin* Minko[1]

[1]National Research University «Moscow Power Engineering Institute», Moscow, Russia

Abstract. Turbulent flow simulations via the LES method place high demands on the accuracy of the utilised numerical schemes. This work considers issues of both the implementation and improvement of numerical schemes and algorithms in the 'in-house' CFD code ANES [1]. A new second-order scheme (MAK) was developed by the authors. Its accuracy was studied by solving well-known test problems, with the aim of extending its use to complex turbulent flow calculations.

1 Numerical algorithms

The ANES code uses a control volume (CV) method. Both structural and non-structural grids with cut-cells and the fine grid algorithm are implemented in the code.

Three well-known and widely used methods of second-order accuracy in space [2, 3, 4] — QUICK, MUSCL and SuperBEE — are also implemented in the ANES code. A primary feature of these algorithms' application for the CV method is using the gradient of the unknown function at the CV nodal points to calculate the total flux of that function through the CV face. The gradient of the function is calculated via two algorithms: 1) the least-squares method with a Gaussian weight model and 2) using interpolation of the function values on the CV face and the gradient computation based on the Ostrogradsky-Gauss theorem.

The authors have proposed an original scheme called MAK for calculating the total flux of the Φ-function on the CV face based on the exact solutions of a one-dimensional steady equation with convective, diffusive and source terms and constant coefficients:

$$\frac{\mathrm{d}}{\mathrm{d}x}\left(\rho u\Phi - \Gamma\frac{\mathrm{d}\Phi}{\mathrm{d}x}\right) = S\,. \tag{1}$$

The analytical solution of equation (1) on the segment that connects the nodal points P and E of two adjacent CVs with the boundary conditions $\Phi(x_E) = \Phi_E$ and $\Phi(x_P) = \Phi_P$ is well known. In the multidimensional formulation of the problem, the value of S will contain fluxes in different directions from the x-direction. To find this value, a linear combination

* Corresponding author: makarovmv2000@yandex.ru

of derivatives at the nodal points $G_P = (\mathrm{d}\Phi/\mathrm{d}X)|_{X=0}$ and $G_E = (\mathrm{d}\Phi/\mathrm{d}X)|_{X=1}$ was used. The result was an exact expression of the density of total flow through the common face 'e' of two CVs with nodal points P and E:

$$J_e = \left(\rho u\Phi - \Gamma\frac{\mathrm{d}\Phi}{\mathrm{d}x}\right)_e = m\Phi_P + Ad(\Phi_P - \Phi_E) + \left(X_e - \frac{1-A}{P}\right)\left(mG_P + Ad(G_P - G_E)\right) \qquad (2)$$

where $X = (x - x_P)/\delta$, $\delta = x_E - x_P$, $m = (\rho u)_e$, $d = \Gamma/\delta$, $P = m/d$, $A = \dfrac{P}{\exp(P)-1}$.

As for time integration algorithm, the second-order, three-level implicit scheme, which is called the Second Order Backward Euler Scheme (SOBES), was implemented in the ANES code.

To calculate the pressure field in the ANES code, a modified version of the SIMPLE + PISO algorithm was used for both structured and unstructured grids, with the step time restricted by the Courant number.

2 Verification of numerical algorithms and schemes

In the first test, a steady flow at an angle of 45^0 to the grid lines was considered. The flow with a density of $\rho = 1$ in a 2D square domain with dimensions $L_x = L_y = 1$ and components of the velocity $u_x = u_y = 1$ was investigated. The equation for the scalar function Φ was solved. The boundary conditions for this equation were as follows: $\Phi = 1$ for the 'left' input boundary (x=0) and $\Phi = 0$ for the 'down' input boundary (y=0). Unlike [5], a small diffusion coefficient in the Φ-equation (Peclet number = 1000) was included for testing of a common influence of both convection and diffusion.

The 'exact' solution was obtained on a 1000×1000 grid. The main calculations were carried out on the 64×64 grid for various schemes. The results indicated that the scheme of the first-order POWER [5] had a high scheme diffusion. The scheme of the second-order SuperBEE gave more accurate results; however, maximum accuracy was obtained when using QUICK, MUSCL and MAK schemes.

The second test was a classical two-dimensional problem of flow in a square cavity with a moving wall at Re = 1000. There is a 'reference' solution for this task [6]. The calculations used the unstructured mesh, with 64×64 nodes. QUICK, MUSCL and MAK schemes showed the best results.

The next problem—the attenuation of two-dimensional vortices of Taylor-Green [7]—was considered. The computational domain was a square with the dimensions of $2\pi L \times 2\pi L$. The periodic boundary conditions were imposed at the opposite boundaries.

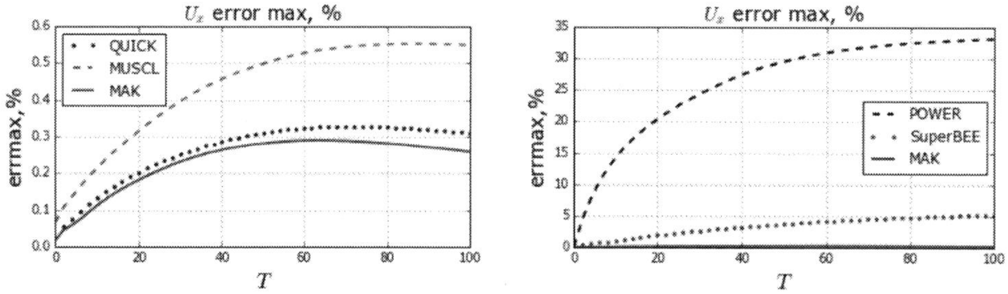

Fig. 1. The maximum error of velocity U_x for different schemes.

The 2D Navier-Stokes equations for this problem have the exact solution of

$$U_x = \frac{u_x}{U_0} = \sin(X)\cdot\cos(Y)\cdot\exp(-\frac{2T}{\mathrm{Re}}),\quad U_y = \frac{u_y}{U_0} = -\cos(X)\cdot\sin(Y)\cdot\exp(-\frac{2T}{\mathrm{Re}}) \quad (3)$$

where $X = x/L = 0\ldots2\pi$, $Y = y/L = 0\ldots2\pi$, $\mathrm{Re} = U_0L/\nu$, $T = tU_0/L$ - dimensionless time, ν - kinematic viscosity.

The calculations were performed at Re = 100 on the uniform 64×64 grid; the step time was 0.05. For integrating in time, we used the second-order SOBES scheme. Fig. 1 shows the maximum estimated error of the velocity U_x, which was normalized by the value $\exp(-2T/\mathrm{Re})$. The maximum solution accuracy was obtained by using the MAK scheme.

The calculations of the Taylor-Green vortex flow were performed in a three-dimensional formulation for Re = 1600, which corresponded to the mode of decay of the isotropic turbulence. The Smagorinsky model for subgrid turbulent viscosity with $C_S = 0.2$ was used on the $192\times192\times192$ grid with a step time of 0.005.

Fig. 2 shows the results of calculations using the SuperBEE schemes and SOBES in comparison with the DNS data [8].

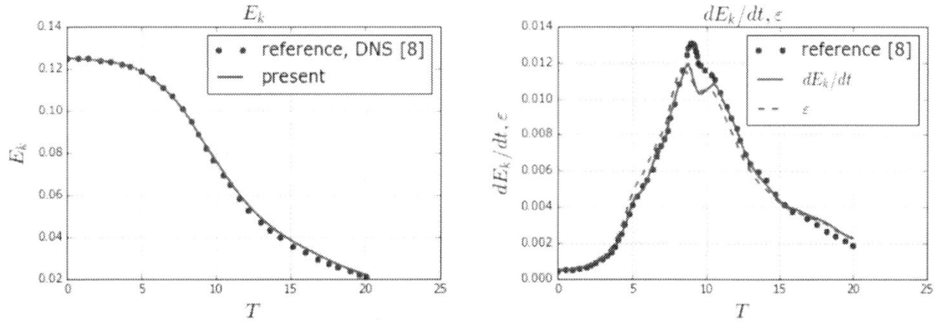

Fig. 2. The kinetic energy E_k and its rate of dissipation ε.

The results of modelling a steady flow in a flat channel for conditions that correspond to the work of R. Moser, J. Kim and N. Mansour (MKM) [9] are shown in Fig. 3.

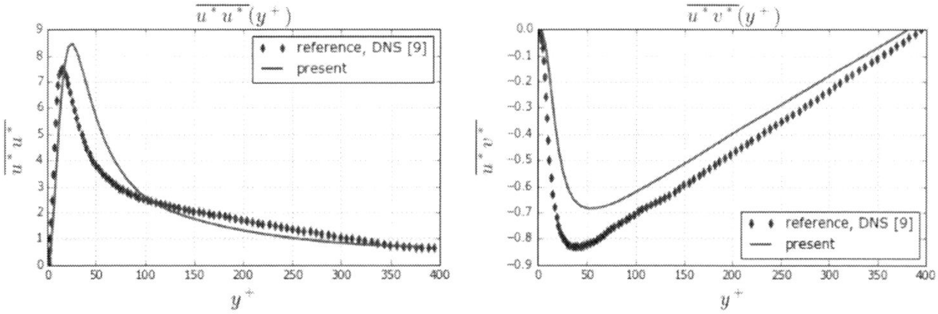

Fig. 3. The intensity of fluctuations of the x-component of the velocity and the LES component of the turbulent shear stress, $\mathrm{Re}_\tau = 395$.

Calculations of the lifting, fully developed turbulent flow with buoyancy forces in a vertical plane channel and a width of 2δ for the conditions of [10] were carried out. On the channel walls, constant temperatures were kept. For the right cold wall $\theta(y=2\delta)=\theta_c$, and for left hot wall $\theta(y=0)=\theta_c+\Delta\theta$. The following dimensionless values were used:

$$u_\tau^* = \sqrt{\tau_w/\rho},\ U^* = u/u_\tau^*,\ \Theta = (\theta-\theta_c)/\Delta\theta,\ \mathrm{Re}_\tau = u_\tau^*\delta/\nu,\ \mathrm{Gr} = g\beta\Delta\theta(2\delta)^3/\nu^2$$

where τ_w - the wall shear-stress averaged on the two walls, g - acceleration of gravity and β - the coefficient of volume expansion.

By integrating the streamwise mean momentum equation with respect to Y, one can obtain the following balance equation:

$$\tau^*(0) - Y = \frac{1}{\mathrm{Re}_\tau}\frac{dU_x^*}{dY} - \overline{u^*v^*} + \frac{\mathrm{Gr}}{\mathrm{Re}_\tau^2}\int_0^Y (\Theta - \Theta_m)dY\,,\ Y = y/\delta\,,\ \tau^*(0) = \frac{\tau(Y=0)}{\tau_w} \tag{4}$$

where $\overline{u^*v^*}$ - dimensionless Reynolds stress and Θ_m - arithmetic mean dimensionless temperature over the channel cross section.

The calculations were performed using Boussinesq approximation via the MAK scheme on a $64 \times 32 \times 32$ grid. The results were compared with the DNS modelling of [10] (Fig. 4).

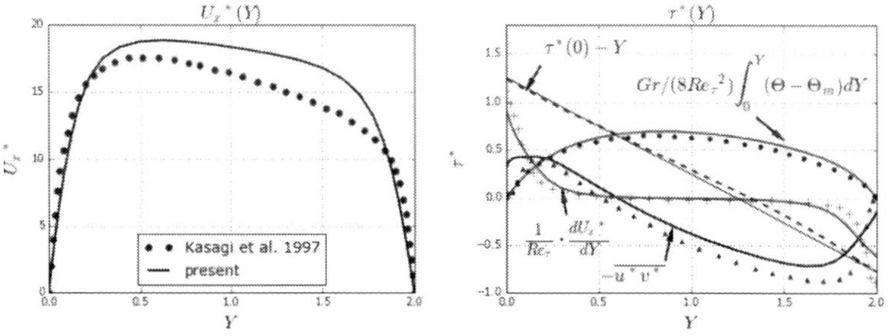

Fig. 4. Dimensionless mean velocity and stresses profiles for $\mathrm{Re}_\tau = 150$ and $\mathrm{Gr} = 9.6 \cdot 10^5$. Markers indicate the DNS data [10].

3 Conclusion

The performances of various schemes were analysed via the LES method with a simple subgrid turbulent viscosity of the Smagorinsky model on the coarse grids. The results showed that the proposed MAK scheme worked efficiently for a wide range of problems, which allowed its use with the LES method for complex turbulent flow.

The authors extend sincere thanks for the financial support of the Russian Foundation for Basic Research (project № 16-08-00981-a, №17-08-01007-a).

References

1. http://anes.ch12655.tmweb.ru
2. B.P. Leonard, Comp. Methods Appl. Mech. **19** 59 (1979)
3. P. K. Sweby, SIAM Journal on Numerical Analysis, **21**, 5 (1984)
4. Ferziger J.H., Peric M. *Computational methods for fluid dynamics* (Springer, 2002)
5. S. Patankar, *Numerical heat transfer and fluid flow* (Hemisphere, 1980)
6. U. Ghia, K.N. Ghia, C.T. Shin. J. Comp. Phys. **48**, 387 (1982)
7. G. I. Taylor and A. E. Green, Royal Society of London Proc. Ser. A, **158** 895 (1937)
8. http://www.as.dlr.de
9. R.D. Moser, J. Kim, N.N. Mansour, Phys. Fluids., **11**, 943 (1999)
10. N. Kasagi, M. Nishimura, Int. J. Heat Fluid Flow, **18**, 88 (1997)

Analytical model of unsteady-state convective heat transfer between the heat carrier and the finite sizes plate adjusted for the thermal relaxation

Danila Makarushkin[1,*], *Yury* Kirsanov[1], *Andrey* Yudakhin[1], and *Alexandr* Kirsanov[2]

[1] Kazan Research Center of the RAS, 420111, Lobachevsky str., 2/31, Kazan, Russia
[2] Kazan National Research Technical University Tupolev, 420111, Karl Marx str., 10, Kazan, Russia

Abstract. A hyperbolic boundary value problem of the thermal conduction of a two-dimensional plate with the third kind boundary conditions is formulated. The transient thermal process in the plate is due to the temperature changes of the external medium over time and along the plate length, and also by a multiple step change of the plate surface heat transfer coefficient throughout the transient process. An analytical solution with improved convergence adjusted for thermal relaxation and thermal damping is obtained for the temperature field in the plate.

1 Introduction

In short-term transient processes, the results of calculations using the classical differential thermal conductivity equation

$$\partial T/\partial\tau = a\,\Delta T \tag{1}$$

are characterized by large deviations from real processes due to neglect of thermal inertia. This specificity of equation (1) was noted by such scientists as Maxwell [1], Onzager [2], Cattaneo [3], Vernotte [4], Luikov [5] et al., and was confirmed experimentally (Fig. 1).

To overcome the limitations of Fourier formula

$$\mathbf{q} = -\lambda\,\mathbf{grad}\,T\,,$$

Luikov [5], Tzou [6] et al proposed to take into account the effect of relaxation phenomena to heat conduction, due to which Fourier formula takes the form of expression, also known as the Maxwell-Cattaneo-Luikov equation [5, 7, 8] or the dual-phase-lag Eq. [6, 9, 10]:

$$\mathbf{q} + \tau_q\,\partial\mathbf{q}/\partial\tau = -\lambda\,\mathbf{grad}\left(T + \tau_T\,\partial T/\partial\tau\right), \tag{2}$$

owing to what equation (1) was transformed into a differential equation of heat conduction of hyperbolic type

$$\partial T/\partial\tau + \tau_q\,\partial^2 T/\partial\tau^2 = a\,\Delta\left(T + \tau_T\,\partial T/\partial\tau\right). \tag{3}$$

* Corresponding author: atpp.danila@mail.ru

Here τ_q is the thermal relaxation time, which characterizes the property of thermal inertia or thermal "elasticity" of the body, s; τ_T is the temperature damping time, which characterizes the reaction of temperature on changing of heat flux, s.

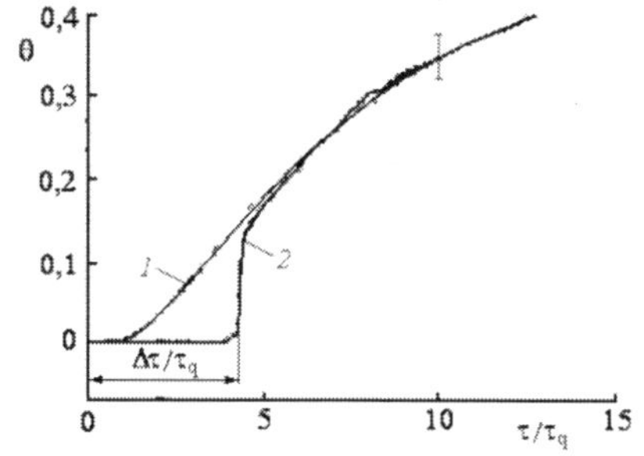

Fig. 1. Comparison of the experimental and calculated transient thermal processes using Fourier hypothesis: *1*-solution data (1); *2* - experimental data [9].

In [7-11] the analytical calculations of transient thermal processes taking into account equation (3) with sufficiently significant simplifications are presented. In [7, 8] models of transient thermal processes in a solid were constructed for a sudden change of the external medium temperature, which was assumed constant with an unchanged heat transfer coefficient during the entire transient process [10], or during individual time intervals (periods) [11].

In this paper a plate of thickness $2h$ and length l is considered. It is assumed, that the heat transfer coefficient of the plate varies in time stepwise, and the medium temperature (heat carrier) in a separate period depends on the time and length of the plate in the polynomial form:

$$\theta_{f,j}(Y,t)=\sum_{k=0}^{k_y} Y^k \sum_{l=0}^{k_t} g_{k,l} t^l ,$$

where $g_{k,l}$ is the coefficient; $t=\tau/\tau_q$ is the relative time; $Y=y/l$ is the relative longitudinal coordinate; k_y and k_t are the counts of members of each row.

2 The third kind hyperbolic boundary value problem

The mathematical formulation of the boundary value problem of the thermal conductivity of a plate with boundary conditions in the form (2) and the differential equation (3) in dimensionless variables:

$$\frac{\partial^2\theta_j}{\partial t^2}+\frac{\partial\theta_j}{\partial t}=\mathrm{Fo}_{q,j}\left[\frac{\partial^2}{\partial X^2}\left(\theta_j+\kappa\frac{\partial\theta_j}{\partial t}\right)+\frac{1}{L^2}\frac{\partial^2}{\partial Y^2}\left(\theta_j+\kappa\frac{\partial\theta_j}{\partial t}\right)\right]; \quad (4)$$

$$\theta_j(X,Y,0)=\begin{cases}1 & \text{for } j=1,\\ \theta_{j-1}(X,Y,t_{p,j-1}) & \text{for } j>1;\end{cases} \quad \frac{\partial\theta_j(X,Y,0)}{\partial t}=\begin{cases}0 & \text{for } j=1,\\ \partial\theta_{j-1}(X,Y,t_{p,j-1})/\partial t & \text{for } j>1;\end{cases} \quad (5)$$

$$\partial\theta_j(0,Y,t)/\partial X=0\,; \quad (6)$$

$$\partial\left[\theta_j(1,Y,t)+\kappa\,\partial\theta_j(1,Y,t)/\partial t\right]/\partial X=-\mathrm{Bi}_{x,j}\left[\theta_j(1,Y,t)-\theta_{f,j}(Y,t)+\partial\theta_j(1,Y,t)/\partial t\right]; \quad (7)$$

$$\partial\left[\theta_j(X,0,t)+\kappa\partial\theta_j(X,0,t)/\partial t\right]/\partial X=\mathrm{Bi}_{y=0,j}\left[\theta_j(X,0,t)-\theta_{f,j}(0,t)+\partial\theta_j(X,0,t)/\partial t\right]; \quad (8)$$

$$\partial\left[\theta_j(X,1,t)+\kappa\partial\theta_j(X,1,t)/\partial t\right]/\partial Y=-\mathrm{Bi}_{y=1,j}\left[\theta_j(X,1,t)-\theta_{f,j}(1,t)+\partial\theta_j(X,1,t)/\partial t\right]. \quad (9)$$

Here $\theta=(T-T^*)/(T_0-T^*)$; T^* is the characteristic temperature, K; T_0 is the initial plate temperature, K; j = 1, 2, … is the period number; $X=x/h$; x is the transverse coordinate relative to the median plane, m; $L = l/h$; $\kappa = \tau_T/\tau_q$; $\mathrm{Fo}_{q,j} = a_j \tau_q / h^2$; $a_j = \lambda_j/(\rho c_j)$; $\mathrm{Bi}_{x,j} = \alpha_j h/\lambda_j$, $\mathrm{Bi}_{y=0,j} = \alpha_{y=0,j} l/\lambda_j$ and $\mathrm{Bi}_{y=1,j} = \alpha_{y=1,j} l/\lambda_j$ are the Bio numbers; α_j, $\alpha_{y=0,j}$ and $\alpha_{y=1,j}$ is the heat transfer coefficient in the j-th period, W/(m^2 K) and $t_{p,j} = \tau_{p,j}/\tau_q$ is the relative duration of the j-th period.

2.1. An analytic solution of the boundary value problem

The solution is obtained using the finite Fourier integral transforms [11, 12]:

$$\theta_j(X,Y,t) = \sum_{n=1}^{\infty} A_{n,j} K_x(\mu_{n,j} X) \sum_{m=1}^{\infty} A_{m,j} K_y(\gamma_{m,j} Y) \theta_{L,j}(\mu_{n,j}, \gamma_{m,j}, t), \tag{10}$$

where $A_{n,j}^{-1} \equiv \int_0^1 K_x^2(\mu_{n,j} X)\, dX$; $A_{m,j}^{-1} \equiv \int_0^1 K_y^2(\gamma_{m,j} Y)\, dY$;, $\vartheta_j^2 = (\mu_{n,j}^2 + \gamma_{m,j}^2 / L^2)$; $\beta_j = 1 + \kappa\, \mathrm{Fo}_{q,j} \vartheta_j^2$

$\zeta_j^2 = \left|\beta_j^2 - 4\, \mathrm{Fo}_{q,j}\, \vartheta^2\right|$; $\theta_{L,j}(\mu_{n,j}, \gamma_{m,j}, t) = C_{1,j}\, f_{1,j}(t) + C_{2,j}\, f_{2,j}(t) + \mathrm{Fo}_{q,j} F_j(t)$; $W_j(t) = \sum_{l=0}^{k_t} d_l t^l + \sum_{l=1}^{k_t} l\, d_l t^{l-1}$

$$f_{1,j}(t) = \begin{cases} t \exp(-\beta_j t/2) & \text{for } \mathrm{Fo}_{q,j} \vartheta_j^2 = \beta_j^2/4, \\ \exp(-a_1 t) & \text{for } \mathrm{Fo}_{q,j} \vartheta_j^2 < \beta_j^2/4, \\ \exp(-\beta_j t/2) \cos(\zeta_j t/2) & \text{for } \mathrm{Fo}_{q,j} \vartheta_j^2 > \beta_j^2/4; \end{cases}$$

$$f_{2,j}(t) = \begin{cases} \exp(-\beta_j t/2) & \text{for } \mathrm{Fo}_{q,j} \vartheta_j^2 = \beta_j^2/4, \\ \exp(-a_2 t) & \text{for } \mathrm{Fo}_{q,j} \vartheta_j^2 < \beta_j^2/4, \\ \exp(-\beta_j t/2) \sin(\zeta_j t/2) & \text{for } \mathrm{Fo}_{q,j} \vartheta_j^2 > \beta_j^2/4; \end{cases}$$

$$F_j(t) = \begin{cases} \int_0^t W_j(\eta)(t-\eta) \exp[-\beta_j (\eta - t)/2]\, d\eta & \text{for } \mathrm{Fo}_{q,j} \vartheta_j^2 = \beta_j^2/4, \\ \frac{2}{\zeta_j} \int_0^t W_j(\eta)\, \mathrm{sh}[\zeta_j (t-\eta)/2] \exp[-\beta_j (\eta - t)/2]\, d\eta & \text{for } \mathrm{Fo}_{q,j} \vartheta_j^2 < \beta_j^2/4, \\ \frac{2}{\zeta_j} \int_0^t W_j(\eta) \sin[\zeta_j (t-\eta)/2] \exp[-\beta_j (\eta - t)/2]\, d\eta & \text{for } \mathrm{Fo}_{q,j} \vartheta_j^2 > \beta_j^2/4; \end{cases}$$

$d_l = \mathrm{Bi}_{x,j} K_{x,j}(\mu_{n,j}) \sum_{k=0}^{k_y} g_{k,l} S_k(\gamma_{m,j}) + \overline{K_{x,j}} \left[\mathrm{Bi}_{y=0,j}\, g_{0,l} + \mathrm{Bi}_{y=1,j} K_{y,j}(\gamma_{m,j}) \sum_{k=0}^{k_y} g_{k,l} \right] \Big/ L^2$; $\quad a_2 = (\beta_j + \zeta_j)/2$;

$\overline{K_{x,j}} = \int_0^1 K_{x,j}(\mu_{n,j} X)\, dX$; $\overline{K_{y,j}} = \int_0^1 K_{y,j}(\gamma_{m,j} Y)\, dY$; $S_k(\gamma_{m,j}) = \int_0^1 Y^k K_{y,j}(\gamma_{m,j} Y)\, dY$; $a_1 = (\beta_j - \zeta_j)/2$;

$K_{x,j}(\mu_{n,j} X) = \cos(\mu_{n,j} X)$ and $K_{y,j}(\gamma_{m,j} Y) = \cos(\gamma_{m,j} Y) + \mathrm{Bi}_{y=0,j} \sin(\gamma_{m,j} Y)/\gamma_{m,j}$ are the transformation kernels; $\mu_{n,j}$ (n = 1, 2, …) and $\gamma_{m,j}$ (m = 1, 2, …) are the the roots of the characteristic Eq.: $\mathrm{Bi}_{x,j} = \mu_{n,j} \tan \mu_{n,j}$, $\mathrm{tg}(\gamma_{m,j})(1 - \mathrm{Bi}_{y=0,j} \mathrm{Bi}_{y=1,j}/\gamma_{m,j}^2) = (\mathrm{Bi}_{y=0,j} + \mathrm{Bi}_{y=1,j})/\gamma_{m,j}$.

The obtained solution is characterized by inhomogeneous convergence of Fourier series, because of which, to obtain the result of an acceptable accuracy, it is necessary to increase the count of Fourier series terms for the longitudinal coordinate in solution (10) for large Bio numbers to several hundred or more. The application of the method for improving the convergence, described for the one-dimensional problem in [12], and for the two-dimensional — in the monograph [13], made it possible to obtain the final solution

$$\theta_j(X,Y,t) = \theta_{f,j}(Y,t) + \sum_{n=1}^{\infty} A_{n,j} K_x(\mu_{n,j} X) \sum_{m=1}^{\infty} A_{m,j} K_y(\gamma_{m,j} Y) \left[\theta_{L,j}(\mu_{n,j}, \gamma_{m,j}, t) - W_j(t)/\vartheta_j^2 \right], \tag{11}$$

where constants $C_{1,j}$ and $C_{2,j}$ are determined from the initial conditions (5). Due to the improvement of convergence, it was possible to reduce the number of the series terms by an order of magnitude or more without loss of the solution accuracy.

The reliability of analytic solutions of heat conduction hyperbolic boundary-value problems can be judged from the results of a comparison of the analytic solution of a hyperbolic boundary-value problem for a one-dimensional body (infinite cylinder) with $\mathrm{Bi}_x = 1$, $\mathrm{Fo}_q = 10$, and a special case of equation (3) ($\tau_T = 0$) with a numerical solution of the same problem, given in the monograph [13 , 208-220 pp.] and shown in Fig. 2.

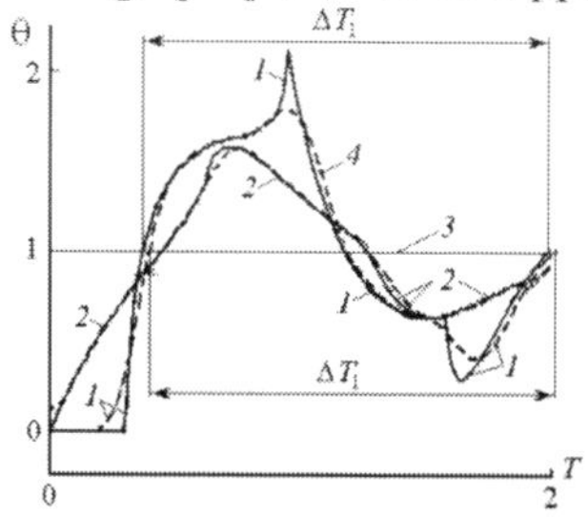

Fig. 2. Transient thermal processes in a cylinder [13]: *1, 4* - on the axis ($X = 0$); *2* - on the surface ($X = 1$); *3* is the temperature of the medium; *1, 2* - analytical solution; *4* - numerical solution.

The graphs in Fig. 2 obviously show that the analytically calculated transient process more closely reflects the harmonics of high (second, third, etc.) orders than calculated by the numerical method, in spite of the fact that in the second case was used a difference scheme of the second order accuracy, both in coordinate and in time.

3 Conclusions

An analytic description of the heat transfer process in a two-dimensional plate was obtained for the first time according to the Maxwell-Cattaneo-Luikov hypothesis, taking into account the dependences of the heat transfer coefficient of the body surface on time and the temperature of the ambient versus time and the longitudinal coordinate.

References

1. J. C. Maxwell, Philos. Trans. Soc. London, **157**, 49 (1867)
2. L. Onsager, Phys. Rev. **37**, 405 (1931)
3. A.V. Luikov, *The theory of heat conductivity* (Moscow, Vysshaya Shkola, 1967) [in Russian]
4. C. Cattaneo, Seminario Matematico e Fisico dell' Universita di Modena. **3**, 83 (1948)
5. P. Vernotte, Compte Rendus. Acad. Sci. Paris. **246**, 3154 (1958)
6. K. Mitra, S. Kumar, A. Vedevarz, M. Moallemi. J. Heat Transfer. **117**, 568 (1995)
7. A.V. Luikov, J. Eng. Phys. Thermophys. **9**, 3 (1965)
8. D.Y. Tzou, Trans. ASME J. Heat Transfer, **117**, 8 (1995)
9. M. Torabi, K. Zhang, Int. J. Heat Mass Transfer, **78**, 960 (2014)
10. Yu.A. Kirsanov, A.Yu. Kirsanov, A. Yudakhin, High Temp., **55**, 1 (2017)
11. Yu.A. Kirsanov, A.Yu. Kirsanov, A. Yudakhin, Trans. of Academenergo, **4**, 32 (2016)
12. E.M. Kartashov, *The analytical methods for theory of heat conductivity of solids* (Moscow, Vysshaya Shkola, 1985) [in Russian]
13. Yu.A. Kirsanov, *Cycle Thermal Processes and the theory of thermal conductivity in regenerative air heater* (Moscow, Fizmatlit, 2007) [in Russian]

Transformation of spectra of refraction index fluctuations in axisymmetric supersonic jet with the increase in the distance from the nozzle

Dmitrii Marakasov[1,*], *Valentina* Sazanovich[2], *Ruvim* Tsvyk[1], and *Andrei* Shesternin[1]

[1]V.E. Zuev Institute of Atmospheric Optics SB RAS, Wave Propagation Laboratory, 634055 Ac. Zuev square, 1, Tomsk, Russia

Abstract. In this paper the results of reconstruction of the radial dependence of refraction index fluctuations spectra in axisymmetric supersonic jet from laser transillumination data are presented. The reconstruction algorithm uses experimental spectra of laser radiation, received behind the jet, and the distribution of time-averaged axial air velocity within the jet. Laser transillumination experiments were performed on the Vertical Jet Unit of ITAM SB RAS (Novosibirsk, Russia). Reconstruction results show the refraction index spectra obey the power-law model, with power index different from that of the Kolmogorov turbulence model (-11/3). Within the first two barrels of the jet the power index is changing from ~-1 near the nozzle to ~-2.7 near the second Mach disk. Also some variations of power index and amplitude coefficient with the distance from the jet axis are found.

Currently the most common contact methods of measurement of pulsation characteristics of the flow at wind tunnels have fatal flaws stemming from their very essence. The direct interaction of the sensors with the supersonic flow leads, on the one hand, to their rapid destruction, and on the other hand, to the distortion of flow structure.

The dynamics of fluid or gas flows can also be studied with the use of noncontact optical methods. Among these are speckle photography and speckle interferometry methods [1], in which the source of information is represented by intensity fluctuations of the laser radiation passed through the flow and a diffuse plate. The widely used methods of shadow visualization (Schlieren photography) do not allow the spatial spectrum of refractive index fluctuations to be determined from the obtained images [2]. Shadow images are integral characteristics of the refractive index in the entire radiation propagation path, and the spatial distribution of the refractive index in different flow layers cannot be reconstructed from them.

Hence it is of interest to study the possibilities of remote diagnostics of the dencity inhomogeneities in supersonic gas flow based on turbulent intensity fluctuations of translucent optical radiation, which does not require the presence of scattering particles.

* Corresponding author: mda@iao.ru

Fig. 1. Vertical jet unit with optical equipment. Source part (right) and receiver part (left).

Based on the algorithm of determining the radial dependence of spatial spectra of the refractive index from time spectra of the received power of laser radiation translucent the axisymmetric jet [3], the transformation of the refraction index spectra accompanied by the change of the air velocity and the turbulence with distance from the nozzle exit was investigated. The original data were obtained in experiments performed in 2016 at the Vertical Jet Unit (VJU) of ITAM SB RAS (Figure 1) also the results of the earlier experiments (2013) were used.

The supersonic jet generated by Witoshinski nozzle with a diameter of 30 mm, went straight up to the room of size 5X5X3 m. On opposite sides of the jet on the optical tables were located laser source and forming diaphragm on one side and the receiving part on the other (Figure 1). The laser beam is converted in a spherical wave by the diaphragm located at a distance of 1 m from the jet axis. At a distance of 1 m behind the jet laser radiation perturbed by inhomogeneities inside the jet was falling on the entrances of the three lightguides with a diameter of 0.15 mm, which carried signals to three PMT. The output of photomultipliers was recorded using oscilloscope TiePie Handyscope HS4diff with a sampling rate of 5 MHz per channel. The record length was 130000 counts. The distance between the lightguides was 2 mm, which corresponded to the spacing of the beams inside the jet 1 mm. Optical table with the laser source and forming the diaphragm moved vertically (along the jet axis), the table of the receiving part moved both vertically and horizontally (perpendicularly to the optical axis). This allowed to vary as the distance of the optical axis from the nozzle exit (42-90 mm) and impact parameter (distance from the beam to the axis of the jet at the point of convergence) up to 23 mm.

Reconstruction algorithm for the radial dependence of the spectra [3] uses a priori information about the time-averaged flow velocity in the corresponding cross section of the jet. The axial component of the average velocity was calculated from the results of numerical simulation of a jet with the ratio of the pressure in the settling chamber to atmospheric $npr = 5$ in a package Fluent 6.3. The radial dependences of the axial velocity for some distances from the nozzle exit corresponding to the second barrel are presented in Fig. 2.

It is seen that in the vicinity of the Mach disk the axial velocity in the inner region of the jet is changed in ~5 times from subsonic values near the axis of the jet to supersonic closely to its external border. In the region between the first and second Mach disks its relative variations over the jet cross section are in the range of 15-20%. Thus, the renormalization of frequencies in the framework of the reconstruction algorithm of the radial dependence of spectra of density fluctuations is essential for sections close to the Mach disk, and does not introduce large refiners for intermediate sections.

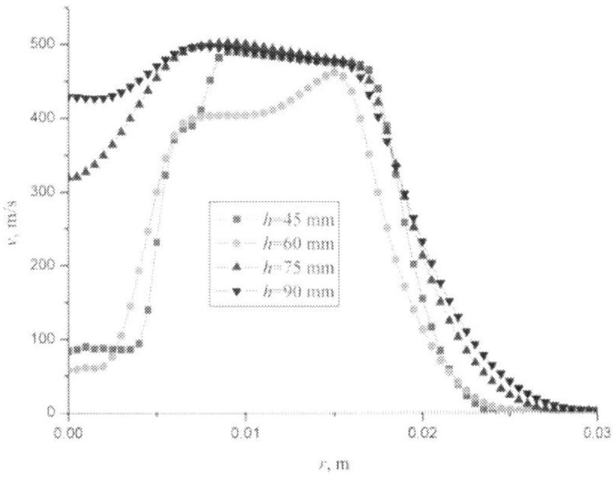

Fig. 2. Axial velocity distribution in the supersonic jet with *npr* = 5.

In the experiments performed in the year 2013 the initial segment (the first barrel) of the jet was transilluminated. Processing of the experimental data obtained at the VJU in 2013, according to the methods developed in [3], showed that the spectra are generally consistent with power law models $\Phi \sim \kappa^{-\nu}$ with index ν varying with the distance from the nozzle and the jet axis. It is seen that the index ν is much smaller than in the Kolmogorov-Obukhov model of developed turbulence (11/3) and gradually increases with the distance from the nozzle exit *h*. The most intense turbulence occurs in the mixing layer at the outer edge of the jet, as evidenced by faster growth of the parameter ν in this area compared to its values at the jet axis. Intense turbulence development leads to the increase of the index ν in this area relatively to its value near the axis (Table.1).

Table 1. Variations of the infex ν inside the first barrel.

	***h* = 3 mm**	***h* = 8 mm**	***h* = 18 mm**	***h* = 28.5 mm**
***r* = 0**	0.45	1.1	1.3	1.65
***r* = 10 mm**	0.5	1.2	1.5	1.85

Experimental data of the year 2016 refer to the next barrel of the jet (heights *h* = 42-90 mm from the nozzle exit). Examples of spectral density of laser radiation passing through the jet at *h* = 42 mm from the nozzle for impact parameters in range 0-20 mm are shown on Figure 3. According to [4], the high frequency asymptotes have the power index -ν-1. Analysis of the results of 2016 showed that the power model, in general, good fits the spectra in the interval of heights 42-90 mm from the nozzle exit also, however, the values of the index ν in the second barrel are noticeably superior to its value in the first barrel but are still less than for the developed atmospheric turbulence (11/3). The second barrel of the values of the index lie in the range 2-2.7 (Table. 2). The index ν appears to be minimal at the jumps of averaged parameters of the jet (internal and external mixing layer, Mach disk). Probably it is the consequence of the presence of spatially coherent sources of turbulence importing whirlwinds with relatively large spatial scales, which degradation into small whirlwinds happens in more distant jet area.

We have to note that the values of the index of the spectrum in the second barrel are close to the model of coherent turbulence [5], describing the decay of large eddies with scales close to the size of the turbulence source. The increase of the index with the increase of turbulence and deceleration of the jet corresponds to the transition to the subsonic flow regime [6].

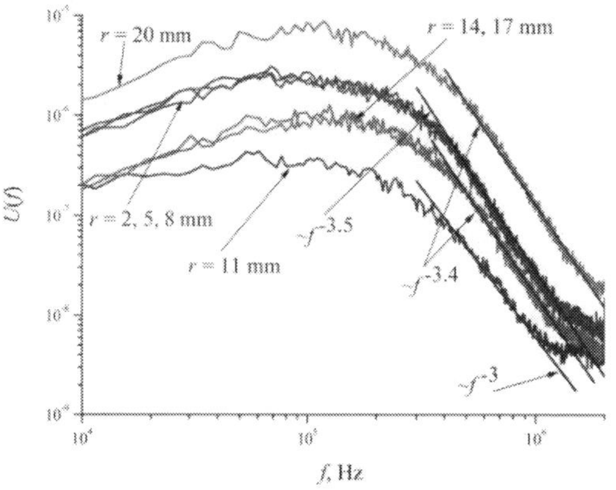

Fig. 3. Transformation of the laser radiation spectra with the increase of impact parameter (h = 42 mm, npr = 5).

Table 2. Variations of the infex ν inside the second barrel.

	***h* = 42 mm**	***h* = 50 mm**	***h* = 60 mm**	***h* = 70 mm**	***h* = 80 mm**
***r* = 0**	2.3	2.0	1.9	2.0	2.4
***r* = 7 mm**	2.3	2.2	2.0	2.3	2.7
***r* = 13 mm**	2.1	2.3	1.9	2.1	2.5

Conclusions

Thus, the reconstruction of the spectra of the refractive index from the data of laser scanning shows that the development of turbulent processes in axially symmetric supersonic jet is described by a power-law model of the spectrum of refractive index with rate ν varying along the jet axis and its radius. The main changes correspond to the increase of the rate as the turbulence of the flow. The result is a small increase in index from the axis to the outer boundary of the jet and its noticeable increase along the axis in the first and second barrels. Small variations appear near the sources of turbulence (Mach disk, mixing layer). Values of power index correspond to the transformation of the coherent turbulence model to the developed turbulence one.

This work was supported by the RFBR grant №15-08-06549 A.

References

1. N.A. Fomin, *Speckle Photography for Fluid Mechanics Measurements. Series: Experimental Fluid Mechanics.* (Springer-Verlag, Berlin, 1998)
2. G. Meier, Exp. Fluids, **33**, 181 (2002)
3. D.A. Marakasov, Proc. SPIE **10035,** 100352B-1 (2016)
4. D.A. Marakasov, V.M. Sazanovich, A.A. Sukharev, R.Sh. Tsvyk, Atmospheric and Oceanic Optics **26**, 11 (2013)

5. V.V. Nosov, P.G. Kovadlo, V.P. Lukin, A.V. Torgaev, Atmospheric and Oceanic Optics **26,** 201 (2013)
6. I.A. Joia, B.J. Uscinski, , R.J. Perkins, G. Balmer, , D. Jordan, E. Jakeman, Waves in Random Media **7**, 169 (1997)

Investigating heat transfer augmentation using gradient heat flux measurement and PIV method

Andrey Mityakov*, *Aleksandr* Babich, *Aleksandr* Bashkatov, *Andrey* Gusakov, *Aleksey Dymkin, Elza* Zainullina, *Sergey* Sapozhnikov, *Vladimir* Mityakov and *Vladimir* Seroshtanov

Deparment of Thermophysics of Power Units, Peter the Great St. Petersburg Polytechnic University, Russian Federation, 195251, Saint-Petersburg

Abstract. The heat transfer and the fluid flow behaviour at the surface of a circular cylinder with and without turbulizing rods, heated by saturated steam rectangular and circular fins were studied for different Reynolds numbers. The proposed approach of simultaneous use of gradient heat flux measurement and PIV method allowed us optimizing the geometry of the system with the rods and increasing heat transfer coefficient up to 20%.

1 Introduction

Up to the present time, it was not possible to combine real-time fluid field with local heat flux in problems of heat flux augmentation. The combining of gradient heat flux measurement [1] and PIV method of fluid visualization [2] gives new opportunities for investigation [3], [4]. Gradient heat flux sensors (GHFS) allow measuring heat flux per unit area practically without inertia because of their response time constant of 10^{-8} to 10^{-9} s. Therefore, we can research both processes simultaneously.

2 Experimental details and results

Experiments were carried out in the closed-circuit wind tunnel. The tube has test section of 870 mm length and contraction cone outlet diameter of 450 mm. The Eiffel chamber is made of plexiglas to allow PIV measurements from outside the chamber. The wind tunnel is equipped with a water-air heat exchanger to maintain constant flow temperature. Free-stream temperature T_∞ and velocity W_∞ were measured by the Testo 435-2 multifunction HVAC and IAQ Meter. The longitudinal turbulent intensity was found to be less than 0.5% for the $10^4 < \mathrm{Re} < 10^5$. In the current study we used the method of simultaneous PIV and heat flux measurement.

To obtain a 3D velocity field stereo PIV of POLIS type was used [2]. The flow was seeded by the smoke with a particle size of about 3 μm in diameter. *ActualFlow* software application was used for data processing.

* Corresponding author: andrey.mityakov@gmail.com

The heat flux was measured by gradient heat flux sensors (GHFS) produced of a single crystal of 99.99% pure bismuth in Peter the Great St. Petersburg Polytechnic University (Russia). The GHFS plan form dimensions were 2 × 2 mm, while its thickness was about 0.2 mm.

The cylinder model, used in the experiments was made of a steel sheet of 0.1 mm in thickness, its diameter was of 66 mm, and its length was of 800 mm. The model was heated by saturated water steam with temperature close to 100°C. Experiments were carried out with constant temperature boundary conditions which were confirmed by the infrared measuring using FLIR P640 infrared camera. The GHFS was installed at the cylinder surface flat with it. The cylinder was turned about its axis by an electric driver which allowed moving the sensor in circumferential direction along the cylinder surface.

Figure 1 illustrates the scheme of our experiment. Attack angle β is defined as an angle between the laser sheet and a cylinder axis (β = 90° for the cross-flow). The GHFS angular coordinate φ is defined as the angular distance between the front stagnation point and GHFS location. Flow velocity component W_z characterizes the flow three-dimensionality. More detailed information about the method could be found in the work [5].

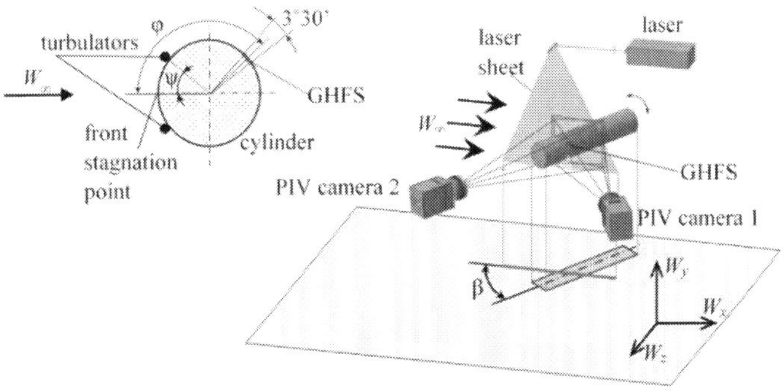

Fig. 1. Scheme of the experiment with cylindrical model

In the first series of experiments the circular cylinder with turbulizing rods was obtained. The rods with diameter d_T were placed parallel to the axis of the cylinder, with a gap (Figure 2, b) δ and without it (Figure 2, a). The rods were placed symmetrically from the front stagnation point at the angle ψ. The values of d_T, δ and ψ were optimized during the experiment.

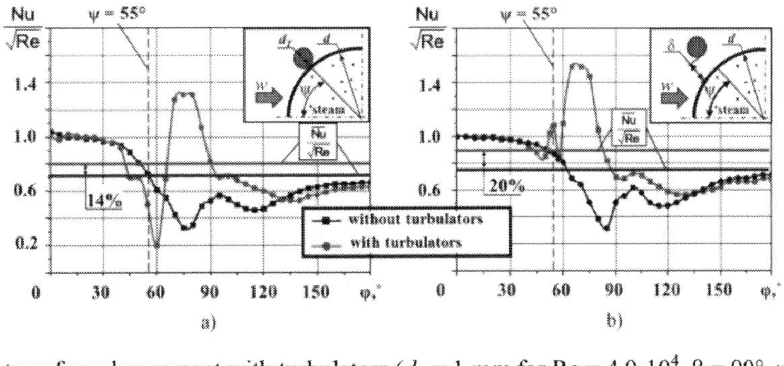

Fig. 2. Heat transfer enhancement with turbulators (d_T = 1 mm for Re = $4.9 \cdot 10^4$, β = 90°, ψ_T = 55°: a — without the gap; b — with the gap, δ = 1.2 mm.

Figure 2 shows curves of heat transfer coefficient when β = 90°. The local heat transfer coefficient near the rods increases 2.5 times, and the average heat transfer coefficient increases by 14 to 20 % in comparison of the cylinder without turbulence rods (black line).

 DOI: 10.1051/matecconf/201711502006

PIV technology visualized the velocity field near the model surface (Figure 3) and allowed finding optimal geometry of the system.

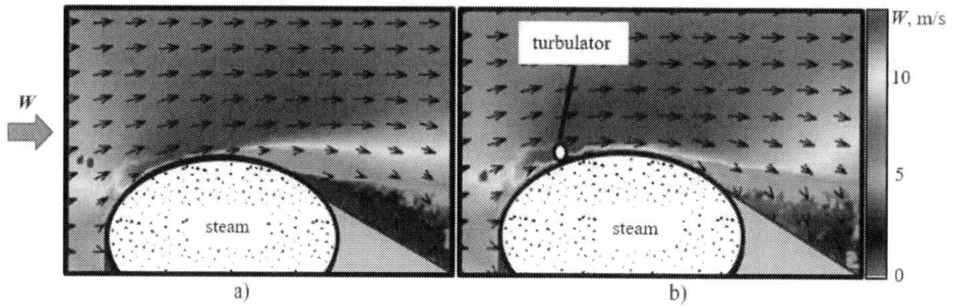

Fig. 3. Visualisation of flow around the cylinder with the turbulizing rods (b) and without them (a). $Re = 5.3 \cdot 10^4$, $\beta = 60°$, $\psi_T = 63°$, $d_T = 1$ mm

Another series of experiments was devoted to investigation of the flow around a circular cylinder at $\beta < 90°$. The turbulizing rods were placed close to the cylinder surface or with some gap from it. It was shown that at different values of β the optimal value of ψ was changed.

The next stage of research was devoted to the flow around isothermal ("ideal") rectangular (Figure 4) and circular (Figure 5) fins which were mounted normally to the circular cylinder surface.

Two GHFSs with sizes 4 × 7 mm and 2 × 2 mm and with the thickness of 0.2 mm were installed at the fins. The volt-watt sensitivity of these sensors was 12.3 mV/W and 21.8 mV/W, respectively. The sensors were installed in different places along the length of the fin. The fin was set in three positions: upstream (Figure 4, a), normal to the flow (Figure 4, b) and downstream (Figure 4, c).

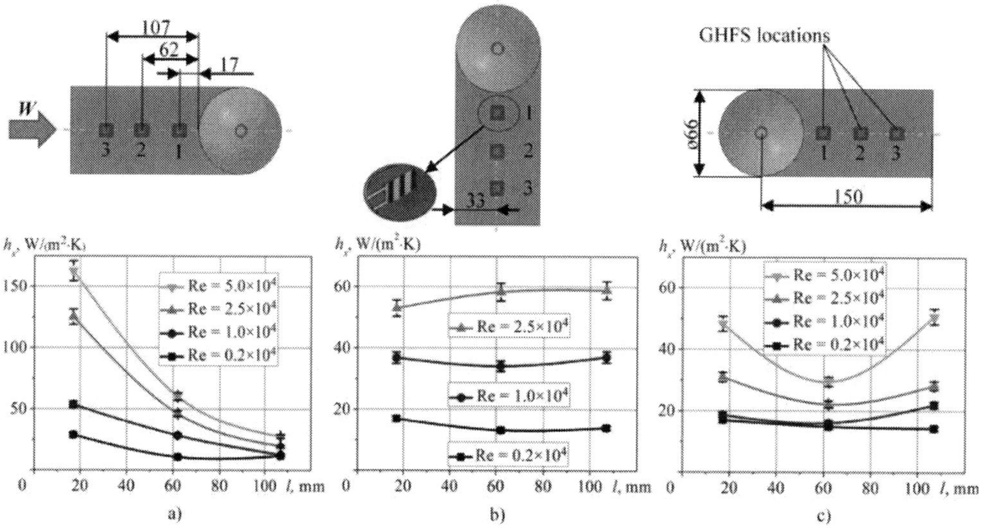

Fig. 4. Rectangular fin orientation within the flow and the local heat transfer coefficients on its surface: a — upstream, b — normal to the stream, c — downstream

The velocity field and heat flux were fixed for some real time moments. Relationship between the local heat transfer coefficient and Reynolds number for fin installed in different positions are shown in Figure 4. The Standard Relative Uncertainty was evaluated in accordance with type B.

Non-monotonic character of the curves in Figure 4 attracts deserves attention and requires further investigation. Next step was to associate heat flux measurement with velocity field.

One of our GHFSs was installed at the surface of the circular fin (Figure 5, a). During the experiment, the model was rotated around the axis of the cylinder. The dependence of the local heat transfer coefficient, azimuthal angle and Reynolds number is presented in Figure 5, b.

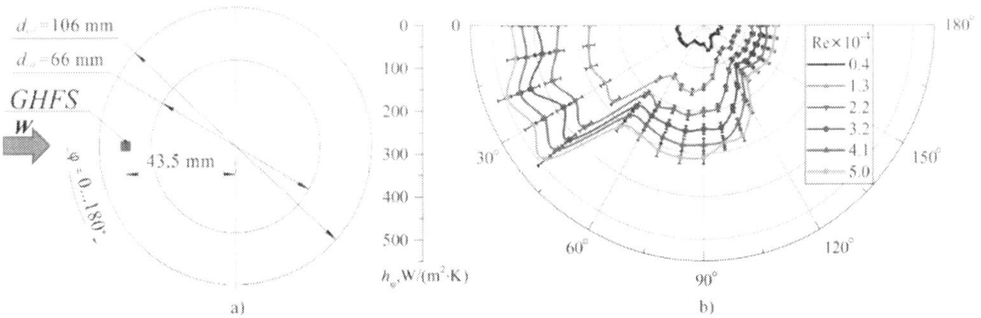

Fig. 5. Location of GHFS at the circular fin (a) and relations between local heat transfer coefficient, azimuthal angle and Reynolds number (b)

3 Conclusion

The combination of PIV method and the gradient heat flux measurement allowed investigating the flow and heat transfer at the surface of circular cylinder with turbulizing rods and with rectangular and circular fins. The geometry of the system with the rods was optimized during the experiment. Moreover, it was the first step in the comparison of heat flux and fluid flow in the same moments of time during the process.

References

1. V. Mityakov, S.Z. Sapozhnikov, V.Y. Mityakov et al., Sensors and Actuators A: Physical **176**, 1 (2012)
2. «POLIS» velocity profile measuring instrument ([website]/Institute of Thermophysics SB RAS, Novosibirsk – access mode: http://eng.polis-instruments.ru/)
3. V.V. Seroshtanov, A.V. Bashkatov, E.R. Zaynullina et al., Industrial and Medical Measurement and Sensor Technology Vehicle Sensor Technology **5**, 26 (2016)
4. A.V. Bashkatov, A.Y. Babich, A.A. Gusakov et al., Industrial and Medical Measurement and Sensor Technology Vehicle Sensor Technology **5**, 15 (2016)
5. S.Z. Sapozhnikov, V.Y. Mityakov, A.V. Mityakov et al., Appl. Mech. Materials **629**, 444 (2014)

Estimating the terms of turbulent kinetic energy transport equation in an unsteady flow on the basis of SIV measurements

Nikolay Mikheev[1,*] and *Ilya* Saushin[1]

[1] Kazan Scientific Center of RAS, Kazan, 420111, Russia
[2] Kazan National Research Technical University named after A.N. Tupolev, Kazan, 420111, Russia

Abstract. Turbulence production and dissipation profiles averaged over a period of forced flow pulsations have been obtained experimentally across the boundary layer, developing on a channel wall under forced periodic freestream unsteadiness. The experiments have been conducted in a wide range of dimensionless numbers β=0.03-0.28 and Strouhal numbers St=2.99-64.2 that characterize relative amplitude and frequency of flow pulsations. The parameters have been estimated based on the instantaneous velocity fields, measured by 2D optical method of Smoke Image Velocimetry (SIV). Turbulence dissipation and production profiles plotted in wall coordinates have been compared to the corresponding steady flow profiles. In the considered ranges of dimensionless numbers, the insensitivity of turbulent dissipation profiles to periodic flow unsteadiness has been established, and conclusion on similar insensitivity of turbulence production profiles drawn on the basis of point measurements and DNS has been confirmed.

1 Introduction

A transport equation of turbulent kinetic energy is a basic equation describing turbulence production, transport and dissipation. In respect to the boundary layer, it is important to have the information on energy balance across the boundary layer. It enables a further insight into complex processes and phenomena, a study laminar-turbulent transition and updating of turbulence models intended for RANS and LES methods.

Turbulence production in the boundary layer is written through a product of the second moments of velocity fluctuations and the spatial gradient of velocity

$$\langle P_{ij} \rangle = \langle u_i' u_j' \rangle \frac{\partial \langle U_i \rangle}{\partial x_j}. \tag{1}$$

This term was estimated in [1-5] for the turbulent boundary layer under forced freestream pulsations of the flow around a plate and in a tube in a wide range of

* Corresponding author: n.miheev@mail.ru

dimensionless numbers. These estimates were obtained using point measurements and included, among others, the data on dynamics of turbulence production profiles over the phase of forced pulsations. Authors [6, 7] analyzed the influence of forced unsteadiness on the production term theoretically on the basis of turbulent fluctuation measurements.

To calculate turbulence dissipation, pairwise correlations of spatial gradients of pulsatory components of all three velocity vector components should be estimated:

$$\varepsilon = \nu\left\{2\overline{\frac{\partial u_i'}{\partial x_i}\frac{\partial u_i'}{\partial x_i}} + \overline{\frac{\partial u_i'}{\partial x_j}\frac{\partial u_i'}{\partial x_j}} + 2\overline{\frac{\partial u_i'}{\partial x_j}\frac{\partial u_j'}{\partial x_i}}\right\}. \quad (2)$$

Obviously, point measurements cannot yield a direct estimation of ε. Authors [5] analyzed the behavior of dissipation integrated over the boundary layer thickness depending on the phase of forced pulsations. The dissipation was estimated by the difference between the turbulent kinetic energy and its production.

Authors [8-10] noted that dissipation cannot be estimated by an optical technique of Particle Image Velocimetry (PIV) with sufficient accuracy as yet due to limited spatial resolution of the method. The potential of 3D particle tracking (LPT) in particle trajectory reconstruction in high seeding density flows was shown in [11] using Shake-The-Box technique. This will obviously allow the direct estimation of the strain rate tensor components and hence dissipation. Yet the development of experimental methods for estimation of dissipation in the boundary layer remains an open problem.

The paper deals with estimation of production and dissipation profiles averaged over the period of forced pulsations in pulsating turbulent boundary layer using the instantaneous velocity fields measured by an optical method of Smoke Image Velocimetry (SIV) [12].

2 Experimental technique and procedure

An experimental setup shown in fig.1 was used to study the boundary layer structure in the pulsating flow. Test section *1* of the experimental setup was a 75×150 mm^2 one meter long rectuangular channel with a 6:1 contraction ratio smooth inlet *10*. Turbulence generating grid *9* with square cells was mounted downstream of the smooth inlet. The grid was made of 1.2-mm thick wire; the distance between the wire axes was 5 mm. A 50-mm long strip of abrasive *12* (ISO 6344-2) was glued onto the channel perimeter. This provided fully developed turbulent boundary layer in the measurement area during the experiments. Channel walls were made of transparent materials (glass and polycarbonate). Stable air flow rate downstream of the test section was provided by a regulating gate *11* and a 1.3 m^3 receiver tank *2*. The flow rate was measured by an ultrasonic flowmeter IRVIS RS4-Ultra *3* mounted downstream of the receiver tank. The relative error in flow rate did not exceed 1%. The average flow rate, frequency and amplitude of forced velocity pulsations were adjusted by a static and rotating flap in pulsator *13*.

To visualize the flow pattern, the air-aerosol mixture (MT-Gravity fluid with medium fog density and average particle size of 0.1…5 μm; Safex aerosol generator *5*) was supplied from the preparation chamber *4* to the channel inlet. The measurement area *6* was illuminated by a continuous diode-pumped solid-state laser KLM-532/5000-h *7*. The flow pattern in the channel symmetry plane at the distance of L = 0.7 m from the turbulence generating grid was recorded by a monochrome high-speed camera Fastec HiSpec *8* with the frame resolution of 665×110 pixel (scaling factor of 0.0625 mm/pixel), frame rate f = 7083 1/s, and recording time of 3 s.

Flow velocity fields were measured by the optical method of Smoke Image Velocimetry (SIV) [12], in which velocity fields are obtained from the analysis of turbulent structure displacements visualized by smoke. Period-averaged frame resolution in y^+-coordinates was 1 pixel = 0.8 y^+, interrogation window size was 16×16 pixel, period-averaged ratio of the resolved scale to Kolmogorov length scale was 1.6.

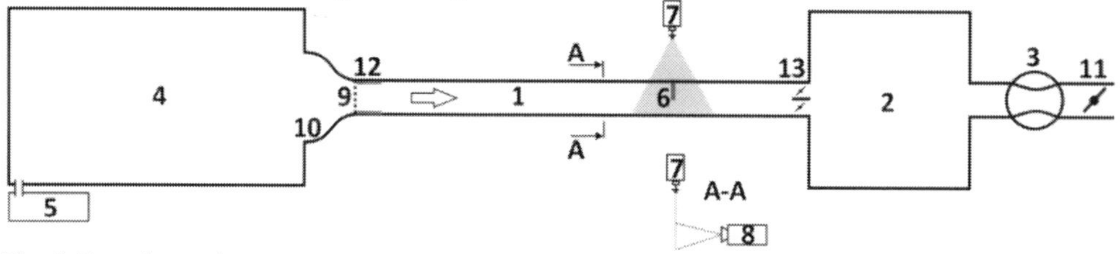

Fig. 1. Experimental setup

3 Results and discussion

Turbulence production and dissipation were estimated by turbulent velocity fluctuations. The latter were obtained from the following relation [13] (for the case of forced periodic freestream pulsations):

$$u_i'(y,t) = U(y,t) - \bar{U}(y) - \tilde{U}(y,t), \tag{3}$$

where U is the instantaneous velocity, $\bar{U}$ is the average instantaneous velocity, $\tilde{U}$ is the fundamental frequency of U oscillogram corresponding to the frequency of forced unsteadiness. Second-order central difference scheme was used to approximate the derivatives. Dissipation was estimated with the assumption of turbulence isotropy in transversal direction.

Fig. 2 shows the period-averaged profiles of turbulence production and generation in wall coordinates. The profiles are normalized by $\nu / \bar{u}_\tau^4$ and illustrate five regimes of pulsating flow and steady flow (St=0) considered in the research. Similar profiles obtained by DNS [14, 15] are plotted with lines. Experimental data for the steady flow case were obtained for two sizes of the measurement area along y-coordinate: 6 and 16 pixel (approximately 5…13 in wall coordinates). As one can see from the figure, the peak of turbulence production curve is significantly "smeared" due to spatial averaging. This indicates that the measurement area should be reduced in the wall-normal direction down to the order of Kolmogorov length scale.

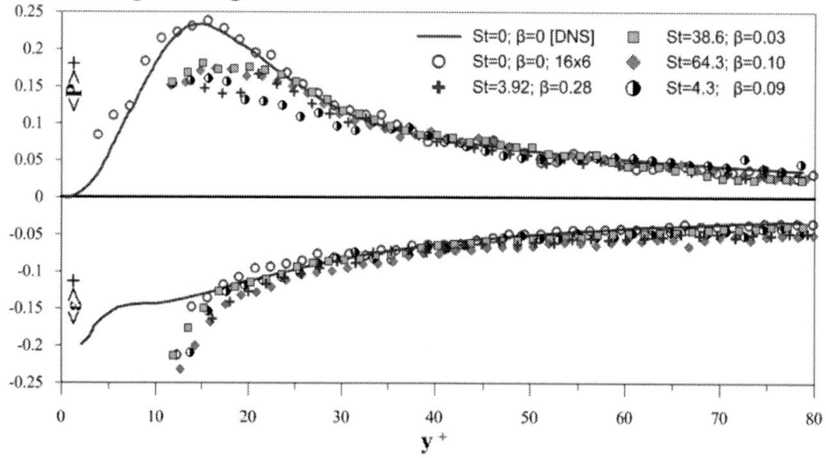

Fig. 2. Period-averaged profiles of turbulence production and dissipation in the boundary layer of steady and pulsating flow; DNS: <P>+ [14], <ε>+ [15]

Fig.2 shows that despite the wide range of dimensionless Strouhal number, St, and relative amplitude of velocity, β, on the channel axis considered in the research, almost identical profiles of turbulence production and dissipation were obtained. As for the turbulence generation profile, this result agrees with conclusions drawn in [1-3, 16] on the insensitivity of $\langle P\rangle^+$ profile to the forced periodic unsteadiness revealed by point measurements and DNS. Dissipation profiles in the turbulent boundary layer in conditions of forced flow unsteadiness have been obtained in the present paper for the first time ever. These profiles appeared to be insensitive to the periodic unsteadiness.

4 Conclusions

Space and time scales of velocity vector fields resolved by SIV technique are sufficient for the estimation of velocity fluctuation derivatives with respect to time and space. Turbulent energy production and dissipation profiles in the pulsating flow in the channel have been estimated from SIV measurements. The insensitivity of turbulent energy production and dissipation profiles to periodic flow unsteadiness has been experimentally established in the considered range of dimensionless numbers.

This work was supported by the Russian Science Foundation (Project no.16-19-10336).

References

1. G.J. Brereton, W.C. Reynolds, R. Jayaraman, J. Fluid Mech. **221**, 131 (1990)
2. G.J. Brereton, W.C. Reynolds, Phys. Fluids A **3**, 178 (1991)
3. R. Akhavan, R.D. Kamm, A.H. Shapiro, J. Fluid Mech. **225**, 395 (1991)
4. M.M. Grigoryev,V.V. Kuzmin, A.V. Fafurin, J. Eng Phys Thermophys **59**, 725
5. M.Hino, M.Kashiwayanagi, A.Nakayama, T. Hara, J. Fluid Mech. **131**, 363 (1983)
6. F.S. Tardu, G. Binder, R.F. Blackwelder, J. Fluid Mech. **267**, 109 (1994)
7. G. Binder, S. Tardu, P. Vezin, Proc. R. Soc. Lond. **451**, 121 (1995)
8. J. Charonko, K. Prestridge, 18th International Symposium on the Application of Laser and Imaging Techniques to Fluid Mechanics, ISBN 978-989-98777-8-8 (2016)
9. J.M. Foucaut, C. Cuvier, M. Stanislas, W.K. George, Progress in wall turbulence 2. Ercoftac series **23**, 429 (2016)
10. J.F.G. Schneiders, F. Scarano, Exp Fluids **57,** 139 (2016)
11. A. Schröder, D. Schanz, R. Geisler, S. Gesemann, 18th International Symposium on the Application of Laser and Imaging Techniques to Fluid Mechanics, ISBN 978-989-98777-8-8 (2016)
12. N.I. Mikheev, N.S. Dushin, Instrum. Exp. Tech. **59**, 880 (2016)
13. A.K.M.F. Hussain, W.C. Reynolds, J. Fluid Mech. **41**, 241 (1970)
14. N.N. Mansour, J.Kim, P. Moin, J. Fluid Mech. **194**, 15 (1988)
15. P. Schlatter, R. Örlü, J. Fluid Mech. **659**, 116 (2016)
16. P.K. Papadopoulos, A.P. Vouros, Int. J. Heat Fluid Flow **58**, 54, (2016)

Mechanism of transition to turbulence in a circular cylinder wake in a channel

Valery Molochnikov[1,*], *Alexander* Mazo[2], *Dmitriy* Okhotnikov[2] and *Anna* Goltsman[1]

[1] Kazan Scientific Center of RAS, Kazan, 420111, Russia
[2] Kazan Federal University, Kazan, 420008, Russia

Abstract. Transition to turbulence in the circular cylinder wake has been studied experimentally and numerically at growing Reynolds number. Good agreement of calculation results with the flow visualization and measurements of instantaneous vector fields of velocity and vorticity has been demonstrated. The growing Reynolds number is shown to make large-scale vortex generation onset move upstream. It also triggers the transition to 3D flow pattern in the cylinder wake. This process is accompanied by non-monotonous behavior of the profiles of velocity and its turbulent fluctuations at equal distances from the cylinder. Non-monotonous behavior of the cylinder drag has been revealed for the Reynolds numbers ranging from 120 to 300.

1 Introduction

The regular Karman street onset in the wake of a cylinder in a cross-flow is attributed (depending on the literature source) to the range of Reynolds numbers $Re = 40 - 100$. Growing Re leads to the increase of the dimensionless shedding frequency (Strouhal number, St), the latter being almost constant in the range of $Re \approx 10^3 - 2\times10^5$. The curve St(Re) has been obtained for the first time by A. Roshko [1]. He showed that the separated shear layer and vortices formed downstream of the cylinder remained laminar in the range of $Re = 40 - 150$. Moving downstream, vortices dissipated due to viscous diffusion. At $150<Re<300$, flow velocity fluctuations were observed in the cylinder wake. They were attributed to the laminar-turbulent transition in the separated shear layer. In this range of Re, there is a non-monotonous dependence of the shedding frequency on the Reynolds number [1]. Authors [2] identified two scenarios of transition to turbulence, each of them in the certain Reynolds number range. Laminar turbulent transition at $200<Re<400$ occurred inside the vortices and resulted from the deformation due to large-scale 3D effects. Growing Reynolds number made the transition onset move upstream; at $Re > 400$ the transition took place in the separated shear layer and resulted from the instabilities of Tollmien–Schlichting waves.

Late laminar-turbulent transition in the cylinder wake has been studied more thoroughly. Williamson [3] showed that the transition to turbulence in the wake had two stages. The first one was observed at $Re = 170 - 180$ and was characterized by the onset of

* Corresponding author: vmolochnikov@mail.ru

vortex loops and their hysteresis. The second one occurred at Re = 230 –260 and corresponded to the changes in small-scale spanwise vortex pattern. Both stages were accompanied by St(Re) discontinuities. The latter were mentioned also in [4] where the transition to turbulence in the cylinder wake at Re = 100 –300 was studied by numerical solution of 3D Navier-Stokes equations. Wavelike shape of large-scale vortices in spanwise direction at Re = 220 was also discovered in [4]. Authors [5] analyzed experimental data and numerical simulation results and established four physically different instabilities in the cylinder wake in the range of Re = 160 –230. One of them was called "vortex-adhesion mode" with Karman vortices that seemed to adhere to a steady or slowly moving point on the cylinder for many periods [5]. The other three instabilities had different wavelengths of 3D global modes. A number of authors note that flow turbulization around the cylinder is accompanied by the transition to 3D wake structure [6].

All available research results dealing with the laminar-turbulent transition in the circular cylinder wake have been obtained for the case of almost unbounded external flow. Besides, the cylinder was always located symmetrically relative to the channel walls. There are no data on the features of transition to turbulence in the cylinder wake in the case when the cylinder is located near the wall. There are only scarce data available on the side wall effect on these processes. However, it is well known that spiral motion of fluid from the side walls of the channel towards its symmetry plane is formed downstream of a semicylindrical rib mounted on the channel wall. This motion governs the processes of laminar-turbulent transition behind an obstacle and formation of large-scale vortices in the mixing layer. There are no data available on existence of such motion past a circular cylinder near the channel wall.

The present work elaborates on experimental results and DNS of the flow structure in the cylinder wake in a cross-flow when the cylinder is located near the rectangular channel wall and the Reynolds number range is $Re_d = 125 - 250$. The flow pattern has been described together with its evolution for the case of transition to turbulence in the cylinder wake.

2 Experimental procedure

The experiments were carried out in a rectangular test section with the width $B = 50$ mm, height $H = 20$ mm and length $L = 250$ mm (fig. 1). The air was pumped through the test section by a vacuum pump. A circular cylinder with the diameter $d = 3$ mm was mounted across the test section parallel to its wide wall at the distance of $l = 100$ mm from the channel inlet. The distance between the cylinder axis and the nearest channel wall was equal to the cylinder diameter, d. Reynolds numbers based on the average velocity in the channel, U_0, and the cylinder diameter were $Re_d = U_0 d/\nu = 135$, 167, 233, and 276. The experiments included flow visualization and measurements of dynamics of instantaneous velocity vector fields in the cylinder wake. Smoke-wire visualization technique was used for flow visualization. The videos were recorded by a high-speed monochrome camera Fastec HiSpec in a light sheet in $x0y$ plane (channel symmetry plane) and in the plane parallel to $x0z$ at $y = 4.5$ mm (tangent to the cylinder generatrix). Instantaneous velocity vector fields in the cylinder wake were measured by SIV technique [6].

DNS was performed on the basis of Navier-Stokes equations for incompressible fluid. The computational domain geometry matched the test section studied in experiments. Calculations were performed in ANSYS Fluent 15.0 software using the second-order finite volume method, and the computational grid was built using locally-structured approach [7]: the domain was filled with a set of structured grid segments, which were connected by non-structured insertions. The grid contained 10^6 cells and was refined near the solid walls and in the wake. Minimal dimensionless grid spacing normal to the cylinder generatrix was 0.02

(all dimensions were normalized by the cylinder diameter). The cell size in the middle wake was 0.15. Minimal grid spacing normal to the wall in the boundary layer of the test section was 0.01. The time step was 0.05. Weightless particles were visualized to imitate the smoke wire used in the experiments.

3 Results and discussion

Good agreement between the calculations, flow visualization and SIV measurements has been established (fig. 1). Main regularities in laminar-turbulent transition in the cylinder wake have been obtained. A closed laminar separation region has been discovered immediately downstream of the cylinder at Re ≤ 167. It consists of two steady symmetric vortices and is followed by a regular large-scale vortex street further downstream. Wake flow remains quasi two-dimensional across almost the whole channel span except for the boundary layer on the channel side walls. The wake flow is laminar: second invariant of the Reynolds stress tensor $II_T = \sqrt{\tau_{ij}\tau_{ji}}$ remains zero throughout the whole wake region, and generated vortices maintain their integrity across the channel span.

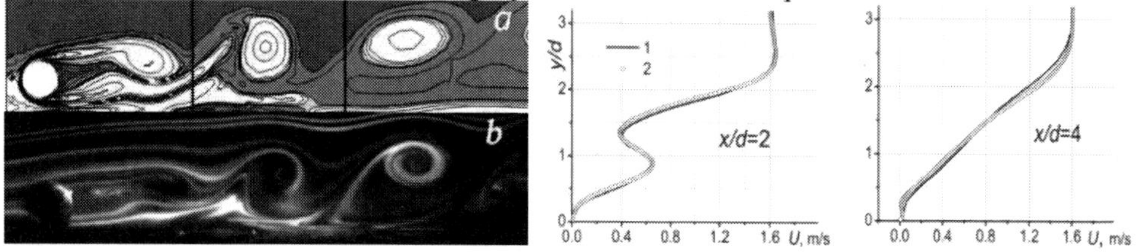

Fig. 1. Velocity field downstream of the cylinder and velocity profiles in the wake at Re = 235: *a* –calculation; *b* – flow visualization; 1 – calculation; 2 – SIV measurements

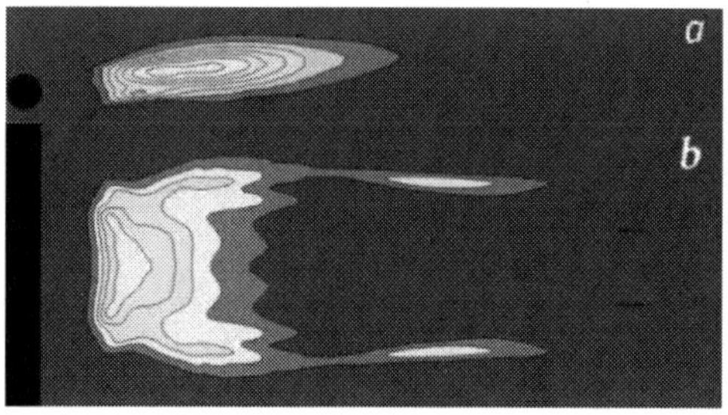

Fig. 2. Second invariant of the Reynolds stress tensor in the cylinder wake at Re = 233: *a* – vertical plane; *b* – horizontal plane

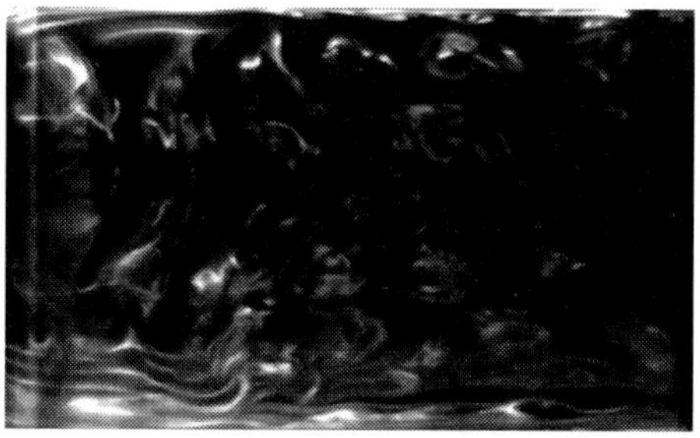

Fig. 3. Cylinder wake visualization at Re = 233 (horizontal plane)

The region of vortex generation onset shifts upstream with the growing Reynolds number. Reynolds stresses are non-zero already at Re = 233 (fig. 2), and the features of 3D flow are observed in the cylinder wake (fig. 3): while moving downstream, vortices split into several vortex clots. However, the near-wake flow (where there are still no Karman vortices) remains laminar.

Further increase of the Reynolds number leads to the filling of high-frequency part of the velocity fluctuation spectrum, which indicates the developed turbulent flow onset. The spanwise profile of velocity is almost uniform at Re = 135 (fig. 4, *a*) but has several local extrema already at Re = 276 (fig. 4, *b*). Turbulent disturbances are localized in the middle wake of the cylinder; spanwise maxima in Reynolds stress are in the wake region.

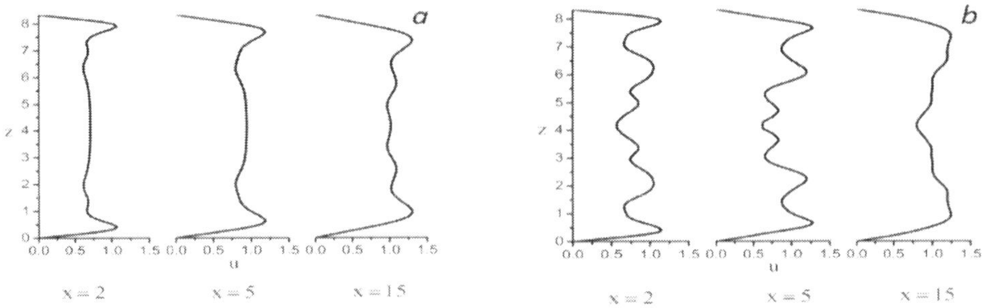

Fig. 4. Spanwise profiles of flow velocity in the cylinder wake: *a* – Re = 135; *b* – 276

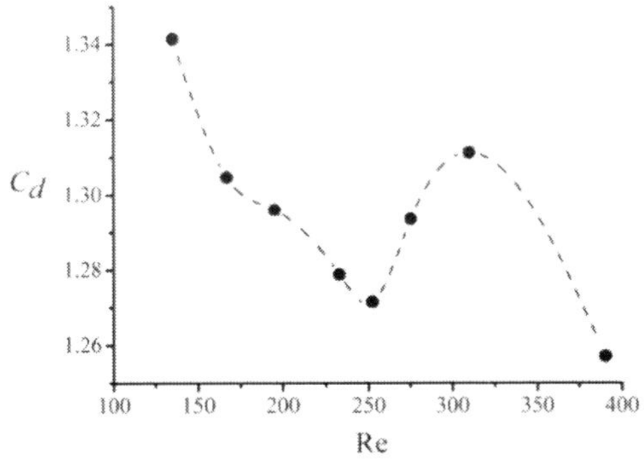

Fig.5. Cylinder drag coefficient vs Re

Thus, the transition to turbulence in the cylinder wake results from the flow instability in the separated shear layer at some distance from the cylinder. Non-monotonous behavior of the cylinder drag coefficient as a function of the Reynolds number has been revealed in the range of Re = 120 – 300 (fig. 5). The latter corresponds to transition to turbulence in the cylinder wake. In the whole considered Reynolds number range there is no spiral motion of fluid from the side walls to the channel symmetry plane.

4 Conclusions

An experimental study and DNS of the pulsating cross-flow past a circular cylinder mounted near the channel wall have been performed in the range of the Reynolds number Re = 135–276. Good agreement between the experimental and simulation data has been obtained. It has been established that a street of regular large vortices is formed downstream of the cylinder surface that is the most distant from the channel wall. Features of 3D flow behavior have been observed in the wake starting from Re ≈ 230. The transition to turbulence downstream of the cylinder has been shown to result from the shear layer instability at some distance from the cylinder. Non-monotonous behavior of the cylinder drag has been revealed in the range of Re = 120 – 300.

This study was supported by the Russian Science Foundation (Project no.16-19-10336).

References

1. A. Roshko, NACA Report 1191 (1958)
2. F.M. S. Bloor, J. Fluid Mech. **19**, 290 (1964)
3. C.H.K. Williamson, Phys. Fluids. **31**, 3165 (1988)
4. H. Persillon, M. Braza, J. Fluid Mech. **365,** 23 (1998)
5. H.-Q. Zhang, U. Fey, B. R. Noack, M. Konig, H. Eckelmann, Phys. Fluids **7**, 779 (1995)
6. N.I. Mikheev, N. S. Dushin, Instr. Exp. Techn. **59**, 882 (2016)
7. A.B. Mazo, D.I. Okhotnikov, Lobachevskii J. Math. **37,** 360 (2016)

Experimental investigation of the influence of the passive porous coating on laminar-turbulent transition of the hypersonic boundary layer of the sharp cone at angles of attack

Sergey Morozov[1,*], *Sergey* Lukashevich[1] and *Alexander* Shiplyuk[1]

[1]Khristianovich Institute of Theoretical and Applied Mechanics SB RAS, 630090 Novosibirsk, Russia

Abstract. The problem of the delaying laminar to turbulent transition by the passive porous coating on the sharp cone at the angle of attack has been studied for the first time. Experiments are conducted in a Transit-M hypersonic short-duration wind tunnel at the Mach number 5.9. The boundary layer laminar-turbulent transition is determined by heat flux distribution on the surface of the cone. It is found that the transition on the porous surface in compare to smooth surface has been delaying on the windward side at the 0.5°, 1° and on leeward side at the 0.5°. There is no effect of the porous coating on the location of the laminar to turbulent transition at angle of attack 1°.

1 Introduction

Research of hypersonic boundary layers stabilization metods are an important problem of the hight speed aerodynamic due to the significant increase of the surface heat flux of the turbulent boundary layer in compare to the laminar boundary layer. The cause of the transition in hypersonic gradientless boundary layers is high frequencies unstable disturbances so called second mode disturbances [1]. On the passive porous coating with microstructure second mode disturbances grows slower than on the smooth surface [2]. The ability to delay transition by passive porous coatings are shown in several works [2 - 4].

The previous experimental studies are shown that it's possible either to reduce or magnify the energy of second mode disturbances using the porous coating with different length on the cone [11]. It indicates that the ability of the porous coating to delay transition is not predictable in case of small varying hypersonic boundary layer parameters. For instance, the stability of the second mode disturbances strongly depends on bluntness of the leading edge [6, 7, **Ошибка! Источник ссылки не найден.**] and angle of attack [9, 10].

Therefore, the present investigations are aimed at an experimental study of the efficiency of laminar to turbulent transition delay by the passive porous coating on a cone at angles of attack.

* Corresponding author: morozov@itam.nsc.ru

2 Experimental setup

Experiments were carried out in the ITAM SB RAS hypersonic short-duration wind tunnel Tranzit-M at the freestream Mach number $M_\infty = 5.9$, stagnation temperature $T_0 = 380 \pm 5$ K and the unit Reynolds number $Re_{1\infty} = 2 \cdot 10^7$ m^{-1}.

The model is 7° half-angle cone with sharp nose tip ($R_{bluntness} = 0.03$ mm). The length of the cone generatrix is 363 mm.

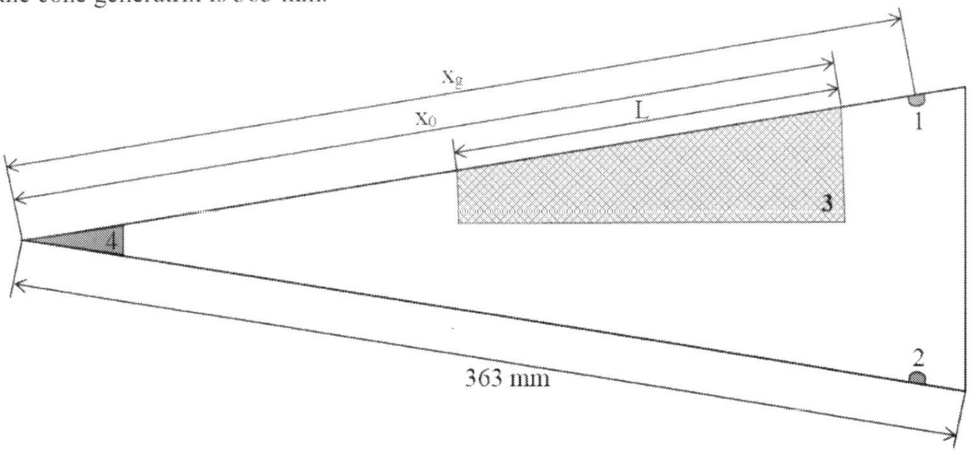

Fig. 1. Schematic of the model. 1, 2 – high frequency pressure gauges, 3 – passive porous coating, 4 – interchangeable nose tip.

The coating is made of three layers of woven mesh made of nylon with a porosity of 44%. The layers are turned with respect to each other by the angle of 45°. The coating thickness are chosen as most effective according to the experimental work [11]. The thickness equals to 0.18 mm. The porous coating length is 210 mm. The cone has been set at angles of attack $\alpha = 0°, 0.5°, 1°$.

The location of the boundary layer laminar-turbulent transition at the cone are determined by heat flux distribution along surface of the cone. Heat flux distributions are calculated from distributions of the surface temperature measured during the wind tunnel run by infrared camera FLIR SC7750L. Temperature measurement sampling rate is 200 Hz.

3 Results

Fig. 2 shows the heat flux distribution along the smooth surface of the cone at zero angle of attack for $Re_{1\infty} = 2 \cdot 10^7$ m^{-1}. The boundary layer is laminar from x = 57 mm to x ≈ 100 mm. The heat flux distribution deviations from linear law in log scales at x ≈ 100 mm indicates the onset laminar to turbulent transition. The end point of the transition $x_{tr} = 258$ mm is determined at the location of the maximum of the heat flux distribution. Therefore, the boundary layer is transition at x = 100 ÷ 258 and turbulent after $x_{tr} = 258$.

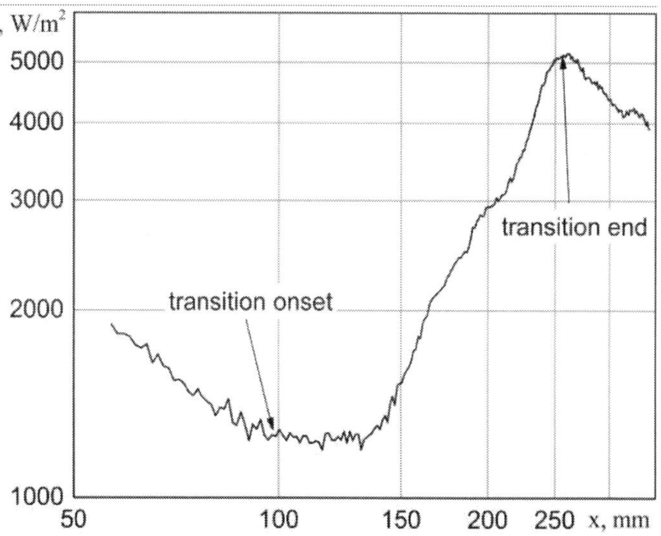

Fig. 2. The heat flux distribution on smooth surface of the cone at zero angle of attack for $Re_{1\infty} = 2{\cdot}10^7$ m^{-1}.

The end point of the transition as function of the angle of attack for the unit Reynolds number $Re_{1\infty} = 2{\cdot}10^7$ m^{-1} are shown in Figure 3. The laminar to turbulent transition on the smooth surface of the cone are shifted upstream on the leeward side and downstream at the windward side with increase of the angle of attack. The end point of the transition on the surface of the cone with passive porous coating are shifted further downstream at angles of attack α = -1; -0,5; 0; 0,5° in compare with smooth surface. There is no transition delay at leeward side at α = 1°. The onset of the transition in this case far downstream (x = 77 mm) then the passive porous coating begins (x = 120 mm). Therefore, the porous coating has no stabilization effect on laminar and early transitional boundary layer.

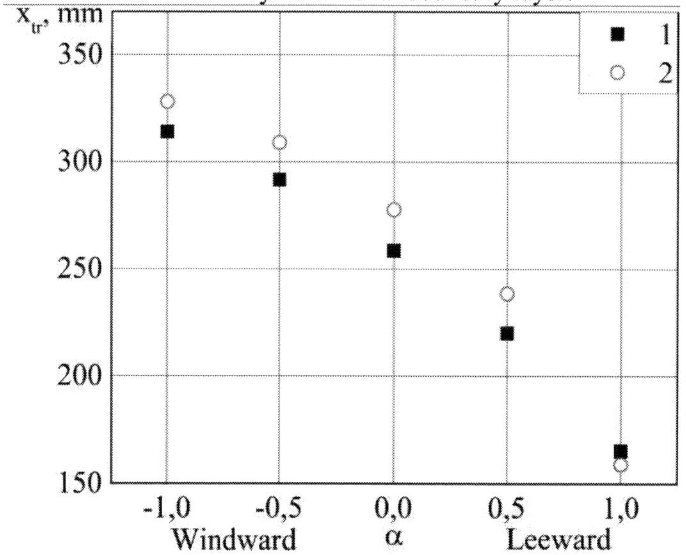

Fig. 3. The end point of the transition as function of the angle of attack for $Re_{1\infty} = 2{\cdot}10^7$ m^{-1}. 1 – smooth surface; 2 – surface with passive porous coating.

4 Conclusions

Experimental investigation of the influence of the passive porous coating on the laminar-turbulent transition location on the sharp cone at angles of attack are performed at the freestream Mach number $M_\infty = 5.9$. The transition location is determined by measured heat flux distributions along smooth surface and surface with passive porous coating. The transition delay by passive porous coating is observed at angles of attack α = -1; -0,5; 0; 0,5°, where minus sign means windward side. There is no transition delay at leeward side at α = 1° because the passive porous coating location in this case is no longer effective.

This work was financially supported by Russian Foundation for Basic Research (Grants No. 16-38-00491 mol_a and No. 16-08-00965).

References

1. L.M. Mack, AIAA J. **13**, 278 (1975)
2. A.V. Fedorov, N.D. Malmuth, A. Rasheed, H.G. Hornung, AIAA J. **39**, 99 (2001)
3. A. Maslov, A. Shiplyuk, A. Sidorenko, P. Polivanov, A. Fedorov, V. Kozlov, N. Malmuth, AIAA Paper **1112**, 1 (2006)
4. A. Wagner, M. Kuhn, J. M. Schramm, K. Hannemann, Exp. Fluids **54,** 1606 (2013)
5. S.V. Lukashevich, S.O. Morozov, A.N. Shiplyuk, J. Appl. Mech. Techn. Phys. **54**, 572 (2013)
6. K.F. Stetson, E.R. Thompson, J.C. Donaldson, L.G. Siler, AIAA Paper **0006**, 1 (1984)
7. I. J. Lyttle, H. L. Reed, A. N. Shiplyuk, A. A. Maslov, D. A. Buntin, E. V. Burov, S. P. Schneider, AIAA J. **43**, 1734 (2004).
8. A.A. Maslov, A.N. Shiplyuk, D.A. Bountin, A.A. Sidorenko, J. Spacecraft Rockets **43,** 71 (2006)
9. K.F. Stetson, E.R. Thompson, J.C. Donaldson, L.G. Siler, AIAA Paper **0492,** 1 (1985)
10. P. Balakumar, L.R. Owens, AIAA Paper **4718**, 1 (2010)
11. S.V. Lukashevich, S.O. Morozov, A.N. Shiplyuk, Techn. Phys. Lett. **38**, 1077 (2012)

Flow around a confined cylinder: LES and PIV study

Egor Palkin[1,2,*], *Mikhail* Shestakov[1,2], *Rustam* Mullyadzhanov[1,2], *Dmitry* Markovich[1,2], and *Kemal* Hanjalić[2,3]

[1]Institute of Thermophysics SB RAS, 1 Lavrentyev Ave., 630090 Novosibirsk, Russia
[2]Novosibirsk State University,2 Pirogov Str., 630090 Novosibirsk, Russia
[3]Delft University of Technology, Bld. 58, Van der Maasweg 9, 2629 HZ Delft, The Netherlands

Abstract. We study the flow over a cylinder placed between two parallel rigid walls using Large-eddy simulations and Particle Image Velocimetry. The Reynolds number based on the inflow velocity and diameter of the cylinder is 3750 corresponding to the subcritical regime with laminar separation. Three-dimensional visualization shows the presence of the horseshoe vortex system prior to the cylinder. The comparison of time-averaged velocity fields and fluctuations shows good agreement between simulations and experiments. Spectral analysis suggests the presence of low-frequency modulations of the recirculating bubble.

1 Introduction

Flows over obstacles in a duct are common in many engineering applications such as cooling systems, bridge piers, heat exchangers, building sections, junctions in wing-body and turbine blade-rotor systems, among others. In such configurations a horseshoe vortex system appears prior to a bluff body increasing the local shear stress and heat transfer [1] while the flow is characterized by periodic shedding of large-scale vortices behind the body that form the Kármán vortex street. Low-frequency modulations of the recirculating zone are detected for various configurations such as a cylinder [2, 3], disk and sphere [4], prism [5], bullet [6], among others. The period is typically $10 \div 100$ times lower compared to the main vortex shedding frequency. In the present work we consider a flow over confined cylinder in a narrow rectangular duct to investigate the effect of walls on the dynamics of the recirculation bubble.

2 Computational and experimental details

We study a water flow over a circular cylinder which is fixed perpendicular to a pair of side walls at the Reynolds number $Re = 3750$ based on the bulk inflow velocity U_b and cylinder diameter D. The inflow velocity distribution represents a steady laminar parabolic profile. The distance between narrow parallel walls is $H = 0.4\ D$. The case is studied using numerical simulations and experiments described below.

*Corresponding author: palkinev89@gmail.com

DOI: 10.1051/matecconf/201711502010

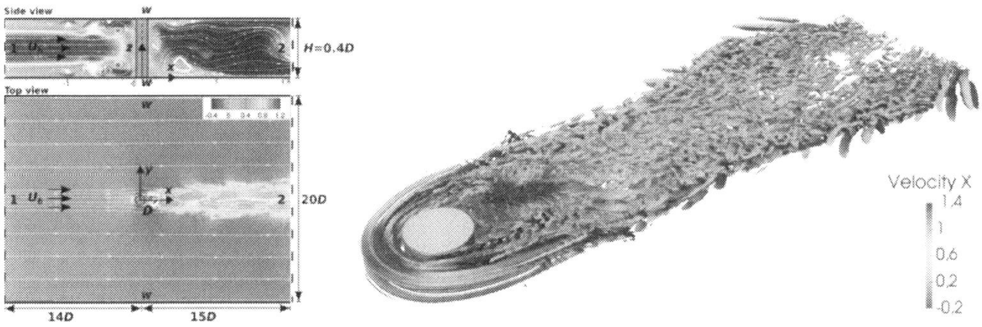

Fig. 1. Left: Schematic view of computational domain with instantaneous streamwise velocity field and streamlines. The boundary conditions correspond to the following marks: '1'– inflow, '2' – outflow, 'w' – no-slip (walls). Right: isosurface of Q-criterion to visualize vortical structures colored with the streamwise velocity.

We perform Large-eddy simulations (LES) using the unstructured finite-volume computational code T-FlowS. The filtered Navier-Stokes and continuity equations for incompressible fluid are closed with the dynamic Smagorinsky subgrid-scale model. The spatial discretization is performed with the second-order central-difference scheme, whereas for the time-marching we use a fully-implicit three-level scheme. The velocity and pressure are coupled with the SIMPLE algorithm. The computational domain shown in Fig. 1 represents a box with a size $x \times y \times z$ = 29D $\times$ $20D \times H$, where x, y, z stand for the streamwise, spanwise and wall-normal directions. The computations were performed on two meshes with 8.7×10^6 and 16.6×10^6 cells, respectively, with no significant differences in the results. Both meshes satisfy wall-resolved LES criteria. In particular, even the 'coarse' mesh corresponds to high resolution since the first cell near the cylinder did not exceed the following limits: $\Delta r^+ < 1$, $(R\Delta\varphi)^+ < 8$ and $\Delta z^+ < 4$, where '+' denotes the wall units and $R = D/2$. The total computational time was around $10^3 D/U_b$ with a non-dimensional timestep 2.5×10^{-3}.

The experiments were performed in a slot channel with the length and width of 38D and $20D$, respectively, where D = 10 mm. In order to provide steady velocity distribution close to parabolic at the inflow, the flow passed through a set of two honeycombs. Velocity fields were measured using Particle Image Velocimetry (PIV) technique. The system consists of a digital PCO camera (1024 × 1280 pix, 500 Hz max. frame rate) and dual cavity Nd:YAG laser (1000 Hz max. rate, 10 mJ max. pulse energy). The camera was located perpendicular to the main channel. The thickness of the laser sheet was equal to 0.7 mm. PIV measurements were performed in a $2D \times$ 2D region behing the cylinder. The averaged characteristics were calculated using 1000 instantaneous velocity fields. The spatial resolution was estimated to be 0.3 mm.

3 Results

The flow regime corresponds to the subcritical one at this relatively low Reynolds number with the separation of the laminar boundary layer and subsequent turbulization of the shear layer. A highly three-dimensional flow appears in the near wake region within the recirculating bubble due to the bounding narrow walls. The hourseshoe vortices decay while interacting with the shear-layer turbulence (Fig. 1). Further downstream the wake becomes fully developed. Figures 2 and 3 demonstrate the time-averaged axial velocity for LES and PIV results together with the comparison at certain axial locations. The length of the recirculating zone is slightly shorter for LES ($\approx$ 1.3 D), while overall results are in a satisfactory agreement with PIV measurements. The influence of horseshoe vortices are

 DOI: 10.1051/matecconf/201711502010

clearly visible on the mean streamwise velocity around $|y| / D \approx 0.7$ showing a pronounced drop. PIV results capture this effect on x/D=1.2 profile of mean U/U_b.

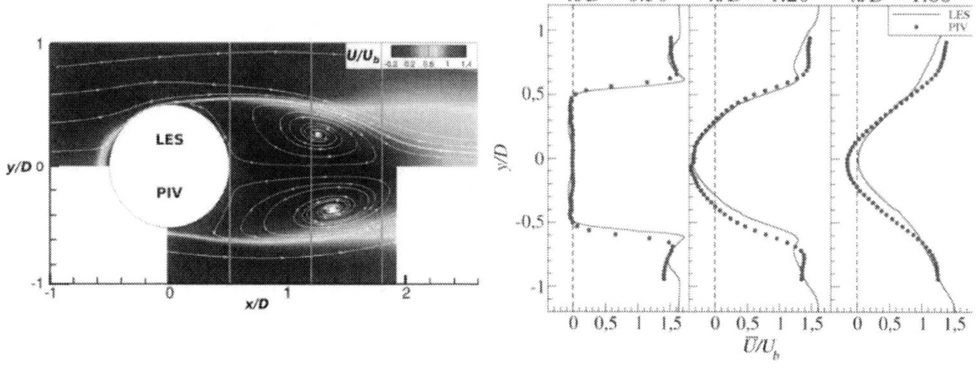

Fig. 2. Left: The field of the time-averaged axial velocity and streamlines from LES (top half) and PIV (bottom). Right: The profiles of the time-averaged axial velocity at three axial locations shown on the left by green vertical lines.

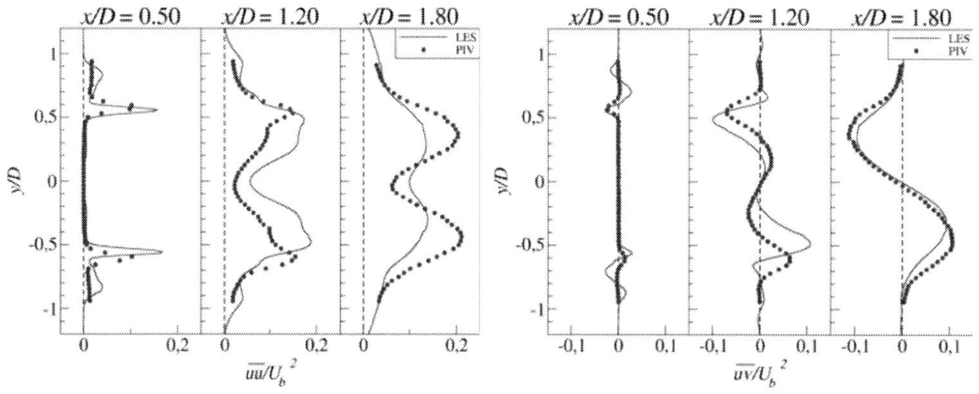

Fig. 3. Profiles of the normal streamwise (left) and shear (right) stresses at the same axial locations as in Fig. 2.

Further we perform spectral analysis of velocity fluctuations to detect characteristic frequencies of the flow, Fig. 4. The comparison of the time-averaged normal spanwise Reynolds stress components shows excellent agreement between LES and PIV. The dominant peak is around $St = fD/U_b \approx 0.23$ corresponding to the vortex shedding frequency. The component mean *ww* was unavailable in PIV and the comparison could not be performed. LES data reveals some low-frequency fluctuations with the peaks around $St \approx$ 0.04. Future work will be devoted to identifying the hydrodynamics behind these low-frequency oscillations.

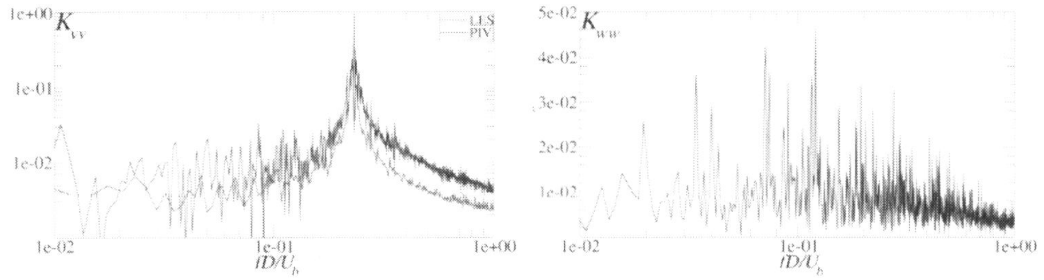

Fig. 4. Power spectra of velocity fluctuations close to the cylinder at the centreline, $(x/D, y/D, z/H)$ = (0.7, 0.0, 0.5). Left: spanwise component from LES and PIV. Right: wall-normal fluctuations.

4 Discussion

We performed LES and PIV of the flow over a confined cylinder in a narrow duct at Re = 3750. Spectral analysis suggests the presence of low-frequency modulations of the recirculating bubble. This will be the topic for the future study. Another issue is the effect of bounding walls on the developed wake. Our observations [7, 8] in confined jets revealed the existence of streamwise meandering vortices in the flow influencing the heat transfer across the channel. It is expected that a similar phenomenon should be present in a confined wake flow.

This work is funded by the Russian Science Foundation grant No. 14-19-01685. The computational resources are provided by Siberian Supercomputer Center SB RAS and Supercomputing Center of the Novosibirsk State University.

References

1. R. Simpson, Ann. Rev. Fluid Mech. **33**, 415 (2001)
2. O. Lehmkuhl, I. Rodriguez, R. Borrell, A. Oliva, Phys. Fluids **25**, 085109 (2013)
3. E. Palkin, R. Mullyadzhanov, M. Hadziabdic, K. Hanjalic, Flow Turbul. Combust. **97**, 1017 (2016)
4. E. Berger, D. Scholz, M. Schumm, J. Fluid Struct. **4**, 231 (1990)
5. J. Miau, J. Wang, J. Chou, C. Wei, J. Fluid Struct. **13**, 339 (1999)
6. V. Gentile, F. F. J. Schrijer, B. W. Van Oudheusden, F. Scarano, Phys. Fluids **28**, 065102 (2016)
7. R. Mullyadzhanov, B. Ilyushin, K. Hanjalic, Int. J. Heat Fluid Flow **56**, 284 (2015)
8. M. Shestakov, R. Mullyadzhanov, M. Tokarev, D. Markovich, J. Eng. Thermophys. **25**, 159 (2016)

Effect of blowing/suction through porous surfaces on boundary layer at supersonic Mach numbers

Pavel Polivanov[1,*]

[1]Khristianovich Institute of Theoretical and Applied Mechanics SB RAS 630090, Institutskaya, 4/1, Novosibirsk, Russia

Abstract. The effect of blowing/suction through perforated surfaces on the turbulent boundary layer was studied experimentally at Mach number M = 1.4. In the experiments, several different porous surfaces were investigated in a wide range of the blowing/suction mass flow rate. The distributions of the integral parameters of boundary layer were obtained depending on the type of porosity and mass flow rate. The integral effect from shock waves on boundary layer depends on set of parameters, which makes search of optimal parameters of porosity difficult.

1 Introduction

Blowing/suction of the gas through the porous surface can be used for various applications, e.g., the laminarization of transonic airfoil and control of flow separation, wall cooling or injection of fuel etc. Many experimental and theoretical studies investigated the effect of blowing/suction through porous surfaces, but mostly for subsonic speeds [1]. Studies performed at Mach numbers M> 1 are usually more complex, so they are rarer and more difficult to systematized. For example in paper [2] the blowing/suction through porous surfaces was used to control of separation flows, but a detailed analysis of the effect of porosity geometry was not performed. In paper [3] the dimensionless dependences of mass flow rate through a porous surface for a wide range of diferent parameters were found. But the absence the data of the effect of the porous geometry on the boundary layers makes compicate the design of optimal control devices for supersonic cases. Modern numerical methods (LES/DNS) allow calculating such flows [4], but they require a lot of computational time. RANS methods are more accessible but require data for verification. It was decided to perform a parametric study of the effect of gas blowing or suction on the supersonic boundary layer, developed on the plate for different porous geometries.

2 Experimental setup and results

The experiments were performed in wind tunnel T-327b (ITAM SB RAS) for Mach number $M_\infty = 1.4$, $T_0 = 298$ K and $P_0 = 1.01\cdot10^5$ Pa ($Re_1 = 15.1\cdot10^6$ m^{-1}). The width of wind

* Corresponding author: polivanov@itam.nsc.ru

tunnel test section is 57 mm, the length is 370 mm, and to compensate the growth of the boundary layer thickness the height of the working part increases from 19 up to 23 mm. The main measurement technique was PIV. The random error of velocity vector measurements was no greater than 1–2%. The porous insert is placed on the wall in the zone of turbulent boundary layer (Fig. 1). CFD RANS (Reynolds-averaged Navier-Stokes) simulation allow to accelerate the process of designing of flow control devices. But for the correct choice of parameters of the semi-empirical turbulent model the experimental data with simplified geometry are required for verification. To simplify the geometry of porous inserts it was decided to use perforated plates with different geometry. The parameters of the porous insert can be found in Table 1, where d is the diameter of the hole, h is the distance between the holes, and β is the angle between the normal to the surface and the axis of the hole. For each perforated surface the flow characteristics were measured at the following fixed mass flow rate of blowing/suction: 3.23, -1.91, -1.21, -0.56, -0.26, 0, 0.26, 0.56, 1.21, 1.91 g/s (sign "-" corresponds to the suction). The maximum mass flow rate was chosen so that for the insert #1 (with a min. total hole area), the velocity of flow through the holes was about the speed of sound. The wind tunnel mass flow rate was ≈210 g/s.

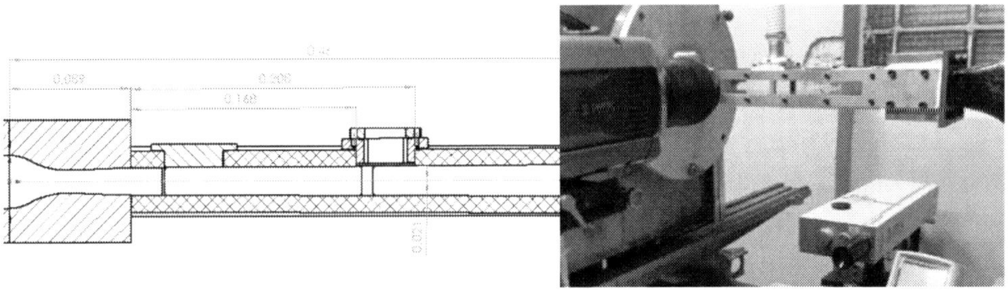

Fig. 1. The draft and photo of wind tunnel

Table 1. Parameters of porous surface

#	d, mm	h, mm	Number of holes	Total square of holes, mm^2	Angle of deviation from the normal	Open area
1	0.8	3.2	9x9	40.7	β=0°	0.05
2	0.8	2.4	11x11	60.8	β=0°	0.09
3	0.8	1.9	14x14	98.5	β=0°	0.14
4	0.8	1.6	16x16	128.7	β=0°	0.20
5	1.0	2.0	14x14	153.9	β=0°	0.20
6	1.4	2.8	10x10	153.9	β=0°	0.20
7	2.0	4.0	7x7	153.9	β=0°	0.20
8, 9	1.0	2.0	14x13	147.9	β=15° (blowing in the reverse and direction of flow)	0.20

Figure 2 shows the velocity fields for the porous insert #4. For a zero mass flow rate, the thickness of the turbulent boundary layer is 3 mm (velocity distribution corresponds to equilibrium turbulence) and almost does not change in the zone of measurement. Blowing leads to an increase in the boundary layer thickness and significant reduces the fullness of boundary layer. The non-uniformity of the flow in the zone of inviscid flow appears, and the flow velocity in the wind tunnel decreases. The reason for this is the shock wave, generated at the leading edge of the porous surface due to the growing displacement thickness. Suction, on the contrary, leads to an acceleration of the flow. In the zone of porous surface there are periodic wavy structures with a period corresponding to a distance between the holes. Wavelike nature of displacement thickness leads to the generation of periodic compression-expansion waves, which can significantly change the flow pattern. One may clearly see the increase of non-uniformity of the mean flow with an increase in the size of the holes (Figure 3, porous insert #7) as compared with the porous insert # 4. For suction, this is explained by a more rapid change of the boundary layer thickness near the holes. This leads to the generation of more powerful compression waves (which generate shock waves) and expansion waves. But for the case of blowing, no periodic structures are found. This can be explained by the increase of the "head" shock wave arising near the beginning of porous surface.

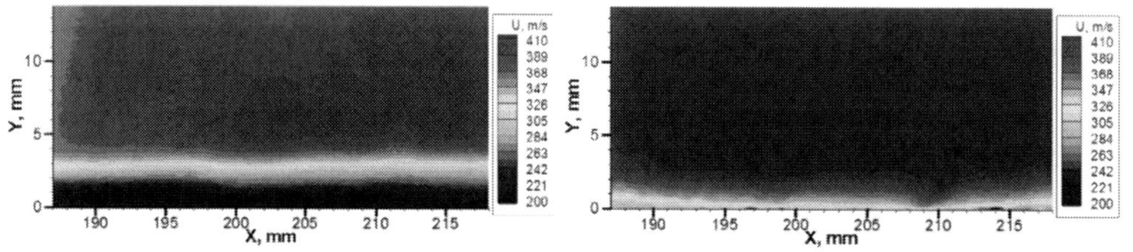

Fig. 2. Velocity fields for case #4, left – blowing at 1.91 g/s, right – suction 1.91 g/s

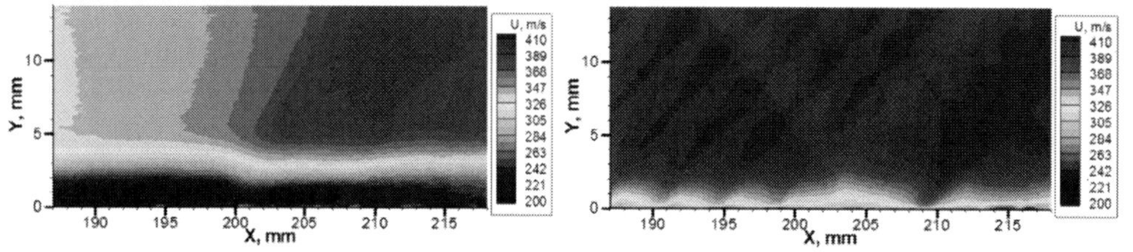

Fig. 3. Velocity fields (case #7) , left – blowing at 1.91 g/s, right – suction 1.91 g/s (Near the coordinate X = 200 mm the perforated part is ends)

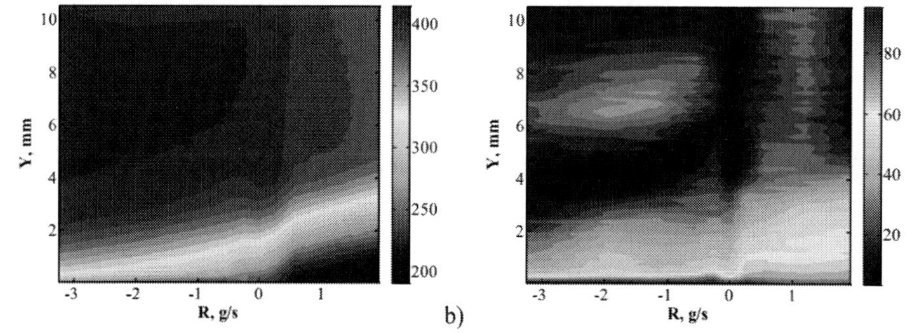

Fig. 4. Distribution of a) mean velocity profiles [m/s] and b) RMS pulsation of velocity [m/s] for various mass flow rate through a porous surface for case #4 at X=215 mm

In Figure 4a one may see the distribution of the velocity profile for porous insert #4 measured behind the porous insert. There is a monotonous variation of the boundary layer thickness with the growing mass flow rate. It is interesting to note that the distribution of pulsations the monotonicity of the data is not maintained especially in the point of zero mass flow rate (Figure 4b). The blowing and suction lead to an increase of pulsations which means that the turbulent boundary layer becomes nonequilibrium behind the porous insert. This should be taken into account when designing devices using a distributed blowing/suction.

Figure 5 shows the distribution of the momentum thickness for all porous surfaces and entire range of mass flow rate through it. The numbering of porous surfaces is selected so that from #1 up to #4 (Table 1) the open areas grow at fixed of diameter of holes. Further from #4 to #7 the hole diameter is increased at a constant value of the open area of porous surface. Case #8 and #9 show the influence of the slope of holes on the parameters of the boundary layer. It is clearly seen that for each porous surface the distribution of momentum thickness is approximately monotonic. But at constant mass flow rate, a variation of one of the porosity parameters demonstrated a significant non-monotonicity of the momentum thickness, which is strongly dependent on the mass flow rate. For example, at strong suction and when the open area is varied (#1-#4) the minimum momentum thickness was found for case #2. This can be explained by the effect of several phenomena on the strength of shock waves, generated near the holes. At low mass flow rate the most important is the size of the holes. But an increase of the mass flow rate leads to a significant distortion of the local flow, which should lead to increased strength of the shock wave.

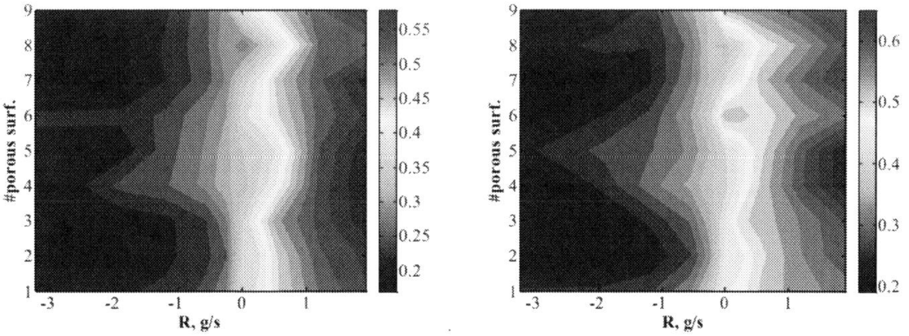

X=192 mm X=215.5 mm

Fig. 5. Distribution of the momentum thickness [mm] for all cases of porosities at various mass flow rate through a porous surface

An experimental study has shown that distributions of integral parameters of the boundary layer for blowing/suction through porous surfaces are non-monotonic. This substantially complicates determination of empirical dependencies. The use of dimensionless data did not allow finding some regularity. The results of the work demonstrate the complexity of the problem, which means that additional experimental and numerical research is needed.

The work was supported by the Grant of President of the Russian Federation for State support of young Russian scientist – Candidate of sciences No Mk- 6682.2016.1.

References

1. V.I. Kornilov, Progress in Aerospace Sciences **76** 1 (2015)
2. Drag Reduction by Shock and Boundary Layer Control: Results of the Project EUROSHOCK II / ed. Egon Stanewsky. Springer-Verlag Berlin Heidelberg. (2002).
3. R.A. Baurle, A.T. Norris, A Source-Term Based Boundary Layer Bleed/Effusion Model for Passive Shock Control. 58th JANNAF Propulsion Meeting, Arlington, VA, USA, P.1-26 (2011).
4. S. Ghosh, J.-I. Choi, J.R. Edwards, AIAA Paper 2009-1330 (2009).

Experimental study of flow around axisymmetric bodies in supersonic flow in case of a local injection into the boundary layer

Evgeny Maslov[1, 2,*], *Irina* Zharova[2], *Evgeny* Kozlov[2], *Valery* Faraponov[1], *Nadezhda* Savkina[1], *Nikolay* Zolotorev[1], *Vladislav* Matskevich[1]

[1] National Research Tomsk State University, Tomsk, Russia
[2] National Research Tomsk Polytechnic University, Tomsk, Russia

Abstract. A technique of experimental determination of the parameters and structure of supersonic air flow around of axisymmetric bodies form in the range of Mach numbers M = (2 ÷ 5) are presented. The results of the investigation of the interaction a supersonic stream flowing around of an axisymmetric model with gas jet which inject into the boundary layer are given.

1 Introduction

Determination of aerodynamic characteristics of bodies in the range of Mach numbers M = (2 ÷ 5) is important for designing aircrafts of various purposes. In this connection the necessity of solving a number of scientific and technical problems arises. This paper presents the technique and results of experimental studies of supersonic flow around an axisymmetric (conical) model in the presence of symmetric and asymmetric gas injection into the boundary layer.

2 Description of the aerodynamic installation and a technique of conducting experimental researches

Experimental researches were carried out on an aerodynamic installation [1, 2]. The range of realized modes of operation of the installation: the Mach number M = (2 ÷ 7), the braking pressure at the nozzle exit section of the diffuser, forming the supersonic flow, P_0 = (0.15 ÷ 0.3) MPa at static pressure P = (0.03 ÷ 0.07) MPa, the braking temperature of the incoming air flow T_0 = (17.5 ÷ 250) °C. Scheme of modernized aerodynamic installation is presented in Fig. 1. The installation consists of the following basic elements: frame, basic valve, 8 cylinders with hand valves, splitter, diffuser, business end, nozzle, forehearth, electrical heater, power source. Before the experiment, the working gas is in 8 cylinders 4 with a total volume of 320 dm^3 at a pressure of up to 15.0 MPa. To start the installation, the basic valve is switched on from the control console and the working gas from the cylinders enters the electrical heater and the forehearth.

* Corresponding author: maslov_eugene@mail.ru

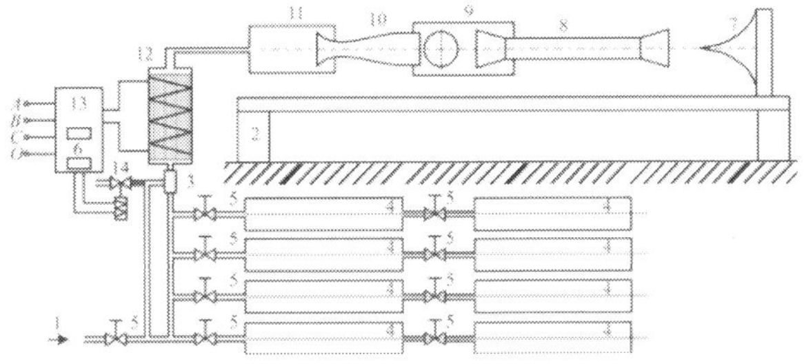

Fig. 1. Scheme of a model aerodynamic installation: 1 – air supply; 2 – frame; 3 – the basic valve; 4 – cylinders with compressed air; 5 – hand valves; 6 – the control panel; 7 – splitter; 8 – diffuser; 9 – working part; 10 – profiled nozzles; 11 – forehearth; 12 – electric heater; 13 – power supply 20 kW; 14 – control valve.

To measure the direct stresses (pressures) on the surface of the test body, holes (diameter 0.25 mm) are made. Drainage holes on the surface of the model were connected to a micromanometer using air-ducts. Each hole in the complex with the air-duct is a receiver (drainage receiver) of static pressure on the surface of the streamlined model. The peculiarity of drainage tests is that defining to their results only that component of the total aerodynamic force, which that is caused by pressure. The frictional forces in this case cannot be measured. The run duration of the aerodynamic installation is (1.0 ÷ 3.0) s. The preheating of the working gas is realized with using a pebble heater. To create a supersonic flow was used axisymmetric profiled nozzles. The installation is provided with registration system of pressure, of temperature, with a system of visualization and with three-component aerodynamic (tensometric) balance. The model aerodynamic unit is equipped with equipment for visualization the processes of gas flow of the studied bodies by the Tepler method with a parallel beam of rays. In Fig. 2 the photography and scheme of the shadow device are given.

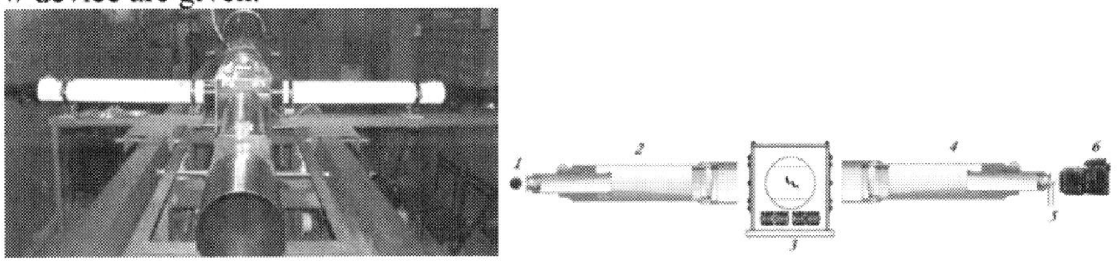

Fig. 2. Photo and diagram of the shadow device: 1 – point light source; 2, 4 – telescope "Tal-100SR"; 3 – working part; 5 – Foucault knife; 6 – video recorder (DVR).

With the help of the Foucault knife, the intensity of the light flux is regulated. The beam enters the lens of the DVR 6, where you can see the image of the body being examined. As a DVR, the high-speed video camera "Phantom v711" is used. The use of a high-speed video camera makes it possible to examine in more detail the picture of the flow around the model by gas, and also allows to correlate the obtained video frames with the readings of sensors, since the initiation of video shooting. The start of the measuring system and the opening of the electromagnetic air valve takes place at a time from one source of the signal.

3 Results of experimental studies

Cone models (Fig. 3) (axial symmetry) with the channels for measuring the pressure on the surface were under tests. Half-angle is 15 °. Drainage holes of the cannels were made in the fixed points of the surface in vertical plane. Shadow method jointly with high-speed videotaping were used during performing-drain and weight tests.

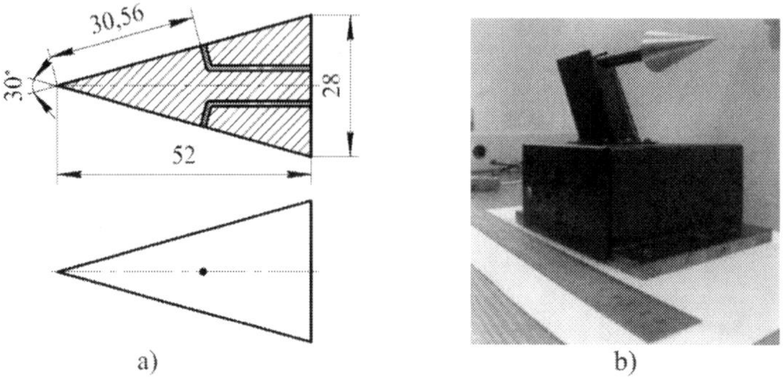

Fig. 3. Scheme (a) and general view (b) of cone model.

Airflow data and force to have an effect on profiling surface by the stream were measured to calculate the model drag coefficient. The results of processing of experimental data of model weight tests for every Mach number are listed in the table 1. The comparison of obtained drag coefficient values with the values [3, 4] are presented in Fig.4.

Table 1. The results of processing of experimental data.

Mach number	C_x	$\overline{C}_x$	ΔC_x	δ, %
2	0.4651	0.4680 ± 0.0036	0.0153	3.27
	0.4750			
	0.4637			
3	0.326	0.3244 ± 0.0014	0.0060	1.85
	0.325			
	0.322			
4	0.2833	0.2830 ± 0.0003	0.0014	0.51
	0.2823			
	0.2833			
5	0.1877	0.1808 ± 0.0038	0.0161	8.93
	0.1748			
	0.1798			

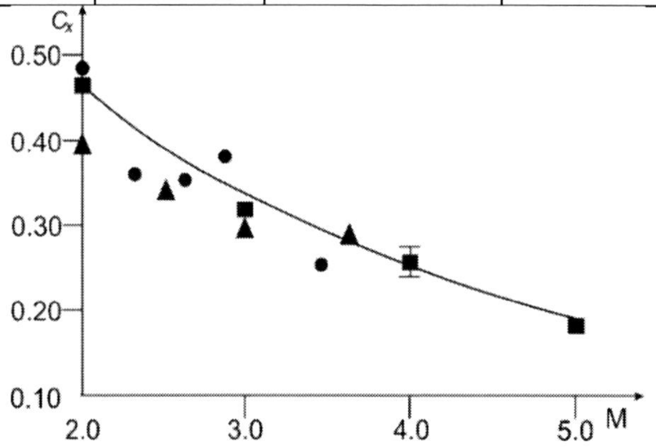

Fig. 4. Coefficient of the drag force for a cone: ■ – results of the experiments; ● – data [3]; ▲ – data [4].

Comparison of the results obtained with the findings of other authors showed their satisfactory agreement. In the range of the Mach numbers M = (2 ÷ 4) the maximum difference between the data of the authors [3] is ~ 8 %, between the data of the authors [4] is ~ 15 %. One of frames of high-speed videotape of supersonic air flow around cone (M = 3) with the organization of a local gas injection into boundary layer through two holes (nozzle-point) located at the top and bottom symmetrically about the model axis, is given in Fig. 5 (a).Conditions of blowing formation with a constant speed of M = 1 are diameter of a drainage hole – 0.25 mm, pressure in channels through which injection of air flow to boundary layer is organized, – (0.5 ÷ 1) MPa. One of video-frames of supersonic air flow around cone (M = 3) at non-symmetrical local injection out through hole in the bottom of model is presented in Fig. 5 (b).

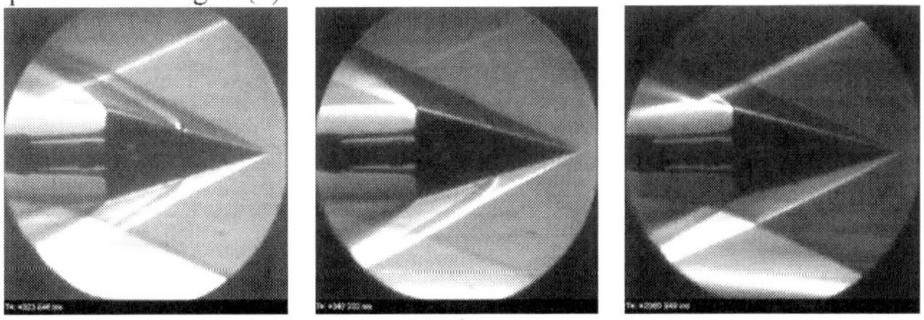

a b c

Fig. 5. Video-frames of supersonic air flow around cone with injection out through two symmetrically located holes (a), out through one hole in the in the bottom of model (b), and without injection (c), M = 3.

Conclusions

The results of experiments when flow around the cone without gas injection in the range of Mach numbers [2] were used as a control-test in experimental studies. The comparison of flow structure when flow around cone with symmetric gas injection and without it showed qualitative change of interaction of supersonic air stream with the cone surface (formation of Mach cone). In presence of injection the angle of Mach cone changes due to interaction between shock wave and injected air flow. It was observed that angles between Mach cone and surface of the model are different when flow around cone with the non-symmetric and with symmetric gas injection. Asymmetric Mach cone is formed when the non-symmetric gas injection. Thus qualitative analysis of the flow structure visualization results when flow around cone with gas injection and without it shows significant impact of injection to the boundary layer dynamics and Mach cone formation, and, consequently, to the change of the dynamic force on the body surface.

This research carried out in 2017 was supported by «The Tomsk State University Academic D.I. Mendeleev Fund Program», grant № 8.2.05.2017.

References

1. V.I. Zvegintsev, *Gasdynamic installations of short-term action* (Parallel, Novosibirsk, 2014) [in Russian]
2. E.A. Maslov, V.V. Faraponov, N. N. Zolotorev, A.V. Chupashev, V.V. Matskevich, S. Yu. Chizhov, MATEC Web of Conferences **92**, 01056 (2017)
3. V.G. Artonkin, P.G. Leutin, K.P. Petrov, Scientific works CAI. **92** 1413 (1972)
4. Yu.V. Sheludko, *Physical and gas-dynamic ballistic studies* (Science, Leningrad, 1980)

Experimental investigation of main rotor wake

Robert Stepanov[1*], and *Sergey* Mikhailov[1]

[1]Kazan National Research Technical University named after A.N. Tupolev, Department of Aerohydrodynamics, 420111, 10 Karl Marx Street, Kazan, Russian Federation

Abstract. In this work, experimental results of rotor wake in hover mode are presented. The experiments were carried out with a rotor rig model in the T-1K wind tunnel in Kazan National Research Technical University (Kazan Aviation Institute). The rotor consisted of four identical blades. The Q-criterion was used to identify tip vortices for a 2D case. The results were then compared with two different wake models.

1 Introduction

Experimental investigation of velocity fields in the wake of the helicopter rotor is a challenging task. Experimental investigation of real rotor wake is very problematic due to complex nature of the flow field and high sensitivity of the rotor due to its operating conditions. The underlying physics of the rotor wake behaviour is very similar to that of a finite wing, but they have differences due to the proximity and mutual interference of vortices trailed from the rotor blades [1].

Much research has been conducted in the past, focusing on the rotor wake characteristics and tip vortices. The rotor wake behaviour is defined by a flight regime in which it operates. McCroskey [2] and Landgrebe [3] carried out comprehensive reviews on different approaches, used to model helicopter rotor wakes.

Being able to predict the rotor's wake plays an important role during the preliminary structural calculations of the rotor blades and overall aerodynamic characteristics of the helicopter during the design phase. At the same time, the underlying physics of the vortex formation and evolution of the tip vortices, trailed from the helicopter rotor blades, is not fully understood and requires further attention [1,4].

Up to the date many physical aspects of rotor tip blade vortex formation and evolution as well as of rotor wake's behaviour are not well understood. This work outlines some of the experimental results, which were obtained using PIV. In this work, the wake of rotor model is investigated in hover mode.

* Corresponding author: rpstepanov@kai.ru

2 Vortex identification

In this work, vortices were identified using Q-criterion, which defines vortices as connected spatial regions, where the Euclidean norm of the vorticity tensor $\boldsymbol{\Omega}$ dominates over the rate of strain tensor $\boldsymbol{S}$ [5]:

$$Q = 0.5[|\boldsymbol{\Omega}|^2 - |\boldsymbol{S}|^2] > 0 \quad (1)$$

Due to the fact that only one camera was used in the experiments, vorticity tensor $\boldsymbol{\Omega}$ and the rate of strain tensor $\boldsymbol{S}$ in equation (1) are reduced to a two-dimensional form:

$$|\boldsymbol{\Omega}|^2 = (\partial u/\partial y - \partial v/\partial x)^2 \quad (2)$$

$$|\boldsymbol{S}|^2 = 2(\partial u/\partial x)^2 + 2(\partial v/\partial y)^2 + (\partial u/\partial y + \partial v/\partial x)^2 \quad (3)$$

Different prescribed wake models exist for obtaining tip vortex positions for hovering flight regimes. Landgrebe's model [6] is based on experimental studies of various rotor models with different number of blades, blade twist and aspect ratio. Later, Kocurek & Tangler [7] derived a prescribed wake model, which was based on a different set of experimental results. Kocurek & Tangler's model was similar to Landgrebe's model, but included blade lift distribution for the modelling of the axial settling rates and did not use inner vortex sheets.

3 Experimental setup

Fig. 1. Experimental setup in the T-1K wind tunnel.

All results were obtained in T-1K wind tunnel at the Kazan National Research Technical University (KNRTU-KAI) named after A.N. Tupolev.

The four-bladed rotor rig (see Figure 1) was used to study its wake in hover mode. It was placed at the center of the test section, so that the distance from the rotor plane to ceiling was 2.6 m, and the distance from the rotor plane to the ground was 2.1 m. The direction of the flow is shown in Figure 2 (b).

The rotor had the diameter of 1640 mm, the blades had a constant chord of 65 mm along their span with rounded tips and no twist. The blades were attached to an articulating hub with three hinges. The collective pitch angle was set at 8°. The rotor's angular speed was fixed at 900 rpm. The blade tip Mach number was M_r=0.23 and Reynolds number at 75% of the rotor radius was $Re_{0.75r} = 2.57\times10^5$.

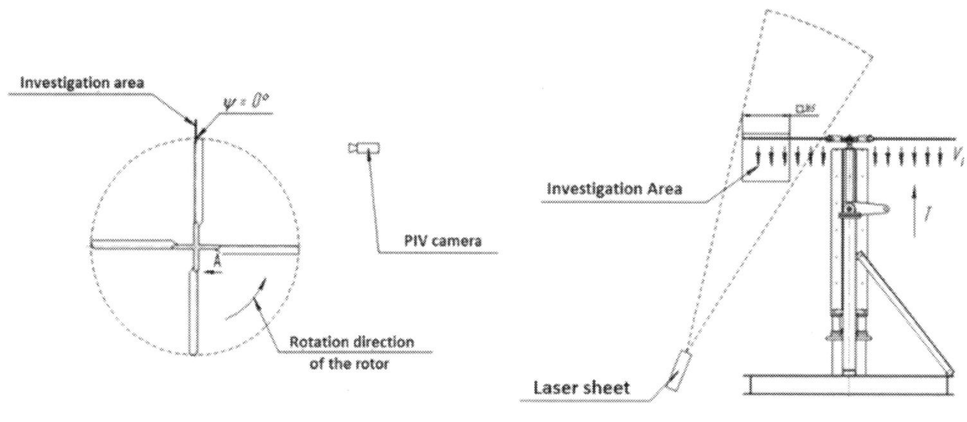

(a) Top view (b) Side view

Fig. 2. Schematic of the experimental setup: (a) top view; (b) side view.

Velocity fields were obtained with Dantec PIV system, which comprised of *Nd-YaG laser Litron 425-10* with maximum output of 425 mJ, exposure time of 6 ns and wave length of 532 nm. Exposure frequency during the experiments was set at 8 Hz. The flow was seeded with olive oil particles using a liquid seeding generator.

The PIV camera *FlowSenseEO 4M*, with the resolution of 2048×2048 pixels, was positioned to capture the tip vortices trailing from the rotor blades, as shown in Figure 2.

4 Results and discussion

As it was already mentioned, Q-criterion was used to identify vortex positions in the wake of the rotor in hover mode. The velocity fields, which were obtained using the PIV system, were used to calculate Q-criterion values for the 2D case using equations (1), (2) and (3).

One of the exposure frames is shown in Figure 3 (a). Here, four distinct vortices can be seen, which are shown as shaded colored regions where $Q > 0$. In addition, the velocity field and streamlines are shown on the same figure. It is worth noting that the positions of the vortices, obtained using the Q-criterion, do not coincide with the points where the resultant velocity is equal to null, but they are rather located closer to the rotor's axis of rotation.

Multiple velocity fields were obtained during the experiment, each corresponding to a certain exposure frame, taken at a different point of time. The collection of each vortex position for each exposure frame is shown in Figure 3 (b) with dots. Those coordinates were then approximated by a polynomial line. The dots on the plot represent points in the vortices with highest Q-values.

Fluctuations of the vortices from their mean values indicate that there is aperiodicity in the wake, also known as vortex wondering. It is also evident, that wake aperiodicity increases with the wake's age.

Comparison of prescribed wake models show that Kocurek & Tangler's model better agrees with experiments, especially in the early age of the wake.

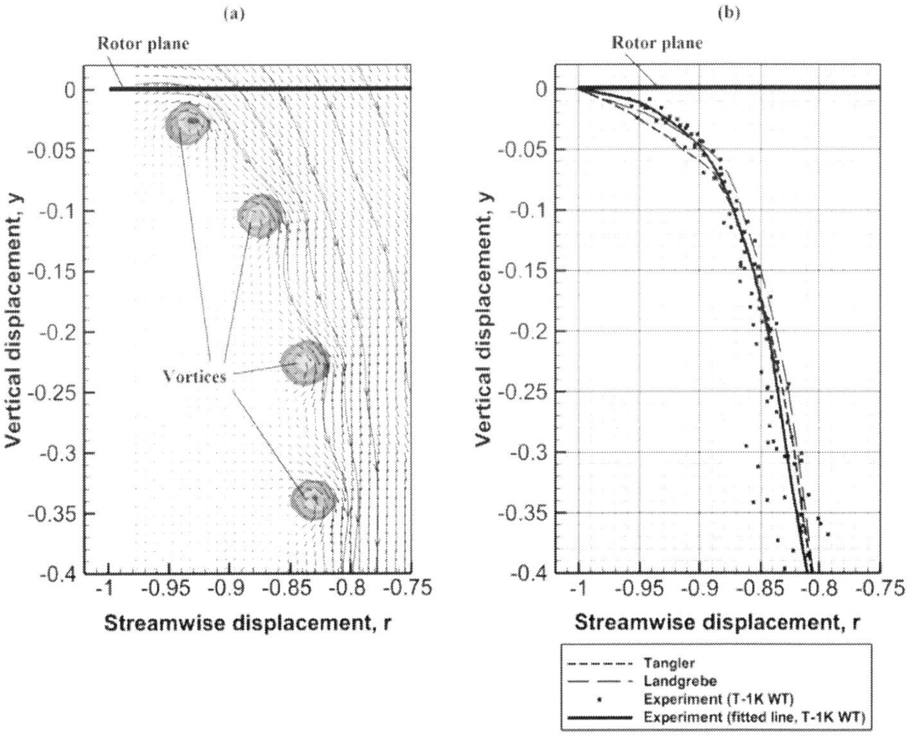

Fig. 3. Experimental results: (a) a typical exposure frame, featuring tip vortices ($Q > 0$); (b) comparison of the experimental data with prescribed wake models.

5 Conclusions

This work presents experimental results of wake investigation using PIV system. Tip vortices were identified using Q-criterion for a two-dimensional case. The wake data is then compared with Landgrebe' prescribed wake model and Kocurek & Tangler's prescribed wake model. The results also reveal that there is aperiodicity inherent to the flow.

This work was supported by the grant 'Numerical and physical modelling of aerodynamic and aeroacoustic characteristics of rotor systems of future concept aircarft' (No. 9.1577.2017/ P Ch) of the Ministry of Education and Science of the Russian Federation.

References

1. G.J. Leishman, *Principles of Helicopter Aerodynamics* (Cambridge University Press, Cambridge, U.K., 2006)
2. W.J. McCroskey, *33rd AIAA ASME*, 95-0530 (1995)
3. A.J. Landgrebe, *2nd Int. Conference on Rotorcraft Basic Research* (1988)
4. J.G. Leishman, A. Bagai, AIAA J. **36,** 415 (1998)
5. G. Haller, J. Fluid Mech. **525** 1 (2005)
6. A. J. Landgrebe, J. American Helicopter Soc. **17,** 2 (1972)
7. J.D. Kocurek, J.L. Tangler, J. American Helicopter Soc. **21**, 24 (1976)

MATEC Web of Conferences 115, 02014 (2017) DOI: 10.1051/matecconf/201711502014
STS-33

Heat transfer at a stagnation point of impinging round air jet at low Reynolds numbers

Vadim Lemanov[1,*], *Ziedillo* Khazhiev[1,2]

[1] Kutateladze Institute of Thermophysics SB RAS, 630090, 1 Ac. Lavrentiev Avenue, Novosibirsk, Russia
[2] Novosibirsk State Technical University, 630073, Karl Marks Avenue, 20, Novosibirsk, Russia

Abstract. This work presents experimental investigation of average and pulsation thermal characteristics at the stagnation point of a round impinging air jet at low Reynolds numbers. In the experiments, the average and pulsation values of heat flux, heat transfer coefficient and Nusselt number were measured. At low Reynolds numbers (Re<4000), in contrast to the known monotonous increase in heat transfer, a non-monotonic change in heat transfer is shown. With an outflow from the tube, a significant increase in Nusselt number is observed in comparison with the case, when the jet flows from a nozzle, and this increase can be 300-500%.

1 Introduction

The impinging jets are widely used for efficient cooling of heat-stressed sections in different fields of engineering [1-2]. However, there are some unresolved issues concerning heat transfer at interaction of the jet with the surface. The complexity of the problem is caused by the large number of parameters influencing hydrodynamics and heat transfer. It is known that heat transfer in the impinging jets depends on Reynolds number (*Re*), Prandtl number (*Pr*), relative distances from the nozzle to the surface (*h/d*), degree of jet turbulence (Tu), and others [1-4]. One of the most important parameters is the Reynolds number ($Re=Ud/\nu$); for the round jets it is determined by superficial velocity U, initial jet diameter d and kinematic viscosity ν. Correlation dependence for calculation of heat transfer in the region of the stagnation point of a flat plate is usually written as

$$Nu_0 = C\,\mathrm{Re}^m\,\mathrm{Pr}^n\left(\frac{h}{d}\right)^p \tag{1}$$

where coefficients C, m, n, p vary in the studies of different authors [1-4]. A monotonous increase in the Nusselt number at a growth of Reynolds number (m=0.4-0.8) follows from formula (1), and this is proved by experimental data of [1-5]. To solve the problem of efficient cooling of heat-stressed sections, the jets with high Reynolds number (Re>10000) at small distances to the obstacles (h/d<6) were studied mainly. Few papers [5-8] deal with

* Corresponding author: lemanov@itp.nsc.ru

the study of heat transfer at low Reynolds numbers (Re<4000) usually at h/d<10. In this regard, the aim of the current work was the study of average and pulsation characteristics of heat transfer in a vicinity of the stagnation point of the impinging round air jet at low Reynolds numbers.

2 Experimental setup and measurement method

The experimental equipment included a setup and measurement system. Working gas was air. The following devices were used as a jet source: a) tube with inner diameter d=3.2 mm, length l=1 m with an inlet of the type "sudden contraction", b) profiled nozzle with inner diameter d=3.2 mm and length of 35 mm with the contraction degree of 20. The distance from the jet beginning to as obstacle was h/d=18-20. The heat exchange section, which receives the impinging jet, is a homogeneous copper plate with the diameter of 190 mm and thickness of 50 mm. A small heat flux sensor (HFS) of the gradient type with the size of 2x2x0.2 mm was mounted at the stagnation point of the plate [9]. The plate was heated using the electric heater, the boundary condition on the copper plate surface was close to T_w = const. To measure the level of turbulence in the initial jet cross-section, DISA 55M heat-wire anemometer was used. In the course of experiment, the following parameters were measured: instantaneous value of the heat flux, gas flow rate through a tube or nozzle, temperatures of the plate T_w (40-50°C) and jet T_j (24-27°C) in the initial cross-section, and barometric pressure. Average Q and mean square q values of the heat flux were determined by the measured time series of instantaneous heat flux values (the sample size of up to $1x10^4$). The average heat transfer coefficients α_0 and mean square pulsations of heat transfer coefficient α_0' were determined respectively by Q and q and temperature difference between the wall T_w and jet T_j. The parameters of air (kinematic viscosity ν and heat conductivity coefficient λ) required for calculation of the jet Reynolds numbers and Nusselt number ($Nu_0 = \alpha_0 d/\lambda$) were determined by the flow temperature in the initial cross-section of jet T_j.

3 Experimental results and discussion

Experimental data for local heat transfer coefficient at the stagnation point of the plate are shown in Fig.1. For the outflow from the nozzle, monotonic dependence of average heat transfer coefficient α_0 on jet Reynolds number is observed in the range of Re=200-20000 (Fig.1a). At the same time, for the outflow from the tube, the behavior of α_0 is significantly different. There are three characteristic zones. The following phenomena were registered in these zones: an increase in α_0 in the first zone (Re<3320), considerable decrease in α_0 in the second zone (Re=3320-3500), and growth of α_0 in the third zone (Re>3500). Thus, a non-monotonic change in the average heat transfer coefficient α_0 was determined in the range of low Reynolds numbers (Re<4000). For the outflow from the nozzle, the level of pulsations is low and approximately constant over a wide range of Reynolds numbers (Fig.1b). At the same time, for the outflow from tube at Re=3320, there is a significant extreme α_0', which correlates with the maximum of α_0.

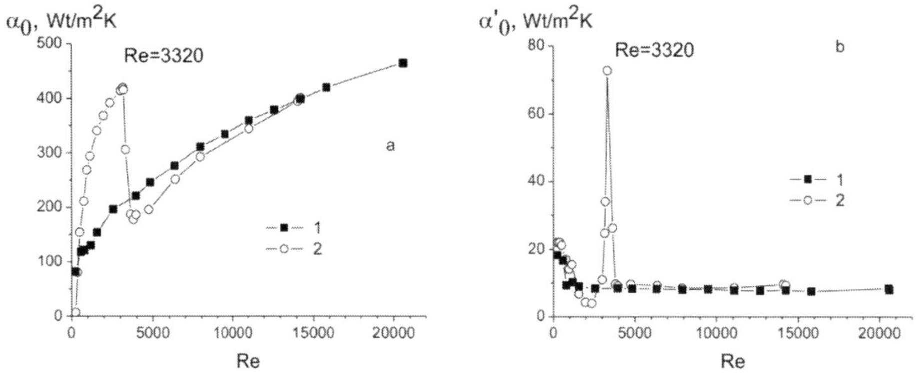

Fig. 1. Average (a) and pulsation (b) value of the local heat transfer coefficient at the stagnation point of the plate: *1* — nozzle, *2* — tube ($h/d = 20$).

The measured level of turbulence in the initial cross-section of the impinging jet is shown in Fig.2a. For the jet outflow from the nozzle, the level of velocity pulsations is low and it is Tu<1% in the range of Re=250-20000. For the jet outflow from the tube, the character of turbulence behavior is different. As in Fig. 1, there are three zones. An insignificant decrease in velocity pulsations (Tu<1%) is registered in the first zone (Re<3300), in the second zone (Re=3300-3500), there is the extreme of velocity pulsations at Re=3320, which is 12%, and in the third zone (Re>3500), there is a monotonous decrease in pulsations (Tu<2%).

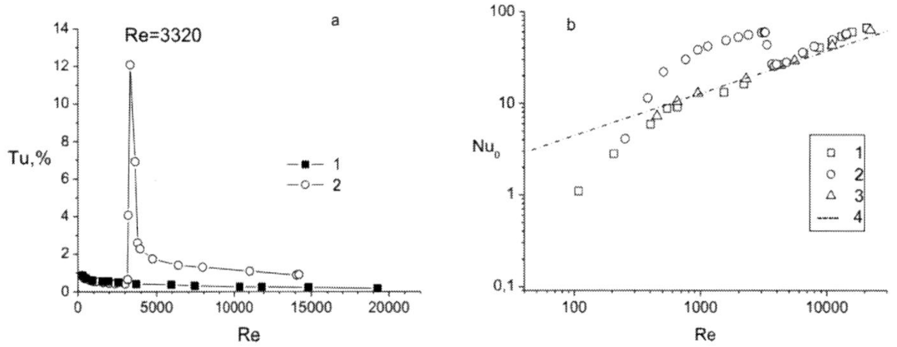

Fig. 2. Dependence of turbulence degree (a) and Nusselt number (b) on Reynolds number: 1 — nozzle, 2 — tube; 3 – data of [7], 4 – calculation by (1).

Data using the Nusselt number for the average heat transfer Nu_0 are shown in Fig.2b. Line 4 in the figure corresponds to dependence (1) with coefficients C=5.25 m=0.5 n=0.33 p=0.77 [1-2]. Data of [7] for heat transfer were obtained in the range of Reynolds numbers Re=450-22000 for air. According to the figure, experiments of [7] are consistent with correlation dependence (1). Our data for the jet outflow from the nozzle are also described by dependence (1) at Re>400, however, for Re<400, there is significant deviation from (1). For the jet outflow from the tube, the character of a change in heat transfer depending on Re number is non-monotonic. As in Figs. 1, 2a, there are three zones. At low Reynolds numbers (Re<400) the data are substantially lower than dependence (1). In the range of

Re=3300-3500, there is a significant increase in Nusselt number in comparison with the case, when the jet flows from a contoured nozzle, and it can be 300-500%. Experimental data are consistent with dependence (1) only in the range of Re>4000. The reason for this behavior of heat transfer at the stagnation point at Re<4000 is the effect of initial conditions of jet formation. For a long tube at Re<3300, velocity distribution in the initial jet cross-section is close to the Poiseuille profile. For the case of a contoured nozzle, the velocity profile at outlet cross-section has a "shock" character with thin boundary layers. In literature, the effect of initial velocity profile for the impinging jets was studied only for high Reynolds numbers (Re>4000). Experiments carried out at IT SB RAS on the range of axisymmetric free jets [10] demonstrate a strong effect of the initial velocity distribution. For the flow from the nozzle, parameters shown in Figs.1, 2 have monotonous dependence on the Reynolds number. At that, a change in the Nusselt number depending on the Reynolds number at Re>400 is consistent with known dependence (1) and experiments of [7].

4 Conclusions

According to the research, heat transfer at the stagnation point of the impinging jet depends not only on similarity criteria (Re, Pr, *h/d*, Tu), but also on the method of jet formation. At interaction of an axisymmetric air jet flowing from a long tube with a flat surface, a non-monotonous change in heat transfer near the front point is observed at low Reynolds numbers (Re<4000). In this case, there is a significant increase in Nusselt number in comparison with the case, when the jet flows from a contoured nozzle, and it can be 300-500%. At Reynolds numbers Re>4000, the difference in heat transfer for two versions of jet formation, decreases asymptotically.

The work was partially supported by RFBR (grant 17-08-00958).

References

1. B.N. Yudaev, M.S. Mikhalkov, V.K. Savin, *Heat transfer at interaction between the jets and obstacles* (Mashinostroenie, Moscow, 1977) [in Russian]
2. E.P. Dyban, A.I. Mazur, *Convective heat transfer at the jet flow around bodies* (Naukova Dumka, Kiev, 1982) [in Russian]
3. K. Jambunathan, E. Lai, M.F. Moss, B.L. Button, Int. J. Heat Fluid Flow **13,** 106 (1992)
4. N. Zuckerman, N. Lior, Adv. Heat Transfer **39**, 565 (2006)
5. V.V. Lemanov, V.I.Terekhov, High Temp. **54**, 454 (2016)
6. G.C. Huang, Proc. ASME J. Heat Transfer **85,** 237 (1963)
7. R. Gardon, J.C. Akfirat, Proc. ASME J. Heat Transfer **88,** 101(1966)
8. D.H. Lee, J.R. Bae, H.J. Park, J.S. Lee, P. Ligrani, Int. J. Heat Mass Transfer **54,** 2408 (2011)
9. S.Z. Sapozhnikov, V.Yu. Mityakov, A.V. Mityakov, *Foundations of gradient heat measurements* (Polytechnical University, St.Petersburg, 2012) [in Russian]
10. V.V. Lemanov, V.I.Terekhov, K.A. Sharov, A.A. Shumeiko, Tech. Phys. Lett. **39**, 421 (2013)

Excitation of localized wave packet in swept-wing supersonic boundary layer

Aleksey Yatskikh[1,2,*], *Marina* Rumenskikh[1,2], *Yury* Yermolaev[1], *Aleksander* Kosinov[1,2] and *Nikolay* Semionov[1]

[1]Khristianovich Institute of Theoretical and Applied Mechanics SB RAS, 630090 Novosibirsk, Russia
[2]Novosibirsk State University, 630090 Novosibirsk, Russia

Abstract. The evolution of the artificial wave packet in swept-wing supersonic boundary layer was experimentally studied at M=2. The localized disturbances were generated by a pulse glow discharge. Measurements were provided by a hot-wire anemometer. The spatial structure of the wave packet was studied. It was found that the wave packet has an asymmetric shape. In addition, the velocity of the propagation downstream of the wave packet was estimated.

1 Introduction

To solve the problem of laminarization the detailed knowledge of laminar-turbulent transition mechanisms is required. At present it is generally recognized that the onset of turbulence is connected with the loss of stability of the initial laminar flow.

The study of the evolution of artificially controlled perturbations allows finding out the wave characteristics of the boundary layers. A commonly used method of the actuation of controlled perturbations in a supersonic boundary layer is a high-frequency electric glow discharge. The experimental confirmation of the main concepts of the linear theory of stability in supersonic zero-pressure-gradient flow on a flat plate was obtained by this method in [1]. Also, this approach is useful in studies of laminar-turbulent transition in inhomogeneous and 3D supersonic boundary layers [2-4]. The approach of periodical disturbances is also used in direct numerical simulation studies (DNS). In [5, 6] results coinciding with experiments were obtained. Also, the method of artificial excitation of periodically pulsations in boundary layers is efficiently used at the hypersonic speeds [7-9].

However, the latter method allows us to study the evolution of narrowband disturbances while natural laminar-turbulent transition is associated with the evolution downstream of fluctuations consisted of a wide range of frequencies and wavenumbers excited by freestream turbulence. Therefore the study of the evolution of broadband disturbance (wave packets) is promising.

The first experimental studies of wave packet propagation in boundary layers were provided at low speed of flow [10]. The evolution of wave packets generated by a short-duration acoustic pulse was investigated by hot-wire measurements. These measurements showed good agreement with the results obtained from the theoretical model which

* Corresponding author: 73.yatskikh@gmail.com

represented the wave packet as a superposition of individual disturbances for all frequencies and spanwise wave numbers of the most unstable linear waves [11]. Further development of the approach of wave packets led to significant progress in understanding the processes of laminar-turbulent transition at a subsonic speed. Thus, *e.g.* in [12] the emergence of wave packets, or forerunners, in the frontal regions of localized structures and their evolution in subsonic boundary layer was experimentally examined. The experiments showed that the forerunners were Tollmien—Schlichting wave packets.

At high speeds of flows the method of the excitation of artificial wave packets is actively developing. The evolution of wave packets in the hypersonic boundary layer of a nozzle was experimentally studied in [13]. The measurements were taken along the axial line of the nozzle using surface pressure transducers. The linear and non-linear growth of wave packets and their transformation into a turbulent spot were found. The evolution and transformation into turbulent spots of wave packets are shown as a continuous process. Also, the method of wave packets is widely used in numerical study of laminar-turbulent transition [14-16].

In [17-19] the technique of pulsed excitation of wave formations was developed and used to study the turbulence origin in a two-dimensional supersonic boundary layer. The evolution of the artificial wave packet in the boundary layer on a flat plate at M = 2 was studied in these works with the help of hot-wire measurements. The wave packets were generated by a pulse discharge. The analysis of the experimental data has shown that for different frequency modes the most unstable waves are oblique. This agrees with the theoretical results.

The main goal of the present work is to carry out the experimental study of the evolution of wave packets in a three-dimensional supersonic boundary layer.

2 Experimental setup

Experiments were performed in the T-325 long-duration blowdown low-noise supersonic wind tunnel, at the Khristianovich Institute of Theoretical and Applied Mechanics SB RAS, at Mach number M = 2 and unit Reynolds number $Re_1 = 6\cdot 10^6$ m^{-1}. A model of swept-wing with a sharp leading edge and a source of controlled disturbances were used (Fig. 1). The swept angle of the leading edge is 40°. The model was installed with zero angle of attack.

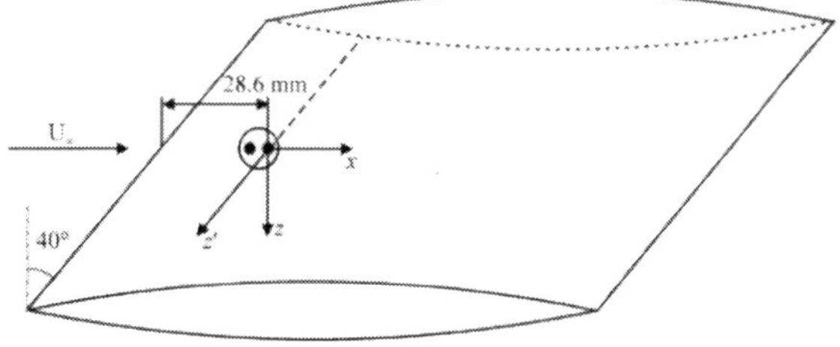

Fig. 1. Swept-wing model with a source of controlled perturbations.

To initiate controlled wave packets an electric discharge was ignited between two copper electrodes separated from each other and the model with an insulator. The insulator and electrodes were mounted flush with the model surface. The electrode centers were located parallel to the direction of the flow (the *x*-axis). The electrode diameter was 0.6 mm, while the minimum distance between them was approximately 0.6 mm. High-voltage (~1 kV) pulses were supplied to the electrodes from the ignition scheme.

The ignition scheme for pulsed electric discharges is based on disrupting the current in the primary circuit of the ignition coil with a powerful bipolar transistor. Disrupting the current in the primary circuit of the ignition coil generates high-voltage pulsing.

In these experiments the duration of discharge was about 20 μs. The energy of a single discharge was about 0.12 mJ.

Pulsations in a supersonic flow were measured using constant-temperature hot-wire anemometry (CTA). A hot-wire probe made of a tungsten wire with a 10 μm diameter was used. The overheat rate of the hot-wire probe was set to 0.8, while perturbations measured by CTA were predominantly mass flux pulsations (*m'*). The measurements were performed at the layer with the most intense natural pulsations of the boundary layer. The probe was moved with the help of traversing gears which the wind tunnel was equipped with. The positioning accuracy of the probe was about 0.1 mm.

To separate perturbations generated by the discharge from the background natural fluctuations of the boundary layer, the measurements were synchronized with discharge ignition. 320 traces were recorded at each measurement point. Data processing included ensemble-averaging 320 traces.

3 Results

The isolines of the mass flux pulsation in plane (z', t) for the cross sections at x = 20, 30 and 40 mm are presented in Fig. 2. The measurements were provided in the area of the maximum level of the boundary layer pulsation. For each cross section the hot-wire probe was moved parallel to the leading edge and at a constant distance to the wall.

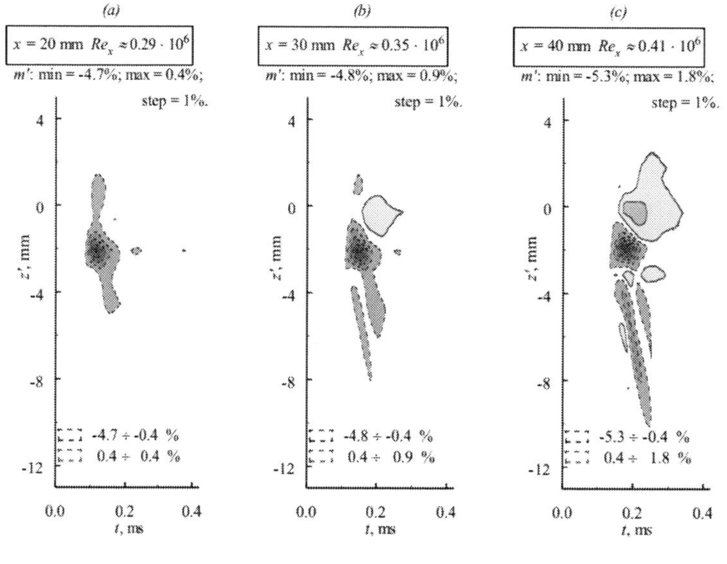

Fig. 2. Contour lines of the mass-flow pulsations of the wave packet in the plane (z', t) at x = 20 (*a*), 30 (*b*) and 40 (*c*) mm.

The disturbances exited by the discharge are asymmetric and shifted to negative values of the z'-coordinate. This result differs from the case of a two-dimensional supersonic boundary layer. This is due to the crossflow in the boundary layer of the swept-wing. Similar results were obtained in the experiments on the evolution of localized disturbances in an incompressible boundary layer [20].

In the initial section, the perturbation from a pulsed discharge is a flow defect localized in space and time. The spatial scales of the wave packet increase downstream. The structures of excess and decrease of the mass flow form at the boundaries of the disturbance from the discharge. At z' = -3÷4 mm wave packet is represented as a vortex in the plane of the model surface. At z' < -3 mm the short-term defects of mean flow are observed.

Moments of time of the wave packet registration are varying downstream. Thus the velocity of propagation downstream of the wave packet can be calculated. It is estimated that the velocities of the wave packet propagation is about $0.5 \div 0.6 \cdot U_{\infty}$.

4 Conclusion

For the first time the evolution of a wave packet in a three-dimensional supersonic boundary layer has been experimentally investigated. The structure of the wave packet has been studied by the hot-wire measurements. It has been found that the structure of the wave packet in swept-wing boundary layer is not symmetric in contrast to the case of a two-dimensional share layer. Also, the velocity of the propagation downstream of the wave packet has been estimated.

The work was financially supported by the Russian Science Foundation (Project No. 17-19-01289).

References

1. A.D. Kosinov, A.A. Maslov, S.G. Shevelkov, J. Fluid Mech. **219**, 621 (1990)
2. Yu.G. Yermolaev, G.L. Kolosov, A.D. Kosinov, N.V. Semionov, Fluid Dynamics **49**, 188 (2014)
3. A.D. Kosinov, G.L. Kolosov, N.V. Semionov, Yu.G. Yermolaev, Phys. Fluid, **28**, 064101 (2016)
4. A.V. Panina, A.D. Kosinov, Y.G. Yermolaev, V.N. Gorev, N.V. Semionov, Thermophys. Aeromech. **21**, 3 (2014)
5. A.N. Kudryavtsev, D.V. Khotyanovsky, Thermophys. Aeromech. **22**, 559 (2015)
6. C.S.J. Mayer, S.Wernz, H.F. Fasel, J. Fluid Mech. **668**, 113 (2011)
7. A.A. Maslov, A.N. Shiplyuk, D.A. Bountin, A.A. Sidorenko, J. Spacecraft Rockets **43**, 71 (2006)
8. D.A. Bountin, A.N. Shiplyuk, A.A. Maslov, J. Fluid Mech. **611,** 427 (2008)
9. A.A. Maslov, A.N. Shiplyuk, A.A. Sidorenko, D. Arnal, J. Fluid Mech. **426,** 73 (2001)
10. M. Gaster, I. Grant, Proc. R. Soc. Lond., A **347**, 253 (1975)
11. M. Gaster, Proc. R. Soc. Lond., A **347**, 271 (1975)
12. M.M. Katasonov, S.H. Park, H.J. Sung , V.V. Kozlov, Exp. Fluids **38**, 363 (2005)
13. K.M. Casper, S. J. Beresh, S. P. Schneider, J. Fluid Mech. **756**, 1058 (2014)
14. C. S. J. Mayer, A.C. Laible, H.F. Fasel, AIAA J. **49**, 67 (2011)
15. J. Sivasubramanian, H.F. Fasel, J. Fluid Mech. **756,** 600-649 (2014)
16. A. Novikov, I. Egorov, A. Fedorov, AIAA J., **54**, 2034-2050 (2016)
17. Y.G. Yermolaev, A.A. Yatskikh, A.D. Kosinov, N.V. Semionov, AIP Conference Proceedings, **1770**, 030037 (2016)
18. A.A. Yatskikh, Y.G. Yermolaev, A.D. Kosinov, N.V. Semionov, Thermophys. Aeromech. **22,** 17 (2015)
19. A.A. Yatskikh, Y.G. Yermolaev, A.D. Kosinov, N.V. Semionov, J. Visualization (2017) doi:10.1007/s12650-016-0414-2 (to be published)
20. G.R. Grek, M.M. Katasonov, V.V. Kozlov, Thermophys. Aeromech. **15**, 549 (2008)

Direct numerical simulation of Mhd heat transfer of the liquid metal in a horizontal pipe with the joint effect of the longitudinal magnetic field and thermo-gravitational convection

Akhmedagaev Ruslan[1,*], and *Listratov* Yaroslav[1]

[1]National Research University «Moscow Power Engineering Institute», 111250 Moscow, Russia

Abstract. The results of the direct numerical simulation (DNS) of the MHD heat transfer in turbulent flow of liquid metal (LM) in a horizontal pipe under the joint influence of a longitudinal magnetic field (MF) and thermo-gravitational convection (TGC) are presented. Discusses the characteristics of hydrodynamics and heat transfer in the longitudinal MF in the absence of TGC. Suppression of turbulent transport is observed with an increase in the Hartmann number: the heat transfer coefficients, the friction coefficient. The authors calculated the effect of TGC in a strong longitudinal MF for a homogeneous heating. Investigated the averaged fields of velocity and temperature, heat transfer characteristics, the distribution of wall temperature along the perimeter of the cross section of the pipe. The effect of TGC on the velocity field is affected stronger than in the temperature field.

1 Introduction

Currently, the world community is working on the creation of an international experimental thermonuclear reactor (ITER), which is being built in Cadarache (France). Despite the water concept of cooling the reactor blanket, it is planned to design separate liquid metal modules for the production of tritium, obtained as a result of lithium irradiation by neutrons. In ITER liquid metal system will work if you have a very strong MF and the research problem of the regularities of hydrodynamics and heat transfer under such conditions at a current LM is very relevant [**Ошибка! Источник ссылки не найден.**].

The purpose of this paper is to simulate the DNS method of a horizontal heated pipe with the combined effect of TGC and longitudinal MF. Calculations for homogeneous heating have been performed. In the future, it's planned to consider one-sided heating in order to study the hypothesis of the occurrence of anomalous temperature pulsations with increasing Hartmann number in the longitudinal MF. This hypothesis is described in detail for a transverse MF [2].

* Corresponding author: akhmedagaev.r.m@gmail.com

2 Physical model

The investigated geometry is shown in figure 1. The length of the investigated section (z) is assumed to be 5d. The radius of the computed area (r) varies from 0 to 1.

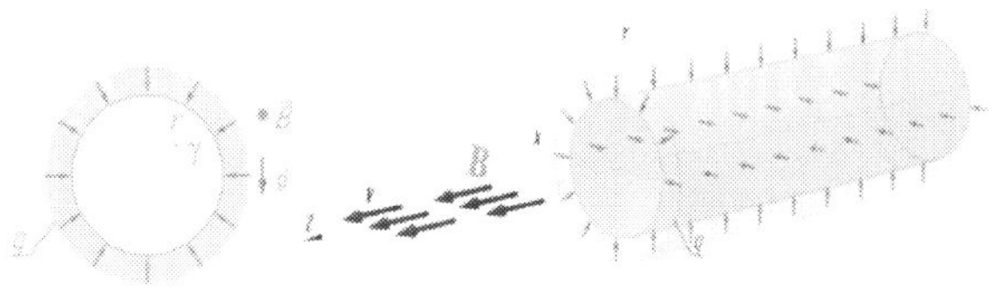

Fig. 1. The flow diagram.

The task is described by means of a system of equations, which includes equations of continuity – eq. (1), of motion – eq. (2) and of energy conservation – eq. (3). Such assumptions as constant properties, incompressible fluid are used, viscous dissipation is not taken into account, low-frequency approximation of electromagnetic hydrodynamics is applied. The final version of the system in a dimensionless form:

$$\nabla \cdot \boldsymbol{U} = 0 \tag{1}$$

$$\frac{\partial \boldsymbol{U}}{\partial t} + (\boldsymbol{U} \cdot \nabla)\boldsymbol{U} = -\nabla p - \nabla \hat{p} + \frac{1}{Re} \cdot \nabla^2 \boldsymbol{U} + \boldsymbol{F}_b + \boldsymbol{F}_L \tag{2}$$

$$\frac{\partial \theta}{\partial t} + (\boldsymbol{U} \cdot \nabla)\theta = \frac{1}{Pe} \cdot \nabla^2 \theta - U_z \cdot \frac{\mathrm{d}\bar{\theta}}{\mathrm{d}Z} \tag{3}$$

The system of equations (1) - (3) is solved by the DNS method, the detailed algorithm of which is presented in [3].

3 Results

Today from experimental data [4] show that the longitudinal MF delays the transition from laminar flow regime to turbulent and it is observed in the growth of the critical Reynolds number, which increases with Hartmann number.

As an example, let us consider the dependence of the friction coefficient on the Reynolds number (fig. 2) and the effect of the longitudinal magnetic field on heat transfer (fig. 3). As the Hartmann number increases, a decrease in friction is observed due to suppression of turbulent velocity pulsations, which corresponds to experiments [4].

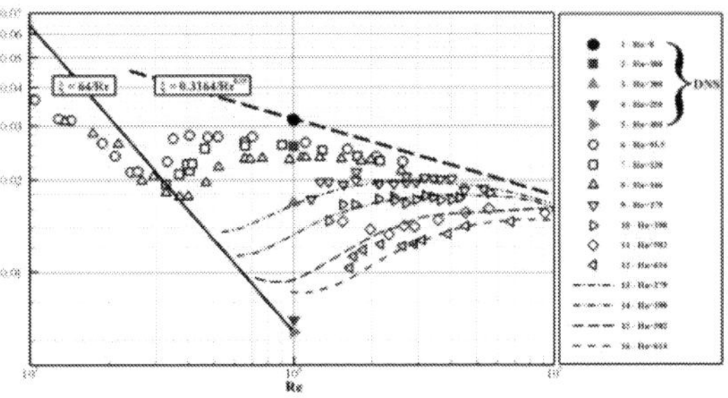

Fig. 2. The dependence of friction coefficient on Reynolds number, Ha=0-614: 1-5 – DNS for $Re=10^4$; 6-12 – experimental points [4]; 13-16 – interpolation lines [4].

A similar situation is observed for heat transfer, which decreases in a strong MF to a laminar Nusselt number. For Reynolds values, approximately equal to the critical Reynolds number for $Ha >> 1$ [5] - eq. (4), the difference between the DNS method and the experimental data increases. This is due to the fact that the hydrodynamic stabilization of the flow was not achieved in the experiments, in contrast to the calculation.

$$Re_{cr} = 30 \cdot Ha \quad (4)$$

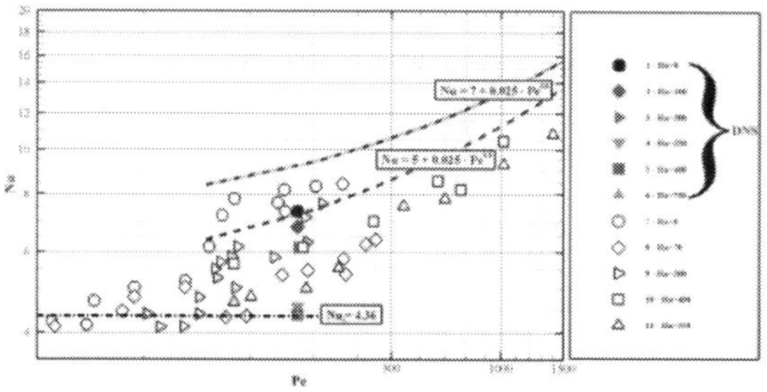

Fig. 3. The dependence of the Nusselt number from the Peclet number, Ha=0-550: 1-6 – DNS for Pe=272; 7-11 – [4].

Now well-known is the effect of TGC on characteristics of hydrodynamics and heat and it depends on the orientation of the investigated area. In the joint conditions in strong MF effects of the TGC may be increase. A characteristic feature of the flow in these conditions is a substantial heterogeneity of the heat transfer coefficients and, therefore, wall temperature on the perimeter of the tube cross-section (fig. 4), which can lead to unacceptably high thermal stresses in the heat exchangers walls. The results of the DNS method are in good agreement with the experiment [6].

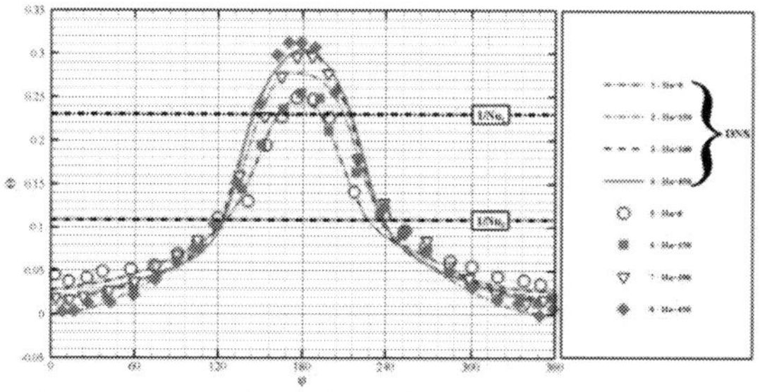

Fig. 4. Dimensionless wall temperature along the perimeter of the horizontal tube cross section, $Re=10^4$, $Gr=0.8 \cdot 10^8$: 1-4 – DNS; 5-8 – [6].

In the horizontal tube affected by longitudinal MF TGC manifests itself in the form of large longitudinal vortices with axes parallel to the vector of the MF induction. MF stabilizes these vortices and, as a result, flow loses axial symmetry (fig. 5 for the temperature). Heat distribution becomes inhomogeneous in the tube cross-section and are formed of a zone of «degraded» heat transfer, which are determined by local Nusselt numbers less than the value for laminar mode – $Nu_L=4.36$.

For the averaged velocity fields (fig. 5 for the velocity), it is seen that TGC in strong MF influences more strongly than on the temperature fields: the maximum velocity is shifted to the lower part of the tube and stretched along the circumference.

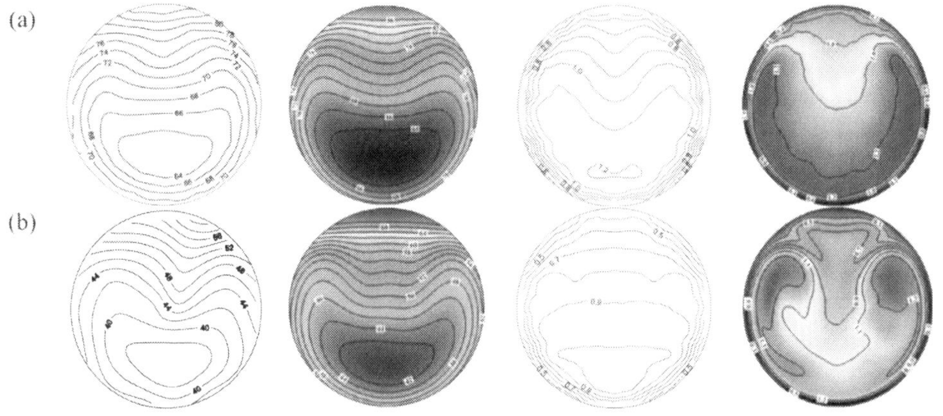

Fig. 5. Fields of average temperature and longitudinal velocity component in the horizontal tube cross section, Re=10^4, Gr=$0.8 \cdot 10^8$: a) Ha=0; b) Ha=300.

4 Concluding remarks

The authors confirmed the well-known fact that in the absence of TGC the longitudinal MF leads to suppression of turbulence with increasing Hartmann number.

In the case of homogeneous heating of the horizontal pipe wall with the combined influence of MF and TGC, the velocity field changes significantly and secondary convective currents appear. This leads to the appearance of inhomogeneity in the distribution of the local heat transfer coefficient along the perimeter of the wall and, as a consequence, the inhomogeneity of the temperature field. These effects can cause high thermal stresses, which should be taken into account when designing the ITER blanket.

The results were obtained during the implementation of the Draft State Order of the Ministry of Education and Science of the Russian Federation in the field of scientific activity No. 13.9619.2017 / 8.9

References

1. C.P.C. Wong, J.F. Salavy, Y. Kim, I. Kirillov, E. Rajendra Kumar, N.B. Morley, S. Tanaka, Y.C. Wu, Fusion Eng. Design **83,** 850 (2008)
2. L.G. Genin, V.G. Zhilin, Y.P. Ivochkin, N.G. Razuvanov, I.A. Belyaev, Y.I. Listratov, V.G. Sviridov, *Temperature fluctuations in a heated horizontal tube affected by transverse magnetic field, in*: Proc. 8th PAMIR Conf. Fund. Appl. MHD, Borgo, Corsica, France, 37 (2011)
3. O. Zikanov, Y. Listratov, V.G. Sviridov, J. Fluid Mech. **720** 486 (2013)
4. L.G. Genin, V.G. Sviridov, *Hydrodynamics and heat transfer of MHD flows in channels* (Moscow, MPEI Publ., 2001) [in Russian]
5. F.W. Fraim, W.H. Heiser, J. Fluid Mech. **33,** 2 (1968)
6. I.A. Belyaev, L.G. Genin, Ya.I. Listratov, I.A. Melnikov, V.G. Sviridov, E.V. Sviridov, Yu.P. Ivochkin, N.G. Razuvanov, Yu.S. Shpansky, Megnetohydrodynamics **49,** 1 (2013)

Large-scale structures in a stratified open channel flow

Konstantin Borynyak[1, *], and *Michael* Hrebtov[1]

[1]Institute of Thermophysics of SB RAS, 630090, 1 Ac. Lavrentiev ave., Novosibirsk, Russia

Abstract. We present the results of direct numerical simulation of stably stratified open channel flow. The simulation was conducted in a frame of reference moving with the mean flow. The chosen setup allowed us to apply temporal filtering without affecting the long-living large-scale laminar and turbulent patches to see the dynamics of the processes on their boundaries. The main goal of the paper was to investigate the vorticity balance in different parts of the flow. It was shown that boundaries between laminar and turbulent patches contain weak large-scale vortex structures sustained by the baroclinic generation.

1 Introduction

Stably stratified flows are known for their fascinating behaviour with large intermittent turbulent and laminar patches. The most common examples of such phenomena are hot airflow over a cooler ocean or night atmospheric boundary layer where the ground cools faster than the air above it. Stratified flows attract increasing attention in scientific community over past decade [1-3].

Stably stratified channel flows are a computationally low-cost model for studying intermittency caused by thermal stratification. Depending on the ratio between the shear and negative buoyancy the flow may become fully turbulent, laminar, or intermittent, with the dynamic alternation of turbulent and laminar spots.

There are two main regions in the stratified boundary layer: the outer region, far from the wall where the shear is low and the flow dynamics is mostly governed by gravity waves [4-5], and the inner region, close to the wall, where the shear is high and the turbulent production opposes the damping effect of negative buoyancy. The outer region is extensively studied and overall well-understood, while the inner region and its dynamics are much more complex, and less-understood, especially the dynamic equilibrium of long-lived turbulent and laminar patches.

The complexity of the phenomena is in large amount of energetic interacting vortices in a strongly anisotropic environment. Not only the flow is anisotropic because of the effect of a wall, but it also becomes anisotropic in horizontal directions due to the intermittency effects. Such highly dynamic configuration is very hard to grasp as a whole so the hope is in statistical processing of the data to simplify the group effects of many vortex interactions in some deterministic way.

* Corresponding author: borisnsk07@gmail.com

In this short paper we investigate the processes on the boundaries between the turbulent and laminar zones with the focus on vorticity dynamics. The vorticity dynamics and especially the large-scale structures that are masked by a numerous smaller eddies generated by the shear stress at the wall may play an important role in sustaining the dynamic equilibrium in the intermittent flow. One way to make these larger structures visible is by changing the reference frame in order to stop the advection and filtering out the small-scale eddies.

2 Computational details

An open channel flow with a low friction Reynolds number ($Re_\tau = u_\tau h / \nu = 180$, where u_τ is a friction velocity, ν is the kinematic viscosity and h is the open channel height) was simulated. The Richardson number ($Ri_\tau = \alpha \Delta T g h / u_\tau^2 = 240$, where α is thermal expansion coefficient, g is gravitational acceleration and ΔT is temperature difference between top and bottom boundaries) was large enough to produce an intermittent turbulent/laminar flow pattern. The Prandtl number value was equal to 0.71.

The computational domain dimensions were $8\pi \times 1 \times 4\pi$ with $512 \times 50 \times 512$ grid points respectively. In longitudinal and transversal directions the grid was uniform, while in vertical direction the grid was clustered toward the wall with the grid step range: $\Delta y^+ = 0.1 - 5.5$.

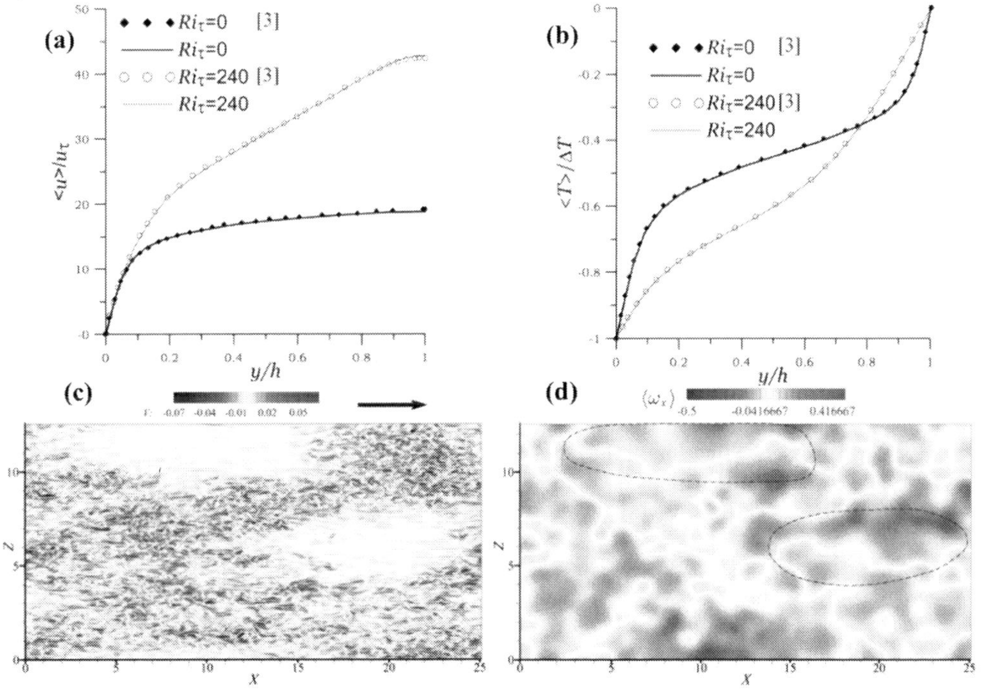

Fig. 1. Comparison of mean vertical profiles obtained in present simulation (lines) and taken from [3] (dots) for cases with stable and neutral stratification (a,b); a typical pattern of laminar and turbulent patches illustrated by vertical vorticity distribution in a cross-section $y/h = 0.125$ (c); Time-filtered longitudinal vorticity distribution at the same cross-section, the lines denote the places of laminar zones in instantaneous fields (d).

We used a DNS solver based on an open-source CFD package called OpenFoam. OpenFoam has been verified for a number of scientific and engineering applications [6,7].

The Navier-Stokes equations of incompressible fluid flow with Boussinesq approximation for buoyancy force were solved numerically using the finite-volume method with second order accuracy in space and time. No-slip and free-slip conditions were set at bottom and top boundaries respectively while at other boundaries the periodic conditions were prescribed.

The simulation was conducted in a reference frame moving with the bulk velocity of the flow. This allowed us to investigate the evolution of laminar and turbulent patches as they appeared almost stationary in the chosen reference frame.

The mean velocity and temperature profiles (together with the Reynolds stress and turbulent heat flux components, not shown here) were compared with the DNS data from [3] (Fig 1a,b). The profiles show good agreement with the literature.

3 Results and discussion

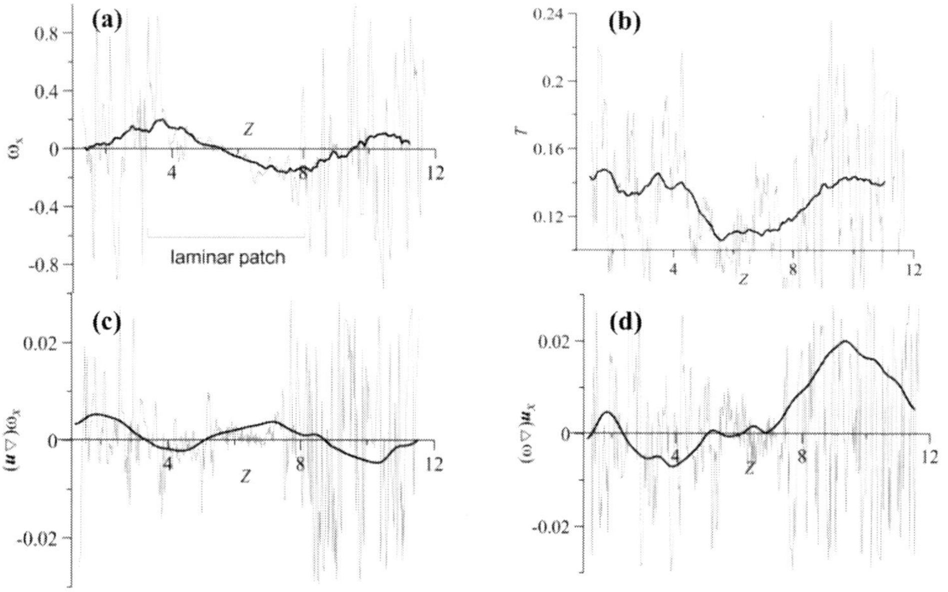

Fig. 2. Transversal profiles at x/h=20, y/h=0.13 for instantaneous (thin line) and time-averaged (thick line) values of: longitudinal vorticity (a); temperature (b); advection of longitudinal vorticity (c); tilting of longitudinal vorticity (d).

Fig. 1c shows a typical pattern of laminar and turbulent patches in the lower part of the channel. In our simulations the laminar spots were relatively stationary due to the chosen reference frame. This allowed us to do time-averaging of the turbulent fields without filtering out the spots structure. By using this procedure it was possible to find some weak large-scale vortex formations which were hidden behind the turbulence. Fig 1d shows the result of time averaging over 6 t_l ($t_l = L/U_{\text{bulk}}$, where L is a streamwise domain length) of longitudinal vorticity component. The dotted ovals show the location of laminar spots. It is clearly seen that after time-averaging there are opposite signs of vorticity on the opposite boundaries of both laminar zones. This vorticity would induce the velocity field which lifts the flow up in the turbulent region and pulls it down in the laminar region, thus contributing to their stability.

Fig. 2a shows the transverse cross-section of the flow intersecting the laminar patch, it is clearly seen that after averaging a pattern emerge with positive and negative vorticity maximums at the boundaries between laminar and turbulent region.

To investigate the source of this longitudinal vorticity we computed separately the components of vorticity balance equation:

$$\frac{\partial \omega_x}{\partial t} = -(u\nabla)\omega_x + (\omega\nabla)u_x - \beta g \frac{\partial T}{\partial z} + \dots \quad (1)$$

From Fig 2c it is clearly seen that the vorticity is at least partially accumulated at the boundary due to advection. Further analysis shows that the main component of advection is longitudinal, that means that the longitudinal vorticity of the needed sign is generated somewhere upstream. The vortex bending term Fig. 2d has a minimum in laminar patch and has large maximums in the turbulent areas, its direction on the boundary between the laminar end turbulent regions is opposite to the existing vorticity sign, so it would work like a sink, meaning that longitudinal vorticity is being bent in some direction. The last term in Eq.1 is baroclinic term. This term describes the production of vorticity by buoyancy gradient. Fig 2b shows the time-averaged temperature profile over the transversal cross-section. It is evident that the laminar patch is cooler than the turbulent spots thus it generate negative buoyancy. So the existing temperature gradients contribute to the vorticity distribution of Fig 1a.

One interesting effect observed in the moving reference frame is that the small velocity fluctuations inside the laminar spots are advected at lower speed than the fluctuations in turbulent patches. Thus, the small eddies from turbulent patches are frequently got into boundaries of laminar zones where they are slowed down and dissipated rapidly. This process creates a constant flow of vorticity toward the boundaries between the laminar and turbulent regions. We speculate that this vorticity flow could be connected with the described large-scale structures.

4 Conclusion

The simulation of stratified open channel flow in a frame of reference moving with the mean velocity allowed us to apply some statistical tools on the laminar and turbulent patches in the flow separately. Weak but large-scale vortex structures were found at the boundaries between the laminar and turbulent regions. These structures are sustained mostly by the baroclinic generation and they may take a part in the supporting of the dynamic balance between laminar and turbulent patches.

This work was supported by the Russian Science Foundation (grant №16-19-00119).

References

1. K. Yeo, B. Kim and C. Lee, J. Turbul., **10**, 17 (2009)
2. M. García-Villalba, J.C. del Álamo, Phys. Fluids, **23** 045104 (2011)
3. P. He, S. Basu, Nonlin. Processes Geophys., **22** 447 (2015)
4. O. Flores, J.J. Riley, Boundary-Layer Meteorol. **139** 241 (2011)
5. E. Deusebio, P. Schlatter, G. Brethouwer, E. Lindborg, J. Phys.: Conf. Ser., **318** 022009), (2011)
6. F. Flores, R. Garreaud, R. Muñoz, Comput. Fluids, **82** 1 (2013)
7. D.A. Hoang, V. van Steijn, L.M. Portela, M.T. Kreutzer, C.R. Kleijn, Comput. Fluids, **86** 28 (2013)

Investigating the round air jet dynamics at low Reynolds numbers

Vadim Lemanov[1], *Konstantin* Sharov[1*], *Andrey* Shumeiko[1], and *Nadezhda* Gorinovich[1,2]

[1] Kutateladze Institute of Thermophysics SB RAS, 630090, 1 Ac. Lavrentiev Ave., Novosibirsk, Russia
[2] Novosibirsk State Technical University, 630073, 20 Karl Marks Ave., Novosibirsk, Russia

Abstract. The laminar-turbulent transition in a round jet flowing from a cylindrical channel with the diameter of 3.2 mm was studied experimentally. In experiments, the range of Reynolds numbers determined by the mean-flow velocity was Re = Ud/ν = 700-12000. The measurements were carried out using a PIV system and one-component hot-wire anemometer. The profiles of average velocities and their pulsations in the zone of laminar-turbulent transition were obtained, as well as axial distributions of longitudinal velocity and pulsations of longitudinal velocity.

1 Introduction

It is known that jets are unstable at low Reynolds numbers Re = 10-30 [1], however, the transition to turbulence occurs at significantly higher values of Re = 30-2000 [2, 3]. The gas jets flowing from the contoured nozzles and at large Reynolds numbers Re>10^4 are studied. It is known that the initial conditions significantly affect the dynamics of the flow development in the jet. The laminar-turbulent transitions occurring in the jet flowing from the nozzle and cylindrical channel differ. This is due to the difference in the initial velocity profiles. There are almost no publications on the jets from the long channels, although in practice, such cases are rather frequent.

The feature of extended channels is that the laminar, turbulent and transient flows can occur there, and this fundamentally determines characteristics of the jets flowing from these channels. By now, the influence of these factors has been studied insufficiently.

2 Experimental methods and facilities

The submerged jet was formed by a round metal channel with diameter d=3.2 mm and length of 100 d. Air from the high-pressure manifold was sent to the channel through a reducer and valve of fine adjustment. The air flow rate was measured by rotameters Krohne DK 800.

* Corresponding author: sharov@itp.nsc.ru

The air jet flowed into the flooded air space (a plexiglass flow chamber of 150x150x400 mm). To visualize the flow and measure the velocity fields, a water aerosol with particle diameter of 1-2 μm was added into the flow. The surrounding space in the flow chamber was also filled with a mixture of air and water aerosol. When using the hot-wire anemometer, the compressed air line, feeding the aerosol generator, was turned off. The automated flow controller EL-Flow of Bronkhorst was included into the compressed air line feeding the main flow instead of the DK 800 rotameter.

The experiments were carried out using a PIV system, including digital camera, solid-state pulsed laser, pulse synchronizer and computer. Resolution of matrix of the digital camera was 1024 × 1360 pixels, laser pulse duration was 5 ns, radiation energy was 50 mJ, laser sheet thickness was 0.2 mm, and minimal time between frames was 20 μs. In experiments with PIV, the measuring region was 15 × 20 mm, and the design area corresponded to the size of 0.53 × 0.53 mm.

Measurements with a hot-wire anemometer were carried out by the DISA 55M system consisting of: single-wire sensor 55P11 with wire diameter of 8 μm, measuring bridge, and voltmeter of mean square values. The signal received from the hot-wire anemometer was converted by a linearizer.

3 Results

3.1 Initial conditions

The initial conditions are very important for the jets, they are distributions of average velocities and velocity pulsations at the outlet of the channel, forming a round jet.

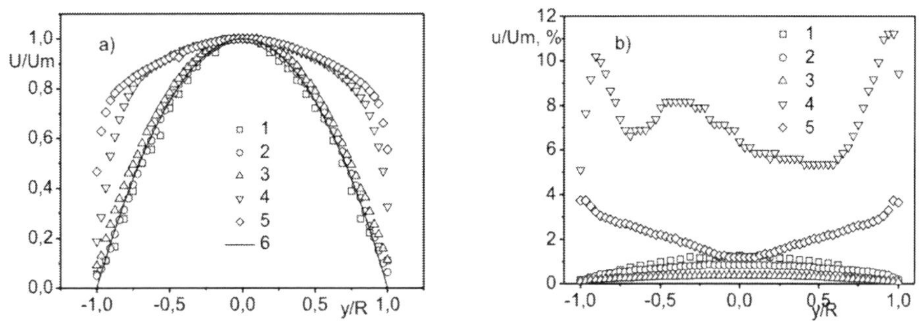

Fig. 1. Profiles of average velocities (a) and turbulent pulsations (b) at the edge of a cylindrical channel: *1* – Re = 700; *2* – 1100; *3* – 3990; *4* – 4060; *5* – 12100; line *6* – Poiseuille profile.

The measurements were carried out by a hot-wire anemometer at a distance of 0.5 mm from the tube edge.

Since the length of the channel, which formed a round submerged jet, is sufficiently large, then for Re <2000 in the initial cross-section of the jet, velocity distribution measured along the cross-section must be close to the known Poiseuille profile (Fig. 1a, line 6) for the flow in cylindrical channels. Measurements on the channel edge (Fig. 1a) showed that the average velocity profile is well described by the well-known Poiseuille dependence, under the experimental conditions of up to Re ~ 4000.

This indicates that the flow in the channel was laminar. The corresponding pulsation profiles are shown in Fig. 1b. It can be seen that for the laminar flow the degree of turbulence does not exceed 1.3%. With a further increase in the Re number, the laminar-turbulent transition starts forming in the channel. The velocity profile becomes smoother

(Fig. 1a, symbols 4), and pulsations increase significantly (Fig. 1b, symbols 4) near the walls and channel axis. At Re ~ 12000 the fully developed turbulent flow is observed on the channel edge.

3.2 Laminar-turbulent transition in a round jet

Distributions of longitudinal average velocities and longitudinal velocity pulsations on the jet axis measured by the hot-wire anemometer are shown in Fig. 2a and Fig. 2b, respectively. For Re < 1000, the average velocity on the jet axis decreases monotonically along the jet. At that, velocity pulsations here increase insignificantly at large distances from the tube edge. This indicates that the jet remains laminar along the entire length. Starting from Re = 1000, there is a noticeable increase in velocity pulsations (Fig. 2b, symbols 2) at a distance of ~ 20 x/d from the cylindrical channel outlet. This indicates the beginning of transient processes in the jet.

With further increase in Re, in spite of the fact that the flow in the channel remains laminar, the laminar-turbulent transition is observed in the jet. There is a distinct peak in distribution of axial pulsations (Fig. 2b, symbols 3), then, there is a rapid drop. Distribution of the average velocity on the jet axis also shows a sharp decrease (Fig. 2a, symbols 3). This behavior is caused by a significant increase in the jet half-apex angle after the zone of laminar-turbulent transition, which causes a decrease in the average velocity on the axis.

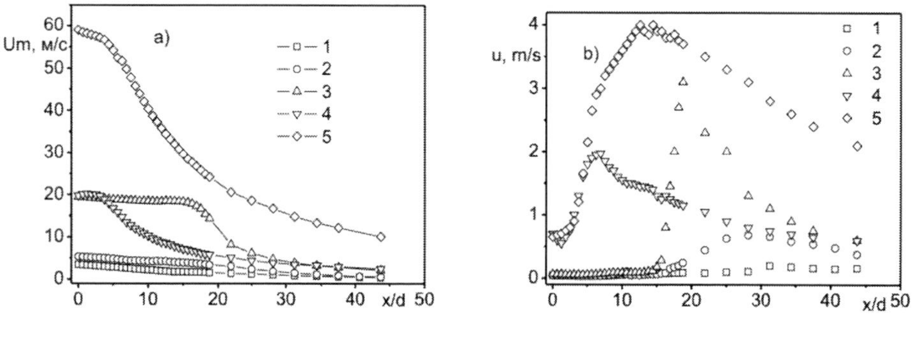

Fig. 2. Distributions of average longitudinal velocities (a) and turbulent longitudinal pulsations (b) on the jet axis along its length: *1* – Re = 700; *2* – 1100; *3* – 3990; *4* – 4060; *5* –12100.

In the range of Re ~ 1000-4000, the laminar-turbulent transition was observed in the jet, although the flow in the cylindrical channel was still laminar. The position of the zone of laminar-turbulent transition in the jet depended on the Re number, and with it's increase it was shifted to the edge of channel. When the Re numbers corresponding to the transient processes in the cylindrical channels were achieved, as it can be seen in Fig. 2a, symbols 4, the laminar region of the jet flow became minimal. Fig. 2b, symbols 4 shows attenuation of pulsations in the laminar flow and their sharp increase during transition.

The transition to turbulence in a cylindrical channel leads to an increase in pulsations from the very edge of the tube (Fig. 2b, symbols 5). At some distance, the pulsations reach a maximum, and their recession begins. The average velocity (Fig. 2a, symbols 5) also shows a drop from the very outlet from the cylindrical channel. In this case, this behavior is caused by the turbulent processes.

The profiles of average velocity and pulsations in the round jets were measured by the PIV system. At Re = 1700, which was determined from the mean-flow velocity, the laminar-turbulent transition in the jet started at a distance of about 20 x/d. The transition point determined as for the hot-wire measurements and corresponding to the maximum of turbulent pulsations was located at the distance of 25 *x/d* from the cylindrical channel edge, where the jet flowed out. In experiments, the measurements were carried out at the site, which covered the distance from 20 to 40 *x/d*. Distributions of the average and pulsation velocity components were obtained during the flow transition from the laminar to turbulent regime.

Distributions of average velocities tend to become smoother to the end of the laminar-turbulent transition. This is explained by an increase in the angle of jet half-apex at transition to the turbulent flow. On the jet axis, the turbulence degree is not large and it makes up several percents. At transition, pulsations on the axis intensively increase, reach a maximum, then there is a small decrease and simultaneously the pulsations profile becomes smoother. Longitudinal and transverse pulsations behave similarly.

The data on Reynolds stresses have been also obtained. They save the zero value on the jet axis along the transition zone and have one extremum in each middle part of the jet mixture layers.

3 Conclusions

The air jet flowing out of a cylindrical channel with large length at low Reynolds numbers has been studied. The effect of initial conditions, such as distributions of average velocities and velocity pulsations at the outlet of the channel, on development of submerged jets has been investigated. Distributions of average axial velocities and pulsations along the jet were obtained in a wide range of Reynolds numbers (Re = 700-12000). The profiles of average velocity, as well as profiles of longitudinal and transverse pulsations in the zone of laminar-turbulent transition in the jet were obtained. It is shown that the laminar-turbulent transition in a jet is associated with the laminar-turbulent transition in a channel. Data on transverse velocity pulsations and Reynolds stresses in the zone of laminar-turbulent transition may be considered as novel.

The work was partially supported by the Russian Foundation for Basic Research (grant 17-08-00958).

References

1. G.N. Abramovich, *The Theory of Turbulent Jets* (M.I.T. Press, 1963)
2. V.V. Lemanov, V.I. Terekhov, K.A. Sharov, A.A. Shumeiko, Tech. Phys. Letters **39**, 421 (2013)
3. V.M. Aniskin, V.V. Lemanov, N.A. Maslov, K.A. Mukhin, V.I. Terekhov, K.A. Sharov, Tech. Phys. Letters **41**, 46 (2015)

Heat transfer in a pulsating flow in a converging channel

Irek Davletshin[1,*], *Anton* Paereliy[1], and *Ilnur* Gazizov[2]

[1] Kazan Scientific Center of RAS, Kazan, 420111, Russia
[2] Kazan National Research Technical University named after A.N. Tupolev, Kazan, 420111, Russia

Abstract. Experimental research of convective heat transfer in a plane converging channel has been conducted. Distributions of heat transfer coefficient on the channel wall in steady and pulsating air flow have been obtained. Quasisteady behavior of heat transfer distributions has been revealed for the pulsating flow case.

1 Introduction

Hydrodynamic and heat processes in gradient flows have their own features. Gradient flows exhibit significant difference between the flow with adverse and favorable pressure gradients. Streamwise pressure gradient in the former case can lead to the flow separation while in the latter case it contributes to the boundary layer stabilization.

Practical relevance of flows in converging channels with favorable pressure gradient determines the need for investigation of processes in such flows [1-4]. It was established experimentally [1, 2] and numerically [3] that favorable pressure gradient promotes flow laminarization. This is accompanied by the reduced hydraulic loss and heat transfer deterioration: the corresponding coefficients drop below the turbulent ones but remain higher than those in zero gradient case. The Kays acceleration parameter is usually taken as a numerical characteristic involved in the analysis of gradient flows. The Kays parameter is estimated by the rate of velocity growth along the streamwise coordinate. Laminar-turbulent transition delay occurs at any value of this parameter. Starting from a certain Kays parameter, the flow remains laminar at any time. Reverse turbulent-laminar transition is even possible at high Kays parameters.

Hydrodynamic and heat processes in channels were studied under the combined effect of the pressure gradient and other factors. In particular, the behavior of flow parameters when different turbulence generating grids were installed in the channel were estimated experimentally [2]. The mesh space and blockage ratio altered the boundary layer and thermal boundary layer and caused earlier transition of heat transfer coefficient from laminar to turbulent values.

Optical methods in hydrodynamics have been developing in the recent years. Kinematic flow structure (fields of velocity, turbulence, vorticity) in different cases is investigated using PIV measurements. In particular, even extensive PIV measurements involving 16 cameras were carried out to obtain the flow structure [4]. However, at present, the

* Corresponding author: davlet60@mail.ru

conducted research is more valuable for the try-out of measurement technique for the considered class of problems.

It should be noted that the majority of available data on gradient flows deal with the steady flow case. Meanwhile, the available information on pulsating flows shows, for example, the changes in the profiles of velocity and shear stress during one period of pulsations [5].

The parameters of separated flow are governed by forced pulsations. Depending on the frequency and amplitude of pulsations, large-scale vortices of different intensity are formed in the separation region. This is accompanied by variation in distributions of pressure, shear stress on the wall and heat transfer coefficient along the separation region. The reattachment length becomes half its initial size, heat transfer enhancement is observed particularly near the obstacle which causes flow separation [6].

Thus, the study of processes (including the heat processes) in gradient flows can yield new information on such complex flows.

2 Experiment

The paper deals with experimental estimation of the heat transfer coefficient in a converging channel with the angle of φ=4.5^0 at steady and pulsating air flow regimes. The test section was a plane channel made of polycarbonate with the length of 1.2 m, width of 150 mm, inflow section height h_0=60 mm and outflow section height h_1=24 mm. This gives the ratio of inflow to outflow areas F_1/F_0 = 0.4 (fig. 1). To turbulize the flow, a turbulence generating grid was installed and an abrasive strip was glued onto the channel perimeter between the channel inlet and the test section. The length of converging section was L=450 mm. The distance between the channel inlet and the converging section inlet was 370 mm. The air flow was provided by a compressor operating in suction mode. Flow pulsations were generated by a special device – pulsator. Its cross section at the channel outlet was periodically blocked by a rotating flap *4*. Adjusting the blockage of the windows (with or without the flap) provided the required air flow rate and the amplitude of velocity pulsations. A receiver tank with the volume of 1.3 m^3 was installed between the test section and the compressor. It damped the velocity (pressure) pulsations further downstream. The air flow rate was measured using an ultrasonic IRVIS RS-4 Ultra flowmeter downstream of the receiver tank.

A 450-mm long heat transfer wall was mounted across the whole channel for heat experiments in converging section. It was a plate (printed circuit board) made of composite epoxy material with the width of 1.6 mm. To eliminate the heat loss, the outer surface of the plate was covered with polystyrene layer. Copper layer on the plate surface was etched to form a single strip. The strip was functionally divided into 33 sections with the length of 13 mm and width of 100 mm. Thus, the copper layer acted as 33 resistance thermometers that measured local temperatures of the wall with the streamwise spacing of 13.6 mm. Total resistance of the plate and the average wall temperature were measured.

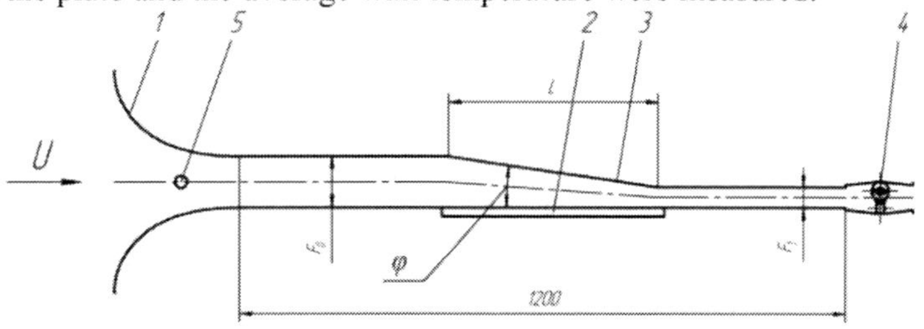

Fig. 1. Experimental setup: *1* – channel inlet; *2* – heat transfer wall; *3* – converging section; *4* – pulsator; *5* – thermometer

The channel was heated by hot air before the measurements. Then, while the wall was being cooled by the air flow at the room temperature, the temperatures of heat transfer walls were measured during 100 seconds with the frequency of 200 measurements per second. Heat balance equations together with these data yielded the distribution of heat transfer coefficient along the channel which was considered time constant (not excepting the pulsating regimes).

The considered ranges of forced pulsation frequency and relative amplitudes of velocity were $f = (6-12)$ Hz and $\beta = A_U/U = 0.3–0.8$, respectively. The frequencies were adjusted by the flap rotation rate. Amplitudes were estimated from velocity, measured by an optical method. The studied ranges of amplitudes and velocities were chosen according to the following considerations. The considered flows are influenced by two factors: pressure gradient due to the converging channel shape and forced flow pulsations. For the considered converging channel, it is in the specified ranges of frequencies and amplitudes that the pressure gradients caused by the channel geometry and wave structure of the pulsating flow are of the same level. Thus, both factors are relevant in the considered ranges of parameters. Under these conditions, the study of heat processes in pulsating gradient flows was of interest in terms of the combined effect of the specified factors.

3 Results

First, heat transfer coefficients were measured in steady air flow in the converging channel section (fig.2, 3). Despite the streamwise acceleration, heat transfer deterioration was observed in the first half of the converging channel, while the second half of the converging channel exhibited typical streamwise heat transfer enhancement (fig. 2). In general, the behavior of these distributions is obviously formed by two factors: the growth of thermal boundary layer further downstream and its thinning under favorable pressure gradient. So, the first factor dominates in the first half, while the second one is prevalent in the second half.

The research of the steady flow was carried out in the range of velocities at the converging channel inlet velocity U_0=(1.5-6.3) m/s and Reynolds numbers $Re_0 = U_0 h_0/\nu = (0.6\text{-}2.5)\times10^4$. Reynolds numbers based on the actual velocity and streamwise coordinate were $Re_x = Ux/\nu = (0.4\text{ -}7.5)\times10^5$, Kays acceleration parameter $K = (3.1\text{-}13)\cdot10^{-6}$, here $K = \dfrac{\nu}{U^2}\dfrac{\partial U}{\partial x} = \dfrac{tg\phi}{Re_0}$ for the plane channel. According to [1-3], laminar flow corresponds to these parameters. However, forced turbulization of flow by the turbulence generating grid and the abrasive provided heat transfer coefficients typical of turbulent flows (fig. 3).

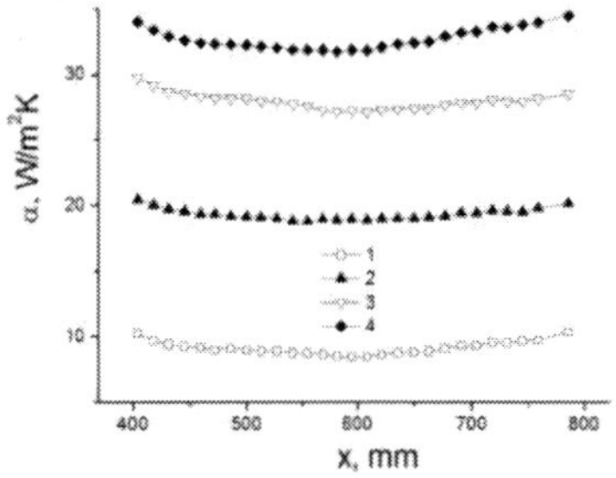

Fig. 2. Heat transfer coefficient in steady flow at inlet velocity U_0: *1*–1.5; *2*–3.2; *3*–4.9; *4*–6.3 m/s.

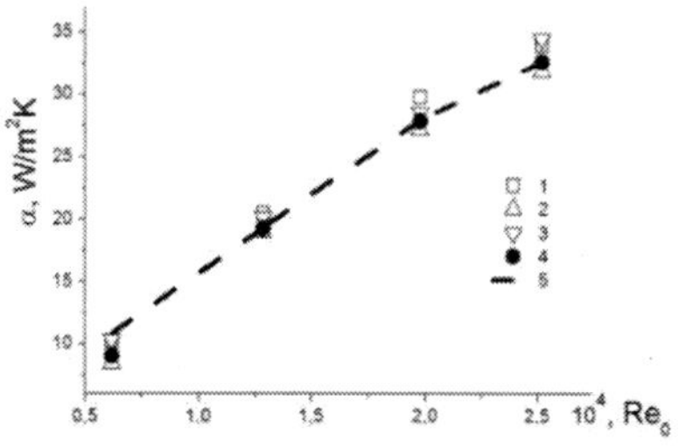

Fig. 3. Heat transfer coefficient: *1* –*x*=34; *2* –225; *3* –416 mm; *4* –wall average; *5* –$\alpha \sim Re_0^{0.8}$

Heat transfer at pulsating flow regimes was studied for a single flow rate (fig. 4) at which the average velocity at the channel inlet was U_0=2.9 m/s ($Re_0=1.2\cdot10^4$). The Stanton number was estimated using the actual values of the average air velocity in the converging channel $St=\dfrac{\alpha}{\rho c U}$, where ρ and c are air density and heat capacity, respectively. The measurements showed that heat transfer coefficient distributions in pulsating flows are quasi steady. Moreover, the values of α appeared to be close to the steady ones with small deviation (up to 5%) towards the lower values. Such decrease can be explained as follows: when the velocity is periodic in time, the inequality $(U^{0.8})_{aver} < U_{aver}^{0.8}$ (subscript "aver" means time average) is true for the curve $\alpha\sim U^{0.8}$ typical of turbulent regimes (exponent less than 1). Hence, in quasisteady approximation, the heat transfer coefficient in pulsating flow should exhibit some decrease depending on the amplitude. Quasi steadiness here means that forced pulsations do not lead to qualitative changes in the flow structure if compared with the steady case.

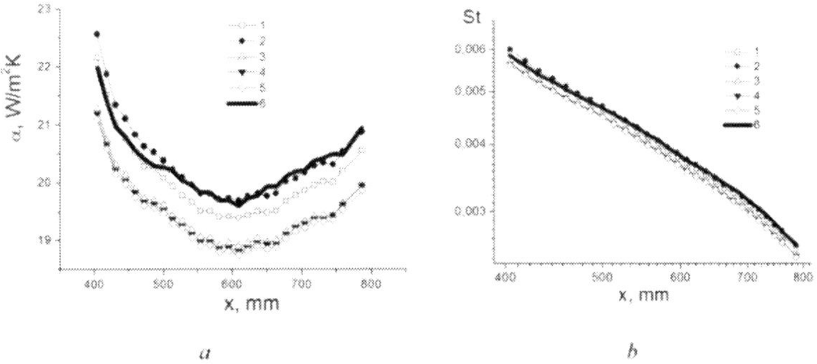

Fig. 4. Heat transfer coefficient (a) and Stanton number (b) at the regimes: 1,2 – 6 Hz; 3,4 – 9 Hz; 5 – 12 Hz; 6 – steady; light symbols – β=0.3; dark symbols – β=0.8.

4 Conclusions

It has been experimentally established that heat transfer coefficient in pulsating flows in converging channels remains close to the steady flow case. This statement is true at least for the flows in which the flow structure can be considered quasisteady.

This study was supported by the Russian Science Foundation (Project no.16-19-10336).

References

1. W.M. Kays, D.W. Kearney, R.J. Moffat, *The turbulent boundary layer-Experimental heat transfer with strong favorable pressure gradients and blowing* (1970)
2. M.F. Blair, M.J. Werle, United Technologies Research Center East Hartford CT, 1981
3. E.P. Volchkov, M.S. Makarov, A.Yu. Sakhnov, Int. J. Heat Mass Transfer **55** 1126 (2012)
4. R. Hain, S. Schamowski, N. Reuther et al. *In Proc. 18th Int. Symp. on the Application of Laser and Imaging Techniques to Fluid Mechanics, Lisbon, Portugal* (2016)
5. A.E. Goltsman, I.A. Davletshin, N.I. Mikheev, A.A. Paereliy, Thermophys. Aeromech. **22** 319 (2015)
6. I.A. Davletshin, N.I. Mikheev, High Temp. **50** 412 (2012)

The effect of hydrodynamic flow unsteadiness on heat transfer enhancement in a channel with square ribs

Nikolay Dushin*, *Nikolay* Mikheev, *Irek* Davletshin, *Anton* Paereliy, *Olga* Dushina

Kazan Scientific Center of RAS, 420111, Kazan, Russia

Abstract. The heat transfer coefficient has been measured in the channel with square ribs for the case of periodic flow rate variation. New technique for heat transfer coefficient measurements has been applied in the study. It allows simultaneous surface heating and estimation of its local temperature. Two different patterns of heat transfer coefficient behavior in the gap between the ribs have been obtained. The first pattern corresponds to enhancement of both local heat transfer coefficient and average heat transfer if compared to its steady flow values. The second pattern exhibits considerable gain in heat transfer in the near-wake of the rib and its deterioration in other measurement regions. The reasons for such effects have been described on the basis of smoke visualization of flows.

1 Introduction

Currently, heat transfer enhancement in channels is based on turbulization of a thin near-wall region of flow, generation of large-scale vortices and boundary layer breakup. This approach allows significant enhancement of heat transfer accompanied by moderate increase in pressure drop. In this case, heat transfer augmentation is attributed to domination of turbulent transport over the molecular one and is governed by the linear scale of vortices and their velocity. Spanwise ribs mounted inside the channels are among the efficient ways to arrange such flows. Optimal rib pitch in the case of steady flow leads to a 26% increase in thermohydraulic performance in the Reynolds number range of $10^4 \ldots 4\times10^5$.

Further augmentation of thermohydraulic performance can be achieved by variation of the rib shape, location of ribs and generation of flow rate pulsations [1 – 2]. Thus, authors [2] obtained a 60% increase in heat transfer (if compared to the steady flow case) in pulsating flow through a single orifice. Besides, they mentioned that the maximum heat transfer coefficient shifted towards the orifice and documented five-fold increase in local heat transfer coefficient near the orifice. The shift of maximum heat transfer coefficient was also mentioned in [3] for the case of acoustic disturbances of different frequency and amplitude in the vicinity of a square rib, but the authors [3] did not achieve thermohydraulic performance comparable to [2]. Periodic blowing-suction of heat-transfer medium through the rib edge is another way to affect the flow pattern. Such an approach

* Corresponding author: ndushin@bk.ru

has been thoroughly studied for the case of a backward-facing step flow [4 – 5]. These papers mainly focused on the control of flow pattern. In particular, the authors on average noted a 30% reduction of the recirculation length. Heat transfer enhancement under periodic forcing of flow through the backward-facing step edge was studied in [6]. A 40% increase in heat transfer coefficient in the near wake of the backward-facing step was obtained at the Strouhal number of 0.275.

The review of research papers on hydrodynamics and heat transfer of separated pulsating flows shows that the majority of results have been obtained from numerical simulation. Meanwhile, experimental results are based mainly on the research of single obstacles in flows with some combinations of unsteadiness parameters. Thus, experimental research of thermohydraulic performance in configurations that are used in real-life engineering devices is a relevant task.

The present paper studies the effect of hydrodynamic unsteadiness of flow on heat transfer enhancement in a ribbed channel. Turbulent air flow in a 0.115×0.15 m^2 rectangular channel at the Reynolds number Re_D = 18200 based on the hydraulic channel diameter was considered. Aluminum square spanwise ribs with the height e = 13.5 mm were mounted in the channel. The rib-to-channel height ratio was e/H = 0.117. The relative rib pitch was p/e = 10, 15, and 20. A smooth inlet section with 6:1 contraction was attached to the channel inlet. The latter was also equipped with a turbulence generating grid and an abrasive with a grit size of about 0.8 mm. Measurements were carried out between the third and fourth ribs because of the attempt to reduce the measurement section length and exclude the possible effect of resonance in the experimental setup on measurement results.

2 Experimental technique and procedure

Local heat transfer coefficients were measured using a technique [7] that allowed simultaneous heating of the heat transfer surface and estimation of its local temperature from the measurement of corresponding electric resistance of a heating element. For this purpose, a composite epoxy plate was embedded into the test section wall. Dimensions of the plate were: length 450 mm, width 200 mm, thickness 1.5 mm. The inwards facing surface of the plate was covered with a continuous zigzag copper strip. The strip width was 0.5 mm. The strip spacing was 0.75 mm (the gap between adjacent parts of the strip was 0.25 mm). The covered area was 420×135 mm^2. Heat transfer coefficients were measured for the regions $b_1\times B$ (measurement sections) equal to 9.5×135 mm^2 by the wires connected to the strip with a given spacing (9.5 mm).

To reduce the side heat losses, through grooves (sealed by a thin film) were made in the plate; and the area of wire surface was minimized. The reverse side of the plate was thermally insulated by two 10-mm thick air chambers. External and dividing walls of the chambers were made of 9-μm thick aluminum foil.

Local heat transfer coefficients were quantified using a constant heat flux method. Taking the heat loss into account, the heat transfer coefficient of the *i*th measurement section of the plate was calculated from:

$$\alpha_i = \frac{\dfrac{U_i U_0}{R_0} - \left[Q_{ti} + Q_{r_i} + Q_{sw_i} + Q_{xi}\right]}{S_i \Delta T}, \qquad (1)$$

where U_0 and R_0 are the voltage drop and resistance of reference resistor, respectively; U_i is the voltage drop at the *i*th measurement section of the plate; S_i is the area of the *i*th measurement section of heat transfer surface.

Heat loss through thermal insulation was calculated for the *i*th measurement section:

$$Q_{ti} = \alpha_{ti}\Delta TS_i, \quad \frac{1}{\alpha_t} = \frac{\delta_1}{\lambda_1} + \frac{\delta_t}{\lambda_t} + \frac{1}{\alpha_c}. \tag{2}$$

Here δ_1 and λ_1 are the thickness and conductivity of the plate, respectively; δ_t and λ_t are the insulation thickness and thermal conductivity coefficient of air, respectively; α_c is the coefficient of heat transfer from the plate surface.

Side heat losses, Q_{swi}, through the plate were estimated by the formula derived for the straight ribs but transformed for the plate. Given that the temperature of heat-transfer medium is equal to the ambient air temperature, it is written as follows:

$$Q_{sw} = \Delta T\sqrt{b_1^2\left(\alpha_c + \alpha_t\right)\left(\lambda_1\delta_1 + \lambda_2\delta_2\frac{b_2}{b_1}\right)}\, th\, mL, \quad m = \frac{\alpha_c + \alpha_t}{\sqrt{\lambda_1\delta_1 + \lambda_2\delta_2\frac{b_2}{b_1}}}, \tag{3}$$

where b_1 and b_2 are the pitch and width of the wires, respectively; δ_1 and λ_1 are the thickness and conductivity of wires, respectively; ΔT is the temperature difference between the measurement sections of the plate and the heat-transfer medium:

$$\Delta T = T_w - T_f = \frac{1}{K}\left(\frac{U_i R_0}{U_0 R_{20_i}} - 1\right) + 20 - T_f. \tag{4}$$

In (4), K is the temperature coefficient of the resistance thermometer; R_i is the resistance of the measurement section strips at the temperature T_i; R_{i20} is the nominal resistance of the measurement section strips at 20°C.

Heat supplied to the ith measurement section with the width Δx from adjacent sections was calculated from:

$$Q_{xi} = -\lambda_1\delta_1 B\left(\frac{T_{i-1} - 2T_i + T_{i+1}}{\Delta x}\right). \tag{5}$$

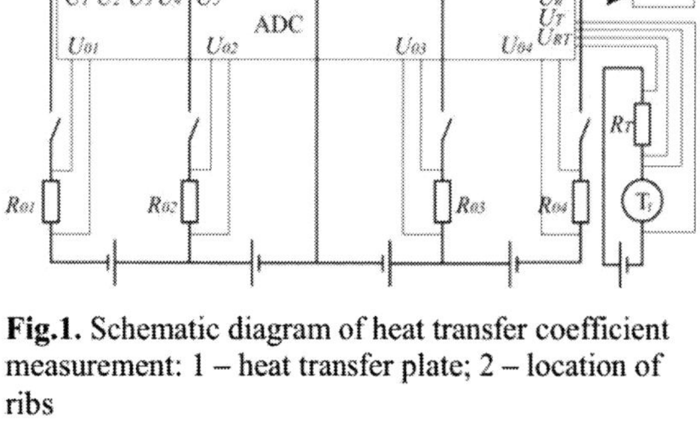

Fig.1. Schematic diagram of heat transfer coefficient measurement: 1 – heat transfer plate; 2 – location of ribs

Heat loss due to radiative heat transfer was not included in (1) due to its negligible value.

The plate was heated by an automotive battery. Ambient air temperature was measured by a platinum resistance thermometer Pt100. Voltage drop was measured by a 14-bit analog-to-digital converter L-card E14-140M (resolution up to 19.5 μV) connected to PC (fig. 1). Air flow rate was controlled by an ultrasonic flowmeter IRVIS-RS4-Ultra. Flow rate fluctuations were generated by a shaped rotating flap mounted at the measurement section outlet. The rotation frequency was controlled by a frequency converter Vacon 10.

3 Results and discussion

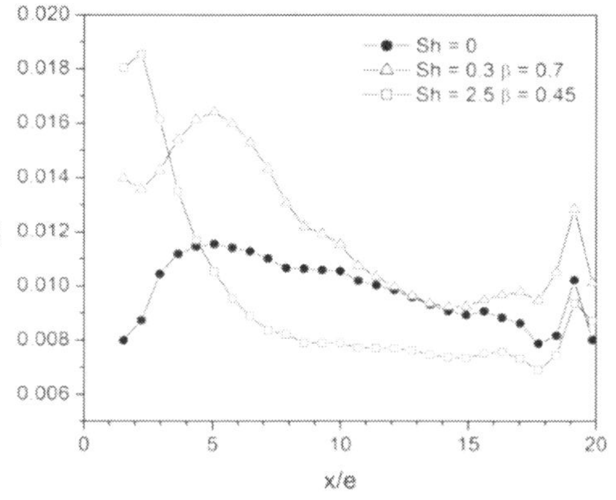

Fig.2. Local heat transfer between the ribs

Fig. 2 shows typical measurements of local heat transfer coefficient at the regimes with heat transfer enhancement and their comparison with the steady flow. Variation of frequency and amplitude of forced pulsations can result in different effects. For example, 44% heat transfer enhancement is observed in separation regions at Sh = 0.3, $\beta = A_u / \overline{u} = 0.7$ (here the Strouhal number, Sh, is based on the steady flow reattachment length). The integrated dimensionless heat transfer coefficient grows by 23%. At the same time, the regime with Sh = 2.5, β = 0.45 is characterized by heat transfer deterioration over the largest part of the gap between the ribs. This results in reduction of the integrated dimensionless heat transfer coefficient by approximately 4%, but significant heat transfer enhancement is observed in the rib near wake (2.12 times), and the peak heat transfer is shifted towards the rib.

To find the reasons for such different effects, the flow was visualized with smoke. The visualization in the first case demonstrated a large vortex that was formed behind the rib in the flow deceleration phase and shed downstream in the acceleration phase. The vortex diameter was 2-2.5 times the rib height. The second case exhibited a vortex with high angular velocity and the diameter comparable to the rib size. In the flow deceleration phase this vortex nestled against the rib and then was shed to the main flow. The air in the remaining part of the gap between the ribs behaved like a solid body almost without mixing with the mean flow.

4 Conclusions

New measurement technique has been employed to estimate the heat transfer coefficient in the flow with periodic flow rate pulsations in the channel with square ribs. Depending on the flow rate frequencies, different flow regimes can occur in the channel, including the ones with enhanced local heat transfer in the whole gap between the ribs and the ones with considerable heat transfer augmentation in the rib near wake and decrease over the remaining part of the gap. Smoke visualization of flow has revealed essential rearrangement of the flow pattern, which is responsible for the discovered effects.

References

1. G. Xie, S. Zheng, W. Zhang, B. Sunden, Applied Thermal Eng. **61**, 289 (2013)
2. I. Davletshin, N. Mikheev, High Temp., **50** 212 (2012)
3. B. Cukurel, C. Selcan, M. Stratmann, Int. J. Heat Mass Transfer **91**, 848 (2015)
4. S. Saric, S. Jakirlic, C. Tropea, Trans. ASME J. Fluids Eng. **127**, 879 (2005)
5. A. Das Gupta, P. Zhao, S. Ray, 54th AIAA Aerospace sciences Meeting (2016)
6. G. Ree, H. Sung, Numerical Heat Transfer A, **37**, 733 (2000)
7. N. Dushin, N. Mikheev, I. Gazizov, I. Davletshin, Russian Aeronautics (2017) (in press)

Flow structure in a flat channel with honeycomb surfaces

Victor Terekhov[1,2], *Jaroslav* Smulsky[1], *Konstantin* Sharov[1], and *Alexey* Zolotukhin[*, 1,2]

[1] Institute of Thermophysics SS Kutateladze SB RAS, 630090, Novosibirsk, 1 Ac.. Lavrentiev Ave., Russia
[2] Novosibirsk State Technical University, 630073, Novosibirsk, 20 Karl Marks Avenue, Russia

Abstract. This paper deals with the experimental study of the gas flow structure at longitudinal flow around a honeycomb surface with hexagonal cells of various sizes and depths. The measurements were carried out using the PIV system for the developed flow in a flat channel with the cross-section 21 × 150 mm and length of 1000 mm. The flow was visualized and the velocity and turbulence components were measured in a smooth channel and in the presence of honeycomb surfaces. It is shown that a boundary layer with high gradient of velocity and turbulent pulsations is formed near the honeycomb.

1 Introduction

An important direction in the development of the heat transfer theory and aerohydrodynamics is the search for new methods of heat transfer intensification with a parity growth of hydraulic losses. In this respect, the surfaces with lunar reliefs are illustrative [1,2]. A number of experimental and numerical studies in this field prove a great potential of such methods.

The honeycomb surfaces considered in this work, used in particular for the purposes of thermal protection of spacecraft elements and as seals for gas turbines, have a number of similarities to the cavity vortex generators. This is evidenced by the results of the first experimental studies of the longitudinal flow around the surfaces with honeycomb coatings [3,4]. Data on acoustic damping [5], which helps to reduce friction, give all the grounds for using such vortex generators as the effective methods for controlling the processes of turbulent transfer. However, due to the lack of detailed experimental information on the flow structure, it is not possible to create an integrated pattern of this complex phenomenon, and therefore its deeper investigation is required.

The development of the flow field and turbulent characteristics along the honeycomb surfaces are studied in this paper. It compares data on the averaged and turbulent characteristics for the honeycombs and channel with smooth walls.

* Corresponding author: zoloav@mail.ru

2 Experimental setup and measurement methods

The working channel of the experimental setup had a rectangular section of 21 × 150 mm, length of 1 m, and height h = 22 mm. It was made of transparent organic glass, 8 mm thick. The transverse dimensions of the inlet channel were chosen to ensure the two-dimensional flow in the central longitudinal cross-section of the channel. At the distance of 580 mm from the channel inlet, the plates were mounted flush with the following dimensions: the first plate of 40 × 200 mm, and the second and third plates of 58.4 × 182 mm with hexagonal cells. The cells were blocked from the outside. Their heights b were 21 mm, 2.4 mm, and 6 mm. The first honeycomb surface was made of textolite with the thickness of cell partitions of 0.2 mm and diameter of a circle circumscribed around a cell of 5 mm. Honeycombs No. 2 and No. 3 were made of plastic and had the cell diameter of 12 mm and wall thickness of 0.5 mm. The cell diameter-height ratio d/b was 0.24 for the first plate, and for the second and third plates, it was 2 and 5, respectively.

The experimental setup includes all elements of the wind tunnel to form a qualitative flow: prechamber, nozzle and working channel. The detailed description of setup and measurement system is given in [6]. In the present experiments, development of the flow along the entire length of the honeycomb plate was studied at three different velocities: 5, 10 and 20 m/s; in the beginning of the plate with honeycombs, the flow was developed. The measurements were performed using the PIV system.

3 Experimental results and discussion

The paper presents the results of measurements of longitudinal velocity profiles, longitudinal and transverse pulsations, and shear stresses in a smooth channel and channel with honeycombs.

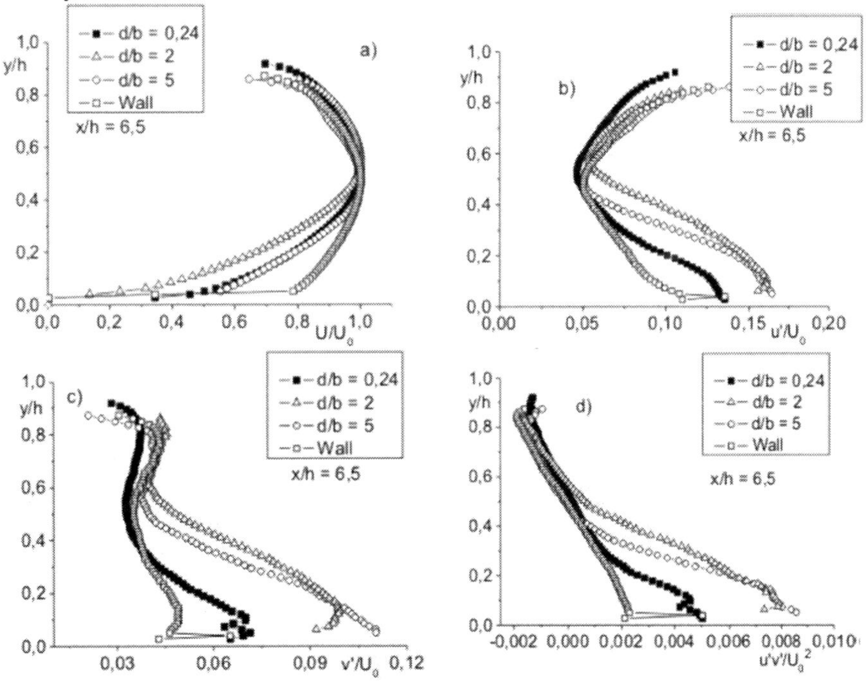

Fig. 1. Distribution along the channel height: a) longitudinal velocity, b) longitudinal, c) transverse fluctuations, and d) shear stresses, working fluid – air, Re = $6{,}4{\cdot}10^4$

All data were obtained for cross-section at x/h = 6.5, x is a distance from the beginning of the surface with hexagonal cells. According to experiments, the flow in a vicinity of the honeycomb surface is fundamentally different from the flow around a smooth wall. In the near-wall region of the honeycomb surface, the velocity profile is more even in comparison with the classical distribution on a smooth wall (Fig. 1a). At the same time, longitudinal and transverse velocity pulsations (Figs. 1b and 1c, respectively) near the honeycomb surface are several times greater than on the smooth surface. A similar tendency exists for distribution of shear stresses across the channel (Fig. 1d). If for the channel with smooth walls, as expected, there is the linear distribution of shear stresses along the height, in the channel with cells at distance y/h ~ 0.2, the stresses increase more than 2-3 times. At the same time, a steady tendency to a decrease in transverse velocity pulsations and turbulent shear stresses is observed directly in the near-wall region.

Such a complex scenario of flow development is caused by interaction of the boundary layer with the honeycomb cells. Indeed, the flow around depressions of various shapes (cylindrical cavity [7], hole [8], transverse trench-groove [9], etc.) on the surface leads to formation of a periodic vortex layer behind them. However, we failed to detect the existence of such structures in the present work.

One of the principal differences between the obtained results and experimental data of [3,4] is the absence of the flow slip along the surface with honeycombs, which, according to these data, could reach about 20-30% of the velocity in the flow core.

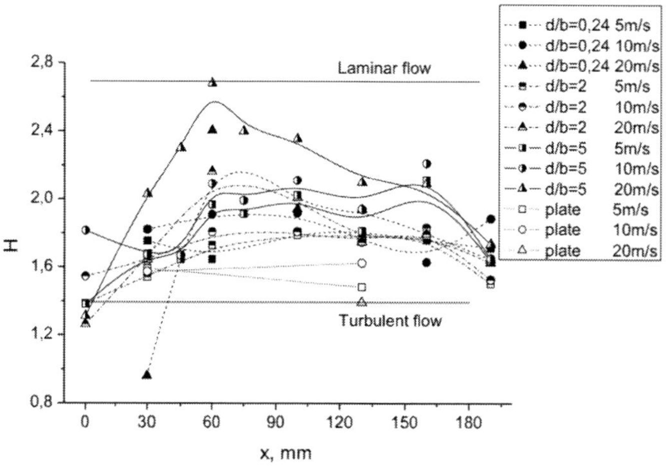

Fig. 2. Change in form parameter H along the honeycomb surface.

Distribution of layer form parameter $H = \delta^{*}/\delta^{**}$ along the honeycombs with different ratio d/b is shown in Fig. 2 for three flow velocities (coordinate $x = 0$ corresponds to the beginning of honeycomb region).

In the diagram, the value H corresponds to the developed turbulent flow; then, we can observe its increase to the value characteristic of the laminar flow regime ($H \rightarrow 2.7$) and subsequent decrease. This behavior of the form parameter is caused, most likely, by the "flattening" of velocity profile with simultaneous increase in turbulence and tangential stresses.

It is not possible now to formulate clearly the reasons for such a behavior of local and integral characteristics of the turbulent flow around a surface with honeycombs, primarily because of multifactority and complexity of the aerodynamic structure, and this will require further detailed studies.

4 Conclusions

The paper presents the results of experimental study of the averaged flow, normal and tangential stresses in gas flowing with different velocities around the honeycomb surfaces with characteristic cell size of ~ 5 mm and ~ 12 mm.

The measurements were performed using the two-component PIV system. The flow was organized in a flat channel of 21 × 150 × 1000 mm.

It was found that the boundary layer on the surface with hexagonal cells is less filled than on a smooth wall. At the same time, the growth of turbulent pulsations and tangential stresses is observed simultaneously. However, there is no effect of flow sliding relative to the cells.

The work was financially supported by the grant of the President of the Russian Federation for the leading scientific schools (project NSh-8780.2016.8.).

References

1. S.A. Isaev, N.V. Kornev, A.I. Leontiev, E. Hassel, Int. J. Heat Mass Transfer **53**, 178 (2010)
2. G.V. Kovalenko, V.I. Terekhov, A.A. Khalatov, J. Appl. Mech. Tech. Phys. **51**, 839 (2010)
3. A.A. Klimov, S.A. Trdat'yan, High Temperature, **41**, 801 (2003)
4. S.A. Trdatyan, A.A. Klimov, Proc. 12-th Int. Heat Transfer Conf. Grenoble. 221. (2002)
5. N.N. Koval'nogov, Energy Problem, № 5-6, 41, (2003) [in Russian]
6. V.I. Terekhov, Ya.I. Smulsky, K.A. Sharov, A.V. Zolotukhin, Thermophys. Aeromech., **21** 719 (2014)
7. M. Hiwada, T. Kawamura, J. Mabuchi, M. Kumada, Bull. JSME. **26,** 1774 (1983)
8. V.I. Terekhov, S.V. Kalinina, J.M. Mshvidobadze, Siberian Phys.-Tech. J. 77 (1992)
9. V.I. Terekhov, N.I. Yarygina, A.V. Shaporin, Industrial Heat Eng. **21**, 22 (1999)

Comparing Ranque tubes of circular and square cross section

Ivan K. Kabardin[1,*], *Vladimir G.* Meledin[1], *Nikolay I.* Yavorsky[2], *Vladimir A.* Pavlov[1], *Margarita Kh.* Pravdina[1], *Dmitry V.* Kulikov[1], and *Valentina I.* Polyakova[1]

[1] Kutateladze Institute of Thermophysics SB RAS, 630090, 1 Academician Lavrentiev Avenue, Novosibirsk, 630090, Russia.
[2] Novosibirsk State University, 2 Pirogova Str., Novosibirsk, 630090, Russia.

Abstract. The efficiency of temperature end energy separation is compared for Ranque tubes of circular and square cross section, with the square side equal to the circle diameter. The "square" tube demonstrates approximately two times less efficiency, yet the separation effect still being evidently presented.

Experimental study of the vortex effect efficiency in Ranque-Hilsh tubes is presented. The insufficience of the available experimental data and theoretical approaches to the understanding of processes inside a vortex tube are well-known [1]. Systematic and detailed experiments are now strongly needed for verification of theoretical models, the techniques being preferable that do not or just slightly disturb the flow. So, the use of optical diagnostic has allowed a considerable progress in the study of inside swirling flow [2-5]. In [2] by means of Gilbert optics the existence of a large-scale vortex structure in the form of a double spiral in the field of optical phase density was for the first time diagnosed. The spiral arises on a plane end cover at the hot exit and propagates along the axis in the cold exit direction. In [3-4] an attempt was made to diagnose this structure by means of LDA. The sufficient nonstationarity of temperature and velocity was showed. While in [5] a circular tube is investigated, in [2-4] the tube was studied with square cross section all along for the convenience and delicacy of optical diagnostic. It was found that in spite of such a modernization the Ranque effect is distinctly manifested though the tube does not show any record features.

At this stage, the authors proceed to investigate the velocity fields inside the same tube by means of optical field diagnostic - PIV [6]. At that the first step is a detailed investigation of tube efficiency in a wide region of regime parameters for the square tube and comparison of results with a circular tube with the diameter defined as that of inscribed circle.

It should be noted that a significant amount of new publications has appeared in recent times with detailed description of certain vortex tube designs and corresponding efficiencies, such as, for example, [7-8]. The data is supposed to help not only to optimize the technical use but also to verify theoretical models. In our case the detailed data is needed as a road map guiding the field measurements inside the tube.

* Corresponding author: kabardin@itp.nsc.ru

Two variants of investigated vortex tubes are shown in Figure 1. The tubes include the two-slit swirl apparatus, the accelerating vortex chamber, the working channel, the diaphragm with connected cylinder channel at the cold exit and the radial diffuser at the hot exit. The hot exit is encased in a buffer volume that may be opened to atmosphere or have an exit to it regulated by the valve.

The compressed air enters the vortex chamber through the two tangential slits of 40 mm^2 square each at the radii $R1$=65 mm. The opened top cover of the vortex chamber is hyperbolic-shaped that keeps the velocity circulation almost constant up to the entrance into the working channel of the tube at $R2$=17 mm. Thus, the tangential velocity at the entrance of the tube is enlarged as compared to the slit exit velocity. The channel length is 450 mm. The circular tube is made of duralumin. Thermo insulation is not involved. The sections of square tube are made of duralumin with two opposite optical glass windows in the measurement part.

The radial diffuser at the hot exit of the channel may be used as a baffler, the precise opening from 0.5 to 1.5 mm being adjusted to regulate the cold fraction that is the ratio of cold flow rate to the inlet one – $\mu=G_c/G_{in}$. The parameter μ could also be regulated by the hot exit valve at maximum opening of 1.5 mm between disks.

At the cold exit, there is a diaphragm with adjusted tube of 10 mm diameter. The square side and the circle diameter in cross sections are of 34 mm.

The flow rates at the entrance and the exits were measured by orifice flowmeters each calibrated by a standard flowmeter PC-4 Ultra from a company "IRVIS", with the error not exceeding 0.5 %. The temperature and pressure values were measured by the probes as follows: DS18B20 temperature sensors (223...398 K, error less than 0.1 K), pressure sensors SDV-I-2,50-1,60-1,00-4- 20 mA (0-10 bar, 0.1 % accuracy), pressure drop sensors Honeywell MPX50100 (0-1 bar, 0.1 % accuracy) and SDV-D-2,50-1,60-1,00-4- 20 mA (0-10 bar, 0.1 % accuracy).

The cold fraction μ, ranged from 0.2 to 0.9. The incoming air overpressure P changed from 1 to 7 Bar. The income flow rate G_{in} came up to 0.07 kg/s.

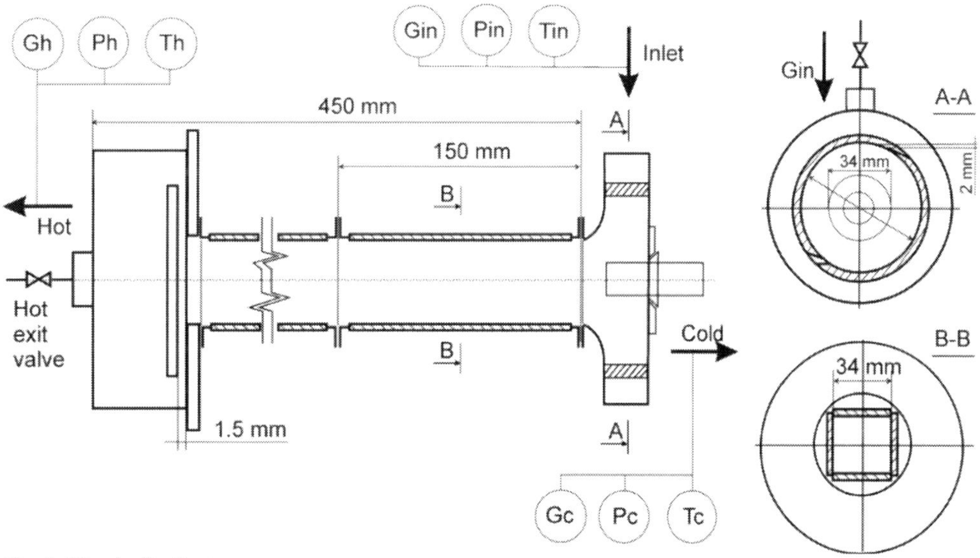

Fig.1. The draft of setup

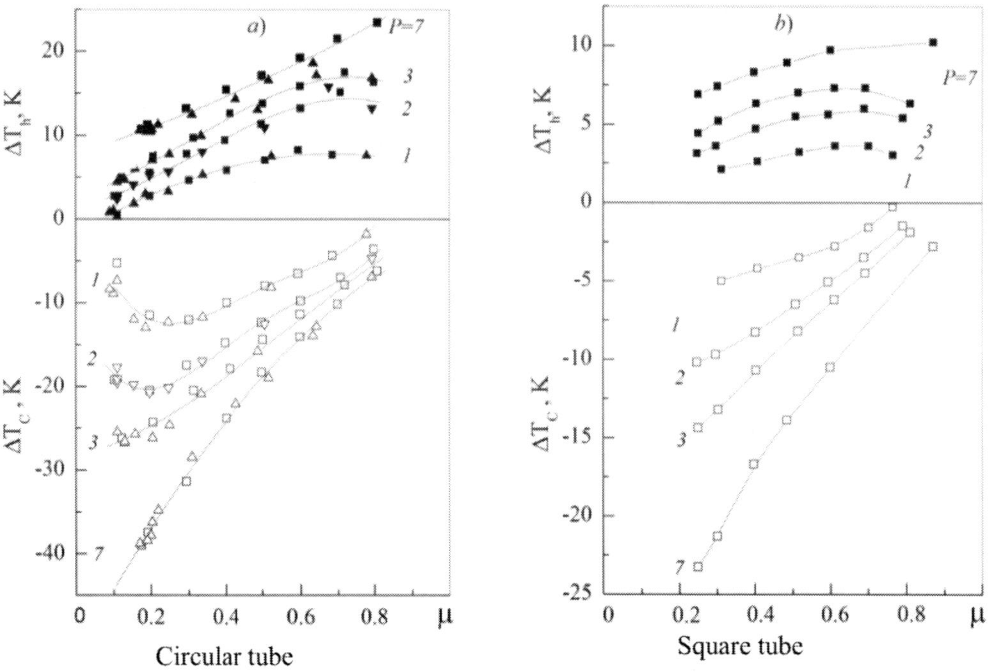

Fig. 2. Temperature separation.

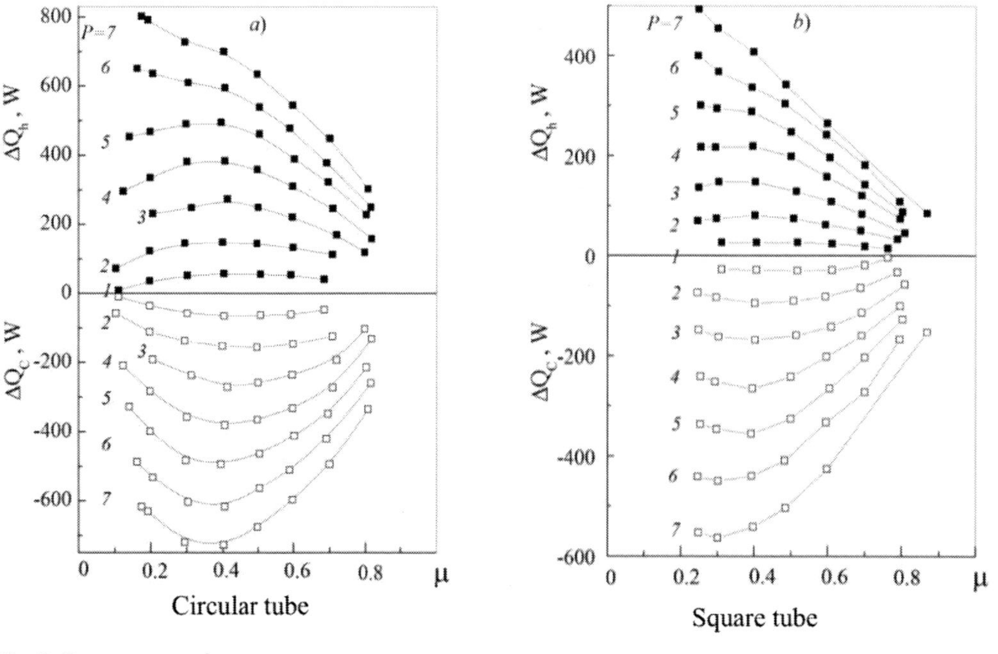

Fig. 3. Energy separation.

The detailed regime maps are executed for circular and square Ranque tubes presenting temperature and energy separation. Figure 2a) shows the circular tube temperature separation data with square points, obtained by regulation of μ by the valve, and with the triangle points, obtained by varying the opening of the radial diffuser at the hot exit. The data show no difference in regulation means. Figures near curves denote the incoming air overpressure (Bar).

In Figure 2b) the experiments on the square tube are presented with only valve regulation of μ. Figures 3a) and 3b) present the energy separation for circular and square tubes with valve regulation of μ. The value of ΔQ (Watts) is obtained multiplying of ΔT, corresponding mass flow rate and specific thermal capacity.

The comparison of the circular and square tubes show almost equal mass flow rate at the same inlet pressure though the temperature separation in the circular tube is about twice as large as that in the square one. For reference the work [9] should be noted where confined swirling flows in the closed polygonal and circular containers were studied. The reduction of temperature separation in the square tube as compared to the circular one with almost no change in hydrodynamic resistance needs a detailed explanation. Yet the amount of separation in the square tube is still evidently presented, that makes it possible and useful to analyze the Ranque effect by aforementioned optical diagnostics. The detailed data on separation features will make a road map guiding the field measurements inside the tube.

The work is financially supported by RFBR-16-08-01120 A and RFBR 15-08-00186 A.

References

1. Y. Xue, M. Arjomandi, R. Kelso, Exp. Thermal Fluid Science **34**, 8 (2010).
2. V.A. Arbuzov, Y.N. Dubnishchev, A.V. Lebedev, M.K. Pravdina, N.I. Yavorsky, Tech. Phys. Lett. **23**, 23 (1997).
3. Y.N. Dubnishchev, V.G. Meledin, V.A. Pavlov, N.I. Yavorsky, Thermophys. Aeromech. **10**, 4 (2003) .
4. I.K. Kabardin, V.G. Meledin, N.I. Yavorsky, V.A. Pavlov, M.H. Pravdina, D.V. Kulikov, V.V. Rahmanov, ICMAR AIP Conference Proceeding 1770 (2016).
5. U. Doll, M. Beversdorff, G. Stockhausen, C. Willert, D. Schlüß, C. Morsbach, 17th Int. Symp. on Appl. of Laser Techniques to Fluid Mechanics. Lisbon, Portugal (2014).
6. M. Raffel, C.E. Willert, J. Kompenhans, *Particle Imaging Velocimetry* (Berlin: Springer-Verlag, 2001).
7. S.N. Jadhav, V.D. Wagh, M.N. Patil, B.V. Kawale, P.P. Patunkar., Int. J. Technology Enchancments Emerging Eng. Res. **3**, 4 (2015).
8. M. Attalla, M. Salem, A. Abo EL-Wafa, Int. J. Mechanical & Mechatronics Eng. **14,** 4 (2014).
9. I.V. Naumov, I.Yu. Podolskaya, J. Fluid Mech. (2017) (to be published).

Numerical simulation of the turbulence effect on heat transfer between fluid and thin plates of a solid material

Nikolay Kozyulin[1,*], *Maxim* Bobrov[1], and *Michael* Hrebtov[1]

[1]Institute of Thermophysics of SB RAS, 630090 Ac. Lavrentiev ave. 1, Novosibirsk, Russia

Abstract. We present the results of numerical simulation of conjugate heat transfer between thin (100-1000 nm) solid plates and periodically heated air flow. Several air inflow conditions were studied with different turbulence intensity and integral scale. The thermal response in the plate was calculated showing that inflow turbulence interacting with the plate generates significant temperature fluctuations in the plate volume. These fluctuations amplitudes grow along the plate and with time for several periods, indicating the existence of some feedback process, with turbulent air flow acting as a medium, transferring thermal signals between different sections of the plate. The described spatial and temporal fluctuations may be used for pyroelectric energy harvesting applications.

1 Introduction

Heat transfer in the boundary layer has great practical importance for many engineering applications. One such application is the conversion of thermal energy into electricity using the pyroelectric effect. In such converters, thin films or plates of ferroelectrics are used with small thin electrodes deposited on their surface. These devices could be installed inside the hot exhaust pipes of automobiles or in gas/diesel electric generators to increase their efficiency.

The hot gas or liquid flowing onto the solid plates can be sufficiently turbulent. In that case, turbulence can affect heat exchange with the wall. There are some works [1,2] investigating the influence of turbulence on the spatially developing mixing layer, but their main focus is to investigate the influence of turbulence on hydrodynamics. The thermal effect is especially interesting if the incoming fluid temperature is alternating in time and the plate thickness is small. The longitudinal roll-instabilities in the boundary layer arising from the interaction of turbulent flow with the wall may affect the heat transfer locally, thus leading to the local increase of temperature oscillations amplitude. This effect in principle could be utilized in pyroelectric harvesting.

The purpose of this work was to investigate the effect of inflow turbulence on the heating and cooling of an array of thin (100 nm – 1000 nm) ferroelectric ceramic plates by an air flow using direct numerical simulation of the conjugate heat transfer process.

* Corresponding author: nikkozyulin@gmail.com

2 Computational details

The single "cell" of the plate-grid was simulated. The computational domain consists of one plate, inlet and outlet sections (Fig. 1a). The computational domain dimensions are 20 mm × 2 mm × 6 mm in longitudinal, vertical and transversal directions respectively. The curvilinear grid of 4.4 × 10^6 hexagonal cells was constructed with refinement in the near-wall region (Δy^+_0 = 0.15). The symmetry conditions were set at the top and bottom boundaries and periodic conditions in the direction transversal to the flow. At the inlet a uniform velocity, ***U***, profile of 15 m/s was superimposed with turbulent fluctuations obtained from separate simulation of isotropic turbulence. The turbulence integral scale was varied between 0.05 and 0.5 mm. The inlet turbulence intensity ($E_{turb}/\boldsymbol{U}^2$) was set to be 6%. The temperature field at the inlet was set uniform and alternating in time to simulate fast changes between two temperature values (273K and 373K) as follows:

$$T_{in}(t) = 323 + 50 \cdot sign(\sin(2\pi\omega t))\left|\sin(2\pi\omega t)\right|^{0.1}, \qquad (1)$$

where ω = 500 Hz is the temperature alternation frequency.

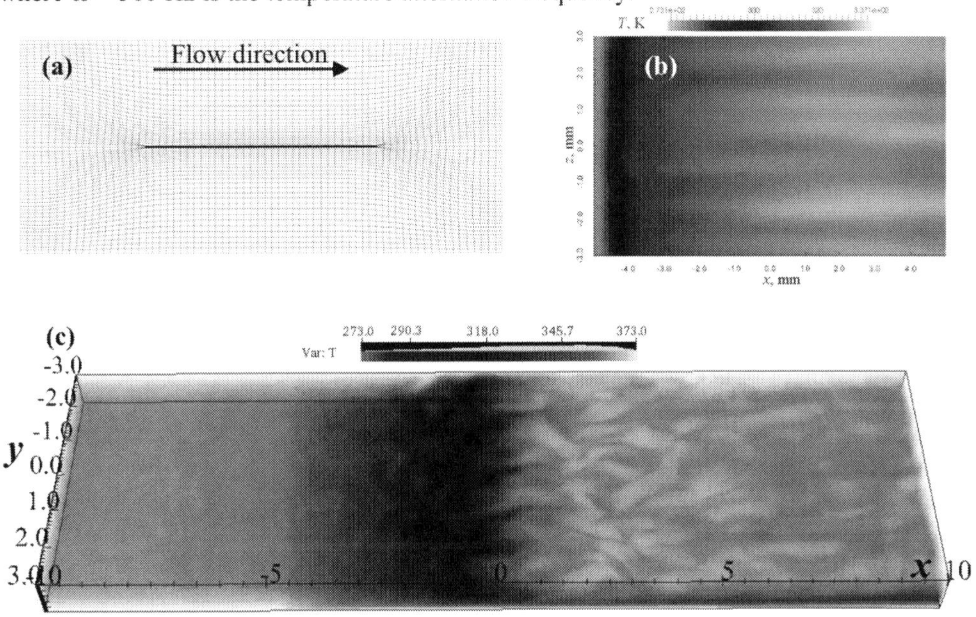

Fig. 1. The schematic of computational domain and mesh, not to scale (a); the temperature distribution at the solid plate for one moment in time (b); the temperature distribution in the fluid with the cool temperature front passing the centre of the domain (c).

The plate thickness was varied between 100 nm and 1000 nm. The plate material was chosen to be barium titanate, which physical properties were taken from [3]. The fluid was chosen to be air (density ρ = 1 kg/m^3, specific heat c_p = 1 kJ/(kg K), thermal conductivity K = 0.03 W/(mK), and kinematic viscosity ν = 2.1× 10^{-5} m^2/s). The temperature dependencies of viscosity and thermal conductivity were neglected at present simulations.

The simulations were done with modified conjugate heat transfer solver from OpenFoam [4] package, second order accurate in space and time. Due to low Reynolds number of the flow (Re~1500, based on the height of computational domain) the Kolmogorov scale is resolved at the chosen grid and the simulation could be considered as

DNS, with no subgrid model used. The preliminary computations on the sequence of refined grids showed that the current setup is sufficient for the flow considered.

3 Results and discussion

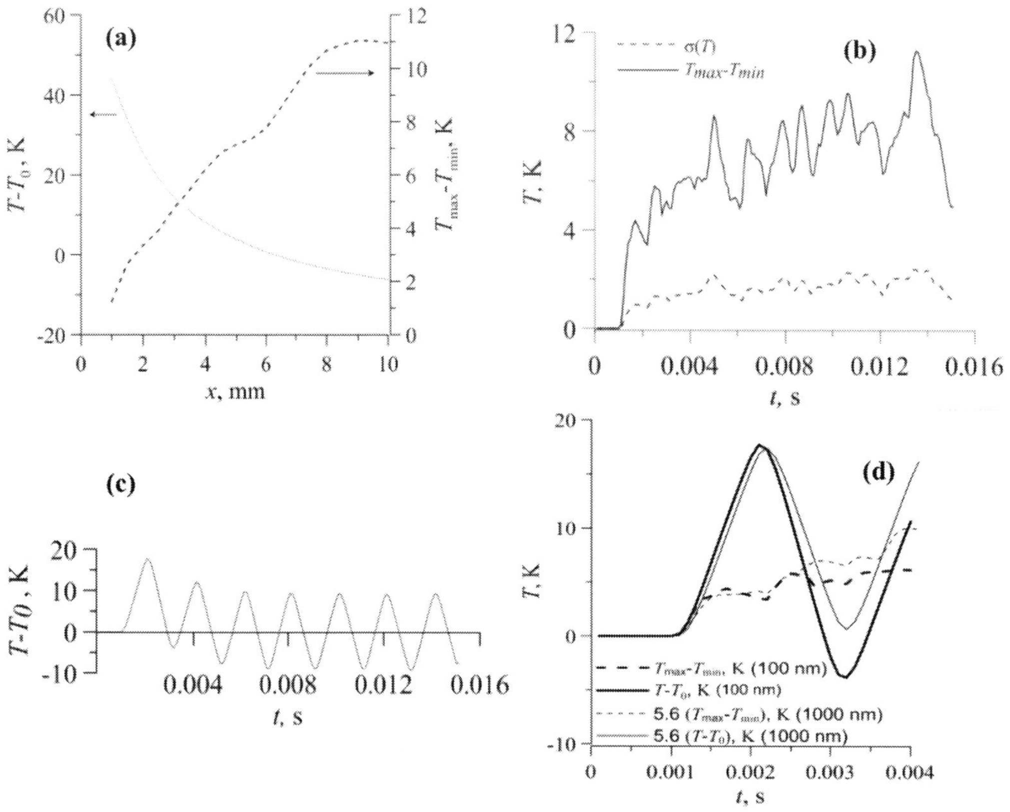

Fig. 2. Plate temperature distribution (averaged over transversal direction) along central longitudinal cross-section together with temperature fluctuations amplitude (at $t = 0.0115$ s) (a); temperature fluctuations time evolution at the end of the plate ($x = 9$ mm) (b); temperature evolution at $x = 9$ mm (c); Temperature and thermal fluctuations evolution for different plate thicknesses (d). .

The effect of turbulent vortices moving with the hot/cold temperature front above the solid surface leads to the formation of longitudinal stripes of hot/cold temperature at the surface (Fig 1b). These stripes are generated due to three-dimensional boundary layer instabilities triggered by inflow turbulence. The longitudinal roll-vortices which are the dominant mode of the instability intensify the heat transfer at some places and at some other places diminish it. The computation of evolution of the 100-nm-thick plate temperature averaged in transversal direction showed that turbulence has very weak effect on the averaged temperature. The mean temperature evolution for 8 cycles was the same with and without inflow turbulence giving only about 0.5% difference at most. However temperature fluctuations inside the plate generated along the transversal direction by the interaction with turbulent flow are quite significant.

We used the difference between the highest and lowest temperature along the transversal cross-sections together with the standard deviation of temperature to measure the intensity of temperature variations.

Fig 2a showing the temperature distribution in central longitudinal cross-section of a 100-nm-thick plate in the first half of seventh heating cycle. It is seen that the mean temperature at the beginning of the plate is much higher that at the plate's end. This is partially due to the loss of heat by the air when it moves along the cooler plate and partially due to the growth of boundary layer thickens. However as it can be seen the temperature fluctuations in the transversal direction are the highest at the plate's end and their magnitude is comparable with the fluctuations caused by the prescribed temperature alteration at the same point (see Fig 2c). What is more interesting, the fluctuations' magnitude at the plate's end is growing with time steadily for at least 6 periods (Fig2 b). This may be explained by some thermal feedback effect of the fluctuations at the plate's beginning on the plate's end. Looking at the evolution of two transversal cross-sections: one at the beginning of the plate and the other at the plate's end it is clearly seen that during the first few cycles the local extrema of the temperature fluctuations are uncorrelated between the cross-sections but then they become more and more correlated (temperature minimums at one cross-section starts to coincide with maximums at the other) and simultaneously the magnitude of the fluctuations at the plate's end starts to grow. This effect needs further investigation and here we only outline our current findings.

The effect of plate thickness on the fluctuations magnitude was also investigated. Fig. 2d shows the normalized mean temperature and fluctuations evolution during the first cycle of simulation for two different plate thicknesses (100 nm and 1000 nm). It could be seen that relative magnitude of fluctuations in comparison with the mean temperature is roughly the same that means that the effect might be reproduced to some extent on the thicker plates. The separate study is needed to find the most effective setup for the generation of strong temperature fluctuations.

Our simulations showed that the integral scale of turbulence is an important parameter for the strong temperature fluctuations to take place, the integral scale should be comparable with the boundary layer rolls diameter (in our case ~0.5 mm) smaller integral scales give much weaker fluctuations (0.25 mm integral scale gives ~4 times less fluctuations magnitude).

4 Conclusion

The described effect of thermal fluctuations generation by the interaction of inflow turbulence with the spatially growing boundary layer conceptually may be used to improve the efficiency of pyroelectric harvesting. The hot/cold temperature "stripes" caused by the roll-vortices are predominantly longitudinal. This allows one to use separate longitudinal electrodes on the surface of ferroelectric material to harvest the fluctuations energy. The width of the electrodes should be the same as the roll-vortices diameter, which could be easily calculated for each situation. Further research is needed to find the optimal parameters for the generation of the strongest fluctuations.

This work was supported by the Russian Science Foundation (grant №16-19-00119).

References

1. T.S. Lund, X. Wu, K.D. Squires, J. Comput. Phys. **140**, 233 (1998)
2. A. Ferrante, S.E. Elghobashi, J. Comput. Phys. **198**, 372 (2004)
3. Y. He, Thermochim. Acta. **419**, 135 (2004)
4. H.G. Weller, G. Tabor, H. Jasak, C. Fureby, Computers in Phys. **12**, 620 (1998)

Influence of the thickness of monolithic compliant coatings on the skin friction drag

Victor Kulik[1,*], *Andrey* Boiko[2,3], and *Inwon* Lee[4]

[1]ITP SB RAS, Academician Lavrentyev Avenue 1, Novosibirsk, 630090, Russia
[2]ITAM SB RAS, Institutskaya Street 4/1, Novosibirsk, 630000, Russia
[3]University of Tyumen, Volodarskogo Street 6, Tumen, 625003, Russia
[4]GCRC-SOP, Busan daehak-ro 63beon-gil, Geumjeong-gu, Busan, 609-735, Korea

Abstract. In this study, the drag reduction capabilities of compliant coatings made from viscoelastic silicon rubbers are verified experimentally. A series of single-layer compliant coatings of homogeneous materials with constant thicknesses were used. The experiments were performed at different flow velocities in a high-speed water tunnel of Pusan National University. They were accompanied by experimental determination of dynamic viscoelastic properties of the coating materials. Then, for each coating, a dynamic compliance in frequency domain was estimated. Assuming that the coatings interact effectively only with dynamic structures of turbulent boundary layer in a frequency band corresponding to the region about the maximum compliance, promising ranges of flow velocity and coating thickness are outlined. The predicted ranges are compared favourably with the experimental observations in the high-speed water tunnel.

1 Introduction

Experimental verification of drag reduction capabilities of compliant coatings has long been of great concern. A comprehensive review of such attempts can be found in [1], where most of researches dealt with 'soft' coatings (either a sponge material covered by a thin film or a gel-like substances). It is known that due to the fluid-solid interaction, such soft materials are subject to different instabilities resulting in surface folding and, hence, in drag increase.

The use of more 'stiff' compliant coatings can prevent the surface folds, thereby being advantageous for the sake of drag reduction applications. The surface of the stiff compliant coating tends to remain hydraulically smooth under the action of a flow, as was shown by [2–5]. In a typical water application at flow velocities up to about 20 m/s, such coatings are characterized by relatively large modulus of elasticity (usually more than 1 MPa) that prevents the formation of folds and enhances the coating. Although the surface deformation is smaller than the thickness of viscous sublayer, the surface velocity is not negligible.

In the present study, the drag reduction capabilities of single-layer compliant coatings made from viscoelastic silicon rubber are verified experimentally in a high-speed water tunnel. The tests were accompanied by experimental determination of dynamic viscoelastic

* Corresponding author: vkulik@itp.nsc.ru

properties of the coating materials to estimate the flow-velocity ranges and the frequency bands for an effective interaction of the coatings with the flow.

2 Experimental setup

The experiments were conducted in the high-speed water tunnel of Pusan National University. The test section of the facility is 2 m long with the cross section being 200×200 mm2. The water tunnel velocities can be set with the accuracy of ±0.05 m/s. In addition, each tunnel section is equipped with differential pressure transducers. The turbulence level in the tunnel was 0.6%. The freestream velocity and the pressure gradient were constant with the deviation being less than 2% throughout measurement time span.

The turbulent boundary layer developed along four walls of the test section. The origin of the streamwise coordinate x was defined at the leading edge of the test section, and no turbulence trip was installed. There were three acrylic windows installed in each of the walls. One of the side windows located downstream was replaced with a floating plate balance unit. The surface of the floating plate balance unit was flush-mounted with the entire side surface of the test section. The upper plate of the floating plate balance unit had 100×50 mm^2 rectangular opening for placing a rectangular insert.

The compliant coatings of varying composition and thickness H were mould in the recess of floating plate inserts. The bed surface of the recess was preprocessed with a special primer to provide a strong silicon adhesion. A detailed inspection indicated that the surfaces of the coatings became concave due to a contraction of silicon rubber with time. To flatten the surfaces, the coatings were carefully grinded to keep the surface smooth. Hence, the actual height of the coatings decreased about 2.5% from the nominal.

Silicone room temperature vulcanizing (RTV) rubbers were used as the compliant coating materials in this study. The RTV 3133 BASE material by *Xiameter* was polymerized by mixing RTV 3081-F catalyst with the mass composition being 5% of mixture. The compound was carefully mixed and vacuumed for 10~15 minutes in order to get rid of air bubbles which might subsequently form unwanted pores in the sample.

The floating plate inserts with the coatings were installed into an immersed balance. Great care was exercised to maintain the gap between the floating plate and the surrounding surface about 30 μm and the level differences less than 5 μm.

3 Measurement results

The main results of frictional force measurements are shown in Fig. 1 for the baseline case (hard insert) and a series of the compliant coatings with varying thicknesses. As seen, the interaction between the coating and the flow depends significantly on the coating thickness. The 4 mm and 12 mm thick coatings approach the drag of the hard surface. The 6 mm, 8 mm, and 10 mm thick coatings show similar quite large levels of drag reduction. The changes of the drag with respect to free stream velocity are monotonous.

Viscoelastic properties of a material can be described by the dynamic complex-valued modulus $E(1\text{-}i\mu)$, where $E(f)$ and $\mu(f)$ are the real-valued frequency-dependent modulus of elasticity and loss factor, respectively. The experimental method to determine the viscoelastic properties of the coatings is described in detail [6, 7]. It is based on measuring the vibration amplitude ratio and phase shift between two ends of cylindrical samples. In the present study, the measurements of the viscoelastic properties of the rubber were performed simultaneously with the water-tunnel tests. A simple interpolation of the obtained data of the viscoelastic properties is as follows

$$E(f) = (3{\cdot}10^4 \ln(f) + 3.67{\cdot}10^5){\cdot}[1-i(0.044\ln(f) - 0.1163)],\ E_0 = 0.42\ \text{MPa}.$$

It is reasonable to assume that one of the necessary conditions for skin friction drag modification is the maximum interaction of the coating with the flow, the condition does not necessarily differentiate drag reduction or drag increase.

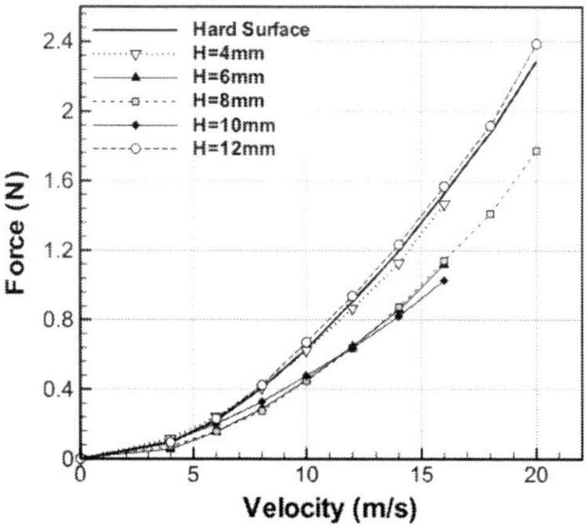

Fig. 1. Friction force for the hard surface and the series of RTV-3133 compliant coatings vs. stream velocity.

The dynamic compliance, i.e. the ratio of dynamic deformation to the applied dynamic mechanical stress (e.g., due to an external periodic forcing), needs to be maximised in order to get the maximum interaction between the surface oscillation and the turbulent pressure fluctuations. It is assumed usually that the normal compliance makes the main contribution to the fluid-coating interaction. The normal compliance C_n scaled with the ratio of the static compliance of a long narrow rod of height H with the static modulus of elasticity E_0 was computed using the measured dynamic properties by the approach described in [2–5] for an infinite plane layer of homogeneous compliant material of height H, density ρ, Poisson's ratio σ, modulus of elasticity E and loss tangent μ. It has been shown [2–5] that if the interaction occurs under the action of pressure wave with wavelength λ traveling at a speed V, the components of the compliance depend only on combinations λ/H and V/C_t, where C_t is the propagation velocities of a transverse shear wave travelling in an infinite space without viscous losses (μ=0) and $f=V/\lambda$ so that $C_t^2=E(f)/[2\rho\,(1+\sigma)]$.

Let us assume that the velocity V of pressure wave induced by a vortical structure in the boundary layer flow is about $0.8U$, where U is the free-stream flow velocity. This is the upper bound of typical convection velocities discussed by [8], which corresponds to large vortices. If the coating interacts with the smaller-scale relatively high-frequency inner layer structures as assumed, e.g., by [9], V needs to be about $0.69U$ according to [10].

The modulus $|C_n|$ for the RTV-3133 coatings of different thickness H computed using the empirical formulas are plotted against frequency f and flow velocity U in Fig. 2. At relatively high velocities (more than about 20 m/s) the compliance is characterized by the presence of a large amplitude ridge stretched parallel to the velocity axis. The maximum compliance occurs at this 'primary' ridge. The position of $|C_n|$ maximum shifts to lower frequencies, as H grows. At velocities about 16~17 m/s the ridge is divided into a pair of 'secondary' ridges with diminishing amplitudes. The location of the secondary ridges is essentially independent of H, i.e. it is determined mainly by the viscoelastic properties of material.

Therefore, two scenarios of interaction of the coating with coherent structures of turbulent boundary layer can be proposed: (i) an interaction in a relatively narrow frequency range of the 'primary' ridge, which depends on the coating thickness; (ii) a broadband in frequency, but less intensive interaction at lower flow velocities corresponding to the position of the 'secondary' ridges and their downhills. To make a compliant coating durable, its modulus of elasticity should be large. However, as seen, this leads to higher velocities for the efficient interaction at the position of the 'primary' ridge of compliance. A way to overcome this shortcoming is to operate at velocities corresponding to the 'secondary' ridges. Despite the compliance is several times smaller at the 'secondary' ridges than the peak value of the 'primary' ridge, the frequency range of interaction is much wider and the phase variations are much smoother that can lead to wider frequency range of the phase lags favourable for the Reynolds stress reduction.

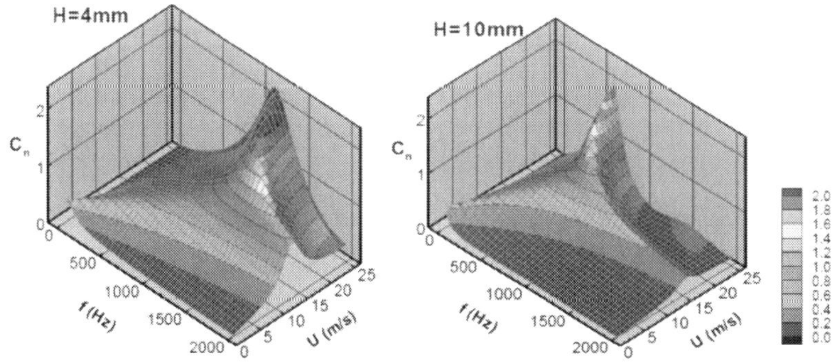

Fig. 2. Computed modulus of normal dynamic compliance of the coatings.

4 Conclusions

It was shown that the efficiency of fluid-coating interaction depends on the thickness of coatings in a special way that supports the proposed simplified approach for the prediction of efficiency of the interaction based on the scenario associated with the lower velocity ridges in the compliance charts.

The experiments were financially supported by the National Research Foundation of Korea grants funded by the Ministry of Science, ICT & Future Planning through GCRC-SOP (No. 2011-0030013) and by the Ministry of Education (No. 2015R1D1A1A01059973). The numerical modelling and data analysis were financially supported by the Russian Science Foundation Grant No. 17-11-01156.

References

1. M. Gad-el-Hak, Appl. Mech. Rev. **49**, 147 (1996)
2. V.M. Kulik, J. Appl. Mech. Tech. Phys. **48**, 221 (2007)
3. V.M. Kulik, Thermophys. Aeromech. **16**, 43 (2009)
4. V.M. Kulik, Int. J. Heat Fluid Flow **33**, 232 (2012)
5. V.M. Kulik, Thermophys. Aeromech. **23**, 505 (2016)
6. V.M. Kulik, B.N. Semenov, A.V. Boiko, B.M. Seoudi, H.H. Chun, I. Lee, Exp. Mech. **49**, 417 (2009)
7. A.V. Boiko, V.M. Kulik, B.M. Seoudi, H.H. Chun, I. Lee, Int. J. Solids Struct. **47**, 374 (2010)
8. B.J. Cantwell, Annu. Rev. Fluid Mech. **13**, 457 (1981)

9. B.N. Semenov, Thermophys. Aeromech. **16**, 219 (2009)
10. S.J. Russell, *Proc. Int. Symp. Seawater Drag Reduction* (Newport, RI, U.S.A., 1998)

Investigation a single-spiral vortex in a swirl flow

Elizaveta S. Gesheva[1,*], *Sergey I.* Shtork[1,2], and *Sergey V.* Alekseenko[1,2]

[1]Kutateladze Institute of Thermal Physics, Siberian Branch of Russian Academy of Sciences, Lavrentyev Ave. 1, Novosibirsk, 630090 Russia
[2]Novosibirsk State University, Pirogova Str. 2, Novosibirsk, 630090 Russia

Abstract. The work is aimed at a detailed study of large-scale helical vortex structures emerging in a high turbulent intensively swirling flow. It was shown that the vortex formed in the chamber by installing diaphragm with shifted outlet is folded into a single-helical vortex. The flow visualization shows that although the vortex axis performs slow oscillations (precession), on average this structure is fixed in space. The velocity fields were measured with the aid of a nonintrusive method of flow diagnostics (PIV). Verification of the calculation results obtained using a LES simulation was performed based on measured experimental data to confirm the correctness of the chosen mathematical modelling approach. It has been asserted that the investigated regimes are in a self-similarity area relative to Reynolds number.

1 Introduction

The application of swirling flows and useful properties of vortex structures has been grounded scientifically and widely used in practice [1]. Longitudinal concentrated vortices increase the efficiency of devices and hardware systems. Stationary helical vortex structures were first observed at the Institute of Thermophysics of the SB RAS [2]. Until now, there is no quantitative information on such vortices that hinders understanding the physics of formation of vortices and developing approaches for mathematical description of the vortex dynamics.

Setting up the experiments with complex swirling flow is a hard task and often it is not feasible to extract enough data, so the numerical modeling plays an important role. The Navier-Stokes equations describing the motion of a viscous fluid are among the most important instruments in hydrodynamics and are used in the mathematical modeling of many natural phenomena and technical processes [3]. Currently, the numerical simulation methods are rapidly developing. Different approaches in the modeling of turbulence make it possible to expand the notion of fluid motion including swirling flows with the formation of complex vortex structures. To verify the simulation results, a joint experimental and numerical study of the vortex structures is used, analogous to [4, 5]

* Corresponding author: gesheva_es@itp.nsc.ru

DOI: 10.1051/matecconf/201711502025

2 Experimental method

To carry out the experiments, a hydrodynamic stand created earlier in the IT SB RAS [2] to study turbulent swirling flow has been modernized. The hydrodynamic stand (Fig. 1) is used basically for experimental study of the vortex structures formed in a working chamber 1 with the tangential flow inlet. Water circulating through a closed hydraulic circuit was used as a working fluid. The liquid from the tank 2 is fed into the working chamber by means of a pump 3, through a system of valves and flow distributors. The liquid flow rate is measured by an ultrasonic flowmeter (5 – electronic module, 6 - measuring section of the flowmeter).

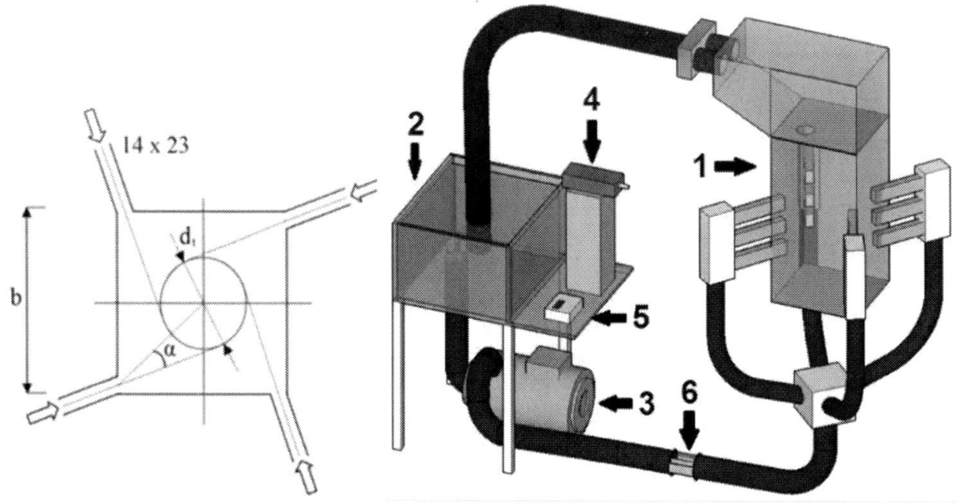

Fig. 1. Scheme of the vortex chamber.

The test section is made of plexiglas as a vertical channel of square cross-section with dimensions 188×188×560 mm. Rectangular nozzles with an output sizes of 14×23 mm are arranged in three tiers and are combined into four corner blocks. The swirling flow was organized by tangential fluid supply to the work area, i.e. by directing the axes of the nozzles along the tangent to some conditional circle with a diameter d_t (Fig. 1).

The flow regimes were characterized by dimensionless parameters, such as the Reynolds number Re and the swirl parameter S. The Reynolds number was determined using the transverse dimension of the channel b and the average velocity V_b in the vortex chamber:

$$Re=bV_b/\nu \quad (1)$$

The swirl parameter was defined through the swirler geometry and varied by changing the angle of rotation of the nozzles α (the diameter of the conditional circle d_t) [4]:

$$S=bd_t/\Sigma_n \quad (2)$$

where Σ_n - area of the inlet nozzles. The experiments included flow visualization, which was organized using a light source and by small air bubbles that were fed into the circuit in front of the working area. The flow pattern was recorded using a high-resolution digital mirror camera with the video recording function at a frequency of 60 frames per second. To obtain quantitative characteristics of the vortex, such as velocity fields, a nonintrusive PIV system (Particle Image Velocimetry) was applied.

3 Characteristics of a spiral vortex

Visualization studies have shown that when the geometry of the chamber changes, various vortex structures are formed. In the chamber with a flat bottom and a centrally located diaphragm orifice, a rectilinear vortex is formed from the bottom of the chamber to the outlet. When the diaphragm is displaced, the vortex filament is bended into a spiral (Figure 2).

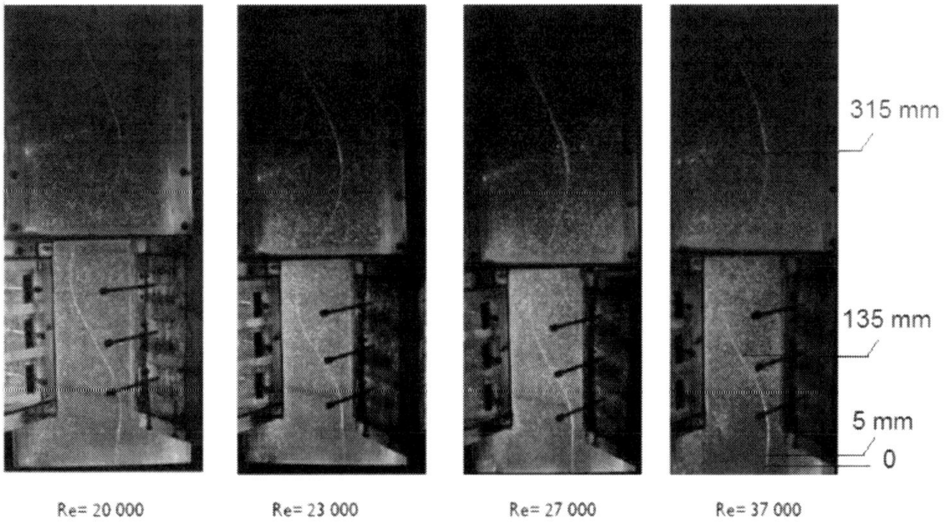

Fig. 2. Visualization of spiral vortex at different Reynolds numbers, S=8,4.

Figure 2 demonstrates that form of the spiral vortex is unchanged for all studied Reynolds numbers. Although analysis of the video recording shows that the vortex axis executes slow oscillations relative to the average position, but in general this structure is stationary. The fluid particles move around the spiral axis, thus making a double spiral motion. Also can be noticed some areas with a reverse flow.

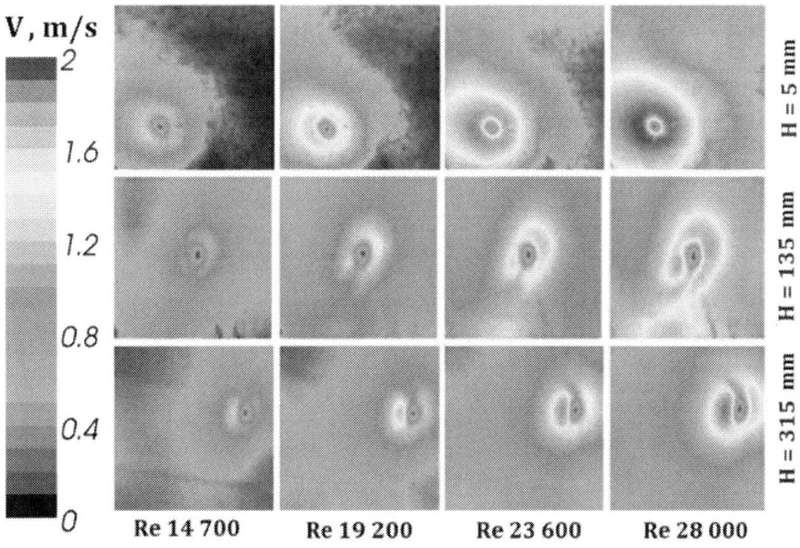

Fig. 3. Velocity fields in the cross sections of the chamber, S=8,4.

Figure 3 shows the velocity fields in and around the vortex core at different horizontal cross-sections of the chamber. The plots demonstrate formation of a localized vortex core at all studied Reynolds numbers.

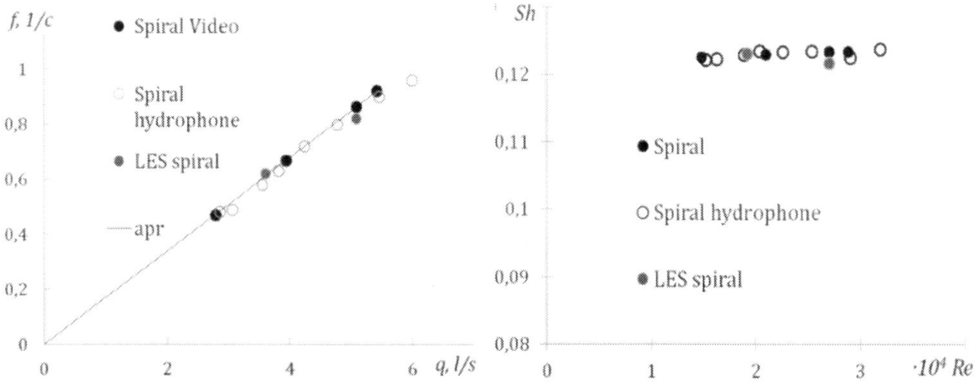

Fig. 4. Frequency dependence of the precession of a vortex on flow rate (left). The dependence of Strouhal number on Reynolds number (right), S=8,4.

Detailed analysis of the video recordings in a horizontal plane allowed to reveal that the mentioned above oscillations of the vortex core axis have a form of precession. Figure 4 shows the dependence of the precession frequency on the fluid flow. The frequency of the vortex precession was measured in several ways: based on the analysis of video recordings; using a hydrophone installed on the wall of the channel and based on the numerical simulation using a CFD package by the LES method [6].

4 Conclusions

As a result of the investigations it was shown that the geometry of the chamber and the boundary conditions have a key influence on the shape of the vortex axis. E.g. displacement of the outlet leads to the formation of a spiral vortex, which is quasi-stationary although makes small regular oscillations in form of precession. Helical vortex is stable and does not change form with Reynolds number. Strouhal number of the precession also does not depend on the Reynolds number that points on the flow self-similarity. The velocity fields measured using PIV technique confirm that spiral vortex has a localized vortex core.

The study was financially supported by the Russian Science Foundation (project № 14-29-00093).

References

1. S.V. Alekseenko, V.M. Dulin, M.P. Tokarev, D.M. Markovich, Springer Proceedings in Physics, **185**, 43 (2016)
2. S.V. Alekseenko, P.A. Kuibin, V.L. Okulov, S.I. Shtork, J. Fluid Mech. **48**, 123 (1999).
3. S. Alimohammadi, E. Fanning, T. Persoons, D.B. Murray, Computers & Fluids **140**, 232 (2016)
4. C.E. Cala, E.C. Fernandes, M.V. Heitor, S.I. Shtork, Exp. Fluids. **40**, 267 (2006)
5. D.J. Ryan, M.J.H. Simmons, M.R. Baker, Chem. Eng. Sci. **163**, 123 (2017)
6. E.S. Gesheva, S.I. Shtork, S.V. Alekseenko, Vestnik of NSU: Physics. **9**, 39 (2014). [in Russian]

A mathematical model of metallized solid propellant combustion under the changing pressure

Vasily Poryazov[1,*], *Aleksei* Krainov[1], and *Dmitry* Krainov[2]

[1]Tomsk State University, Department of Physics and Engineering, 634050, Russia, Tomsk, 36 Lenin Ave.
[2]Tomsk Politechnical University, Institute of Power Engineering, Department of Theoretical and Industrial Heat Systems Engineering, 634050, Russia, Tomsk, 30 Lenin Ave.

Abstract. This paper presents the mathematical model describing a non-stationary combustion of metallized solid propellant. The model takes into account the heat transfer, the oxidizer decomposition and gasification of the solid propellant, two-phase, dual-velocity, two-temperature reactant flow over the propellant surface. The conditions on the surface perform the conservation of energy and mass fluxes. The model is based on the research works [1, 2]. Our research provides data of the non-stationary burning rate depending on the Al powder dispersion and the pressure drop value.

1 Mathematical model

The mathematical formulation of the problem in the coordinate system associated with the propellant surface has the following form:

The heat transfer and burn-out equations for condensed phase:

$$\rho_1 c_1\left(\frac{\partial T_1}{\partial t}+u\frac{\partial T_1}{\partial x}\right)=\lambda_1\frac{\partial^2 T_1}{\partial x^2}+Q_1k_1\rho_1(1-\eta)e^{-\frac{E_1}{RT_1}}, \quad (1)$$

$$\frac{\partial \eta}{\partial t}+u\frac{\partial \eta}{\partial x}=k_1(1-\eta)e^{-\frac{E_1}{RT_1}}. \quad (2)$$

The system of equations for reactant flow over propellant surface:

$$\rho_2 c_2\left(\frac{\partial T_2}{\partial t}+v\frac{\partial T_2}{\partial x}\right)=\lambda_2\frac{\partial^2 T_2}{\partial x^2}+\frac{dp}{dt}+Q_2k_2\rho_2 Ye^{-\frac{E_2}{RT_2}}+4\pi\alpha r_3^2 n(T_3-T_2), \quad (3)$$

$$\frac{\partial Y}{\partial t}+v\frac{\partial Y}{\partial x}=D\frac{\partial^2 Y}{\partial x^2}-k_2 Ye^{-\frac{E_2}{RT_2}}, \quad (4)$$

$$\frac{\partial \rho_2}{\partial t}+\frac{\partial \rho_2 v}{\partial x}=-G, \quad (5)$$

*Corresponding author: poryazov@ftf.tsu.ru

$$p = \rho_2 RT/\mu, \tag{6}$$

$$c_3\rho_3\left(\frac{\partial T_3}{\partial t} + w\frac{\partial T_3}{\partial x}\right) = -4\pi\alpha r_3^2 n\left(T_3 - T_2\right) + \frac{2\mu_{Al}}{3\mu_O} GQ_{Al}, \tag{7}$$

$$\frac{\partial\rho_3}{\partial t} + \frac{\partial(\rho_3 w)}{\partial x} = G, \tag{8}$$

$$\frac{\partial w}{\partial t} + w\frac{\partial w}{\partial x} = -\tau_{fr}, \tag{9}$$

$$\frac{\partial n}{\partial t} + \frac{\partial(nw)}{\partial x} = 0, \tag{10}$$

The boundary conditions are:

$$\lambda_1\frac{\partial T_1(x_s,t)}{\partial x} = \lambda_2\frac{\partial T_2(x_s,t)}{\partial x},\ T_1(x_s,t) = T_2(x_s,t),\ \frac{\partial T_1(-\infty,t)}{\partial x} = 0,\ \frac{\partial T_2(\infty,t)}{\partial x} = 0,$$

$$\frac{\partial Y(\infty,t)}{\partial x} = 0,\ \alpha_{Al}\rho_1 u = \rho_3 w(x_s,t),\ \rho_2(x_s,t) = P/RT_2(x_s,t)$$

$$n(x_s,t) = \rho_3(x_s,t)\Big/\left(\tfrac{4}{3}\pi r_{Al,0}^3\rho_k\right), \tag{11}$$

$$(1-\alpha_{Al})\rho_1 u = (\rho_2 vY)\big|_{(x_s,t)} - D\rho_2(x_s,t)\frac{\partial Y(x_s,t)}{\partial x},\ (1-\alpha_{Al})\rho_1 u = \rho_2(x_s,t)v(x_s,t).$$

The initial conditions are:

$$-\infty < x < x_s:\ T_1(x,0) = T_0,\ \eta(x,0) = 0, \tag{12}$$

$$x_s < x < \infty:\ T(x,0) = T_{ig},\ Y(x,0) = 0,\ v(x,0) = 0,\ p(0) = p_0 \tag{13}$$

$$\rho_2(x,0) = p_0\mu/RT_{ig},\ \rho_3(x,0) = 0,\ w(x,0) = 0,\ n(x,0) = 0.$$

The particle interaction force with the gas is defined by the formula:

$$\tau_{tr} = F_{tr}\Big/\left(\tfrac{4}{3}\pi r_3^3\rho_k\right),\ F_{tr} = C_R S_m\rho_2(w-u)|u-w|/2. \tag{14}$$

The drag coefficient is defined by the empirical formula [3]:

$$C_R = \frac{24}{\mathrm{Re}}\left(1 + 0{,}15\,\mathrm{Re}^{0{,}682}\right),\ \mathrm{Re} = 2r_k\rho_2\,|u-w|/\eta. \tag{15}$$

The heat transfer coefficient is determined as:

$$\alpha = \frac{Nu\lambda_2}{2r_k},\ Nu = 2 + \sqrt{Nu_l^{\,2} + Nu_t^{\,2}}, \tag{16}$$

where $Nu_l = 0{,}664\,\mathrm{Re}^{0{,}5}$, $Nu_t = 0{,}037\,\mathrm{Re}^{0{,}8}$.

The mass-change rate of particles during combustion:

$$G = \frac{3\mu_O}{2\mu_{Al}} n\rho_k 4\pi k_{Al}a^{0.9}r_{Al}^{1.5},\ k_{Al} = 2.22\cdot10^{-5}\ m^{1.5}/s. \tag{17}$$

The radius of unburnt aluminum particle r_{Al} and the radius of a whole particle:

$$r_{Al} = \left[\frac{2\mu_{Al}}{3\mu_O}\left((1+\frac{3\mu_O}{2\mu_{Al}})r_{Al,0}^3 - \frac{\rho_3}{(4/3)\pi n\rho_k}\right)\right]^{1/3},\ r_3 = \left[r_{Al}^3 + \frac{\mu_{Al} + 3/2\,\mu_O}{\mu_{Al}}\left(r_{Al,0}^3 - r_{Al}^3\right)\right]^{1/3} \tag{18}$$

The pressure drop value ranges from p_0 (at $t_{p,n}$ - start drop time) to p_k at the end of the drop ($t_{p,k}$). The linear combustion rate u is the rate of conversion level $\eta = 0.99$ motion.

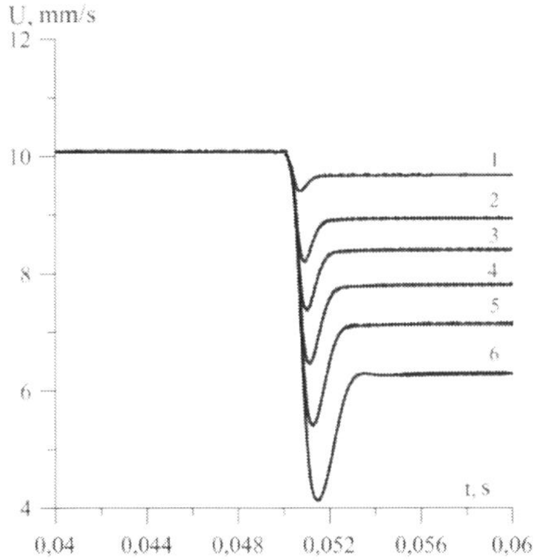

Fig. 1. Burning rate of N powder with micron sized aluminum particles in time. Pressure drop ranges from 100 atm to: 1 - 80, 2 - 60, 3 – 50, 4 – 40, 5 – 30, 6 – 20 atm. Pressure drop rate 10^{10} Pa/s, $r_{Al} = 5\ mkm$

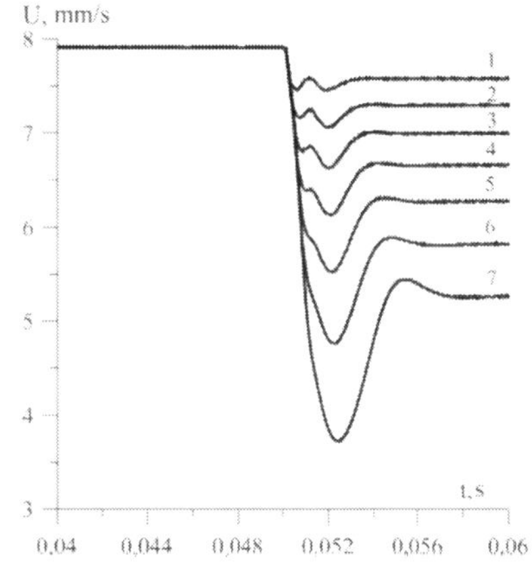

Fig. 2. Burning rate of N powder with micron sized aluminum particles in time. Pressure drop ranges from 100 atm to: 1 - 80, 2 - 70, 3 – 60, 4 – 50, 5 – 40, 6 – 30, 7 – 20 atm. Pressure drop rate 10^{10} Pa/s, $r_{Al} = 15\ mkm$

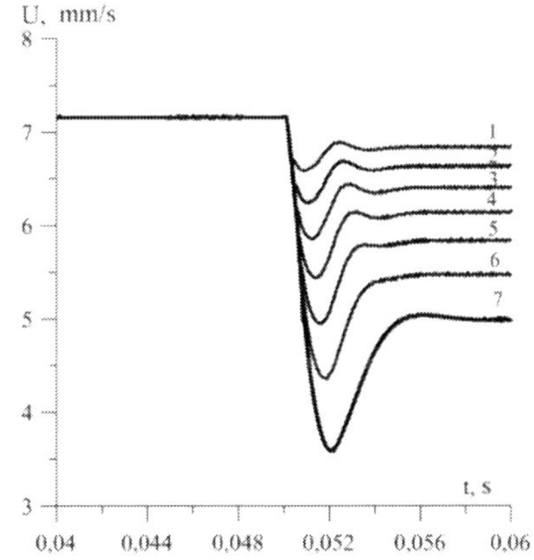

Fig. 3. Burning rate of N powder with micron sized aluminum particles in time. From 100 atm to: 1 - 80, 2 - 70, 3 – 60, 4 – 50, 5 – 40, 6 – 30, 7 – 20 atm. Pressure drop rate 10^{10} Pa/s, $r_{Al} = 30\ mkm$

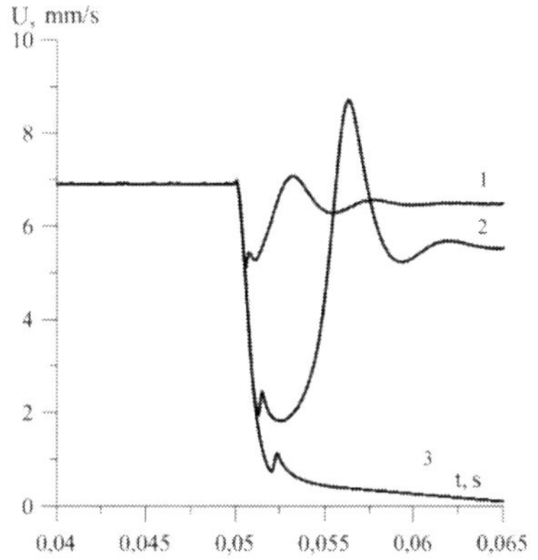

Fig. 4. Burning rate of N powder in time. Pressure drop ranges from 100 to: 1 - 80, 2 - 40, 3 – 20 atm. Pressure drop rate 10^{10} Pa/s

Here: t is the time, x is the coordinate, x_s is the coordinate of combustion surface, ρ_1, ρ_2, ρ_3, ρ_{Al}, ρ_k is the density of solid propellant, gas, reduced density of particles, aluminum, particle in the flow of combustion product, T_1, T_2, T_3 is the temperature of solid propellant, gas phase, aluminum particles in gas phase, T_0, T_{ig} is the initial temperature of solid and gas phases, η is the conversion degree of the condensed-phase material, u is the linear combustion rate, v, w is the gas velocity and particle velocity, n is the number of particles in a unit volume, r_{Al}, r_3 is the radius of unburnt particle of aluminum and the whole radius of a particle, c_1, c_2, c_3 is the specific heat of solid propellant, gas and particles at a constant pressure, λ is the thermal conductivity, Q_1 is the thermal effect of reaction in the solid phase, Q_2 — thermal effect of reaction in the gas phase, Q_{Al} — heat of aluminum combustion, k is the pre-exponential factor in the Arrhenius law, E is the energy of activation, R is the universal gas constant, α_{Al} is the is the aluminum mass fraction in solid propellant, p is the gas phase pressure, p_0, p_k is the initial and final pressure in gas phase , $t_{p,n}$, $t_{p,k}$ is the start and end time of pressure drop, Y is the oxidant concentration in the gas phase, D is the diffusion coefficient, μ , μ_{Al}, μ_O is the molar masses of gas phase, aluminum particles and oxygen, G is the massis thechange rate of particles during combustion, τ_{fr} is the particle interaction force with the gas, α is the heat transfer coefficient. The quantities related to the condensed phase are denoted by index 1, to the gas is the 2, to particles is the 3, to the initial conditions is the 0.

We have solved the system of equation numerically using the methods described in the papers [1,2]. The calculations of N powder combustion of with micron sized aluminum particles were carried out at the following values of the thermophysical and kinetic quantities from [1,2]

We have conducted the calculation of the non-stationary burning rate during the pressure drop over the propellant surface. The composition combusts stationary under the pressure of 100 atm, at the time of 40 *ms* the pressure over the surface begins to drop in linear fashion at the rate of 10^{10} Pa/s . The values of the final pressure are 80, 70, 60, 50, 40, 30 and 20 atm. The size of Al particles varies from 5 to 30 *mkm* (Pic.1, Pic.2, Pic. 3).

The combustion rate during the pressure drop is less than under the final pressure. Pic. 4 shows the combustion rate of non-metallized N powder. One can see in the picture that the combustion attenuates, when the pressure drops to 20 atm. In comparison to the pattern for metallized N powder, the addition of Al particles increases the combustion stability and the particle size growth reduces the time of transient interval.

Conclusion

We developed the mathematical model of metallized solid propellant non-stationary combustion. The statement of the problem includes the heat transfer, the oxidizer decomposition and gasification of the solid propellant, two-phase, dual-velocity, two-temperature reactant flow over the propellant surface. The model takes into account the diffusion and the exothermal reaction in gas phase, heating and combustion of Al particles in the flow, the particle velocity lag relative to the gas. The conditions on the surface perform the conservation of energy and mass fluxes. The paper provides data of non-stationary combustion under decreasing pressure. The work investigates transient-state conditions depending on the pressure change and the size of Al particles in solid propellant.

This work was carried out with financial support from President of Russian Federation grant MK-1763.2017.8..

References

1. A.Y. Krainov, D.A. Krainov, V.A. Poryazov, J. Eng. Phys. Thermophys. **89**, 458 (2016)
2. A.Y. Krainov, V.A. Poryazov, Combust., Explosion and Shock Waves **51**, 664 (2015)

3. Spravochnik po teploobmennikam. V dvukh tomakh. **1**. M .: Energoatomizdat, 1987. [in Russian]

MATEC Web of Conferences 115, 03002 (2017) DOI: 10.1051/matecconf/201711503002
STS-33

Combustion of solid propellant with micron-sized Aluminium under the acceleration force

Vasily Poryazov [1,*], and *Dmitry* Krainov[2]

[1]Tomsk State University, Department of Physics and Engineering, 634050, Russia, Tomsk, 36 Lenina avenue
[2]Tomsk Politechnical University, Institute of Power Engineering, Department of Theoretical and Industrial Heat Systems Engineering, 634050, Russia, Tomsk, 30 Lenina avenue

Abstract. The paper presents a physical-mathematical model for metallized solid propellant combustion under acceleration directed along the normal to the burning surface. The model takes into account the thermal effect of decomposition of the condensed phase, convection, diffusion, the exothermic chemical reaction in the gas phase, heating and combustion of Al particles in the gas flow, the flow of combustion products, the particle velocity lag relative to the gas and the field effect of acceleration on the motion of Al particles. The effect of the Al particle size and mass fraction, emitted from the burning surface, on the burning rate is also taken into consideration. The impact of the Al particle size, emitted from the burning surface, on the linear burning rate has been investigated under acceleration. The study results showed that with increasing acceleration the burning rate increased. It was also revealed that the larger was the size of aluminum particles emitted from the burning surface, the higher was the response of the burning rate to acceleration. The results showed that increasing the mass fraction of aluminum in the propellant composition led to an increase in the response of the relative burning rate. It was also found that the relative burning rate sensitivity increased with pressure above the burning surface. The results obtained are in a qualitative agreement with those described in the scientific literature.

1 Mathematical model

The paper presents a physical-mathematical model for metallized solid propellant combustion under acceleration directed along the normal to the burning surface. The model takes into account the thermal effect of decomposition of the condensed phase, convection, diffusion, the exothermic chemical reaction in the gas phase, heating and combustion of Al particles in the gas flow, the flow of combustion products, and the particle velocity lag relative to the gas and acceleration forces. To model the burning out of aluminum particles in a flow of powder combustion products, we used the experimental results [1] showing that under pressures above 20 atm the rate of aluminum particle combustion in an oxidant flow do not depends on pressure, ignition of an aluminum particle occurs when it reaches a

*Corresponding author: poryazov@ftf.tsu.ru

certain temperature. Mathematical model of solid propellant combustion with micron-sized aluminum under the acceleration forces is based on [2,3].

Under the admitted assumptions, the mathematical model is well fitted by the following equations: the energy conservation equations for the gas and aluminum particles, the mass conservation equations of the gas and particles, the particle number concentration equation, the equation of the reagent burning-out in the gas phase, the equation of particle motion in a gas flow, and the gas state equation. The system of equations written in the Cartesian axes has the following form (the origin of the coordinates is on the combustion surface):

$$c_2\rho_2\left(\frac{\partial T_2}{\partial t}+u\frac{\partial T_2}{\partial x}\right)=\lambda_2\frac{\partial^2 T_2}{\partial x^2}+Y\rho_2 k_0 Q_2\exp\left(\frac{-E_2}{R_y T_2}\right)+4\pi\alpha r_k^2 n\left(T_3-T_2\right), \tag{1}$$

$$\left(\frac{\partial Y}{\partial t}+u\frac{\partial Y}{\partial x}\right)=D_2\frac{\partial^2 Y}{\partial x^2}-Yk_0\exp\left(\frac{-E_2}{R_y T_2}\right), \tag{2}$$

$$\frac{\partial \rho_2}{\partial t}+\frac{\partial(\rho_2 u)}{\partial x}=-G, \tag{3}$$

$$c_3\rho_3\left(\frac{\partial T_3}{\partial t}+w\frac{\partial T_3}{\partial x}\right)=-4\pi\alpha r_k^2 n\left(T_3-T_2\right)+GQ_{Al}\frac{2\mu_{Al}}{3\mu_O}, \tag{4}$$

$$\frac{\partial \rho_3}{\partial t}+\frac{\partial(\rho_3 w)}{\partial x}=G, \tag{5}$$

$$\frac{\partial w}{\partial t}+w\frac{\partial w}{\partial x}=-\tau_{tr}-a, \tag{6}$$

$$\frac{\partial n}{\partial t}+\frac{\partial(nw)}{\partial x}=0, \tag{7}$$

$$P=\rho_2 R T_2=const, \tag{8}$$

Here Eqs. (1) and (4) are the energy equations for the gas phase and particles, Eqs. (2), (3), and (5) are the mass conservation equations for the gas phase, gaseous reaction products, and particles, respectively, Eq. (6) is the particle motion equation, Eq. (7) is the particle number concentration equation, and Eq. (8) is the ideal gas law.

The calculations of the N powder combustion were carried out under the following values of the thermophysical and kinetic parameters from [3].

According to the Belyaev–Zel'dovich model, the linear combustion rate of the powder is defined by the formula [4] : $V_k=K_\upsilon\exp\left(-E_1/2R_y T_S\right)$, where K_υ is the empirical constant.

The boundary conditions at $x=0$ are the mass and energy conservation laws:

$$(1-\alpha_{Al})\rho_1 V_k=\rho_2 u\big|_{x=0},\ \alpha_{Al}\rho_1 V_k=\rho_3 w\big|_{x=0},\ \rho_1 V_k Y=D_2\rho_2\,\partial Y/\partial x\big|_{x=0}+\rho_2 uY\big|_{x=0},$$
$$\lambda_2\,\partial T/\partial x\big|_{x=0}=\rho_1 V_k\left(c_2 T_2\big|_{x=0}-Q_1-c_1 T_{1,0}\right),\ T_3\big|_{x=0}=T_2\big|_{x=0}, \tag{10}$$
$$\rho_2\big|_{x=0}=P\Big/RT_2\big|_{x=0},\ n\big|_{x=0}=\rho_3\big|_{x=0}\Big/(4/3)\pi r_{Al,0}^3\rho_k,\ P=const.$$

The following boundary conditions are specified on the boundary $x=\infty$:

$$\partial T/\partial x\big|_{x=\infty}=0,\ \partial Y/\partial x\big|_{x=\infty}=0. \tag{11}$$

Here $r_{Al,0}$ is the radius of an aluminum particle coming out from the N powder combustion surface into the gas flow and $T_{1,0}$ is the initial temperature of the powder, $T_2\big|_{x=0}\equiv T_S$.

The initial conditions:

$$T_2(x,0)=T_{ign},\ T_3(x,0)=T_{ign},\ Y(x,0)=0,$$

$$\rho_2(x,0)=P\Big/RT_2(x,0),\ \rho_3(x,0)=0,\ w(x,0)=0,\ n(x,0)=0. \qquad (12)$$

The particle interaction force with the gas is defined by the formula:

$$\tau_{tr}=F_{tr}\Big/\tfrac{4}{3}\pi r_k^3\rho_k,\ F_{tr}=C_R S_m\rho_2(w-u)|u-w|\Big/2, \qquad (13)$$

where S_m is the midsection area; F_{tr} is the drag force of a sphere; and C_R is the drag coefficient.

According to the empirical formula [5], the drag coefficient is:

$$C_R=24\left(1+0,15\,\mathrm{Re}^{0,682}\right)\Big/\mathrm{Re},\ \mathrm{Re}=2r_k\rho_2|u-w|\Big/\eta,$$

where η is the dynamic viscosity coefficient.

The heat transfer coefficient is equal [6]:

$$\alpha=\frac{Nu\,\lambda_2}{2r_k},\ Nu=2+\sqrt{Nu_l^{\,2}+Nu_t^{\,2}}, \qquad (14)$$

where $Nu_l=0,664\,\mathrm{Re}^{0,5}$, $Nu_t=0,037\,\mathrm{Re}^{0,8}$, Nu - Nusselt number.

The mass change rate of a particle during the combustion is determined as in [3]:

$$G=\frac{3\mu_O}{2\mu_{Al}}n\rho_k 4\pi r_{Al}^{1,5}k_{Al}a_{ox}^{0,9},\ k_{Al}=2.22\cdot10^{-5}\ м^{1.5}/c, \qquad (15)$$

The radius of unburnt aluminum particle r_{Al} is:

$$r_{Al}=\left[\left((1+{}^{3\mu_O}\!/_{2\mu_{Al}})r_{Al,0}^3-{}^{\rho_3}\!/_{(4/3)\pi n\rho_k}\right)^{2\mu_{Al}}\!/_{3\mu_O}\right]^{1/3}. \qquad (16)$$

We investigated the influence of acceleration forces on the combustion rate of N powder with micron-size aluminum particles. The acceleration, particle radius and pressure in the calculations ranged from -50*g* to 50*g*,from 1 to 30 *μm*, and from 20 to 100 atm, respectively. The aluminum mass concentration was equal to 9%. Pic. 1 shows the calculations result of combustion rate depending on acceleration forces directed along the normal to the burning surface.

Research of dependance of non-dimensional quantity U on aceleration value *a*, where $U=V_k/V_{k,0}$, V_k is the combustion rate at specified value of accelaration, and $V_{k,o}$ is the combustion rate without accelaration, calculated under each value of pressure and different aluminum dispersity. Actually, the resalt provides relative velocity gain with rising acceleration for each composition under set pressure near the combustion surface.

This dependaces are of some practical interest.

As Pic.1 shows, the more acceleration rises, the more combustion rate changes. The increase of Al particle size leads to the rising of relative velocity gain at the same acceleration value.

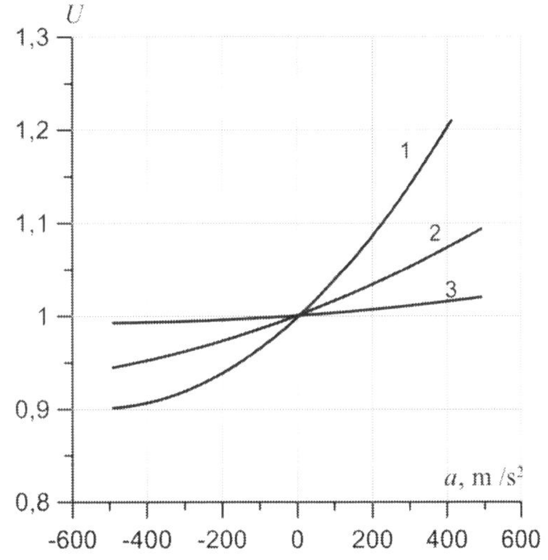

Fig. 1. The N powder combustion rate – acceleration curves for different Al particle size $r_{Al,0}$. $P = 40$ atm . 1 - $r_{Al,0}$ =15 μm, 2 - $r_{Al,0}$ =10 μm, 3 - $r_{Al,0}$ =5 μm

The combustion rate at certain aceleration value grows more significantly at high pressure. Thus we observe the increase of relative velocity gain near the combustion surface.

The growth of Al mass concentration up to 20% of propellant resulting in the rising of the combustion rate at the same acceleration values.

The reaserch showed that the combustion rate of solid propellant under acceleration forces directed along the normal to the burning surface increases due to the presence of Al particles which stay close to the combustion surface during the acceleration. The Al particle burning increases the gas temperature along the powder surface, the conductive heat flux grows, the temperature of powder surface rises, and therefore linear combustion rate of the propellant goes up. The obtained results are in a qualitative agreement with those described in the scientific literature.

This work was carried out with financial support from President of Russian Federation grant MK-1763.2017.8.

References

1. A.F. Belyaev, Yu.V. Frolov, A.I. Korotkov, Combust., Explosion and Shock Waves **4**, 182 (1968)
2. Ya.B. Zel'dovich, J. Exp. Theoret. Phys. **12**, 498 (1942) [in Russian]
3. V.A. Poryazov, A.Yu. Krainov, D.A. Krainov, J. Eng. Phys. Thermophys. **88**, 94 (2015)
4. A.Yu. Krainov, V.A. Poryazov, Combust., Explosion and Shock Waves **52**, 177 (2016).
5. Spravochnik po teploobmennikam. V dvukh tomakh. V.1. M .: Energoatomizdat, 1987. [in Russian]
6. A.F. Belyayev, Goreniye, detonatsiya i rabota vzryva kondensirovannykh system, Moscow, Nauka, 1968 [in Russian]

Studying solid fuel ignition by CO_2-laser

Alexander Korotkikh[1*], *Konstantin V.* Slyusarskiy[1], *Ivan V.* Sorokin[1]

[1]Tomsk Polytechnic University, Institute of Power Engineering, 634050, Tomsk, Russia

Abstract. The ignition data on Siberian bituminous coal and lignite by a continuous CO_2-laser were presented in current work. Measurements were carried out at the heat flux density range of 90–150 W/cm^2 in air. Dependences of ignition delay time on heat flux density were obtained along with the surface temperatures at the moment of fuel sample ignition. Ignition delay time was found to be in the range 50-520 ms for lignite. For bituminous coal ignition the delay time values are larger 1.5-2 times. However, fuel surface temperature in the moment of ignition is the same for both fuels, i.e. 640-680 °C.

1 Introduction

Solid fuel is one of the most common among fossils which are used to produce energy worldwide [1]. However, coal-fired plants are the main source of pollution and greenhouse gas, like carbon, sulfur and nitrogen oxide, as well as solid particles (ash, slag) into environment [1]. Decreasing of the temperature in intense chemical reaction area during the solid fuel combustion is one of the methods for enhancing combustion efficiency and pollution mitigation (nitrogen oxide, in particular [1]). Lower temperature in fuel combustion area causes chemical reaction rate changing and, as a result, alters energy equipment characteristics. That's why studies on dependences and characteristics of gasification, ignition and combustion of widespread in Siberia coals are up-to-date.

Many studies on ignition of single coal particles [2] and coal dust in oxidizer flow [3] were presented in scientific literature recently. Majority are dedicated to conductive [4] or convective [5] heating. This creates the need for using criterial equations and heat-transfer coefficient to obtain heat flux. But in existing coal-fired boilers the main heat transfer mechanism in furnace is radiation [3]. There are significantly less articles on solid fuel ignition and combustion under conditions of radiant heating. In articles [5, 6] the data on ignition of coal powders with different disperse composition under conditions of radiant heat flux. In articles [7, 8] the results on radiant ignition of single coal particle are presented. In all mentioned above publications the impulse lasers with low exposure times, while in real conditions the emission is continuous.

The current study presents the experimental data on ignition by continuous radiant heat flux tests in air medium for two samples of energy coals with grade T and 2B

* Corresponding author: korotkikh@tpu.ru

2 Experimental section

2.1 Fuel samples

The powders of T-grade bituminous coal and 2B-grade lignite of Kuznetsk and Cansk-Achinsk deposits were studied. Initial fuel was ground in a ball mill and sieved through a sieve with 80 μm mesh size. Dispersed characteristics of investigated fuel samples were defined using laser-analyzer HELOS. The average diameters and characteristic sizes of coal powders are given in Table 1.

Table 1. Average diameter and characteristic size for studied fuel samples.

Sample	Average diameter, μm					Characteristic size, μm		
	d_{10}	d_{20}	d_{30}	d_{32}	d_{43}	x_{10}	x_{50}	x_{90}
Coal T	36.9±0.2	45.4±0.2	52.1±0.2	69.2±0.1	79.2±0.1	6.45±0.1	31.6±0.1	75.3±0.2
Coal 2B	38.9±0.1	47.3±0.1	53.6±0.1	69.7±0.1	79.1±0.2	6.80±0.2	34.9±0.2	77.5±0.1

The proximate and elemental analyses of two investigated samples are carried out according to ISO 17246:2010 methodic and using Euro EA 3000 analyzer. Analysis results are given in Table 2.

Table 2. Proximate and ultimate analysis data for studied fuel samples.

Sample	Proximate analysis, wt. %				Ultimate analysis, wt. %				
	M	A	V	FC	C	H	N	S	O
Coal T	3	21	20	66	68	4	3	<1	4
Coal 2B	7	16	32	45	50	4	2	<1	28

Here M, A, V and FC are the moisture, ash, volatile matter and fixed carbon content, respectively.

T-grade coal has higher fixed carbon and ash content while moisture and volatile matter content are higher for 2B-grade coal. Tested samples were pressed into tablets with 10 mm diameter under pressure 310 MPa. The density of tablets of T-grade coal was 1200 kg/m^3 and 1100 kg/m^3 for 2B-grade coal.

2.2 Fuel ignition

Study on solid fuel sample ignition was carried out using experimental setup for radiant heating based on continuous CO_2-laser with 10.6 μm wavelength and 200 W power. The front end of samples was visually controlled for pore, crack and cavity absence. Scheme of experimental setup is given in Fig.1.

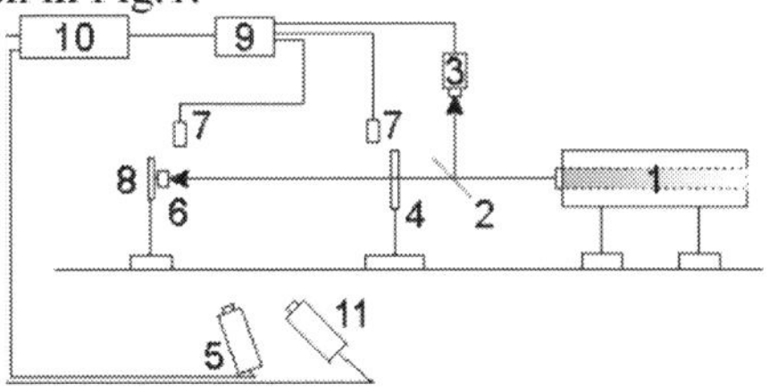

Fig. 1. Scheme of experimental setup based on CO_2-laser: 1 – CO_2-laser; 2 – semi-transparent; 3 – thermoelectric sensor of radiation power; 4 – electromagnetic shutter; 5 – video camera; 6 – fuel sample; 7 – photodiode; 8 – holder; 9 – ADC; 10 – PC; 11 – thermal imaging camera.

Studied fuel sample (6) was attached to sample holder (8). After opening of shutter (4) the laser emission was directed to studied fuel sample (6). Signals from photodiode (7) and emission power measurer (3) were directed to personal computer (10) through ADC of signal L-card E-14-440 (9). Later they were processed via LGraph2 software. The fuel ignition delay time t_{ign} was defined as a difference between signals from photodiodes (7): one registered the appearance of flame on the surface of sample, another one registered the opening of the shutter.

The surface temperature of samples during heating, gasification, ignition and combustion of fuel was registered with thermal imager Jade J 530 SB with frequency 50 Hz in the temperature range of 260–1800 °C. Thermo and video-imaging data as well as photodiode signal were synchronized by the signal of shutter opening. The radiant heat flux density was measured using Ophir FL400A thermoelectric sensor.

3 Results and discussion

3.1 Sample ignition delay

The dependencies of ignition delay time on radiant heat flux density for fuel samples are given in Fig. 2.

The presented data reveal that the ignition delay time for T-grade bituminous coal sample is 1.5–2.0 times higher than for 2B-grade lignite sample. Increasing of the heat flux density from 90 W/cm^2 to 150 W/cm^2 decreases difference between the ignition times from 400 to 50 ms for both studied samples. It may be connected either with lower activation energy of the gasification process for lignite and to lower conductivity coefficient [9].

Experimental data was approximated using equation:

$$t_{ign} = A \cdot q^{-B}$$

where t_{ign} is the solid fuel ignition delay time, ms; q is the heat flux density, W/cm^2; A, B are approximation constants.

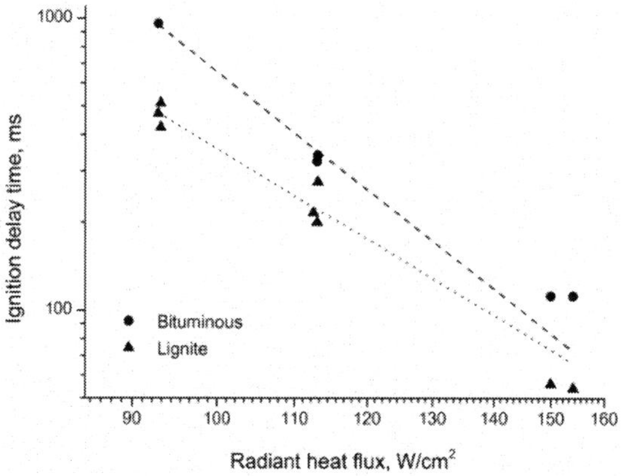

Fig. 2. The ignition delay time of solid fuel samples vs. radiant heat flux density.

Approximation constant values as well as determination coefficient R^2 are given in Table 3.

Table 3. Approximation constants and determination coefficient values.

Sample	A	B	R^2
Coal T	$9.61 \cdot 10^{12}$	5.8	0.99
Coal 2B	$2.20 \cdot 10^{10}$	3.9	0.96

3.2 Sample surface temperature

The temperature distribution on the sample surface at the moment of flame appearance is presented in Fig. 3. The maximal and averaged surface temperatures of tested samples are presented in Table 4.

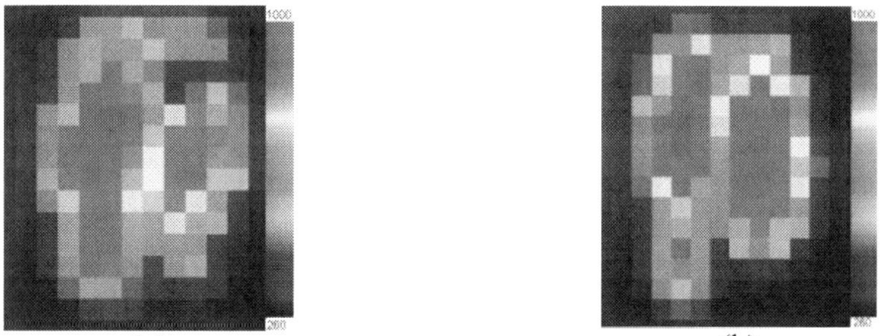

(a) (b)

Fig. 3. Thermal imager data in the ignition moment at $q = 113$ W/cm^2 for fuel samples: (a) – T-grade coal; (b) – 2B-grade coal.

Table 4. Fuel sample surface temperatures at the ignition moment.

Sample	Coal T	Coal 2B
Maximal temperature, °C	1038	1066
Average temperature, °C	644	677

The heating and ignition processes for tested solid fuels could be divided into two stages: intense heating with intense gasification and appearance of flame with transition into stable combustion. The obtained temperatures of sample surface in the moment of ignition of T-grade and 2B-grade coals are close to each other.

4 Conclusion

The ignition process of T-grade bituminous coal and 2B-grade lignite samples was studied using an experimental setup based on CO_2-laser in air at normal conditions. The ignition delay time values for fuel samples were obtained in the radiant heat flux range of 90–150 W/cm^2 and were in the range of 50–950 ms. It was found that the ignition delay time for T-grade bituminous coal sample is 1.5–2.0 times higher than for 2B-grade lignite sample. The average surface temperature for fuel samples is equal to 640–680 °C during initiation by radiant heat flux $q = 113$ W/cm^2.

The reported study was supported by the Russian Scientific Foundation, Grant 16-19-10316.

References

1. Inventory of U.S. Greenhouse Gas Emissions and Sinks: 1990-2014. April 15, 2016
2. A.C. Sarroza, T.D. Bennet, C. Eastwick, H. Liu, Fuel Proc. Tech. **157**, 1 (2017)
3. W. Cao, Q. Qin, W. Cao, Y. Lan, T. Chen, S. Xu, X. Cao, Powder Tech. **310**, 17 (2017)
4. A.G. Korotkikh, K.V. Slyusarskiy, A.A. Ditts, Russ. J. Phys. Chem. B. **10**, 576 (2016)
5. J.C. Chen, M. Taniguchi, K. Ito Fuel. **75**, 323 (1995)
6. T.H. Dubaniewicz Jr., K.L. Cashdollar, G.M.J. Laser Appl. **15**, 184 (2003)
7. M. Qu, M. Ishigaki, M. Tokuda, Fuel **75**, 1155 (1996)
8. B.A. Wong, G.R. Gavalas, R.C. Flagan, Energy Fuels **9**, 484 (1994)
9. V.G. Lisenco Fuel. *Rational combustion, control and technological application* (Teplotehnik, Moscow, 2003) [in Russian]

Physical and mathematical modeling of ignition of methane-air mixture in the presence of coal microparticles

Aleksandr Fedorov and *Dmitry* Tropin*

Khristianovich Institute of Theoretical and Applied Mechanics SB RAS, 630090, 4/1 Institutskaya Str., Novosibirsk, Russia

Abstract. The results of numerical investigation of the ignition of a stoichiometric methane-air mixture in the presence of carbon particles with diameters of 20-52 μm in the temperature range 950-1150 K and pressures of 1.5-2.0 MPa are presented. The calculated data of the ignition delay times of coal particles in the coal particles/air mixture and of the ignition delay times of methane and coal particles in the methane/coal particles /air mixture are compared with the experimental ones. A satisfactory agreement of the data on the coal particles ignition delay times and methane ignition delay times in all the mixtures considered is shown.

1 Introduction

Investigation of the ignition and combustion of disperse systems consisting of a gas suspension of reactive particles in a mixture of gaseous fuels and oxidizer is relevant from the point of view of explosion and fire safety, for example in the coal mining industry. A vast experimental material on the ignition characteristics of single relatively large (with a radius of more than 20 μm) coal particles under conditions of weak convection of the oxidizing gas in the reaction chamber is contained in the monograph [1]. A lot of works are aimed at forecasting the conditions of ignition and flame propagation. The main attention was paid to the formation of a cloud of solid particles and initiation of its combustion due to local ignition and propagation of the combustion wave in methane. As shown in [2, 3], hybrid mixtures are more easily flammable than each of the components individually. A lot of work has been devoted to a theoretical study of the problem of ignition of coal suspensions (see [4-7] and references therein). The studies were carried out within the framework of point models of ignition of coal particles, the provisions of the Semenov's thermal explosion theory were widely used to formulate the critical ignition / extinction conditions.

Investigation of this phenomenon is also interesting from the point of view of controlling the process of ignition of the gas mixture, which can become the scientific basis for creating new combustion regimes for traditional fuels.

* Corresponding author: d.a.tropin@itam.nsc.ru

2 Physical and mathematical model

Let us consider a space filled with the gas suspensions of coal microparticles with the diameters from 26 to 52 microns in air or in methane-air mixture, with high initial state parameters (temperature ranging from 850 to 1150 K and pressure ranging from 1.3 to 2.0 MPa). We assume, as in [5, 6], that the processes of thermal destruction of coal particles occurs simultaneously with the release of volatiles into the gaseous phase, the ignition and combustion of volatiles in the gaseous phase, the heterogeneous reaction of oxidation of solid carbon (coke), the ignition and combustion of methane/air mixture. The mass rates of the pyrolysis reaction, oxidation reaction of solid carbon will be described in the form of diffusion-kinetic dependences presented in [5, 7]. Kinetic constants for these dependences are presented in [5, 8]. To describe the kinetics of ignition and combustion of gaseous volatile and methane/air mixtures we use the simplified detailed kinetic scheme [9]. Earlier, with the help of this kinetics model, we described the processes of propagation, attenuation and suppression of detonation in methane/oxygen and methane/hydrogen/oxygen mixtures [10], as well as the experimentally observed ignition delay times of gas suspension of iron particles in methane/air mixture [11].

2 Calculation results

The ignition of coal/air and methane/coal/air mixtures was calculated using the mathematical model [5-11]. As the particle ignition delay time we consider the time needed to reach the maximum particles temperature rise in time – $\max\left(\frac{dT_2}{dt}\right)$ follow [11-12].

2.1 Coal particles/air mixture

Fig. 1 shows the experimental [13] and calculated dependences of the ignition delay time of coal particles in air on the gas temperature at the end of the compression tact (the pressure at the end of the compression tact varies from 1.5 to 1.77 MPa). It is seen from fig. 1 that the increasing of the gas temperature results in decreasing the ignition delay time of the coal particles. In addition, in the whole pressure range ($p \in (1.5, 1.77)$ MPa) and temperatures ($T \in (897, 1123)$ K) of the mixture under consideration, we have a good agreement between the experimental and calculated values of the ignition delay of coal particles.

2.2 Methane/coal particles /air mixture

Fig. 2 shows the experimental [13] and calculated dependences of the coal particles ignition delay times and the methane ignition delay times in the methane/coal particles /air mixture on the temperature of the gas mixture at the end of the compression tact. It can be seen that there is a satisfactory agreement between the experimental [13] and calculated ignition delay times in all considered pressure ranges ($p \in (1.6, 1.82)$ MPa) and temperatures ($T \in (989, 1052)$ K) of the mixture.

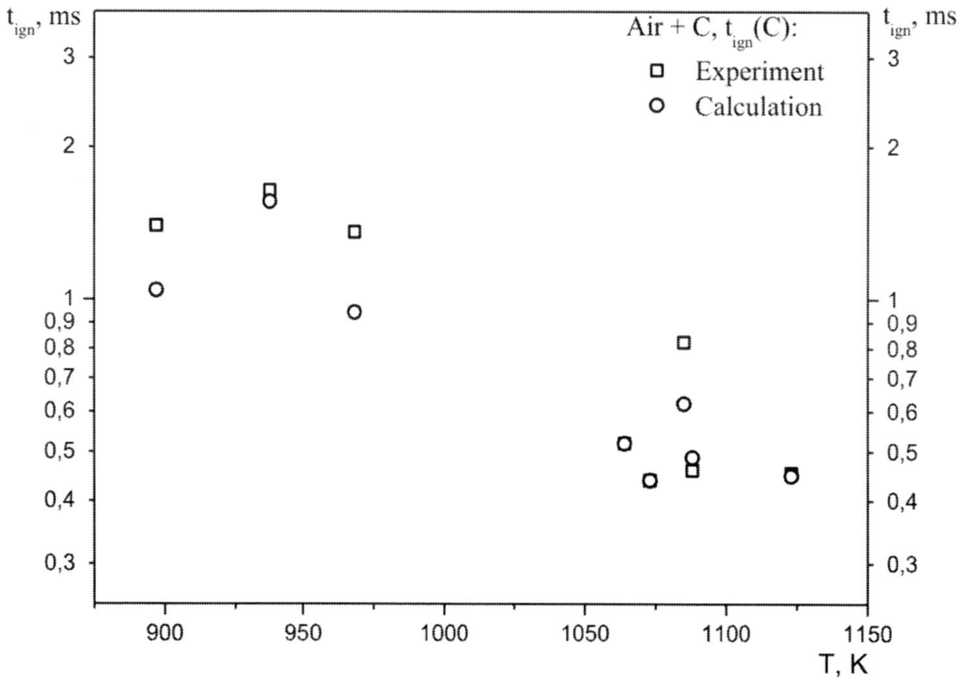

Fig. 1. Dependences of ignition delay time of coal in coal/air mixture on mixture temperature. Comparison the calculated data with experimental one [13].

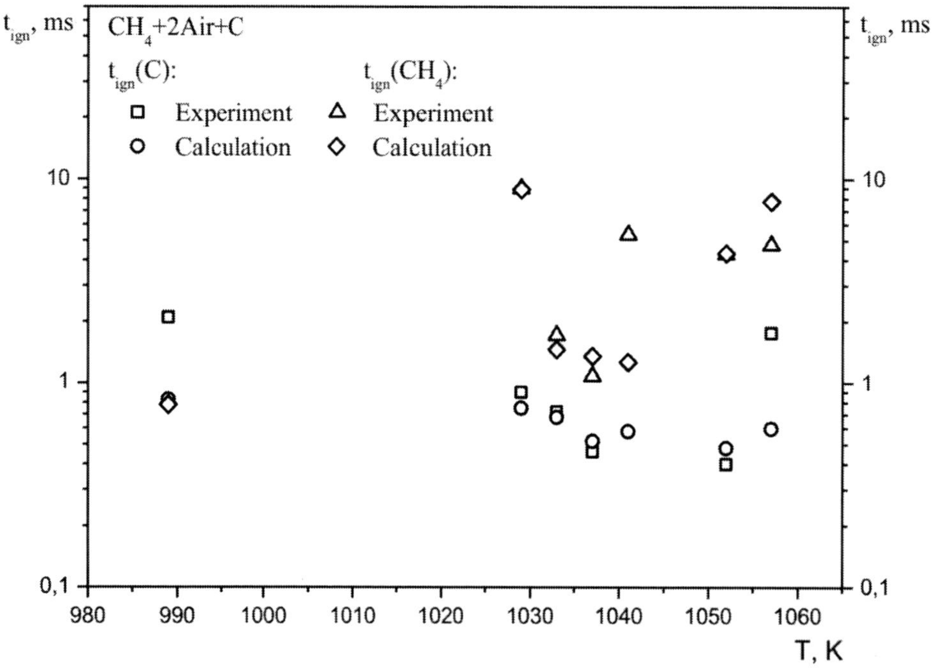

Fig. 2. Dependences of ignition delay times of methane and coal particles in methane/coal particles /air mixture on mixture temperature. Comparison of the calculated data with experimental one [13].

3 Conclusions

Physical and mathematical model of ignition and combustion of methane/air mixture in the presence of coal microparticles, which takes into account the detailed kinetics of oxidation of methane/hydrogen/air mixture and the thermal destruction of coal particles with the volatile (methane and hydrogen) release into the gas phase, ignition and combustion of volatile in the gas phase and the heterogeneous carbon oxidation reaction is proposed.

The calculated data of the ignition delay times of coal particles in the coal particles /air mixture and of the ignition delay times of methane and coal particles in the methane/coal particles/air mixture are compared with the experimental ones [13]. A satisfactory agreement of the data on the coal particles ignition delay times and methane ignition delay times in all considered mixtures is shown.

The work was supported by the Russian Foundation for Basic Research (grant No. 16-58-00052-Bel_a).

References

1. V.I. Babiy, Yu.F. Kuvaev, *Combustion of coal dust and calculation of a pulverized coal flare* (Energoatomizdat, Moscow, 1986). [in Russian]
2. J. Nagy, D.W. Mitchell, *Experimental coal-dust and gas explosions*, Washington, U.S. Department of the Interior, Bureau of Mines, 1963)
3. P.R. Amyotte, K.J. Mintz, M.J. Pegg, Y.H. Sun, Fuel **72**, 671 (1993)
4. A.V. Fedorov, Combust. Explos. Shock Waves **34**, 418 (1998)
5. A.V. Fedorov, V.M. Fomin, Yu.A. Gosteev, *Dynamics and ignition of gas suspensions* (NSTU, Novosibirsk, 2006) [in Russian]
6. Yu.A. Gosteev, A.V. Fedorov, Combust. Explos. Shock Waves **37**, 646 (2001)
7. Yu.V. Kazakov, A.V. Fedorov, V.M. Fomin, Archivum Combustionis. **7**, 353 (1987)
8. A.V. Fedorov, T.A. Khmel, Combust. Explos. Shock Waves **38**, 700 (2002)
9. C.K. Westbrook, P. Urtiew, Combust. Explos. Shock Waves **19**, 65 (1983) [in Russian]
10. D.A. Tropin, A.V. Fedorov, Combust. Explos. Shock Waves **50**, 542 (2014)
11. D.A. Tropin, A.V. Fedorov, O.G. Penyazkov, V.V. Leschevich, Combust. Explos. Shock Waves **50**, 632 (2014)
12. A.V. Fedorov, D.A. Tropin, Combust. Explos. Shock Waves **44**, 552 (2008)
13. V.V. Leschevich, O.G. Penyazkov, S.Yu. Shimchenko, Combust. Explos. **9**, 3 (2016)

Numerical and experimental study of supersonic combustion chamber at entrance Mach number 4. Part 1: Non-reacting flow

Natalya Fedorova[1,*] and *Marat* Goldfeld[1]

[1]ITAM SD RAS, 4/1 Institutskaya Str., Novosibirsk, 630090, Russia

Abstract. Results of joint experimental and numerical study are presented for a non-reacting 3D flow in a channel with backward-facing steps (BFS) located on top and bottom walls. Experiments were conducted in a hot-short wind tunnel at entrance Mach number 4, total pressure 9.5 bar and total temperature 2120K. Argon was injected into the channel from eight round holes, placed on channel walls upstream of the channel expansion. Simulations performed in a 3D RANS approach with k-ω SST turbulence model give a reasonable agreement with experimental data and allow evaluating the pressure flowfield and argon mass concentration in the channel.

1 Introduction

The organization of ignition of fuels at a supersonic flow velocity at the inlet to the combustion chamber channel seems to be one of the most difficult and important problems in the scramjet technology development. From technological considerations, the ignition and stabilization of the flame is the first and foremost problem for the scramjet application since ensuring the stabilization of the flame with minimum losses of the total pressure and a high level of heat release are the key to obtaining high engine characteristics [1]. The process of ignition in a supersonic stream depends on many factors, such as the mixture residence time, the local excess fuel ratio, the configuration of the flame stabilizer, the ignition energies, and the physical and kinetic properties of the fuels [2, 3].

Despite the fact that ignition and flame stabilization at supersonic flow rates are given great attention, however, the physical and chemical mechanisms, the characteristics of ignition, the processes of flame propagation and heat release in the channel have not been sufficiently studied to reliably predict the processes in the scramjet combustion chamber, especially at high flow rates at the channel inlet.

A decisive role in solving this problem is played by experimental methods. Together with this, the joint use of numerical modeling and experimental methods allows significantly expanding the possibilities of studies of complex supersonic flows in a channel with a mass supply and chemical reactions. In present paper, results of non-reactive

* Corresponding author: nfed@itam.nsc.ru

flows investigations are presented aimed at experimental and computational study of argon and air mixing in a channel at M=4.

2 Experimental setup and numerical procedure

The model of the combustion chamber is a flat channel of 100 mm wide consisting of an insulator of 200 mm long, injector and expanding sections, the outline of which is shown in Fig. 1. As a flame stabilizer, the backward facing steps (BFS) of height $h = 25$ mm, located on the upper and lower walls of the channel were used. The experiments were conducted in a hot-short wind tunnel IT-302 (ITAM SD RAS) with a working time of 80 ÷ 150 ms. The incoming flow parameters at the entrance of the injector section were as follows: Mach number M = 4, the total temperature T0 = 2120K, and the total pressure P0 = 0.95 MPa. As the injected gas, at this stage of the investigation, argon was selected, which was fed in the direction of the core of the flow at an angle of 45 degrees from 8 round holes centered at a distance of 8 mm from the edge of the BFS. The distance from the line of symmetry to the center of the injection hole is 12.5 mm, the distance between the holes is 20 mm, the jet-to-freestream momentum ratio $J = (\rho U^2)_j/(\rho U^2)_0$ varies from 2.35 to 8. In the experiments, the distributions of static pressure and heat fluxes on the model walls, the Pitot pressure at the inlet and outlet of the combustion chamber, and the flow rate of air and injected gas according to the outflow conditions from the closed volume were measured. High-speed shadow visualization was used to determine the flow pattern.

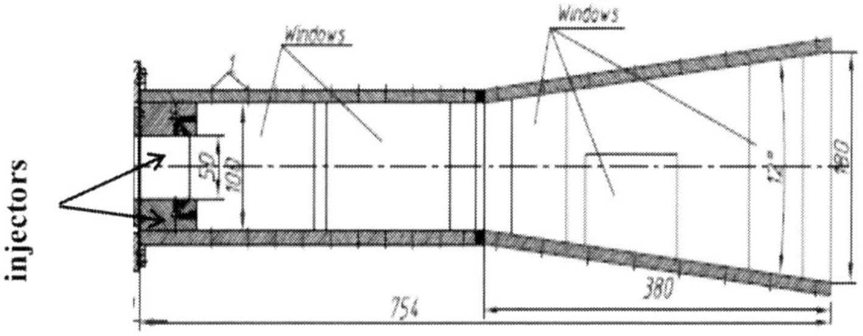

Fig. 1. Outline of injector and expanding sections of experimental model.

Mathematical simulation of the flow was performed under the same inflow conditions based on the Reynolds averaged Navier-Stokes equations supplemented by the k-ω SST turbulence model using the commercial software ANSYS Fluent 17.0. The results of calculations with the same approach performed earlier to model 2D turbulent supersonic flows in channels, including various gas injections, have shown good agreement with the experimental data [4, 5]. The present calculations were carried out in 3D taking into account the symmetry of the channel in the vertical and transverse directions. The calculation domain had the same dimensions as the experimental model, and was limited by inlet and outlet sections from the left and right sides, by symmetry planes from the top and front sides and by bottom and back walls. In the computational domain, a structured hexa-mesh condensed toward the walls was constructed. The final grid contained about 4.5 million cells. Non-dimensional coordinate y^+ did not exceed 10 at the first cell layer. At the inlet sections, the pressure-inlet type of boundary conditions was applied with the profiles of total and static pressure, temperature and turbulent parameters obtained in the preliminary computation for nozzle and isolator part of the channel. On the walls, the cold wall temperature conditions Tw=300K were used.

 DOI: 10.1051/matecconf/201711503005

3 Results

As a result of the studies, the structure of the non-reactive flow in the channel was determined, the distribution of pressure and heat fluxes on the model walls, flow parameters as well as argon concentration in the channel were obtained.

Figure 2 shows the static pressure fields in the channel in the symmetry plane z=0 for the case without blowing (above) and blowing the argon jet (below). The top picture demonstrates the structure typical for supersonic flows behind BFS. At the inlet to the injector part, the static pressure is $P_{inf} = 0.85$ bar. After the steps located on the upper and lower walls of the channel, rarefaction waves are formed, in which the pressure is reduced to the level of 0.15 P_{inf}. The interaction of rarefaction waves in the central part of the channel reduces the static pressure to the level of 0.06 P_inf. Behind the steps, the separation zones are formed with a length of about $2h$, closing by tail shocks. The shocks meet on the top symmetry line at $x \approx 0.3$m, which leads to a significant pressure increase up to $1.5 \cdot P_{inf}$. Due to reflection from the walls and interference, a symmetrical system of shock waves and rarefaction waves is formed in the channel. The intensity of the shock waves decreases downstream, which is due to the expansion of the channel and the interaction of shock waves with rarefaction waves. For the case with injection, due to presence of the jets pressure maximum increases significantly up to $4 \cdot P_{inf}$, and the whole wave structure moves upstream. Local static pressure maximum can be observed on the symmetry line in the region of intersection of the shocks generated before the injection holes. By the end of the channel, the pressure level is approximately the same in both cases.

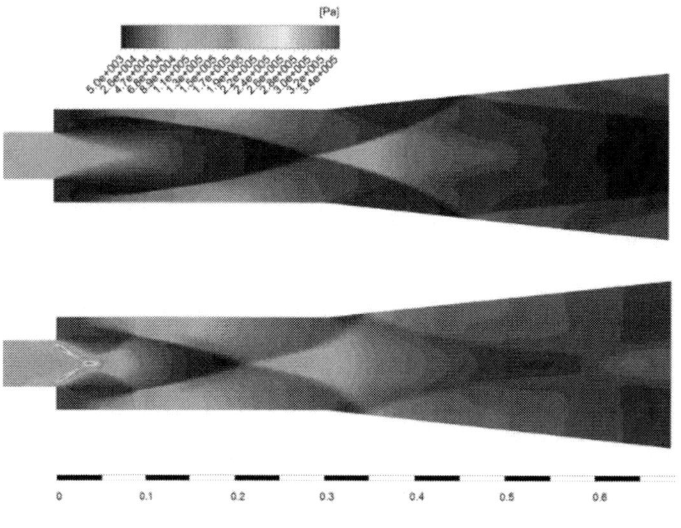

Fig. 2. Static pressure fields in z = 0 plane for the case without blowing (above) and with blowing the argon jet (below).

In Figure 3, computed and experimental static pressure distributions along the wall at the symmetry plane are compared for the cases without and with argon injection. Despite the fact that injection orifices are shifted from the symmetry plane, a sharp pressure rise before the BFS is observed which is caused by the jet-induced shock. It should be noticed that presence of the jet increases the pressure twice in the separation zone and additional pressure maximum exists at x=0.14. Figure shows a satisfactory agreement of numerical and experimental results, nevertheless the pressure maximum for the second case is slightly underpredicted in the computations.

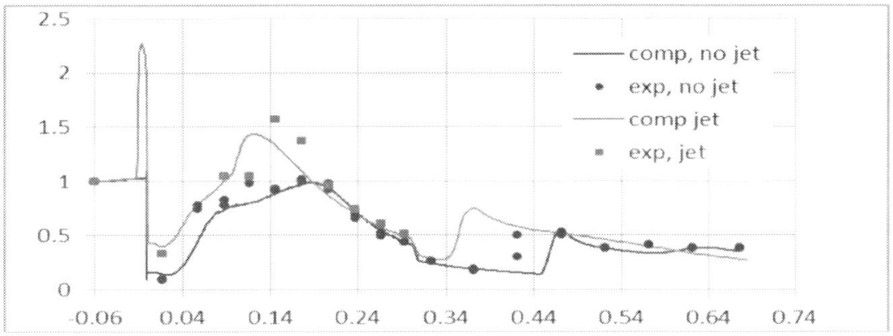

Fig. 3. Experimental and computed static pressure distributions along the wall.

Figure 4 presents the computed argon mass fraction on the bottom wall and in several cross-sections along the channel. The jets do not reach the top symmetry plane, and even at the exit sections, a middle part of the channel does not contain substantial amount of the injected gas. Simultaneously, a rather high argon concentration is observed in the separated region after the BFS. The significant 3D effect in the shape of contours and mean values of argon concentration far downstream from the injection could be seen due to the back wall presence.

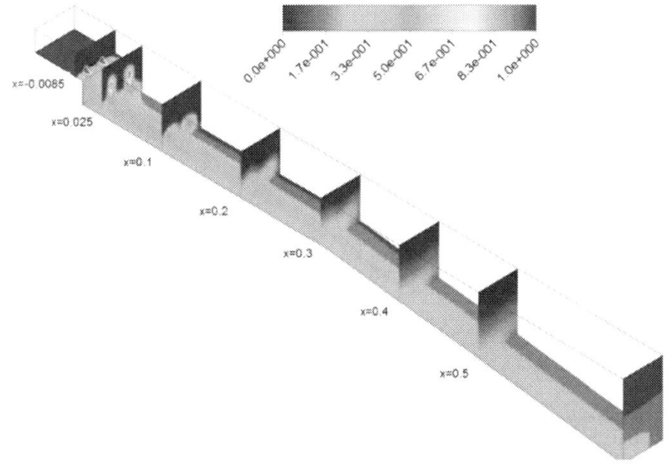

Fig. 4. Computed argon mass concentration at various *x*-cross-sections.

The work was supported by the Russian Foundation for Basic Research grant 17-08-01158 A.

References

1. E.T. Curran, J. Propuls. Power **17**, 6 (2001)
2. C.C. Rasmussen, J.F. Driscoll, K.Y. Hsu, J.M. Donbar, M.R. Gruber, C.D. Carter, J. Propul. Power **21** (2005)
3. D.J. Micka, J.F. Driscoll, Combust. Flame **159**, 1205 (2012)
4. M.A. Goldfeld, Yu.V. Zakharova, N.N. Fedorova, Thermophys. Aeromech. **19**, 4 (2012)
5. N.N. Fedorova, I.A. Fedorchenko, A.V. Fedorov J. Appl. Mech. Tech. Phys. **54**, 2 (2013)

Flow structure diagnostics in a four-vortex furnace model using PIV-method

Evgeniy Shadrin[1,*], *Anatoliy* Papulov[1,2]

[1] Institute of Thermophysics SB RAS, Novosibirsk, Russia
[2] Novosibirsk State University, Novosibirsk, Russia

Abstract. In this paper the flow structure in a four-vortex furnace of a pulverized coal boiler has been studied. The results of an experimental study of inner aerodynamics carried out on a laboratory isothermal model of a combustion device are presented. Using the particle image velocimetry method, flow velocity distributions have been obtained in a number of sections. A spatial flow structure consisting of four stable conjugate vortices with vertical flow rotation axes has been visualized.

The perspectives of coal energy development necessitate the creation of new ways of efficient and environmentally safe use of solid fuels. The limited reserves of high-quality solid fuels lead to the need to involve low-grade ("non-project") coals in the fuel and energy balance. The solution of this problem requires a significant increase in the characteristics of steam boilers of TPPs and the development of new types of combustion devices that satisfy modern standards of energy efficiency and environmental safety. One of the promising technologies in this field is fuel combustion in the vortex flow. The flow swirling provides for a more complete fuel burnup due to better mixing and longer residence time of fuel particles in the combustion chamber.

In this paper, the flow structure in a perspective combustion device [1] with a four-vortex solid fuel combustion scheme is experimentally studied. Four vortical structures with vertical axes and opposite directions of rotation are formed there. The furnace is designed for the use at thermal power plants for burning brown slagging coal. The furnace is equipped with a shielded rectangular combustion chamber. Two diagonally directed blocks of multilevel burners are mounted on the sidewalls; they are close to the axis of symmetry of the sidewalls so that the ratio of distance between the burners to the burner height is 0.8–1.2. In the center of the front and rear walls, there are nozzles of secondary and tertiary inflow, made in the form of vertical near-wall slots, whose height is equal to the height of the burner unit and oriented relative to each other in opposite directions along the walls, where they are located. Primary fuel combustion occurs in the direct-flow part of the side burner flame. Due to optimal positioning of the side burners, efficient interaction of burner jets mixing and ignition is achieved via intensive flue gas suction in the inter-burner space. The colliding straight-flow parts of the flame, directed from the sidewalls towards each other, turbulize the flow, and due to the nozzles of secondary and tertiary inflow, form four vortex flows with vertical rotation axes (Fig. 1-a).

* Corresponding author: evgen_zavita@mail.ru

The four-vortex scheme of pulverized coal combustion [1] was developed for boiler BKZ-320-140, station No. 18 at Krasnoyarsk CHP-1 with its transfer from liquid to dry slag removal. It was also implemented at reconstructing furnaces of boilers BKZ-640-140-TP at the Gusinoozerskaya TPP (Russia). According to the practice, burning the high-ash coals in the furnaces of BKZ boilers leads to intense slagging of the furnace and steam superheaters, whereby the boiler load is reduced. Evaluating the effectiveness of boiler reconstruction has identified a number of shortcomings in boiler operation, requiring its further modernization. To improve the technical, economic and environmental performance, it is necessary to optimize operating and design parameters of the four-vortex furnace of the pulverized-coal boiler on the basis of a detailed study of its aerodynamics on laboratory models.

To study the flow structure in the four-vortex furnace model, an experimental stand was developed and debugged. The scheme of this stand is shown in Fig. 1-b. The main elements of this stand are [2]: compressed air supply system with control and regulating devices; four-vortex furnace model; system for seeding a stream by tracer-particles; and measuring system.

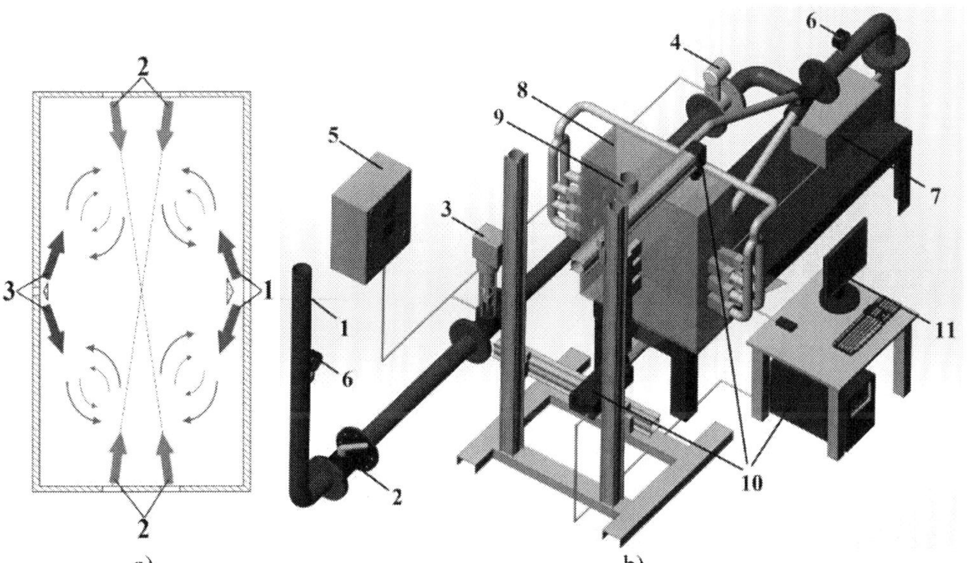

Fig.1. Four-vortex fuel combustion scheme (a): 1 – dust-air mixture flow through the front nozzles; 2 – dust-air mixture flow through the side nozzles; 3 – tertiary air; (b): 1 – compressed air supply line, 2 – stop valve, 3 – flow control valve, 4 – flow meter, 5 – control box, 6 – manometers, 7 – fog generator, 8 – four-vortex furnace model, 9 – ventilation, 10 – PIV measuring system, 11 – computer.

The model is made of 10 mm thick plexiglass at a scale of 1:25 (internal dimensions 290 x 880 x 730 mm). On the side walls at three levels, there are two diagonally directed nozzles at an angle of 6° (dimensions 28 x50 mm), and the axes of the side nozzles are directed to the center of the furnace. On the front and rear walls at three levels (locations of side nozzles are marked), there are also two central nozzles (dimensions of front nozzles 23 x 66 mm, and rear ones – 11 x 64 mm) directed towards the side walls at an angle of 20°.

Study of the flow structure in the four-vortex furnace model was carried out using the particle image velocimetry (PIV) method, which is the more productive in comparison with the method of laser Doppler anemometry, also used for diagnostics of flow pattern [3–7].

PIV is a non-contact optical panoramic method, allowing measurements with high performance. Measurements of the instantaneous flow velocity field in a given section are based on the measurement of the motion of impurity particles (tracers) located in the plane of the section under study, for a fixed time interval. This displacement is determined based on the application of correlation methods to tracer pictures, using regular partitioning into elementary cells.

In this work the PIV-system "POLIS" (development of IT SB RAS) was used; it consists of the CCD camera Videoscan 4021 (with a resolution 2048 x 2048 pixels, the frequency of shooting up to 1.25 Hz, and an exposure time of 28 ms) and a pulsed Nd: YAG laser QuantelEVG (wavelength of 532 nm, energy in the pulse up to 145 mJ, and a pulse duration of 10 ns). To obtain the average velocity field, a series of 100 pairs of images was conducted in each measuring section. The regime under which the velocity was 5 and 3 m/s in the outlet from the side and central nozzles, respectively, was investigated. In horizontal sections, the flow is symmetrical with respect to the vertical plane passing through the center of the central walls of the furnace. Therefore, in order to increase the spatial resolution and save the time of the experiment, measurements were made only in the half area of the investigated horizontal cross-sections (290 x 365 mm).

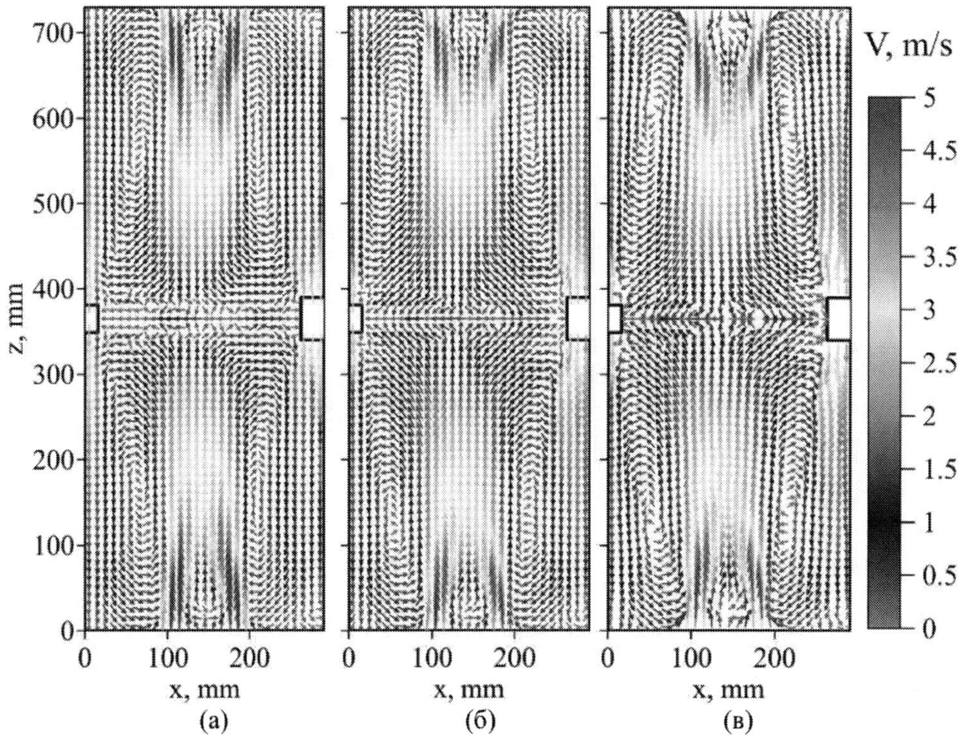

Fig.2. The time-average flow velocity field in a four-vortex furnace model in horizontal sections at the marks of the centers of three levels of nozzles (in the figures the spatial grid spacing is 1 cm): (a) the lower level; (b) the middle level; (c) the upper level.

The obtained results show the complex structure of the flow in the investigated furnace model, consisting of four conjugate stable vortices. The jets, which flow from the nozzles located on the side walls at a distance of three nozzle calibers, merge and spread as a single stream to the center of the furnace. Then the stream turns back and, interacting with the jets which flow from the central nozzles, is directed parallel to the front and rear walls. In the corners, the stream turns and flows to the bottom of the side burners, thereby forming four closed vortices with vertical axes and opposite direction of rotation.

With this flow structure, there is an intensive "washing" of the furnace screens (maximum tangential velocity is near the walls), which in practice contributes to the improvement of convective heat transfer. With such an aerodynamic scheme, a long torch length is provided; there is no direct surge of the unburned fuel particles on the screens, which reduces the danger of intense surface slag typical for the frontal arrangement of the burners. The mounting of tertiary blast nozzles is a necessary condition for the formation of a four-vortex flow structure and, at the same time, allows organizing a gradual combustion of fuel in order to reduce emissions of nitrogen oxides.

The results of the research are useful for verification of developing mathematical models applied in full-scale numerical simulations of furnace processes [8, 9].

The study was carried out with the financial support of the Russian Science Foundation (project No. 14-29-00093).

References

1. S.V. Sryvkov, M.Ya. Protsailo, A.A. Dekterev, S.G. Kozlov, M.S. Pronin, A.M Kovalevsky, V.P. Popov, RF Patent No. 2032853 (1995) [in Russian]
2. I.S. Anufriev, D.O. Glushkov, A.P. Papulov, E.Yu. Shadrin, MATEC Web of Conf. **92,** 01059 (2017).
3. I.S. Anufriev, O.V. Sharypov, E.Yu. Shadrin, Tech. Phys. Lett. **39** 466 (2013)
4. Yu.A. Anikin, I.S. Anufriev, E.Yu. Shadrin, O.V. Sharypov, Thermophys. Aeromech. **21,** 807 (2014)
5. I.S. Anufriev, D.V. Krasinsky, E.Yu. Shadrin, O.V. Sharypov, Tech. Phys. Lett. **40,** 879 (2014)
6. I.S. Anufriev, P.A. Strizhak, M.Yu. Chernetskii, E.Yu. Shadrin, O.V. Sharypov Tech. Phys. Lett. **41,** 727 (2015)
7. I.S. Anufriev, P.A. Kuibin, E.Yu. Shadrin, D.K. Sharaborin, O.V. Sharypov, Thermophys. Aeromech. **23,** 621 (2016)
8. M.Yu. Chernetskiy, V.A. Kuznetsov, A.A. Dekterev, N.A. Abaimov, A.F. Ryzhkov Thermophys. Aeromech. **23,** 591 (2016)
9. D.V. Krasinsky, O.V. Sharypov, J. Eng. Thermophys. **24,** 348 (2015)

On the shape of self-sustained evaporation front in a metastable liquid

Oleg Sharypov[1,*], *Denis* Krasinsky[1]

[1]Kutateladze Institute of Thermophysics SB RAS, 630090 Novosibirsk, Russia

Abstract. A theoretical model which takes into account new experimental data is proposed for the description of a stationary surface of the self-sustained evaporation front propagating in a layer of superheated liquid flowing over a flat heater. An approximate analytical dependence of the vapour layer thickness on the coordinate and physical parameters is obtained and found to satisfactorily agree with the experimental results. A dimensional parameter is introduced that allows the description to be presented in an invariant dimensionless form.

1 Introduction

Experiments on the heat exchange in boiling liquids have revealed a regime of interest for both basic science and technical applications, in which a vapor layer propagates over a cylindrical heater under conditions of liquid superheating above the boiling temperature at a given pressure [1-4]. Models proposed previously [1, 5-9] aimed at approximate description of the dependence of evaporation front propagation velocity on some set of physical parameters. Some of these models [5, 7, 9] suggest that a "frontal stagnation point" exists at the interface. However, some recent experimental data did not confirm the existence of the frontal stagnation point [2], which implies the need for developing a new approach to the description of these phenomena. In addition, the aforementioned models do not describe the shape of a stationary interface. In this context, the present work aims at developing the analytical description [10] of a stationary shape of self-sustained evaporation front propagating in a layer of superheated liquid flowing over a flat heater.

2 Description of the model

Reported shadow images [2] showed that the thickness of a vapor layer monotonically increases along the heater surface. In the frontal region, the interface is initially smooth. Thus the model assumptions are as follows. The flow is laminar, the flow velocity varies insignificantly, the curvature of the interface is small. Under these conditions, we can ignore the dynamic and surface pressure and assume that the liquid flows at a constant velocity over an almost flat free surface $y = f(x)$ with velocity V (in the frame of reference connected with evaporation front). The proposed model also ignores the following factors:

* Corresponding author: sharypov@itp.nsc.ru

hydrostatic pressure; compressibility and viscosity of phases; buoyancy, wettability, and thermocapillary effects; finite thickness of the interface; heat transfer between the vapor and the heater. It is also assumed that the evaporated liquid is subsequently not condensed. In the steady-state problem formulation, we will ignore variations of heater temperature T_W and heated-layer thickness δ_T (within the time interval under consideration) and consider the vapor layer as semi-infinite. It is taken into account that the ratio $\rho_V/\rho_L = \varepsilon$ is small ($\varepsilon << 1$), subscripts L, V refer to the liquid and vapor phases, correspondingly. Using Bernoulli integral, conditions of continuity at the interface (for material flow, energy flux, normal and tangential components of momentum flux), and also the balance between vapour production and vapour flow, the analytical solution (1) has been derived [10, 11], which describes the vapor layer thickness f depending on the coordinate x and main physical parameters ($\varphi = mx$, $g = mf$):

$$\varphi = (1 + g/2)\sqrt{g + g^2/4} - 2\,\mathrm{Arsh}\left(\sqrt{g}/2\right). \quad (1)$$

The solution is presented in an invariant dimensionless form with the unique parameter of the model: $m = \mathrm{Ja}^{-2}\pi V/a_L$, here $\mathrm{Ja} = c_p \Delta T/\varepsilon L$ is Jakobi number, a_L and c_p are the thermal diffusivity and heat capacity of the liquid, L is the evaporation heat, ΔT is overheating of the metastable liquid. According to the solution (1), the derivative $f_x \to \infty$ as $x \to 0$. The presence of this singularity does not mean that the model is principally inapplicable, since the integral fluxes of heat and vapour in any finite region of the interface remain finite. The dimensionality of the parameter m corresponds to inverse length V/a_L, and it characterizes the curvature of the interface along the coordinate x.

3 Discussion

The obtained analytical solution (1) doesn't take into account the curvature of the interphase surface along the transversal direction (z), which is about R^{-1} (R is radius of the cylindrical heater). Nevertheless, the solution (1) is applicable if R is large enough. Physically this condition means $R >> 1/m$. The plot of $f(x)$ calculated according to (1) for the case $m^{-1} = 20\,\mu\mathrm{m}$ (this value corresponds to the conditions of experiment [2] with Freon-21, $\Delta T = 76$ K, $V = 3.9$ m/s, $R = 1.5$ mm) is presented in the Fig. 1.

The comparison shows that solution (1) well describes the experimental data [2], except for some initial region of the interface with vapour layer thickness $g < 1$. It may be assumed that the stagnation zone with the closed stream lines is formed near the forward point $x = 0$. This is confirmed by the results of numerical modeling of the steady-state incompressible isothermal liquid flow (Fig. 2) along the non-deformable bottom ($y = 0$ at $x < 0$ and $y = f(x)$ at $x \ge 0$). The derivative of tangential velocity component along the normal direction at the bottom equals to null.

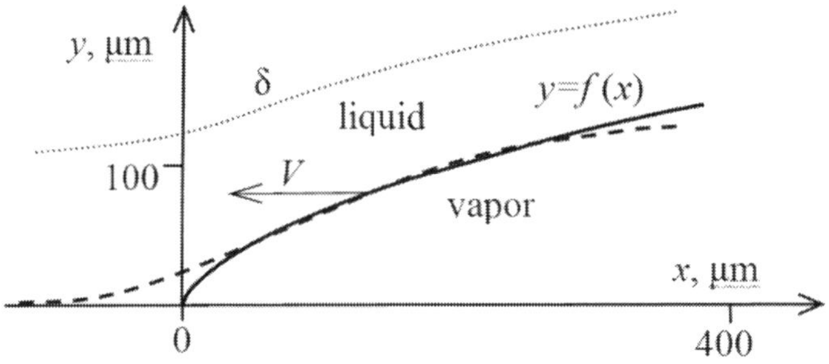

Fig. 1. Analytical solution $f(x)$ (solid line) and experimental data [2]: the interphase surface (dash line) and the thermal layer (δ).

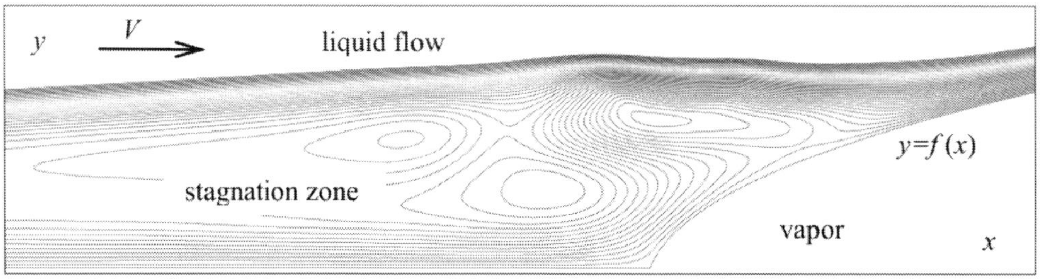

Fig. 2. Stream lines in the stagnation zone in liquid at the forward part of the interphase surface (numerical modeling), hydrodynamic parameters correspond to the conditions of the experiment [2].

The analytical solution (1) also provides good correspondence to the experimental data [12] with acetone ($\Delta T = 65$ K, $V = 7.1$ m/s, $1/m = 0.4$ mm $R = 1.25$ mm), see Fig. 3.

Fig. 3. Photo of the propagating vapor cavity – experimental data [12] and the analytical solution (1) (curves) with $1/m = 0.4$ mm.

The conditions of the experiments [13] with acetone also satisfy the inequality $R > 1/m$, and the obtained solution is close to approximations of the experimental data (here $\Delta T = 119$ K, $V = 25.5$ m/s, $1/m = 0.4$ mm $R = 1.25$ mm), see Fig. 4. According to [13], the thickness of the vapor cavity is proportional to $A_1 t^{0.66}$ ($t = x/V$) and the radius of the initial bubble is proportional to $A_0 t^{0.66}$ (approximate dependences). The analytical solution (1) has the same dependence: $g = (3\varphi/2)^{2/3}$ at the forward part, where $g << 4$. If we rewrite this in the form $f = Bt^{2/3}$, then the numerical value of the coefficient $B = 8.36$ $\mathrm{mm/ms^{2/3}}$ exceeds the value $A_1 = 5.14$ $\mathrm{mm/ms^{0.66}}$, but exactly equals to A_0.

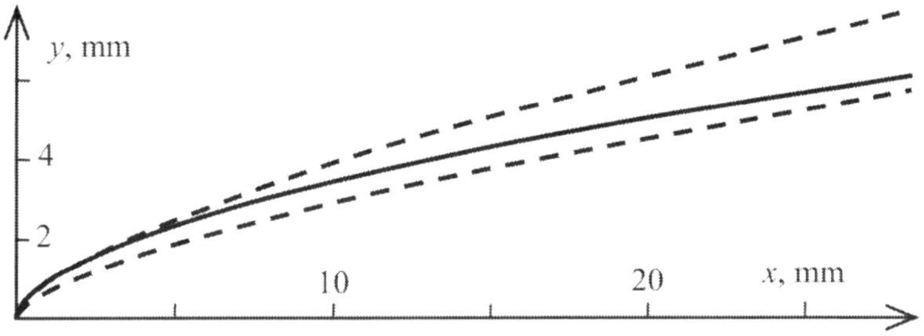

Fig. 4. The analytical solution (1) with $1/m = 0.4$ mm (solid line) and the approximations of experimental data [13] for the thickness of vapor cavity at the upper and lower parts of the surface of cylindrical heater (broken lines).

The solution (1) noticeably exceeds experimental data in case $R < 1/m$, when the curvature of the interphase surface in z direction can not be neglected comparatively with the curvature along x-axis.

4 Conclusions

The analytical dependence (1) of the vapor-layer thickness on the coordinate (derived in the framework of proposed approximate theoretical model of stationary propagation of the self-sustained evaporation front in a layer of superheated liquid flowing over a flat heater) satisfactorily agrees with the known experimental data for cylindrical heaters under the condition $R > 1/m$. The governing parameter m in physical sense plays the role of characteristic curvature of the interphase surface in the direction of evaporation front propagation. This stationary solution can be used as the base to study the problem of weak perturbation stability at the liquid-vapor boundary.

This work was supported by the Russian Foundation for Basic Research, project no. 15-08-01359.

References

1. B.P. Avksentyuk, V.V. Ovchinnikov, V.Ya. Plotnikov, Izv. Sib. Otd. Akad. Nauk, Ser. Tekh. Nauk. 17 (1989) [in Russian]
2. A.N. Pavlenko, E.A. Tairov, V.E. Zhukov, A.A. Levin, M.I. Moiseev, J. Eng. Thermophys. **23**, 173 (2014)
3. A. Stutz, J. R. Simoes-Moreira, Int. J. Heat Mass Transfer **56**, 683 (2013)
4. V. E. Zhukov, D.V. Kuznetsov, M. I. Moiseev, Innov. Nauka, 76 (2016)
5. B.P. Avksentyuk, Russ. J. Eng. Thermophys. **5**, 1 (1995)
6. B.P. Avksentyuk, V.V. Ovchinnikov, High Temp. **34**, 799 (1996)
7. A.N. Pavlenko, V.V. Lel', Thermophys. Aeromech. **6**, 105 (1999)
8. A.S. Moloshnikov, I.I. Shmal', High Temp. **38**, 53 (2000)
9. S.P. Aktershev, V.V. Ovchinnikov, J. Appl. Mech. Tech. Phys. **49**, 194 (2008)
10. O.V. Sharypov, Tech. Phys. Lett. **43**, 379 (2017)
11. O.V. Sharypov, Thermophys. Aeromech. **24**, (2017) (to be published)
12. B.P. Avksentyuk, V.V. Ovchinnikov, Thermophys. Aeromech. **15**, 267 (2008)
13. B.P. Avksentyuk, V.V. Ovchinnikov, Thermophys. Aeromech. **11**, 609 (2004)

DOI: 10.1051/matecconf/201711503008

Study of the two-stage gasification process of pulverized coal with a combined countercurrent and concurrent flow system

Viktor Kuznetsov[1,2], *Mikhail* Chernetskiy[2,3,*] *Nikolay* Abaimov[3], *Alexandr* Ryzhkov[3]

[1]Siberian Federal University, 79 Svobodny, Krasnoyarsk, 660041, Russia
[2]Kutateladze Institute of Thermophysics, Novosibirsk, 630090, Russia
[3]Ural Federal University, 19 Mir Street, Yekaterinburg 620002, Russia

Abstract. The paper presents a numerical study of an advanced two-stage gasifier with a combined countercurrent and concurrent flow pattern and dry fuel feed system EAGLE. The Reynolds Stress model was used for the numerical simulation of turbulent flow. We have conducted studies on the influence of upper burner's inclination angle on heat and mass transfer processes in the gasifier as well as on coal conversion. It is shown that the increase in the upper burner's inclination angle in the horizontal plane allows intensifying the process of two-stage gasification.

1 Introduction

Use Gasification of coal is one of the most strategic pathways of coal processing. Market analysis of gasification technologies shows that the most popular are flow gasifiers, whose share is about 80% [1].

The analysis of technical solutions used in new designs of gasifiers in terms of efficiency and economy of their application in integrated gasification combined cycle (IGCC) has shown that the use of two-stage principle of the fuel conversion process is one of the upcoming trends in technology modernization. According to [2], the transition from single-stage process in the flow gasifier to two-stage process can increase the efficiency of the IGCC by almost 1%.

The EAGLE-type two-stage gasifier, developed in the frameworks of the Japanese project "Coal Energy Application for Gas, Liquid and Electricity" (EAGLE) [3] is one of the original technical. In this paper, on the basis of numerical studies, using the Kuznetsk coal as a fuel, we examine the influence of several operating parameters of the EAGLE gasifier on the heat and mass transfer parameters in the gasification chamber as well as the efficiency of the gasification process.

[*] Corresponding author: micch@yandex.ru

2 Problem statement and research methods

The operational principle of the EAGLE gasifiers is based on a hydrodynamic flow separation, providing a combined countercurrent and concurrent flow pattern (Fig.1). One advantage of this solution is reducing the required height of the reaction zone, while increasing the residence time of coal particles in the reactor. One flow, with oxygen to coal ratio close to stoichiometric value, is delivered to the lower part of the gasification chamber. While combusting, it creates the necessary temperatures for efficient proceeding of the gasification reactions. The second flow, with the lack of air, is delivered into an upper part of the chamber, where it interacts with the high temperature up flow of combustion products. Due to the lack of oxygen for complete combustion, the carbon residue transforms into gasification reaction products through the conversion process. In this work, the Kuznetsk coal was used as a gasifier fuel. Analysis of Kuznetsk coal: W^a=5.4%, A^d=22.3%, V^d=34.7%, V^{daf}=44.7%, C^{daf}=75.57%, H^{daf}=5.66%, N^{daf}=1.78%, S^{daf}=0.55%, O^{daf}=16.44%, Q_s^{daf}=7086kcal/kg. The estimated coal consumption is 1700 t/day. The dimensions of the gasification chamber are presented in Fig. 2.

For the numerical simulation of turbulent flow of an incompressible liquid we used the Reynolds equations taking into account the interfacial interaction. The Reynolds equations were closed using the Reynolds stress model (RSM). Abandoning the isotropic eddy-viscosity hypothesis, the RSM closes the Reynolds-averaged Navier-Stokes equations by solving transport equations for the Reynolds stresses, together with an equation for the dissipation rate. Since the RSM accounts for the effects of streamline curvature, swirl, rotation, and rapid changes in strain rate in a more rigorous manner than one-equation and two-equation models, it has greater potential to give accurate predictions for complex flows [4]. In the present work, to describe the particles motion we used the Lagrange method. Combustion the gas phase is represents the combination of kinetic combustion model of the gas components with the eddy break-up model. The coal particle burning process was represented as the following consecutive steps: evaporation of moisture from the fuel, devolatilization and the combustion of volatile components, and combustion of the coke residue. The combustion rate of coke residue was calculated in accordance with the provisions of the classical diffusion-kinetic theory. To describe gasification processes, we included the reactions of steam-air conversion of coal into the mathematical model. Reactions and kinetic constants are presented in Table 1.

The proposed model and solution methods were previously tested for solving problems of pulverized coal combustion and gasification [5] and showed a satisfactory agreement with the experimental data on the basic process parameters in the combustion chamber.

Table 1. Specific reaction rates of combustion and gasification.

	Reaction	Ks = A exp(-E/RT)		References
		m/s	J/mol	
1.	$C + O_2 \rightarrow CO + CO_2$	$3.3 \cdot 10^4$	135756	D.M. Khzmalyan (1977)
2.	$C + H_2O \rightarrow CO + H_2$	$1.6 \cdot 10^4$	181427	V.F. Yudin (1957)
3.	$C + CO_2 \rightarrow 2CO$	$3.5 \cdot 10^4$	140365	V.A. Parilov (1971)

3 Results and discussion

As was shown earlier [6], the use of the RSM turbulence model enables more accurate description of aerodynamics in the twisted flow pulverized-coal chamber. The aerodynamics features in the combustion chamber strongly affect heat and mass exchange processes and pulverized coal fuel gasification. The gas flow direction influences the motion trajectory and residence time of the coal particles in the chamber. Several variants of the two-stage gasification process organization were investigated. The influence of upper burner's inclination angle in the horizontal (angle α) and vertical plane (angle β) on heat and -mass exchange processes, coal particles conversion and syngas composition is researched (Fig. 3.4). The initial calculated data are presented in Table 2.

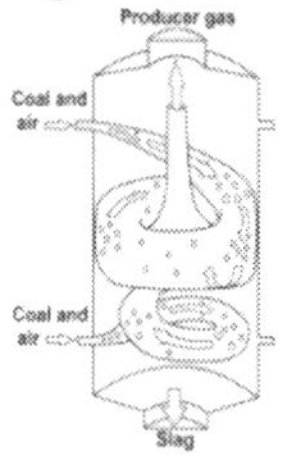

Fig. 1. EAGLE gasifier

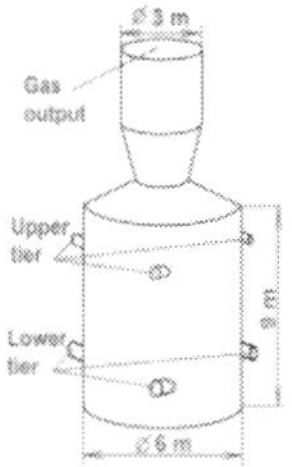

Fig. 2. Feeding diagram and gasifier dimensions at a flow rate of 1700 t/h

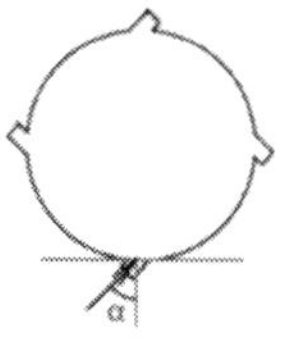

Fig. 3. Inclination angle of burners in the horizontal plane of the upper tier

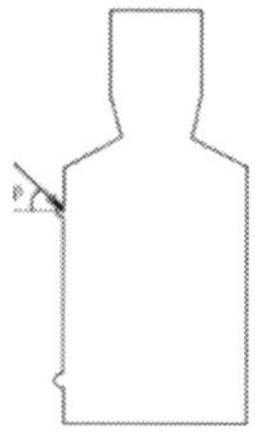

Fig. 4. Inclination angle of burners in the vertical plane of the upper tier

Table 2. Gasification process options in the chamber.

Option	1	2	3
Inclination angle of upper burners α, °	30	45	60
Inclination angle of upper burners β, °	0	0	0

Below are the results of calculations. It can be seen that there is a significant effect of the upper burner's inclination angles in the horizontal plane on the heat-mass transfer processes in the gasifier. The increasing of upper burner's inclination angle leads to a decrease of the maximum temperatures in the gasifier (Fig. 5). This is explained by the difference of downstream flow velocity near the gasifier walls. Comparison of the axial velocity profiles (Fig. 5a) across the gasifier at 2.5 m height shows that angle α = 30°, in comparison with the angle α of 45 and 60°, leads to less intensive gases motion down along the gasifier walls. The downward flow velocity at an angle α = 30° does not exceed 4 (m/s), and at an angle α = 60° it reaches 8-9 (m/s). The low temperatures in the chamber at α = 60° (Fig. 5d-5f) can be explained by more intense gas flow downward from the upper stage of the burners, promotes their better mixing with combustion products of the lower burners and more intensive gasification endothermic reactions.

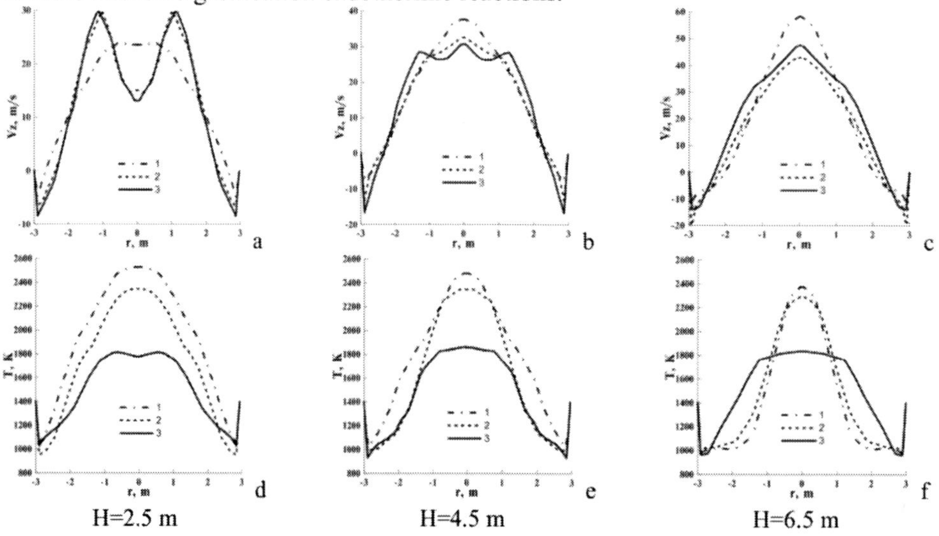

Fig. 5. Radial distribution of axial velocity and temperature over the chamber height.

Comparison of the carbon particles concentration across the gasifier at a 2.5 m height shows that at the angle $\alpha = 60°$, in comparison with the angles α equal to 30 and 45°, a larger amount (0.8 kg/m^3) of fuel near the walls is observed (Fig.6a) . In the upper part of the gasifier, for an option at $\alpha = 30°$, the concentration of coal particles is higher - 4.5 kg/m^3 (Fig.6c). Thus, an increase in the upper burner's inclination angle results in a more intensive descending movement of coal fuel from the upper burners to the lower part of the gasifier.

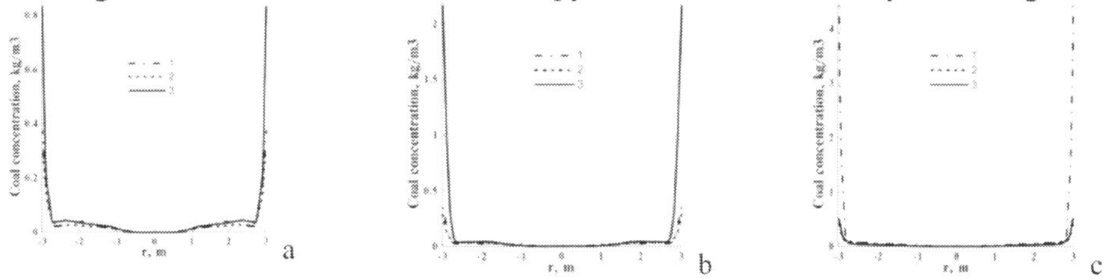

H=2.5 m H=4.5 m H=6.5 m

Fig. 6. Radial distribution of coal concentration over the chamber height.

Comparative analysis of the calculation results showed that an increase in angle α leads to increase of carbon conversion from 82.51% to 99.05% and increase of chemical efficiency (cold gas efficiency) from 80.60% to 87.10%.

Conclusion

Numerical studies of advanced two-stage flow gasifier have been carried out for Kuznetsk coal gasification.

It is shown that the increase of the angle of rotation of the burners of the upper tier in a horizontal plane increases coal conversion and chemical efficiency.

The increase of the angle α leads to intensification of the two-stage gasification process and allows reducing the gasifier height without decreasing the syngas heating value.

Problem statement and results analysis were carried out at the Ural Federal University and financially supported by the Russian Science Foundation (project no. 14-19-00524).
The development of methods for mathematical modeling and calculations was carried out with the financial support of the Ministry of Education and Science of the Russian Federation under the Grant Agreement No. 14.607.21.0150 (Unique identifier of the RFMEFI60716X0150)

References

1. C. Higman, *Proceedings of the International Pittsburgh Coal Conference*, Beijing, China, September 16–19, 2013
2. C. Higman, M. Burgt, *Gasification* (Elsevier Science, 2008)
3. F. Kiso, T. Akiyama, A. Morihara, K. Takahashi, E. Kida, J. Iritani, et al., *Proceedings of the 25th International Conference on Coal Utilization & Fuel Systems, Clearwater, FL*, USA, 2000
4. B.E. Launder, G.J. Reece, W. Rodi, J. Fluid Mech. **68,** 537 (1975)
5. N.A. Abaimov, A.F. Ryzhkov, Thermal Eng. **62,** 767 (2015)
6. V.A. Kuznetsov, N.S. Chernetskaya, M.Yu. Chernetskiy, J. Phys.: Conf. Ser. **754,** 112006 (2016)

Investigation of combustion and gasification mechanically activated coal fuel of various degrees of metamorphism on the 5-MW heat setup

Evgenii Butakov[1*], *Anatoly* Burdukov[1], and *Artem* Kuznetsov[1]

[1]Kutateladze Institute of Thermophysics, SB RAS, Novosibirsk, Russia

Abstract. The technology of mechanochemical activation of combustion and gasification of coals is of unquestionable scientific and technical interest; an increase in chemical activity of coals at their mechanically activated grinding is associated with an increase in the rate of reaction of the coal substance. To study the combustion and gasification process, the reactor model with tangential scroll input of coal-air suspension and cylindrical reaction chamber was used at the 5-MW thermal power plant. The experiments were carried out with coals of G and SS grades of the Kuznetsk deposit. Coal, ground after the boiler's standard mill, is fed by a feeder to the disintegrator; then, it enters the scroll inlet of the reactor burner with transport air. The suspension is ignited by a gas igniting device with the power of 50 kW. In experiments on combustion and gasification of fine coal performed at the temperature in the reaction chamber of 1000-1300°C and air excess $\alpha = 0.5\text{-}0.7$, the data on concentrations of CO and H_2 were obtained: for coal of grade G, concentration of H_2 was 6.3% and concentration of CO was 15.3%; for coal of SS grade, concentration of H_2 was 9.5% and concentration of CO was 15.6%.

1 Introduction

Russia is the largest coal country, and one third of the world's resources is concentrated in its interior. At the same time, the quality of coal supplied to power plants is constantly deteriorating. At that, burning of low-grade coals at the boiler plants is associated with considerable difficulties, when igniting a coal flame, stabilizing its combustion, and fuel burn-out with simultaneous reduction of ecological parameters of power plants [1,2].

The use of coal in power engineering meets various difficulties, leaving great potential for improving and optimizing the efficiency of combustion technologies. To improve the ignition and stabilization of combustion of the pulverized coal flame, the methods of increasing the grinding fineness, high heating of the air mixture and secondary air, using dust of high concentration, and joint use of fuel oil and coal are usually used. However, these methods have some disadvantages and only partially solve the above-mentioned problems.

* Corresponding author: e_butakov@mail.ru

Recently, plasma technologies of ignition and lighting of a pulverized-coal flame have been actively developed [2]; however, due to relatively low service-life of plasmatrons, the technology is more applicable to the ignition processes at the boiler plants.

The fact of an increase in the chemical activity of coals at their grinding in highly stressed disintegrator mills has been established at the Institute of Thermophysics SB RAS. A number of investigation results related to ignition and combustion of mechanically activated fine coal of different stages of metamorphism have been published [3, 4]. Based on obtained data, a new technology is proposed to replace high-reaction gases and fuel oil by fine coal at the power plants.

2 Experimental apparatus and procedure

The experiments were carried out at a pilot setup with the thermal power of 5 MW. During the experiment, the temperatures were measured along the entire length of setup, and gas analysis was performed at the outlet of the combustion chamber.

Coal powder with the size of up to 1 mm or dust after the ball-tube mill (BTM) is fed into the hopper, then through a dispenser, it enters the grinding mill and after grinding it is sent to the funnel, through which it is fed by ejector to the burner with the transport air. The pulverized coal mixture is formed before entering the scroll swirler of the fine-coal burner, where ignition and combustion of the dust-air mixture are subsequently initiated with the help of ignition safety device (ISD). It burns under the lack of air in a muffled combustion chamber, i.e., ignition and partial gasification of coal with the release of volatiles (CO and H_2, etc.) occur there. The highly reactive fuel mixture, after partial gasification, enters the afterburner chamber, where it mixes with the secondary air, fed tangentially.

The combustion process can be observed through the inspection windows located along the combustion chamber and afterburner. The temperature control of combustion is performed by eight thermocouples located along the combustion chamber and afterburner. The setup includes the nozzles for sampling the gaseous products of combustion and gasification for gas analysis of their composition. The flow rates of transport, primary, and secondary air are controlled by thermal anemometers and regulated by means of valves and air fans on the ejector.

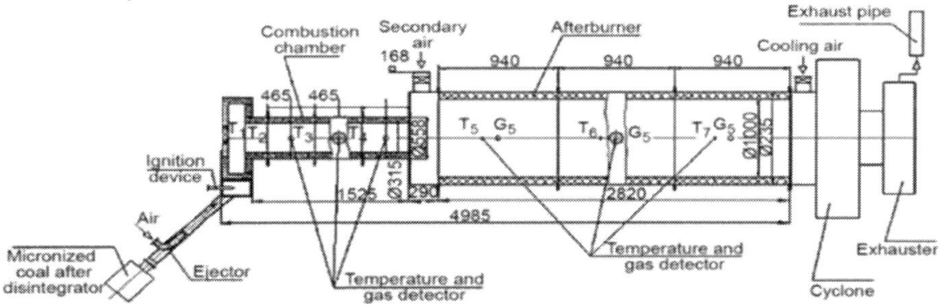

Fig. 1. Experimental setup of up to 5 MW.

3 Results and discussion

Experiments on ignition, combustion and gasification of Kuznetsk deposit coals (Siberia) of grade G and S (gaseous and sintering), crushed at the disintegrator mill with productivity of up to 2 t/h, were carried out. Characteristics of the studied Kuznetsk coals of grade G: moisture Wrl - 19.0%, ash A_r - 14.3%, volatile content - 43.5%, sulfur S_{dl} - 0.30%, high heat value Q_{daf} - 7327 kcal/kg; S: W_{rl} - 7.0%, ash A_r - 13.5%, volatile content - 27.5%, sulfur S_{dl} - 0.30%, and high heat value Q_{daf} - 8180 kcal/kg.

 DOI: 10.1051/matecconf/201711503009

Experiments were carried out with air excess α of 0.5 - 0.7 and power of the gas ignition device of 50 kW. During the entire experiment, the temperature was measured and gas analysis of the combustion products was carried out. The gas intake tube was located at the end of the combustion chamber, and sampling was performed in the chamber center. During the experiments, the minimum time, required to achieve autothermal combustion, was determined; the time was measured from the beginning of feeding the mechanically activated coal into the combustion chamber prior to disconnecting the lighting-up propan-butane.

The first experiments on studying ignition, stable autothermal combustion and air gasification of coals of two metamorphic stages (G and S), are presented in this work.

The experiment was as follows: ~ 100 s after the beginning of supply of dust-gas coal mixture to the chamber using working ignition gas device (IGD), gas supply was stopped and combustion of coal-dust suspension occurred autothermally; when working with SS coal, autothermal combustion started in ~ 240 s. After combustion stabilization by the temperature, the main temperature and gas analytical measurements were carried out in the combustion chamber.

Experimental results on ignition and gasification of coals of grade G and S are shown in Fig. 2.

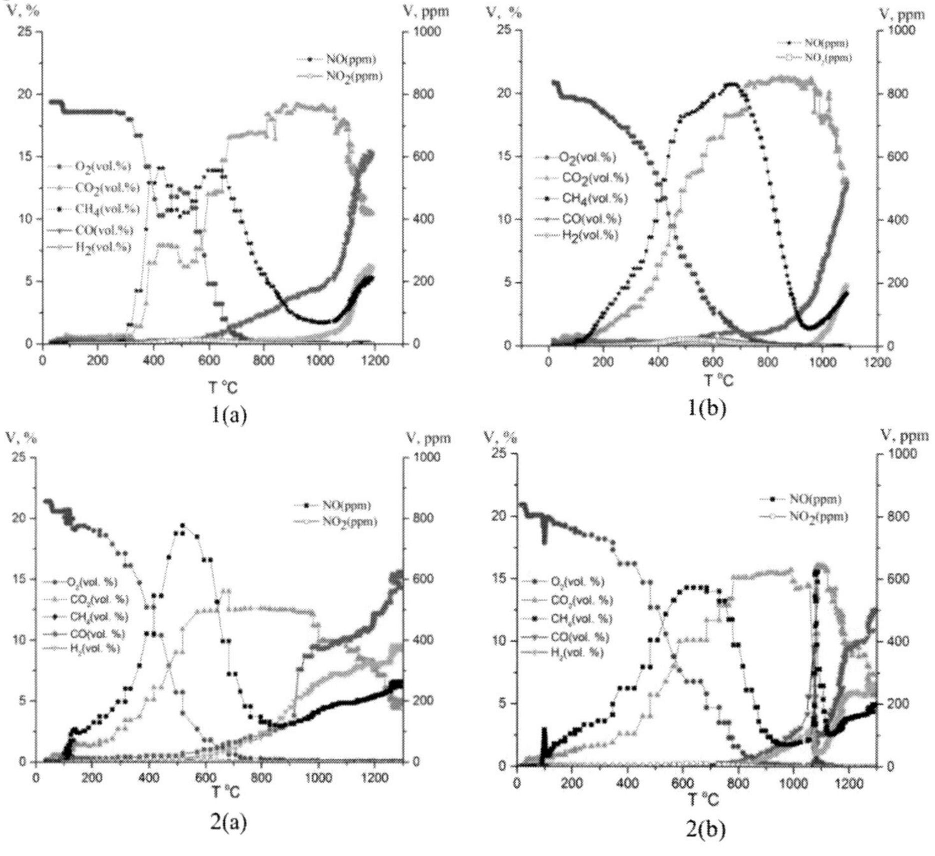

Fig. 2. Temperature dependence of concentrations of gas phase components at gasification of mechanically activated coal of grade: 1 – G, 2 – S; a – α = 0.5; b – α=0.7.

In all cases, complete burnout of O_2 occurs, i.e., along 3 calibers, there is burnout of 50 - 70% of coal fuel. When O_2 reaches ~ 3% concentrations of CO and H_2 start increasing. Concentration of CO_2 increases and has a maximum at complete burnout of O2; subsequently it starts decreasing due to reaction of gasification in carbon dioxide [5, 6].

A change in the excess air factor α from 0.7 to 0.5 leads to an increase in maximal concentration of H_2 and CO, in the case of coal of grades H, concentration of H_2 is 4.8%-6.3% and concentration of CO is 12.9%-15.3%; for coal of grade SS, concentration of H_2 is 6.6% - 9.5% and concentration of CO is 12.5% - 15.6%.

4 Conclusion

Stable autothermal combustion of the dust suspension of Kuznetsk coals of two grades G and SS has been achieved as a result of their mechanically activated grinding and ignition by a gas burner.

Investigation results on air gasification of these coals confirm the possibility of using the processes of mechanically activated grinding to obtain acceptable values of H_2 and CO concentration.

For obtaining the real high-efficient processes and devices for gasification of fine coal, further studies are needed to optimize the operation parameters: temperature levels, air excess, time of dust suspension stay in the reaction chamber, etc.

Data on ignition and combustion of mechanically activated coals of grade G and SS serve to recommend the technology of microgrinding of these coals for the systems of oil-free firing of the boiler plants.

The work was financially supported by the Russian Foundation for Basic Research (grant - 17-08-00445 A)

References

1. M.Yu. Chernetskiy, A.P. Burdukov, E.B. Butakov, I.S. Anufriev, P.A. Strizhak, Combust., Explosion and Shock Waves **52**, 326 (2016)
2. A.P. Burdukov, E.B. Butakov, V.I. Popov, M.Y. Chernetskiy, N.S. Chernetskaya Thermal Science **20**, 23 (2016)
3. A.P. Burdukov, V.I. Popov, G.V. Chernova, M.Yu. Chernetskiy, A.A. Dekterev, N.S. Chernetskaya, V.M. Markova, V.N. Churashev, T.S. Yusupov, Thermal Eng. **60**, 889 (2013)
4. V.E. Messerle, A.B. Ustimenko, O.A. Lavrichshev, Fuel **164**, 172 (2016)
5. N. Couto, V. Silva, E. Monteiro, Appl. Thermal Eng. **78**, 185 (2015)
6. R. Mota, G. Krishnamoorthy, O. Dada, S. Benson, Appl. Thermal Eng. **90**, 13 (2015)

Experimental investigation of open cavity as flame holder of supersonic combustor

Marat Goldfeld[1,*]

[1]ITAM SB RAS, 4/1 Institutskaya, Novosibirsk, 630090, Russia

Abstract. The results of a study of combustion stabilization in a supersonic combustor for Mach numbers 2, 2.5, and 3 are presented. It was shown that the choice of the fuel injection scheme has decisive influence on the mixing efficiency and flame holding. It has been established that the efficiency of combustion stabilization depends essentially on Mach number and total temperature at the channel entrance.

1 Introduction

Recent advances in the design of scramjet have shown the technical feasibility of creating and cost-effective application of a hypersonic vehicle. Together with this, the development and testing of the propulsion system is a key problem in the implementation of hypersonic flights. The relevance of such studies is confirmed by national and international programs [1, 2]. A key element of the scramjet is the combustion chamber with supersonic flow speed at the entrance. Achieving effective ignition and stable combustion in high-speed flows remains a serious problem in designing the combustor. Due to the low static temperature and the incomplete mixing of fuel and air, there is a tendency of the flame to break, even if stabilizers such as a cavity or a backward facing step, artificial ignition with a pilot flame or plasma discharge are used. The problem can be solved by providing the required level of mixing and the corresponding flow structure in front of the flame holder in order to increase the residence time of the mixture in the recirculation region of the flame holder. The aim of the present work was the research of fuel injection and flow structure in an open cavity to ensure self-ignition and stable combustion of kerosene in a relatively short supersonic combustor (0.4-0.8m).

2 Model and facility

The model consists of an insulator, an injector section and an experimental channel with an expansion angle of 6 ° (Fig. 1). Open cavity with ratio L/h = 5.5 was used as flame holder.

Fuel supply was carried out from the top wall of the model. Different schemes were used for fuel injection: into the flow core in front of the step at different angles to the main flow (AI), into the cavity against the flow from the rear wall of cavity(CI) or perpendicular to the flow from bottom wall of cavity (NI). The combination of these ways of injection

* Corresponding author: gold@itam.nsc.ru

allowed studying 6 variants of fuel supply. The system for fuel injection consisted of fuel tanks, pipe lines, locking valves and synchronization system. During the tests, fuel and air flow rates were measured in order to determine the fuel-air equivalent ratio.

Model tests were performed in the mode of attached pipeline. Hot-shot wind tunnel with operation duration of 80-150ms which is the effective tool for such research [3] was used as a source of high-enthalpy air. Experiments were made at the following range of flow parameters: Mach numbers at combustor entrance from 2 to 4, total temperature from 1500K to 2700K, static pressure from 0.06 MПa to 0.25MПa, and equivalence ratio from 0.45 to 1.5. Distributions of static pressure and heat flux on model walls and Pitot pressure at entrance and exit of combustor were measured. Shadow visualization was carried out for the purpose of determining the ignition and position of reaction zones.

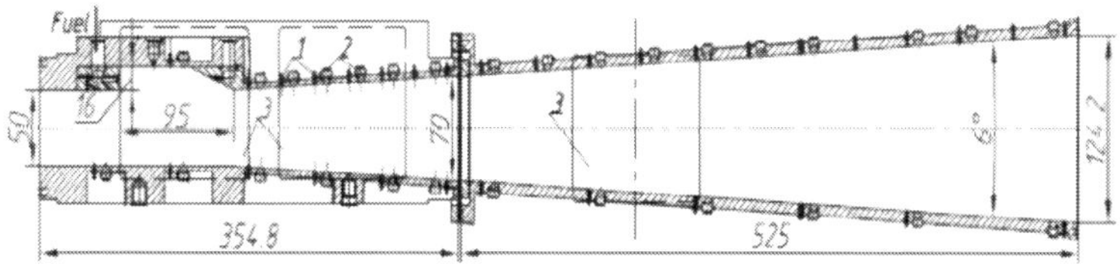

Fig. 1. Model layout: 1- pressure gages, 2-heat flux gages, 3-windows.

3 Results and discussion

The initial air temperature at the combustor entrance has decisive influence on the efficiency of ignition. It was found that at total temperature below 1500K self-ignition of kerosene does not occur even at Mach number 2, if the kerosene injection was carried out according to the AI scheme. There was only a slight increase of the static pressure and heat fluxes in the channel. The maximum rise (about 50%) of the parameters was achieved at the initial part of the channel at a length of up to 200 mm. Steady combustion throughout the combustor was obtained at kerosene injection according to the scheme AI +CI (against flow in the cavity) and at an increase of the total temperature up to 2100K (Fig. 2).

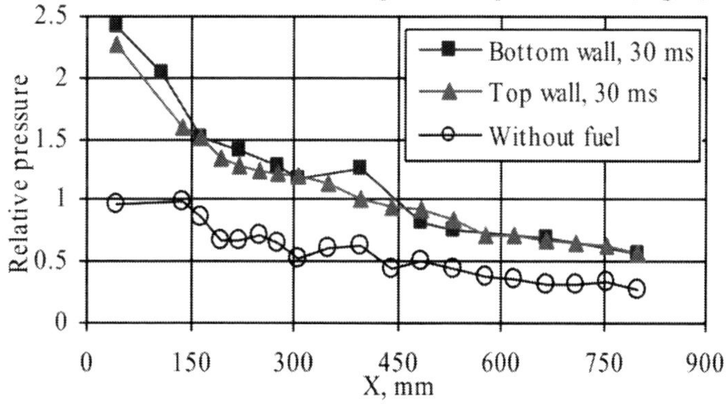

Fig. 2. Pressure distribution at kerosene combustion: M=2.

Static pressure increased approximately two to three times along the entire length of the channel and evenly on top and bottom walls. The interaction of the counter jet of kerosene with the air flow led to good spraying and mixing. High temperature behind the shock waves provided self-ignition and stabilization of the combustion process. The results of heat flux measurements (Fig. 3) and flame visualization data confirm the presence of combustion reactions and their intensity throughout the channel volume. Combustion continued for 70 ms. Breaking of combustion occurred as a result of the excessive enrichment of the mixture. At Mach numbers at the combustor entrance of 2.5 and 3, stable combustion of kerosene did not occur if the AI + CI injection scheme was used. Increasing the total temperature up to 2500 K did not also lead to intensive combustion due to small residence time of the mixture in cavity.

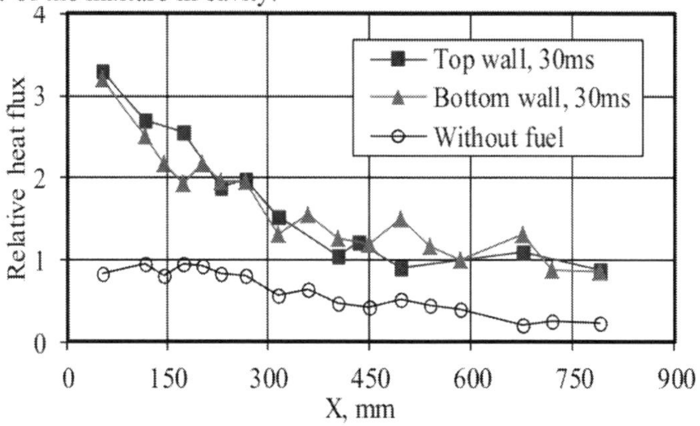

Fig. 3. Heat flux distribution at kerosene combustion: M=2.

Therefore, the fuel injection according to the scheme AI + CI + NI (normal injection in the cavity) was tested. The equivalence ratio in the cavity remained unchanged. As a result, intense combustion at Mach number 2.5 was realized and accompanied by a higher increase in pressure than in Mach number 2 (Fig. 4).

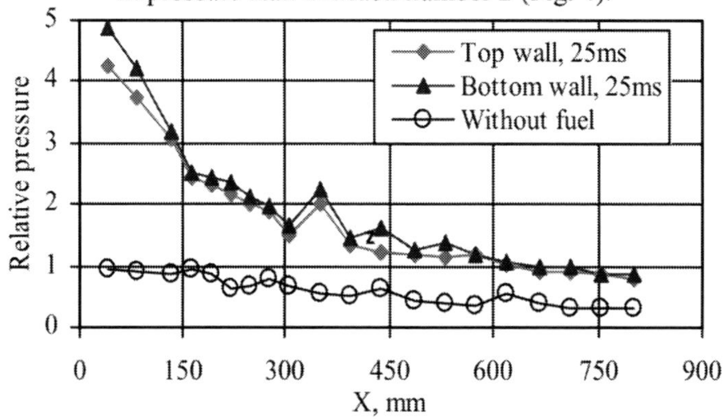

Fig. 4. Pressure distribution at kerosene combustion: M=2.5.

The levels of heat flux in the channel walls increased accordingly (Fig. 5). Visualization data in the visible range also confirmed the presence of intense combustion throughout the combustion chamber. It should be noted that the application of the AI + NI scheme at Mach number 2.5 and 3 did not lead to self-ignition of the mixture in the cavity due to the low level of mixing (highly lean mixture).

As a result, the cavity turned out to be ineffective as ignition source and flame holder. In this case, as in the case without fuel supply into cavity, there was slight increase of pressure in the cavity and immediately behind the one, without flame propagation into the flow core.

The intensity of combustion and the efficiency of heat release at Mach number 3 were lower than at Mach number 2.5, despite the increase in the total temperature at the channel entrance up to 2700K. Therefore, the fuel injection according to AI + CI scheme and with the pylon on BFS before the fuel jets was tested in order to ensure its penetration into the flow core. The use of such injection scheme resulted in sustained combustion of kerosene throughout the entire volume of the channel and significant increase of pressure along the channel length (Fig. 6).

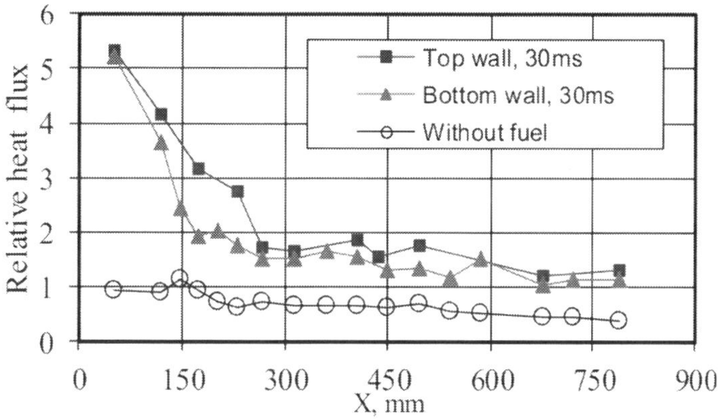

Fig. 5. Heat flux distribution at kerosene combustion: M=2.5

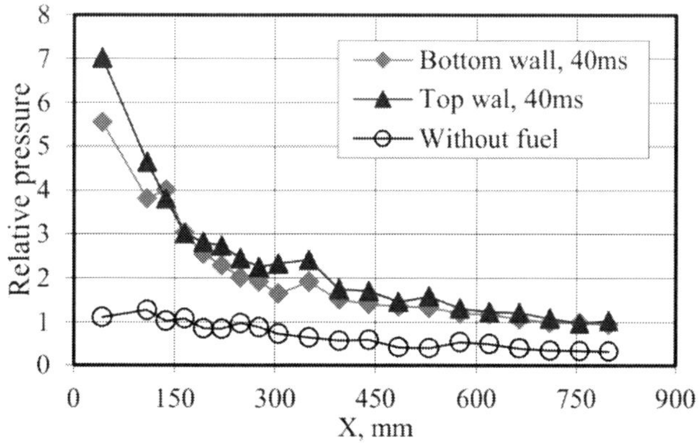

Fig. 6. Pressure distribution at kerosene combustion: M=3.

In the cavity and on the initial section of the channel, the pressure increased more than six times in comparison with the test without fuel. The distribution of heat fluxes in this case had the same features. An increase of fuel-air equivalence ratio was accompanied by a decrease of the duration of supersonic combustion. And then, the transition to subsonic combustion, propagation of the flow disturbance upstream, and channel blocking occurred.

4 Summary

Thus, the performed investigations have shown that an open cavity can be used as a source of self-ignition and an effective flame holder. The condition for the efficiency of flame stabilization is the correct choice of the fuel injection scheme, which depends on Mach number and total temperature at the combustor entrance. With an increase in the Mach number, it is necessary to increase the fuel angle injection or (and) use the pylons to ensure fuel supply and efficient mixing in the flow core.

References

1. J.P. Drummond, AIAA J. **52**, 465 (2014)
2. K.R. Jackson, M.R. Gruber, S. Buccellato, J. Propulsion Power **31**, 36 (2015)
3. M.A. Goldfeld, A.A. Maslov et al., AIP Conf. Procs. **1770**, Paper 030020 (2016)

Catalytic combustion and steam reforming of hydrocarbons in microreactor

Sergey Dimov*[1], *Ol'ga* Gasenko[1]

[1]Kutateladze Institute of Thermophysics SB RAS, 630090 Novosibirsk, Russia

Abstract. Catalytic combustion of fuel gas using a platinum catalyst was experimentally investigated in the slit microreactor. The composition of the exhaust gases was determined depending on temperature and time of contact. Data of methane steam reforming were received in that reactor with rhodium catalysts depending on temperature for three samples with different composition of doping substances.

1 Introduction

The development of catalytic combustion technology and steam conversion of natural gas and other hydrocarbon fuels is relevant for the development of environmentally friendly catalytic gas turbine plants, heating devices, hydrogen and syngas production. Catalytic combustion of hydrocarbons has several advantages over flaring. The decrease in the process temperature leads to a significant decrease in the formation of nitrogen oxides, the creation of heat fluxes with lower process temperatures [1]. The importance of studying the oxidation of hydrocarbons in small concentrations is due to the need to solve environmental problems for purification of industrial emissions [2]. A promising solution for hydrogen production was the creation of catalytic reactors based on microchannel technology [3]. Micro-reactors have a number of advantages: high ratio of catalytic surface to volume, high rate of heat and mass transfer, and short contact times. The steam reforming of hydrocarbons and hydrocarbon derivatives in microchannels was experimentally and numerically studied in [4-7]. At the same time, one of the main problems is the search for suitable catalysts for the oxidation of hydrocarbons and steam reforming. These catalysts should have high activity and stability for carrying out reactions at the lowest possible starting temperature and high thermal stability (up to 1350C). The same problem is associated with the search for effective doping additives to increase the activity and stability of the catalyst. Platinum is commonly used to carry out the oxidation of hydrocarbons [2,8]. For steam reforming, the choice of catalyst is more problematic, since the process takes place at high temperatures with the absorption of a large amount of heat. The main part of the heat is spent on evaporation of water, heating the steam. According to the data of [9, 10], rhodium is the most effective catalyst for steam reforming of methane.

The purpose of this work was to study the oxidation of the propane-butane mixture and the methane steam reforming in slit reactor whose walls are coated with a nanoscale catalyst.

* Corresponding author: dimov@itp.nsc.ru

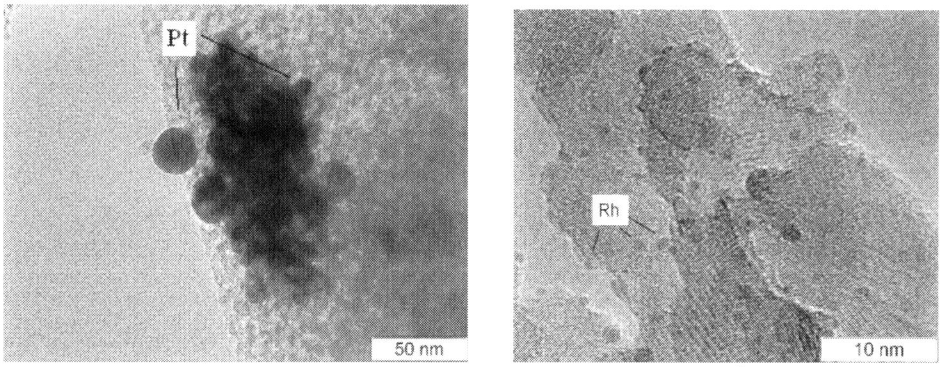

a b

Fig. 1. TEM pictures of the catalyst for the oxidation of hydrocarbons (a), steam reforming (b).

2. The experimental part

2.1 Preparation of catalysts

The use of noble metals as a catalyst enables oxidation of main components of the fuel gas at a temperature of 250-450 ° C and at a temperature of 750-850 ° C, steam reforming of methane. The multilayer catalytic coating is based on platinum for the oxidation of hydrocarbons and based on rhodium doped with lanthanum and barium oxides for steam reforming.

The catalyst support, γ-Al_2O_3, was impregnated with solutions of nitrate salts of lanthanum and barium, subjected to high-temperature calcination at 1050 ° C. It was then impregnated with a solution of rhodium nitrate, calcined at 500 ° C. The metal rod was treated with acetone for degreasing, etched in 1M hydrochloric acid at 60 ° C. It was then densely coated with titanium oxide from a complex organic solution to prevent oxidation of the metal surface at high temperatures of the steam reforming process [11]. The aerosol spraying procedure of the catalyst was repeated many times until the required weight was reached. In order to avoid cracking of the applied layer, the final calcination was carried out at a rate of temperature rise of not more than 3C / minute. The procedure for preparing a catalyst for the oxidation of hydrocarbons includes the steps of impregnating alumina with an ammonium complexes of platinum, drying, calcination. The application of the catalyst to the metal rod is similar to the above. Figure 1 shows photographs of the catalyst particles obtained on a transmission electron microscope for combustion (a) and steam reforming (b).

2.2 Experiment: Oxidation

Experiments on the oxidation of the propane-butane mixture were carried out using the set-up described in [6] for three contact times of 32, 98, and 161 ms. The mixture of fuel gas and air, set by the Bronkhorst gas controllers, was supplied to the preheating chamber and then to the slit annular reactor. In the course of the experiments, gas flow rates, reactor temperatures, and the composition of the inlet and outlet gas mixtures were measured. The temperature of the reactor was measured by a type K thermocouple. The composition of the output and input gases was analyzed by a chromatograph with a mass selective detector Agilent, as well as a Maestro chromatograph with a thermal conductivity sensor. The necessary temperature of the reactor was created by the degree of heating of the electric

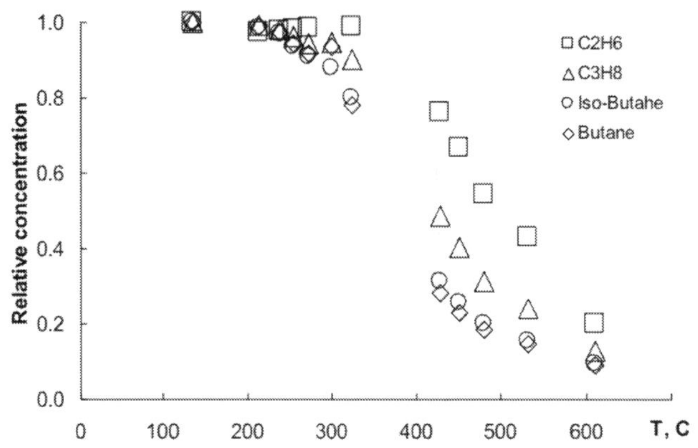

Fig. 2 Dependence of the relative concentrations of output gases normalized to the input from the temperature of the reactor at a contact time of 98 ms.

heater. Figure 2 shows the composition of the exhaust gases after the reactor at different temperatures for a contact time of 98 ms.

The conducted experiments showed high catalytic activity and stability of platinum catalysts on a modified multilayer γ -Al_2O_3 support. It is found that the main components of the fuel gas (propane and butane) are oxidized at a temperature of 450-500 C with a contact time of 98 ms. An increase in the contact time up to 161 ms weakly affects the composition of the exhaust gases, and a decrease of up to 32 ms results in a three-fold deterioration in the oxidation of the fuel gas at the same process temperatures.

2.3 Methane steam reforming

The study of methane steam reforming was carried out in an annular microchannel reactor, with a nano-sized catalyst, deposited on the surface of the inner cartridge-cylinder. The technology of the preparation of the catalytic surface is described in [6,7]. In the experiments carried out, the amount of doping additives La_2O_3 and BaO in the catalyst support γ-Al_2O_3 was changed. The obtained catalyst samples were studied by electron microscopy. In the process of studying chemical transformations, the temperature of the reactor, the contact time, and the composition of the mixture changed. Experiments on methane conversion were carried out with a contact time of 30, 60, and 120 ms. The results of the degree of methane conversion for three samples with different composition of doping substances at one contact time (60 ms) depending on the reactor temperature are shown in Fig. 3. Reducing the concentrations of lanthanum and barium oxides below 1.5 and 1%, respectively, leads to a decrease in the thermal stability of the catalyst. At high operating temperatures, γ - Al_2O_3 conversion occurs in high-temperature modifications of the α- and θ- Al_2O_3, leading to sintering and destruction of the catalytic layer. The catalyst with a higher content of lanthanum and barium (3 and 2%, respectively) is characterized by a less uniform dispersion of rhodium particles, which also leads to a decrease in catalyst activity due to the sintering of rhodium particles.

3 Conclusions

The degree of oxidation of hydrocarbons on a platinum catalyst supported on γ alumina in a microchannel was experimentally obtained for different reactor temperatures and contact

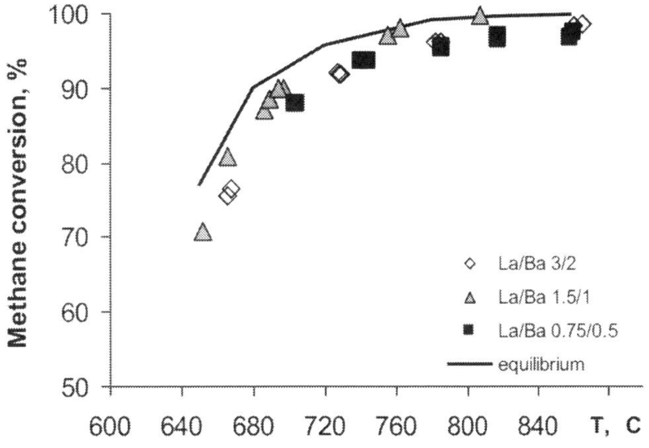

Fig. 3. Steam conversion of methane as a function of reactor temperature with a contact time of 60 ms. The line corresponds to the equilibrium degree of methane conversion.

times. It is shown that there is an optimum degree of doping of alumina with lanthanum and barium oxides for effective steam reforming of methane on a rhodium catalyst.

The research of chapter 2.3 was carried out with the partial support of the Program E2 of the Department of Power Engineering, Mechanical Engineering, Mechanics and Control Processes of the Russian Academy of Sciences.

References

1. Z.R. Ismagilov, M.A. Kerzhentsev, Cat. Rev: Sci. Eng. **32,** 51 (1990)
2. T.G. Alkhazov, L.Ya. Margolis, Moscow, Khimiya Publ. (1985)
3. W. Ehrfeld, V. Hessel, H. Lowe, *Microreactors: New technology for modern chemistry* (Weinheim:Wiley-VCH, 2000)
4. V.V. Kuznetsov, O.V. Vitovsky, J. Eng. Therm. **17**, 191 (2008)
5. V.V. Kuznetsov, O.V. Vitovsky, O.A. Gasenko, Theor. Foun. Chem. Eng. **48**, 376 (2014)
6. V.V. Kuznetsov, S.P. Kozlov, J. Eng. Therm. **20**, 229 (2011)
7. V.V. Kuznetsov, S.P. Kozlov, J. Eng. Therm. **17**, 53 (2008)
8. I.E. Beck, V. I. Bukhtiyarov, I. Y. Pakharukov, V. I. Zaikovsky, V.V. Kriventsov, V.N. Parmon, J. Cat. **268**, 60 (2009)
9. M. Maestry, D.G. Vlachos, A. Beretta, G. Groppi, E. Tronconi, J. Cat. **259**, 211 (2008)
10. V.V. Kuznetsov, O.V. Vitovsky, O.A. Gasenko, J. Eng. Therm. **18,** 187 (2009)
11. V.V. Kuznetsov, O.V. Vitovsky, O.A. Gasenko, Pat. #2549619 Rus. Fed. (2015)

MATEC Web of Conferences 115, 03012 (2017) DOI: 10.1051/matecconf/201711503012
STS-33

Plasma-thermal electric furnace for gasification of carbon-containing waste

A.S. Anshakov[1,2], *E.S.* Butakov[1], *P.V.* Domarov[1,2*], *A.E.* Urbakh[1], and *V.A.* Faleev[1]

[1]Kutateladze Institute of Thermophysics SB RAS, Novosibirsk, Russia
[2]Novosibirsk State Technical University, Novosibirsk, Russia

Abstract. The plasma-thermal electric furnace for gasification of various carbonaceous wastes (domestic, biological, agricultural, and other organic waste) has been created for the first time. Its constituent parts are: hydraulic drive for supplying the packed waste into the reaction zone; gas burner with the thermal power of 42 kW; electric-arc plasmatron with a power of 50 kW; chamber for ash residue melting. The test operation of the electric furnace showed that plasma gasification of carbon-containing materials produces synthesis gas suitable for the needs of heat and electric power industry. The results of thermodynamic calculations are in satisfactory agreement with the experimental data.

According to the data of [1], the volume of carbon-containing (organic) wastes is significant. In the world, their annual quantity is over 3 billion tons of solid waste and up to 700 km^3 of liquid domestic waste and industrial effluents. In the developed countries, from 300 to 750 kilograms of communal and from 1000 to 2500 kilograms of industrial waste are generated per capita per year.

The increasing volumes of waste lead to the search for the new, more efficient ways of their processing. The thermal methods of processing come to replace the open landfills and earth fills, whose exploitation causes environmental pollution; these ways are incineration; pyrolysis; and gasification.

Combustion is one of the main methods of eliminating waste in the world, since it contains a large amount of organic fuel. Since the process of waste combustion occurs at the temperature of 800 - 900°C, the exhaust gases include the substances poisoning the atmosphere (dioxins, furans), high molecular toxic compounds, etc.

Currently, waste recycling/utilization by plasma technology is one of the safest ways. All over the world, there are only a few small plants testing the plasma technology, whose main disadvantage is insignificant resource of plasmatron electrodes (up to 100 h). In Russia, the plasma plants are not used for waste disposal. In Russia, there are the single experimental samples of plasma plants for processing waste with various degrees of hazard.

The use of plasma technology [2-5] in application to organic waste processing allows the following: reducing significantly the volume of gas that is subjected to purification; melting and vitrifying the inorganic fraction of waste in the reactor (to obtain an inert slag that can be used in construction); excluding formation of nitrogen oxides due to the

* Corresponding author: Domaroff@yandex.ru

reducing medium in the reactor; obtaining the synthesis gas with calorific value of up to 10 MJ/m^3, suitable for its subsequent use in power generating devices. The technology is based on high-temperature (1200 ÷ 1600°C) plasma action and complete decomposition of high-molecular organic compounds and gasification of recyclable products using the arc plasma to produce a useful product: synthesis gas, which is a mixture of hydrogen and carbon monoxide, and inert slag, which can be used as a building material.

A schematic diagram of the experimental plasma-thermal setup with the capacity of up to 20 kg/h, designed to adjust the technology for high-temperature (plasma) recycling of renewable carbon-containing wastes of various origins (sawdust, rags, polyethylene), is shown in Fig. 1. Based on the results of testing the pilot plant, recommendations for the construction of pilot and industrial plants for processing the organic materials and waste and solving the environmental and energy problems will be developed.

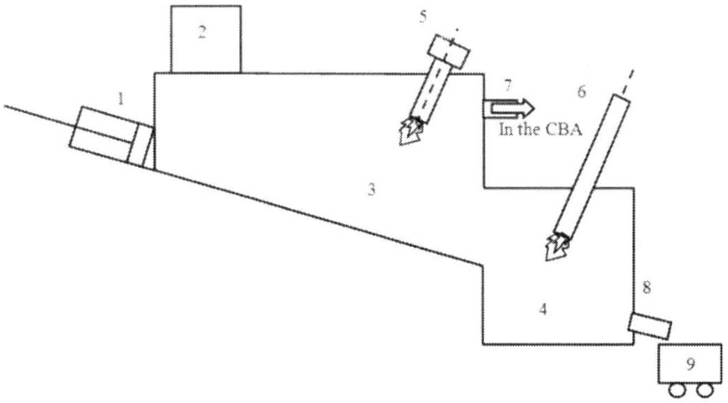

Fig. 1. Scheme of electric-plasma furnace.

Carbonaceous wastes with a bulk weight of 150-200kg/m^3 are delivered in the boxes with overall dimensions of 200 * 200 * 250 mm and are fed into the working space of the furnace through loading device 2. The loading device is equipped with a lock chamber, which prevents the escape of flue gases into the atmosphere from the working chamber of the plasma furnace at an excess pressure and suction of the atmospheric air into the working chamber, when it is vacuumized. The waste moves along the inclined sole in gasification zone 3 by means of hydraulic pusher 1 connected to the oil station and the control unit. Preheating of the lining and gasification zone is performed by gas burner 5 with the power of 42 kW and plasmatron 6 with the power of 50 kW. Then, the temperature of 1200°C at gasification is maintained only by the plasmatron. The mean mass temperature of the air plasma flowing from the plasmatron is 4000K. The synthesis gas obtained in the zone of gasification (fuel gas) is fed to a centrifugal-bubbling apparatus (CBA) for quenching the plasma chemical reactions and dust removal. The ash residue formed by waste gasification, which contains the unburned particles and carbon residue, falls into combustion zone 4 (Fig. 1), where carbon is burned under the action of a plasma jet and the mineral part of waste is re-melted into inert slag. After the CBA, the fuel gas is supplied to the after-burner chamber, where it burns down to CO_2. Before the release into the atmosphere, the gas passes into the mixer for subsequent cooling by mixing with the required amount of atmospheric air.

The source of energy for gasification of the organic part and melting the inorganic part is the plasmatron, as well as the chemical energy of carbon oxidation reactions. Oxygen enters the furnace through the plasmatron with plasma-forming gas. The slag is periodically poured off through tap hole 8 into trolley 9.

Furnace lining has the fire-resistant and heat-insulating layers made of chamotte and mullite-siliceous bricks; the chamber fastenings and vaults are made of heat-resistant concrete. On the lateral walls of gasification chamber, there are the nozzles for air supply, viewing window, and hatch for inspecting the lining. In the upper part of gasification chamber, there is a pipe branch for synthesis gas removal from the furnace; on this pipe branch, there is a device for sampling gases for analytical control.

The operating parameters of the electric furnace are determined by stable operation of its constituent elements. The plasmatron is the main of them. The results of its experimental characteristics are presented below.

To maintain a high temperature in the furnace chamber, an electric arc plasmatron is used. The trouble-free operation of setup required reliable long-term operation of the plasmatron. In this connection, the plasmatron with the power of 50 kW was designed, manufactured, and tested out with different flow rates of plasma-forming air $(3 \div 6) \cdot 10^{-3}$ kg/s with current varying from 120 A to 200 A.

The current-voltage characteristic of the arc is the most important electrophysical and energy characteristic of the arc plasmatron. It determines the region of stable arc burning at a change in the determining parameters: current strength, flow and type of plasma-forming gas, pressure of the medium, and geometric dimensions of plasmatron discharge chamber. The parameters of power supply source that ensure reliable continuous operation of the plasmatron are determined by the shape of the current-voltage characteristic and level of attainable voltage and current strength of the arc.

The experimental voltage-current characteristics of arc are shown in Fig.2 depending on the flow rate of plasma-forming gas.

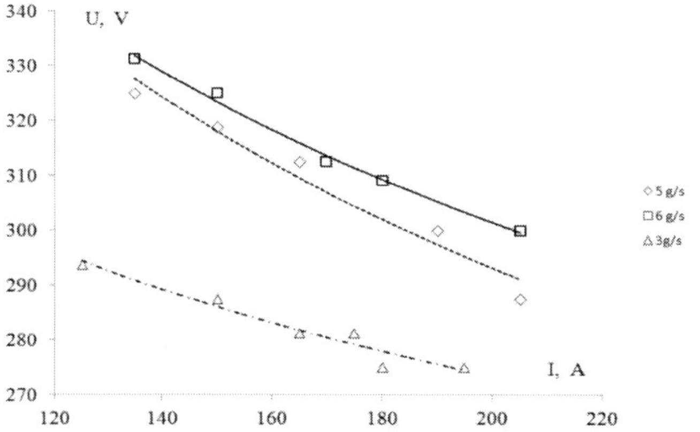

Fig. 2. Voltage-current characteristics of plasmatron arc.

One of the main parameters that determine the plasmatron efficiency is its thermal efficiency. In electric arc plasma generators, the electrodes perceive heat from arc radiation, convective heat exchange between arc discharge, plasma-forming gas and electrode walls, conductive heat transfer, and, finally, the cathode and anode arc spots. Heat from the outside of electrodes is removed with cooling water. Plasmatron efficiency is determined by the measured heat fluxes. At that, the integral heat fluxes entering the electrodes and other water-cooled elements of the structure are determined without dividing them into components. The method of heat flux measuring is standard: it is necessary to know the flow rate and temperature drop of water at the inlet and outlet of plasmatron.

The obtained experimental data are shown in Fig. 3. As it can be seen, the value of thermal efficiency at the certain low airflows is about 80%.

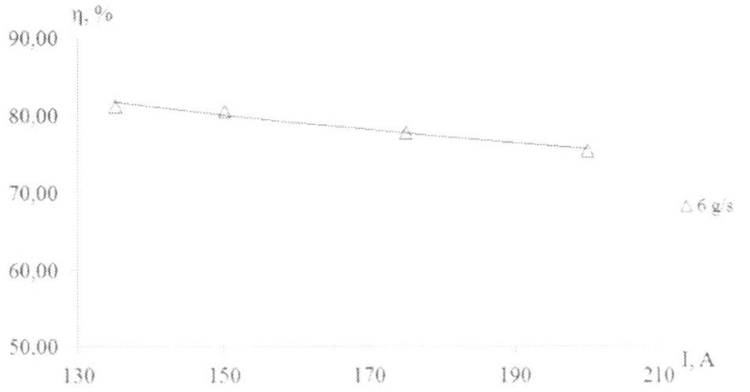

Fig. 3. Thermal efficiency of plasmatron depending on arc current.

Thermodynamic calculations of the process of gasification of model waste in the form of polyethylene pellets, textiles and sawdust were carried out, when the oxidizing gas in the amount stoichiometric for gasification, was fed into the reaction zone. These calculations were aimed at the following experimental studies of the plasma furnace and electric-plasma technology in general. According to investigation results, it was obtained that the synthesis gas in the volume from 3 to 6.75 nm^3 and caloricity from 6 to 12 MJ/m^3 is produced per 1 kg of polyethylene production waste depending on the oxidizing gas (air, steam, oxygen). The synthesis gas in the volume from 1.22 to 1.5 nm^3 with calorific value from 9 to 11.5 MJ/m^3 is produced per 1 kg of textile waste depending on the oxidizing gas (air, steam, oxygen). The synthesis gas in the volume from 0.9 to 1.6 nm^3 with calorific value from 7 to 11.5 MJ/m^3 is obtained per 1 kg of sawdust depending on the oxidizing gas (air, steam, oxygen).

Conclusions

1. A plasma-thermal electric furnace for gasification of carbonaceous wastes with productivity of up to 20 kg/h was developed, made and tested for the first time.
2. Experimental studies of electric-arc plasmatron showed the possibility of continuous operation of its electrodes during 500 hours and longer.

The work was financially supported by the Ministry of Education and Science of the Russian Federation according to the Subsidiary Agreement No.14.607.21.0118 (unique project identifier RFMEFI60715X0118)

References

1. V. Lotosh, *Processing the waste of environmental management* (Yekaterinburg: Poligrafist 2007)
2. A. Anshakov, V. Faleev, A. Danilenko, et al, Thermophys. and Aeromech., **14,** 607 (2007)
3. A. Anshakov, A. Valiliev, A. Pinaev, V. Faleev, Thermophys. and Aeromech., **17(4),** 595 (2010)
4. A. Bratsev, V. Kuznetsov, V. Popov, A. Ufimtsev, High. Temp., **49,** 244 (2011)
5. V. Messerle, A. Mosse, A. Ustimenko, Thermophys. and Aeromech., **23,** 637 (2016)

Burning of substandard liquid hydrocarbons with steam gasification

Igor Anufriev[1,*], *Sergey* Arsentyev[1,2], *Mikhail* Vigriyanov[1], *Evgeny* Kopyev[1], and *Oleg* Sharypov[1]

[1] Kutateladze Institute of Thermophysics SB RAS, Novosibirsk, Russia
[2] Novosibirsk State University, Novosibirsk, Russia

Abstract. The characteristics of burning of substandard liquid hydrocarbons in a direct-flow evaporative burner with controlled forced jet feeding of superheated steam are studied by the example of waste transmission automobile oil. The influence of the steam regime parameters on the temperature distribution in the flame, the thermal power and the composition of the combustion products is determined. The regimes providing simultaneously a low concentration of CO and NOx and a high power of heat release have been found.

At present, high attention is paid to environmental safety. 2017 is declared as the year of the ecology in Russia. An important problem is the disposal of a large volume of liquid hydrocarbon wastes (waste oil and lubricating fluid, remains of substandard fuels, and refinery waste, etc.) accumulated in enterprises of thermal energy sector and other industries. These wastes have not found wide practical application and carry a significant danger of the environmental pollution. The wide use of such hydrocarbons for replacing high-quality organic fuels is hampered by the non-compliance between the existing technologies and environmental requirements, their insufficient energy efficiency, problems associated with fuel ignition, coking of surfaces, etc. In this regard, of interest is to search for and scientifically substantiate new ways of burning substandard liquid hydrocarbons, and to develop and create power plants, effectively utilizing such fuel to obtain thermal energy.

One of the promising technologies is burning of low-quality liquid hydrocarbon fuels in the presence of superheated steam [1, 2]. Preliminary studies [3, 4] conducted in the IT SB RAS by the example of burning diesel fuel in original autonomous burner devices of evaporative type with 10-50 kW power [5, 6] have shown that when feeding superheated steam jet into the combustion zone, the combustion of liquid hydrocarbons sharply intensifies, providing steady ignition at high fuel consumption. In the process of gasification, intermediate components CO and H_2 are formed. It is shown that soot is practically absent in the combustion products, and its content in the combustion products is 35 mg/m^3 [7]. The use of autonomous burners during the experiments provides constant values of the steam state parameters [5, 6], but does not allow obtaining data on the regularities necessary for physical justification of effective control over the process.

* Corresponding author: anufriev@itp.nsc.ru

Developing previous studies [3, 4], this paper experimentally investigates the combustion of liquid hydrocarbons with controlled forced supply of the superheated steam jet to the combustion zone by the example of waste transmission automobile oil. The aim is to study the influence of regime parameters (flowrate and temperature of steam) on the main characteristics of the process.

The research was carried out on an automated fire stand (Fig. 1), equipped with an original direct-flow evaporative-type burner (up to 10 kW), an electric steam generator (average power consumption of 1.5 kW), a plunger dosing pump (flow rate up to 1.6 l/h), a steam generator control system, a heating and fuel supply system, electronic scales for controlling the consumptions of water and fuel, as well as the necessary instrumentations.

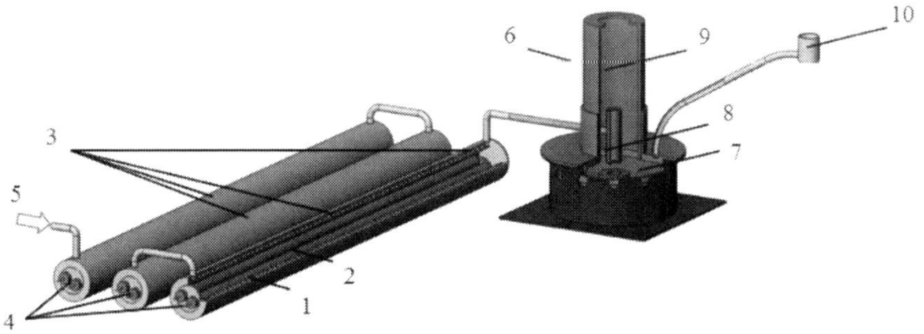

Fig. 1. Scheme of burner device and steam generator: 1 – steam generator; 2 – electric heater; 3 – thermocouples; 4 – heater power terminals; 5 – water supply; 6 – burner; 7 – combustion chamber; 8 – steam injector; 9 – gas generation chamber; 10 – fuel receiver with fuel pipeline.

The main elements of the burner are: a cylindrical body, a combustion chamber, a steam nozzle and a gas generation chamber. All elements are made of 12X18H10T steel. There are openings for natural air inflow from the atmosphere in the lower part of the combustion chamber. Since the burner is of evaporative type, it is necessary to ensure high-viscosity fuels preheating for their stable ignition. For this, a fuel heating system (up to 60 °C) based on electric heaters is used. Fuel is transferred to the combustion chamber through the fuel line. Stable fuel consumption is set by the fuel injector and pump. The steam nozzle is installed coaxially above the combustion chamber at the base of the gas generation chamber and is oriented vertically (output diameter of 0.5 mm). The steam nozzle is connected to an external steam generator. The laboratory electric steam generator allows obtaining superheated steam in the range of temperatures $T = 150\text{-}550$ °C and consumptions $F = 0.25\text{-}1.5$ kg/h (at pressures up to 10 atm). The scheme of the combustion process is similar to an autonomous burner [3]. The chemical composition of the used fuel is as follows: mass fractions of carbon – 84.24%, hydrogen – 13.48%, oxygen – 1.55%, and nitrogen – 0.73%.

With constant fuel consumption $F_f = 0.675$ kg/h, the value of $\gamma = F / F_f$ was varied from 0.31 to 1.38. At these values of γ, steady combustion was provided (Fig. 2). In previous studies of diesel fuel combustion, the influence of the steam superheating temperature on CO and NO_x concentration was not observed in combustion products. Therefore the measurements were carried out at a constant temperature of the superheated steam $T \sim 250$ °C (at the pressure of 2.5–8.0 atm).

Using platinum-rhodium/platinum-rhodium thermocouple, temperature profiles were obtained in the outer flame along the vertical axis of the burner (Fig. 3). Almost in all regimes, the maximum temperature in the flame reached 1400 °C and decreased with decreasing γ. The characteristic longitudinal size of the luminous zone ("length of the torch") varies from 0.27 m to 0.1 m with the increase in γ from 0.31 to 1.38 (Fig. 2). With

DOI: 10.1051/matecconf/201711503013

increasing γ and reducing flame length, the region with the maximum temperature approached the burner outlet (Fig. 3).

Fig. 2. Combustion of waste transmission automobile oil in the direct-flow evaporative burner at various superheated steam consumption ($T \sim 250$ °C): (a) $\gamma = 0.31$; (b) $\gamma = 0.54$; (c) $\gamma = 0.75$; (d) $\gamma = 0.97$; (e) $\gamma = 1.38$.

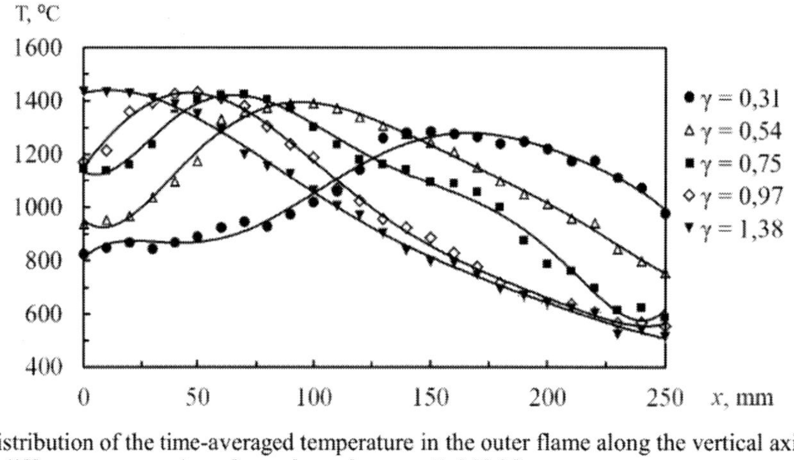

Fig. 3. Distribution of the time-averaged temperature in the outer flame along the vertical axis of the burner at different consumption of superheated steam, T=250 °C.

The thermal power of the burner device in the investigated range of steam parameters was measured with a flow calorimeter [3]. The specific heat and power were determined under stationary experimental conditions from a difference between the coolant (water) temperatures at the inlet and outlet and the consumptions of the coolant and fuel. The results showed that within the measurement error, the specific heat release (per 1 kg of fuel) did not depend on steam consumption and was $q = 45$ MJ/kg at a power of 8.5 kW. When calculating the specific heat, the energy received by the water in the steam generator was subtracted.

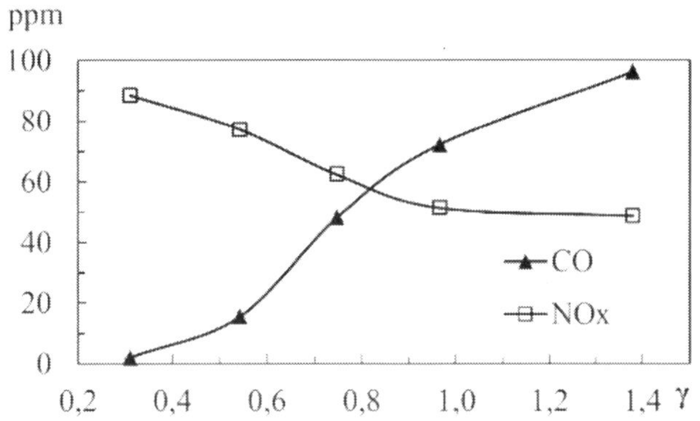

Fig. 4. CO and NOx concentrations at different ratio of steam and fuel consumption, *T*=250 °C.

Gas analyzer TESTO 350 was used to control the composition of the combustion products (O_2, CO, NO, NO_2, SO_2, CO_2). The studies were conducted simultaneously with the calorimetric measurements. The reaction products (cooled to ambient temperature) were sampled at the calorimeter outlet. For all studied regimes, the CO and NO_x content in the flue gases was significantly lower than the maximum allowable concentration level (Fig. 4). For the given fuel consumption, the value of steam consumption ($\gamma \approx 0.8$) provides low concentrations of CO and NO_x simultaneously with high thermal output.

Research was supported by the Russian Science Foundation (Project No. 17-79-20209).

References

1. S.V. Alekseenko, S.E. Pashchenko, V.V. Salomatov, J. Eng. Phys. Thermophys. **83**, 729 (2010)
2. M.S. Vigriyanov, V.V. Salomatov, S.V. Alekseenko, RF Patent **2219435** (2003)
3. S.V. Alekseenko, I.S. Anufriev, M.S. Vigriyanov, E.P. Kopyev, O.V. Sharypov, Combust. Expl. Shock Waves **52**, 286 (2016)
4. S.V. Alekseenko, I.S. Anufriev, M.S. Vigriyanov, V.M. Dulin, E.P. Kopyev, O.V. Sharypov, Thermophys. Aeromech. **21**, 393 (2014)
5. M.S. Vigriyanov, S.V. Alekseenko, Anufriev, O.V. Sharypov, RF Patent **2579298** (2016)
6. M.S. Vigriyanov, S.V. Alekseenko, Anufriev, O.V. Sharypov, RF Patent **2523591** (2014)
7. I.S. Anufriev, A.M. Baklanov, O.V. Borovkova, M.S. Vigriyanov, O.V. Sharypov, Combust. Expl. Shock Waves **53** 140 (2017)

Studies of ignition and combustion of coals subjected to electrochemical activation

Artem Kuznetsov[1*], *Evgenii* Butakov[1], *Anatoly* Burdukov[1], *Egor* Yaganov[2]

[1] Institute of Thermophysics SB RAS, Novosibirsk, Russia
[2] Open joint-stock company Sibtechenergo, Novosibirsk, Russia

Abstract. Coal is one of the most important energy sources in the world. According to forecasts, by 2020, the share of coal in the global energy sector will reach 50%, primarily due to the very likely reduction in oil and gas consumption, as well as the revision of the policy on the development of nuclear energy. For thermal power plants with pulverized-coal boilers, a problematic economic and technological issue is ignition and lighting, carried out with high-reaction fuel - gas, fuel oil, and diesel fuel. The cost of this fuel is much higher than the cost of coal itself, which means that the introduction of new technologies that allow excluding petroleum products from the energy processes that take place at the CHP plant is topical [1]. In this paper, the technology of electrochemical ignition, which is an alternative to gas and fuel oil ignition of power boilers, is studied.

The purpose of this work was to detect the influence of the consumption-dynamic parameters of the dust-air mixture on the efficiency of ignition of coal dust of Kuznetsk coal, and, to determine the optimal design and layout parameters of the ignition system.

The experiments were carried out on a large fire stand with a thermal power of 5 MW. The technological scheme of the fire stand is shown in Fig. 1. The main units and auxiliary systems of the fire stand are functionally related as follows [2]. Vibrant feeder of coal dust with a frequency converter with a capacity of up to 290 kg/h provides fuel to the air ejector with air capacity of 100 m^3/h. As a working medium, compressed air with a pressure of 6 bar is used. This ejector is designed for stable supply of coal dust to the electric arc block. Further, the primary dust-air mixture enters the second ejector with a capacity of up to 600 m^3/h, in which it mixes with air from a blower with a frequency converter. The second ejector is needed to regulate the flow of primary air to the arc block [3,4]. After this ejector, the dust-air mixture enters the electric arc block of the system, in which its primary ignition occurs. The ignited particles of coal dust enter the first stage of the reactor, where its active burning takes place. Solid admixtures of combustion products (ash and unburned coal dust) are separated in a cyclone, after which the gases are released by the smoke exhaust into the atmosphere.

* Corresponding author: temkansu@yandex.ru

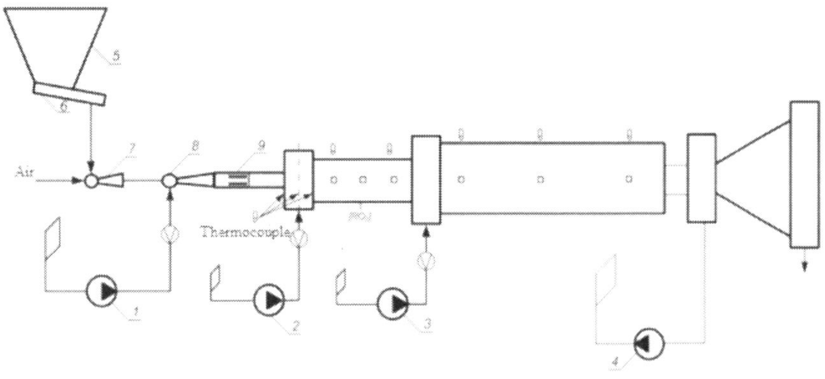

Fig.1. Technological scheme of the stand: 1,2,3 – blowers, 4 – smoke exhauster, 5 – coal dust batcher, 6 – vibrator feeder, 7,8 – ejectors, 9 – electric arc block (EAB).

The procedure for carrying out the experiments is determined by the sequence of preparation and measurement of the parameters. In the cold purge mode, the air flow to the first and second ejectors was measured. The inputs necessary for the experiment, as measured by the Prandtl tube and the differential manometer TESTO-435, were calculated and adjusted. Debugging of the gas and temperature analysis systems before the beginning of the experiment, testing the temperature sensors and the gas analysis system for the breakage of the measuring circuits, checking the contamination of the thermocouples and the purging of the gas pipelines from the measurement point to the gas analyzer were performed. Dust from the feeder was fed into the furnace using an ejector and a compressed air system. The initiation of the ignition of a pulverized-coal flame was carried out using an electric arc block (Fig.2).

Fig.2. Electrochemical fuel system a) working system of electrochemical ignition b) high and low voltage transformers

During the experiments, the temperature of the internal lining of the muffle, the combustion chamber and the afterburner was measured. Combustion products were analyzed with the help of the gas analyzer. After the afterburning chamber, the fly ash was taken-abductions according to the isokinetic sampling method, as well as analysis of the flue gas composition with the help of an additional TESTO-340 gas analyzer. The main selection of ash was made from the cyclone bunker in front of the smoke exhauster.

The efficiency of ignition of the dust-air mixture flow was studied as a function of the mass flow of dust. Condition - constant volumetric flow of primary air. In experiments with a constant flow of primary air at a level of 200 m3 / h, the flow rate of coal dust supplied for combustion was changed from 42 kg/h to 290 kg/h. The concentration of dust in the flow thus varied from 0.17 to 1.13, and excess air varied from 0.89 to 0.13. A significant difference between the excesses measured in the experiments in the chamber compared to the coefficients calculated based on the measured flows of fuel and air indicates an incomplete reaction (mixing) of oxygen with coal dust during its combustion. Only a part of the supplied air participates in the combustion of coal fuel after its ignition in EAB. Analyzing the obtained data, it should be noted that nitrogen oxides decrease with electrochemical combustion, with an increase in excessively organized air supply. This is because when feeding more fuel into the EAB, the process of ignition intensifies - the torch in the combustion chamber is formed instantaneously. A greater number of carbon particles falls into the area of the electrodes and passes through electric arcs and their electric fields. The amount of fuel supplied determines the amount of heat released during combustion. The increase in heat in the experiments leads to an increase in the rate of heating of the lining and the combustion chamber itself-conditions are created for the thermal heating of coal particles, the escape of volatile and combustion of carbon. The content of combustibles in the entrails practically does not change in the experiments, with electrochemical combustion and with rectilinear fuel supply only 40- 50% of the carbon contained in coal dust burns into the combustion chamber. The reason for this is the lack of conditions for the oxidant to mix with the carbon. In electrochemical combustion, combustion occurs due to the interaction of carbon particles with transport air at the same velocities of both. Initiation of the same combustion occurs in the EAB [5] of the system without thermal preparation of fuel.

The optimal position of electrodes for effective ignition of coal dust was determined. In the experiments, the position of the electrodes in the EAB was changed. The layout of the electrodes in the EAB is shown in Fig. 3. With a constant flow of primary air at 215 m^3/h, the flow rate of coal dust supplied for combustion was changed from 50 kg/h to 220 kg/h. The concentration of dust in the stream varied from 0.19 to 0.87, while excess air varied from 0.81 to 0.18. When coal dust was injected into the EAB, the ignition was instantaneous. However, if compared with the previous location of the electrodes in the experiments, the glow of electric arcs with coal dust became worse. Visually, the dust was clearly trailing along the bottom of the quartz tube. The quartz camera itself was quickly absorbed. Burning in the reactor and combustion chamber is stable from the moment of coal dust supply, and the set of temperatures is stable.

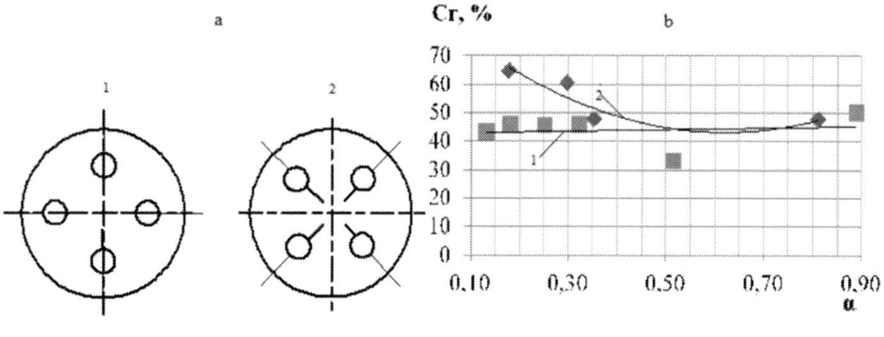

Fig.3. a) scheme of the location of electrodes, 1 – "Rhombus", 2 - Square" (45-degree rotation) b) dependences of the change in the content of combustibles in the entrainment on the coefficient of excess air in the layout of the electrodes "Square" and "Rhombus"

Comparing the results obtained with respect to the content of combustibles in entrails with previous experiments, it should be noted that with reduced excess air, in conditions of supplying more fuel, the changed scheme of the location of the "Square" electrodes worsens fuel more than in the previously tested "Rhombus" scheme. So, under the "Rhombus" scheme, the content of combustibles in the entrainment did not exceed 45% in the whole range of the investigated excess air, and in the "Square" scheme the range of this value increased to 65%.

Conclusion

1 The maximum possible speed of the dust-air mixture in the cross section of the electrode installation (electric arc block) is 20 m/s. Further increase in speed is impossible due to the breakthrough of coal particles of electric arcs.

2 The minimum possible speed of the dust-air mixture in the cross section of the electrodes is 6 m/s. A further reduction in speed leads to separation of coal dust in the electric arc block and its heating.

3 The minimum fuel consumption for combustion is not less than 50 kg/h for Kuznetskiy coal with a maximum excess of supply air of not more than 0.8.

4 When coal dust is supplied to the operating electric arc unit, the ignition of the fuel takes place instantaneously, regardless of the amount of primary air flow.

5 Preheating the flue gases to the maximum temperature at a distance of half a meter occurs at a speed of 30-40 °C/min. Stabilization of the combustion regime - the output to a constant oxygen content occurs within 2-5 minutes, depending on the excess air.

6 Burnout of coal dust is also incomplete. Visually the ash has a dark color, which indicates a significant underburning of the fuel. The relative decrease in the yield of volatiles in the selected samples was 50-75%, and the reduction of carbon was 30- 50%.

7 When the electrodes are positioned according to the "Square" scheme, the burnout of the fuel is worse than under the "Rhombus" scheme. So, under the "Rhombus" scheme, the content of combustibles in the entrainment did not exceed 45% in the whole range of the investigated excess air, and in the "Square" scheme the range of this value increased to 65%.

The work was financially supported by the Russian Ministry of Education and Science, according to Subsidiary Agreement No. 14.607.21.00150 (unique project identifier RFMEFI60716X0150).

References

1. V.N. Churashev, G.V. Chernova, *Problems of substitution of fuel oil in heat power engineering, New technologies of fuel compression: their current state and future use*: Proc. All-Russia. Scientific-techn. Seminar. Moscow 262 (2001)
2. A.P. Burdukov, E.B. Butakov, V.I. Popov, M.Yu. Chernetskiy, N.S. Chernetskaya, Thermal Science. **20,** 23 (2016)
3. A.S. Koshkarov, Y.I. Naumov, S.F. Nikolayev, A.A. Shinkarev, RF Patent No. **2498159** (2013) (in Russian)
4. Y.I. Naumov, S.F. Nikolayev, A.A. Shinkarev, Proc. IX All-Russian conferences with international participation «Combustion of solid fuel »Novosibirsk, November 22-25, 2015.
5. M.F. Zhukov, E.I. Karpenko, V.S. Peregudov, Low-temperature plasma **15,** 304 (1995) [in Russian]

Influence of volatiles oxidation processes on kinetics of sawdust combustion

Igor Donskoy[1,*], *Aleksandr* Kozlov[1], and *Vitaliy* Shamanskiy[1]

[1]Melentiev Energy Systems Institute, 130 Lermonova Str., Irkutsk, 664033, Russia

Abstract. The paper considers influence of the devolatilization on the combustion of a small woody particle at TGA conditions. Differences between woody char and raw sawdust combustion kinetics are evaluated. Gas formation kinetic parameters are estimated.

1 Introduction

Most of the studies devoted to the modeling of wood particle combustion assume simplifications related to the impact of kinetics of gas-phase reactions on heterogeneous reactions of fuel. For example, it is often assumed that kinetics of gas-phase reactions can be defined by a small number of kinetic equations for overall reactions that involve model substances (methane or a conventional surrogate component [1]). The reaction capability of a wood char in this case normally does not depend on the history of its formation. However, owing to the high rate of chain reaction spread in the gas medium space, the flame front of volatiles combustion can influence kinetics of heterogeneous oxidation. Active particles formed in the flame penetrate into the surface layer where they can both activate part of solid phase of already pyrolized fuel, and accelerate pyrolytic destruction of still remaining macromolecules and their fragments [2, 3]. At the same time thermal destruction of wood macromolecules occurs primarily according to a radical mechanism, and transport of radicals from internal regions to the particle surface may be of a relay-race nature, thus forming their diffusion flow. This process in turn triggers chain mechanisms in the near-surface flame and increases the rates of gas-phase reactions.

2 Experimental setup and results analysis

To study the phenomena that occur at thermochemical conversion of wood particles we used a synchronous thermal analyzer manufactured by the firm NETZSCH-Geratebau. It consists of thermal analysis unit STA 449 F1, quadruple mass spectrometer QMS 403 AEOLOS and a pulse thermal analysis unit Pulse TA. The obtained products of solid fuel pyrolysis and combustion were recorded in a range of mass numbers from 1 to 200, with 70 eV electron impact ionization. The mass-spectrometer signal was calibrated by pulse supply of respective calibration gases.

* Corresponding author: donskoy.chem@mail.ru

The initial wood composition was determined by thermal analysis: moisture content – 3.47%, ash content – 1.40%, carbon content – 54.28%, hydrogen content – 5.20%, oxygen content – 40.12%, nitrogen content – 0.40%, and the lowest calorific value – 20.52 MJ/kg. Lignin content in the organic matter makes up 34.0%, tar – 5.3%. The detailed technique is presented in [4].

The research involved 11-12 mg of sawdust, about 0.2 mm in size. A representative sample of sawdust (pine) prepared by a standard quartering method was placed in the alumina crucible with low walls. Such a form of the crucible facilitates a uniform access of oxidizer to fuel particles. The crucible was placed in the furnace of thermal analyzer and heated from a room temperature up to 900°C at a rate of 10°C/min in the argon flow with a flow rate of 80 ml/min. The sawdust pyrolysis was accompanied by removal of volatiles and formation of wood char particles. Then, the sample was cooled to 100°C, without taking it from the furnace, and further the obtained char was completely burnt in the air flow at the same heating rate. The air flow rate was 70 ml/min. The other sample was heated in the air flow until complete combustion under the indicated parameters. The obtained thermogravimetric curves are presented in Figure 1. The results were calculated on a dry and ash-free organic wood matter basis.

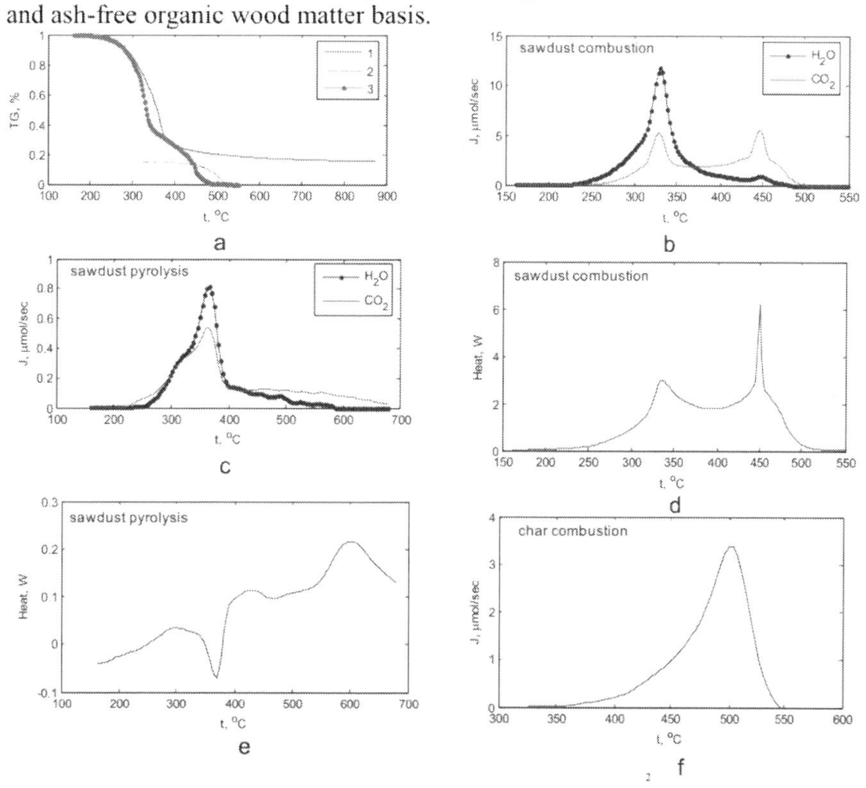

Fig. 1. Experimental curves for sawdust and char combustion.

Figure 1*a* shows that the oxidative pyrolysis and further combustion of wood char without clearly identified pyrolysis stage occur more intensively even at lower temperatures than on the curves with preliminary pyrolysis in an inert medium. The first bend on thermogram 3 apparently corresponds to a transition from combustion of mainly carbohydrate components to combustion of lignin, and the second one corresponds to complete combustion of wood char. This is confirmed by the dynamics of H_2O and CO_2 formation in sawdust combustion, which is shown in Figure 1*b*. Figure 1*b* illustrates that the maximum intensity of water formation is observed in a temperature interval of 320-

350°C, whereas the intensity of CO_2 production has two peaks which can be referred to the destruction of aliphatic and aromatic structures: 320-350°C and 430-560°C, respectively. Similar phenomena were observed in [5-8].

In order to assess the potential impact of organic wood matter pyrolysis on the intensity of gas-phase reactions we will consider temperature dependence of the CO and CO_2 formation intensity in the course of pyrolysis in an inert medium (Fig. 1*c*). In this case the peaks of the CO and CO_2 formation intensity occur in an interval of 350-400°C, whereas in the oxidative pyrolysis as soon as this temperature is reached the burnout of cellulose components of the organic wood matter finishes. An analysis of the total thermal effect at oxidation of volatile components (Fig. 1*d*) shows that after 250°C their oxidation is accompanied by the formation of CO_2. Pyrolysis in an inert medium (Fig. 1*e*) in the region of volatile yield peak values is a low-energy process. It is however worth noting that in this case we record a total effect of exothermic intramolecular oxidation with molecule restructuring and formation of volatile components, and energy consumed for the related disruption of chemical and intermolecular bonds. We can assume that the development of oxidative processes in gas phase (most probably in the near-surface layer of a fuel particle) under such a relatively low temperature is possible only in the case where generation of active particles (radicals) occurs at the destruction of macromolecules, i.e. inside the particle, where this energy consumption is compensated for by the heat of internal oxidation. Figure 1*e* shows that it is precisely in the area near 250°C the total thermal effect transitions to the positive value region. The experiment that proves the emergence of radicals in gas phase at biomass pyrolysis is described in [2].

An analysis of wood char combustion after pyrolysis is made on the basis of temperature dependence of carbon dioxide formation in this process (Fig. 1*f*). The elementary kinetic equation that describes this process can be written as follows:

$$J = \frac{dn_{CO_2}}{d\tau} = k \cdot \exp\left(-\frac{E_a}{RT}\right) \cdot S \cdot \left(n^0 - n\right) \cdot C_{O_2} . \qquad (1)$$

Here: E_a is activation energy, J/mole; S is specific area of a sample, m^2/mole; n is wood char quantity, mole; C is oxygen concentration in gas phase, mole/m^3; and k is the matching factor. Assuming that at the beginning of the process the area of a sample reaction surface changes little, since the total decrease in the geometrical surface can be accompanied by an increase in the surface concentration of active centers, and taking into account that oxygen concentration in the oncoming reaction flow does not change, (1) can be transformed to the form: $\ln J = a - E_a/RT$. Figure 7 presents a diagram of relationships between $\ln J$ and $1/T$. Linear part belongs to an ascending branch of curve in Fig. 8. This is a range of constant activity of wood char surface (T~360÷500°C). The activation energy determined from the diagram equals 119 kJ/mole.

There is a similar section in the interval of temperatures T~270÷320°C on the curve constructed for oxidative pyrolysis (Fig. 8), which also corresponds to an ascending branch of a carbon dioxide formation curve for the first peak in Fig. 1*b*. In this case the process rate rises approximately 250 times (while in reality even more, since the combustion temperature in the case of oxidative pyrolysis is lower). Since the parameters of external gas phase (i.e. beyond the diffusion surface layer) in both cases differ only in temperature, such an increase in the rate can be explained by a higher energy of carbon atoms in the initial wood. However, the actual equality between the activation energies of CO_2 formation at wood char combustion and at the beginning of the oxidative pyrolysis needs to be explained. We can suppose, in particular, that the internal oxidation, i.e. oxidation involving atoms of oxygen in the wood, which undoubtedly occurs at pyrolysis, hardly takes place for carbon in this case. Most likely, oxygen of carbohydrate and related molecules is consumed for water formation. Intensive and synchronous with CO_2 release of water in this range of temperatures is seen in Figure 1*b*. After removal of water molecules from macromolecules, for example, cellulose molecules, the areas of carbon matrix that differ from steady allotropic structures of free carbon, i.e. have a higher Gibbs energy, are formed in the wood. It is precisely these areas that are attacked by the molecules of oxygen or oxygen containing radicals from a surface layer. Certainly such a mechanism does not refer, for example, to the carboxyl carbon or carbon of extractive components (tar). However, in percentage terms the concentration of such carbon is low: no more than several per cent of the total amount. In the case of pyrolysis in an inert medium (Fig. 1*c*), it is seen that carbon dioxide yield is almost 10 times less than in an oxidative medium. Note that nearly 65% of organic wood matter (Fig. 1*a*), i.e. its carbohydrate constituent, burns out before the temperature reaches ~ 330°C.

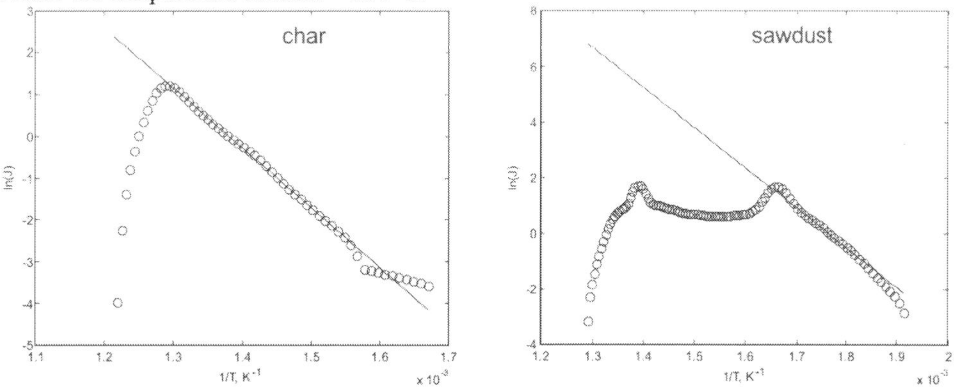

Fig. 2. Arrhenius plot for CO_2 in char and sawdust combustion

An in-depth consideration of the above qualitative suppositions on the wood pyrolysis and combustion mechanisms is possible involving the tools of detailed chemical kinetics [9]. The study on the sawdust pyrolysis will further the detailed study of chemical kinetics, which will create preconditions for research into large particles where thermal inhomogeneity leads to simultaneous processes of oxidative (at the surface) and inert (intraporous, protected by Stefan flows) pyrolysis.

The work is supported by Russian Foundation for Basic Research (project No. 16-08-00694).

References

1. J. Yu, M. Zhang, J. Zhang, Proc. Comb. Inst. **32,** 2037 (2009).
2. A. Kozlov, D. Svishchev, I. Donskoy, V. Shamansky, J. Therm. Anal. Calorim. **122**, 1098 (2015).
3. B. Goshayeshi, J.C. Sutherland, Comb. Flame. **141**, 1900 (2014).
4. A. Kozlov, D. Svishchev, I. Donskoy et al., J. Therm. Anal. Calorim. **122**, 1213 (2015).
5. S. Munir, S.S. Daood, W. Nimmo et al., Bioresource Tech. **100**, 1413 (2009).
6. K. Cheng, W.T. Winter, A.J. Stipanovic, Polymer Degradation Stability **97**, 1606 (2012).
7. W. Wu, Y. Mei, L. Zhang, R. Liu, J. Cai, Fuel **156**, 71 (2015).
8. N. Aniza, S. Hassan, M. Int. J. Mech. Eng. Sci. **10**, 1943 (2016).
9. E. Ranzi, T. Faravelli, F. Manenti, Adv. Chem. Eng. **49**, 1 (2016).

Ignition study of high-energy materials containing Al, B, AlB_2 and TiB_2 powders

Alexander Korotkikh[1,2*], *Vladimir* Arkhipov[2], *Konstantin* Slyusarskiy[1], and *Ivan* Sorokin[1]

[1]Tomsk Polytechnic University, Institute of Power Engineering, 634050, Tomsk, Russia
[2]Tomsk State University, Research Institute of Applied Mathematics and Mechanics, 634050, Tomsk, Russia

Abstract. Boron and its compounds are considered as possible high efficiency fuel components for high-energy materials (HEMs), since they have high gravimetric and volumetric heats of oxidation in comparison with aluminium powder. This work aims to modify a commercial 99.5% pure boron and AlB_2 or TiB_2 powders to improve the ignition and combustion characteristics for solid propellants. The ignition parameters for the model HEM sample based on ammonium perchlorate, ammonium nitrate and energetic binder, containing Al, B, AlB_2 and TiB_2 powders are presented. It was found that the greatest effect at the laser ignition is achieved at a full replacement of Al by B powder, at which the ignition time and the activation energy of HEM are reduced by 2.1–2.7 and 1.8 times, respectively. When replacing Al by AlB_2 powder in HEM the ignition time is reduced by 1.7–2.4 times, and the activation energy is increased by 1.4 times.

1 Introduction

Boron and its compounds are considered as possible high efficiency fuel components of solid propellants for solid fuel rocket and ramjet engines [1–3]. Boron hydrides and organic derivatives have high burning rate, varying in a wide range at changing the air relation and pressure in the combustion chamber. The combustion heat of 1 kg boron in the oxygen more than 1.9 times in comparison with the combustion heat of aviation kerosene and 1 kg aluminum.

Note that solid propellants are widely used, which contain up to 22 wt.% aluminum micropowder. According to the study [4, 5] the aluminum oxidation during the combustion of composite solid propellants strongly influence by the presence on the burning surface the refractory layer containing the aluminum oxide particles and carbonaceous resudues. Melting point of alumina significantly above the melting point of aluminum. The aluminum particles combustion possible at the high temperature gradient of high-energy materials (HEMs) reaction layer near the burning surface with occurrence of cracks and destruction of the oxide layer resulting in the oxidation of active metal. It is well known [6] that the destruction of the aluminum oxide layer on the particle surface may react with the carbon particles to form aluminum carbide.

* Corresponding author: korotkikh@tpu.ru

However, HEMs, containing aluminum micropowders, have a long delay ignition time and low burning rate compared to HEMs, containing nanopowders of aluminum or other metals [7–10].

This paper presents the experimental dependences of the ignition time on heat flux density, values of the ignition temperature and activation energy for the model HEMs based on ammonium perchlorate and nitrate, energetic binder, containing powders of aluminum, boron, aluminum diboride and titanium diboride.

2 Experimental

2.1 HEM samples

To study the main ignition characteristics we used four tested compositions of HEM. The first – the basic composition, containing ammonium perchlorate (fraction with particle size 160–315 μm), ammonium nitrate (fractions with particle size less than 50 μm), the energetic binder of MPVT-ASP type, and 30 wt.% Al micropowder of ASD-4 type [7]. In other tested compositions the Al powder is completely replaced by the amorphous boron, aluminium diboride and titanium diboride powders (Fig. 1).

Fig. 1. SEM images of amorphous boron (a) and aluminium diboride (b) powders.

2.2 Ignition of HEM

The ignition process was studied with the use of setup for the radiant heating on the basis of CO_2-laser with the wavelength of 10.6 μm and power of 200 W (Fig. 2). Prior to testing the samples were cut into tablets of 5 mm in height.

The test HEM sample (6) was attached to the substrate of recoil force transducer (8) to register the gasification products outflow from the burning surface. When opening the shutter (4) the radiation was focused by the sodium chloride lens (5) to the HEM sample (6). Signals from the recoil force transducer (8) and photodiodes (7) were transmitted to the L-card-E 14-440 ADC (9) and recorded in the personal computer (10), and then processed with the software application LGraph2. The time delay of start gasification t_{gas} of HEM sample was determined as time interval between the moments of signals change of photodiode near the shutter (7) (or a thermocouple installed in the laser beam behind the shutter), and the recoil force transducer (8). Photodiode (7) registered the moment of opening the shutter, transducer (8) recorded the appearance of recoil force signal of gasification products flowing from the front (irradiated) sample surface. The ignition time t_{ign} of HEMs was determined by difference between the moments of signals from two photodiodes (7), one of which registered the appearance of flame near the end surface of HEM sample. The relative error of delay times measuring of t_{gas} and t_{ign} was equal 5–12 % at the value of confidence probability 0.9.

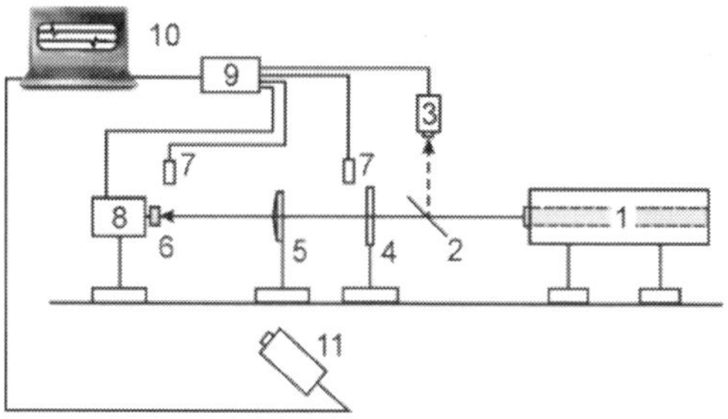

Fig. 2. The scheme of experimental setup based on CO_2-laser: 1 – CO_2-laser; 2 – beam-splitting mirror; 3 – thermoelectric sensor of radiation power; 4 – shutter; 5 – lens; 6 – HEM sample; 7 – photodiodes; 8 – transducer; of recoil force; 9 – ADCs; 10 – PC; 11 – thermal imager.

The radiation power and heat flux density of CO_2-laser beam was measured by the thermoelectric sensor of radiation power (3). The maximum radiation power was defined in the center of the laser beam using a diaphragm with diameter 2 mm. The diameter of the laser beam incident on a HEM sample was 10 mm.

3 Results and Discussion

The ignition time of tested HEM samples was determined in atmospheric conditions. The approximation dependences for HEMs ignition time vs. the heat flux density were determined (Fig. 3).

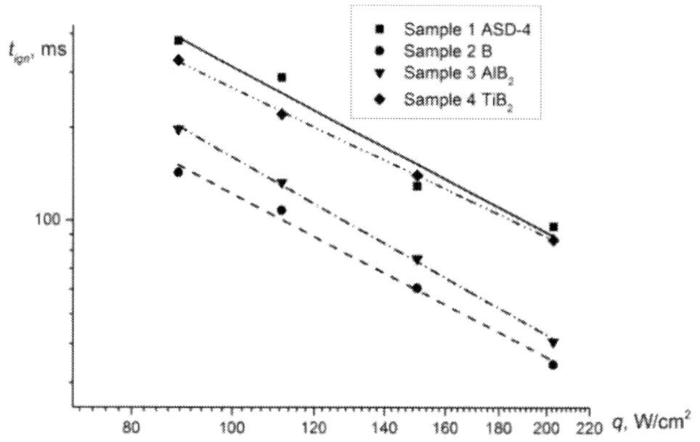

Fig. 3. The ignition time vs. the heat flux density for tested HEMs containing powders:

Al – t_{ign}=9.90·10^5·$q^{-1.75}$; B – t_{ign}=2.02·10^5·$q^{-1.61}$; AlB_2 – t_{ign}=7.39·10^5·$q^{-1.83}$, TiB_2 – t_{ign}=4.71·10^5·$q^{-1.62}$.

It was found that a complete replacement of aluminum by boron powder in HEM leads to a decrease of the ignition time by 2.1–2.7 times in the heat flux density range of 90–200 W/cm^2. Replacement of aluminum by aluminum diboride micropowder in the HEM basic composition reduces the ignition time of sample by 1.7–2.4 times. At the same time, the ignition time of the HEM containing TiB_2 is 1.05–1.2 times lower in comparison with the basic composition.

The kinetic constants and ignition temperature of HEM were determined according to [11] and with use of the t_{ign} and T_{ign} equations. The ignition data for tested HEM samples are presented in Table 1. Here there are experimental values of the average $\bar{T}$ and the maximum T_{max} surface temperature on the HEM reaction layer at appearance of visible flame which obtained using a thermal imager Jade J530 SB at q = 114 W/cm^2.

Table 1. The activation energy, specific heat of reaction, ignition temperature and surface temperature on the reaction layer for tested HEMs.

Parameter	HEM sample containing metal powder			
	ASD-4	B	AlB_2	TiB_2
E_a, kJ/mol	219	120	307	136
Qz, W/g	$8.32 \cdot 10^{10}$	$3.77 \cdot 10^{8}$	$4.81 \cdot 10^{18}$	$3.84 \cdot 10^{7}$
T_{ign}, K	1203	859	935	1139
$\bar{T}$, K	1040±30	949±110	1009±90	897±20
T_{max}, K	1516±80	1531±120	1600±170	971±30

4 Conclusions

The ignition parameters for the HEMs based on ammonium perchlorate, ammonium nitrate and energetic binder of MPVT-ASP type, containing powders of aluminium, boron, diboride aluminum or titanium diboride are presented. It was found that the greatest effect at the laser ignition is achieved at a full replacement of the Al micropowder by 99.5% pure boron, at which the ignition time and the activation energy of HEM sample are reduced by 2.1–2.7 and 1.8 times, respectively, in the heat flux density range of 90–200 W/cm^2.

When replacing the Al micropowder by the AlB_2 powder in HEM the ignition time is reduced by 1.7–2.4 times, and the activation energy is increased by 1.4 times. For HEM containing TiB_2 powder the ignition time is slightly reduced by 1.05–1.2 times. Thus boron powder and its compounds can be used to improve the ignition characteristics for HEMs.

The reported study was supported by the Russian Scientific Foundation, Grant 16-19-10316.

References

1. A. Gany, Y.M. Timnat, Acta Astronautica **29**, 3 (1993)
2. T.L. Connell, G.A. Risha, R.A. Yetter, C.W. Roberts, G. Young, J. Propulsion Power **31**, 1 (2015)
3. K.-L. Chintersingh, M. Schoenitz, E.L. Dreizin, Comb. Flame **173**, 288 (2016)
4. P.F. Pokhil, A.F. Belyaev, Yu.V. Frolov, V.S. Logachev, A.I. Korotkov, *Combustion of Powdered Metals in Active Media*, (Science, Moscow 1972) [in Russian]
5. V.E. Zarko, O.G. Glotov, Science and Technology of Energetic Materials, **74** (2013) [in Russian]
6. A.A. Gromov, A.G. Korotkikh, A.P. Il'in, L.T. DeLuca, V.A. Arkhipov, K.A. Monogarov, and U. Teipel, *Nanometals: Synthesis and Application in Energetic Systems, Energetic Nanomaterials: Synthesis, Characterization, and Application* (Eds. V.E. Zarko and A.A. Gromov, Elsevier Inc. 2016) [in Russian]
7. V.A. Arkhipov, S.S. Bondarchuk, A.G. Korotkikh, V.T. Kuznetsov, A.A. Gromov, S.A. Volkov, L.N. Revyagin, Combustion, Explosion, and Shock Waves **48**, 5 (2012)
8. A. Abraham, H. Nie, M. Schoenitz, A. Vorozhtsov, M. Lerner, A. Pervikov, N. Rodkevich, E. Dreizin, Combustion and Flame **173** 179 (2016)
9. V.A. Arkhipov, A.G. Korotkikh, V.T. Kuznetsov, A.A. Razdobreev, I.A. Evseenko, Russian J. Phys. Chem. B **5**, 4 (2011)
10. T.R. Sippel, S.F. Son, L.J. Groven, S. Zhang, E.L. Dreizin, Comb. Flame **162** 846 (2015)
11. V.A. Arkhipov, A.G. Korotkikh, Comb. Flame **159** 409 (2012)

Mathematical modelling of sparkplug ignition of a coal-dust monodisperse suspension in a methane-air mixture

Alexey Yu. Krainov and *Ksenia M.* Moiseeva*

Tomsk State University, Tomsk, 634050 Russia

Abstract. The paper provides the numerical simulation of a coal-dust monodisperse suspension sparkplug ignition in a methane-air mixture. The methane-air mixture is not stoichiometric. The aim of the research is a design-theoretical determination of the minimum ignition energy of coal-dust-methane-air mixture (CDMAM) depending on the methane volume fraction. Simulation has shown that the increase of the methane volume fraction leads to the decrease of the CDMAM minimum ignition energy, and with the expansion of coal particles the minimum ignition energy rises.

1 Introduction

The research is devoted to sparkplug ignition of the coal-dust monodisperse suspension in a methane-air mixture. The solution to the problem provides applied data dealing with fire and explosion safety in coal industry. The research [1] describes the igniting mechanism and explosion in mines. It offers the explosion emerging scenarios and proposes assumptions of the influence the coal-dust content (particle size, mass content of reactive particles, and presence of the inert state) and reacting gas content (percentage of methane and impurities in the gas) on the possibility of the mixture to ignite.

The research [2] has drawn a conclusion about the ignition peculiarities and flame propagation in reacting gas mixture. The work provides research efforts of the coal-dust suspension explosive risk in air and in methane-air mixture. The reactive gas-coal dust-air mixture is capable of exploding under the low concentration of the coal dust. The paper [3] has shown theoretically that the coal dust burning in the volume becomes possible after the addition of a small amount of methane or after the heating of the volume walls.

The presence of the coal dust significantly effects the ignition characteristics of reactive mixtures. The minimum ignition energy of the coal dust depends on its mass concentration [4, 5]. The "minimum sparkplug ignition energy-coal dust mass concentration" curve has been plotted in [5] and its minimum value has been determined. The work [6] defines the minimum energy dependence on the parameters of the coal dust. The gas suspension motion due to its heating has not been taken into account. The small particles decrease the minimum ignition energy. The larger particles increase the minimum ignition energy when the mixture is close to the stoichiometric condition, and decrease when the mixture is far

* Corresponding author: Moiseeva_KM@t-sk.ru

from stoichiometric. The large particles do not influence the minimum ignition energy; it is equal to the mixture ignition energy. The researchers in [7] have investigated the problem of the thermal expansion influence on the minimum sparkplug ignition energy of the gas. According to the obtained results the thermal diffusion model of the gas sparkplug ignition energy decreases the value of the minimum energy more than two times.

2 Mathematical model

The monodisperse suspension of the coal dust is uniformly distributed in volume. The mass concentration of the coal dust is m_{dust} and the methane volume concentration is a_{CH4}. The mixture is not stoichiometric. The instantaneous filiform ignition source is located in the center of the volume. The outer boundary of the computational domain is supposed to be infinitely far from the ignition source. The diffusion and the heat transfer coefficients depend on the temperature [8]. There are two parallel reactions in the mixture, the first is the exothermic chemical reaction in the gas, and the second one is the heterogeneous reaction on the particle surface. The reaction rate in the gas is determined by the second-order kinetics (the first-order kinetics for methane and the first-order kinetics for oxygen). The heterogeneous reaction rate is defined taking into account the diffusion of the oxidizer to the particle surface [9]. The chemical reaction rate constants depend on the temperature under the Arrhenius law. It is considered the interphase interaction between gas and particles. The electrode heat waste is neglected.

The mathematical model is based on the dual-velocity two-phase model of the reacting gas-dispersion medium [10]. The system of equations includes the mass-conversation equation, the impulse-conversation equation, the total energy-conversation equation of the gas and the particles taking into account the thermal conductivity and the chemical reactions in the gas and on the particle surface, the mass-conversation equation of the mixture gas components considering the diffusion and the combustion and the particle radius and concentration equations. The initial conditions were set in the same way as in [11].

The problem was solved numerically by S.K. Goduvov's method [12]. The summands on the right side of the gas energy-conversation equation and the oxygen mass balance equation, which determine the diffusion and the heat transfer processes, were explicit approximated by the three-point stencil. To solve the equations for the particle mass balance, the particle impulse–conversation and the particle energy-conversation the method from [13] was used. The time step was calculated by Courant's stability criterion. To verify the program code the adiabatic temperature after the explosion was calculated and compared with theoretical one, and the balance ratios were reviewed. The calculation error was equal to 0.03 %.

3 Results and discussion

The problem of the sparkplug ignition was solved under thermophysical and kinetics parameters of the coal dust taken from [8]. The dimension of the ignition energy was determined as *J/m*. To define the energy in *J*, the anode-cathode spacing should be taken into account, which is in the range of $4\cdot10^{-3} - 6\cdot10^{-3} m$ [4].

Fig. 1(a,b) shows the minimum spark ignition energy – the methane volume fraction curves. The curve in fig. 1b fairly agrees with the results in [4]. The distinction in the results is due to the absence of necessary data about coal dust dispersity. The fundamental result of the calculation corresponds to the data in [4] – the increase of the methane volume fraction in the mixture leads to the decrease of the minimum spark ignition energy of the

CDMAM. For example, the 1% methane volume fraction in the coal dust suspension decreases the minimum spark ignition energy 5 times.

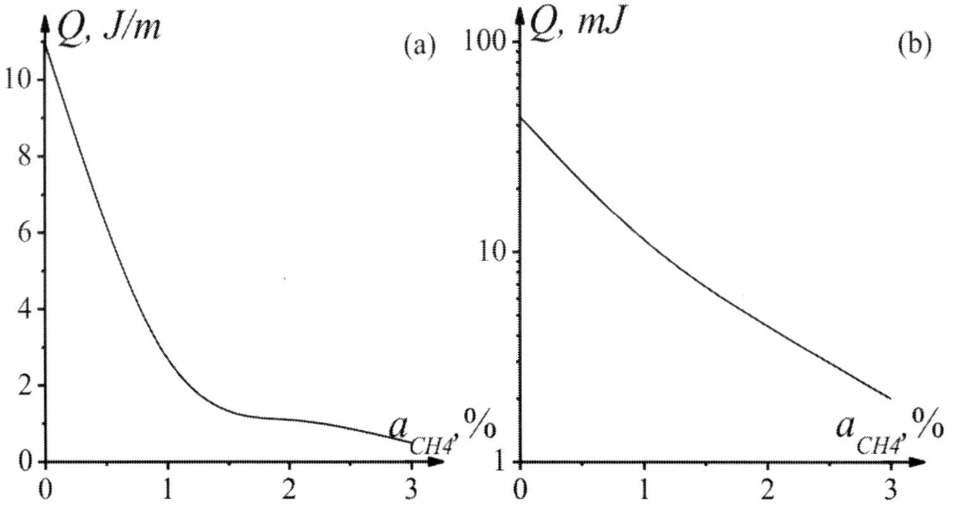

Fig. 1. The minimum spark ignition energy of CDMAM – the methane volume fraction curve. $m_{dust} = 0.4$ kg/m^3, $r_k = 5\cdot10^{-7}$ m. (a) –dimension [J/m], (b) – the curve, when the anode-cathode spacing is $4\cdot10^{-3}$ m

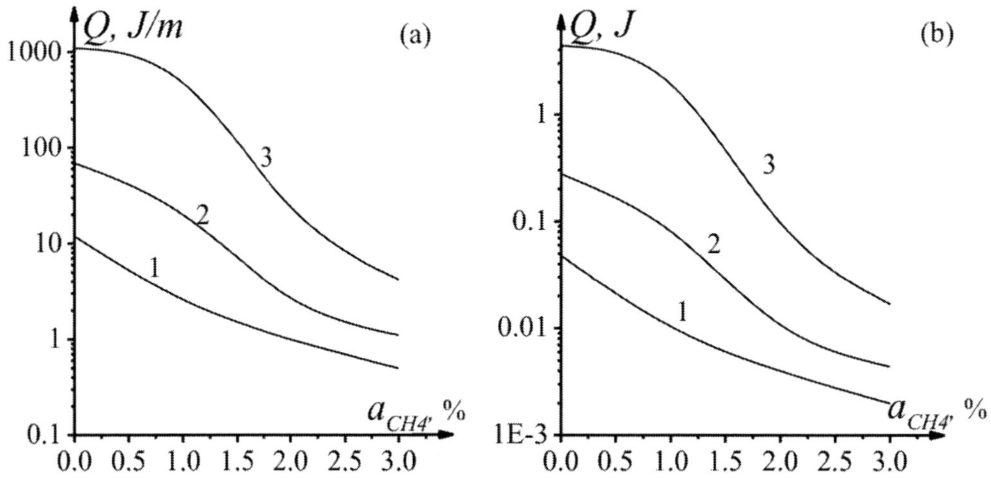

Fig. 2. The dependences of the minimum spark ignition energy of CDMAM on methane content in the mixture. $m_{dust} = 0.4$ kg/m^3, rk = 5·10-7 m (curve 1), 10^{-6} m (curve 2), $2\cdot10^{-6}$ m (curve 3). (a) – dimension [J/m], (b) – the curve, when the anode-cathode spacing is $4\cdot10^{-3}$ m

The particle size of the coal dust influences the minimum spark ignition energy. The specific surface area decreases with the increase of the particle radius. The warm-up time of the particle rises. The ignition energy of the spark spreads in the gas during the warm-up time. The particles do not heat up to the inflammation temperature, the ignition does not appear. The calculation results of the particle size influence on the minimum ignition energy of CDMAM are presented in fig.2. The calculations were held for the particle radii: $r_k = 5\cdot10^{-7}$ m, 10^{-6} m, 2 10^{-6} m. With addition of the 1% methane volume fraction in the coal dust suspension the minimum spark ignition energy decreases 5, 3 and 1.1 times, respectively. The increase of the methane volume fraction leads to the decrease of the minimum ignition energy.

4 Conclusions

The mathematical modeling of sparkplug ignition of CDMAM with low content of methane has been carried out. The influence of coal dust particle size and methane volume fraction on the minimum spark ignition energy has been obtained. The larger particles are the more energy is required to ignite the coal dust suspension in air or in the reactive gas. With the increase of the methane volume fraction in the mixture, the minimum ignition energy of the mixture with large particles decreases faster than for the mixture containing small particles.

This work was supported by the Russian Science Foundation grant No. 16-33-60091 mol_a_dk.

References

1. A.A. Vasil'ev, A.V. Pinaev, A.A. Trubitsyn, Combust., Expl., Shock Waves **53,** 8 (2017)
2. S. Kalyakin, V. Kostenko, E. Zav'yalova and L. Shtrokh, *Aktualne problemy zwalczania zagrożeń górniczych: II konferencja naukowo-techniczna (Brenna)* **7-9,** 176 (2012) [in Russian]
3. D. Bradley, M. Lawes, M.J. Scott, N. Usta, Combust. Flame **124**, 82 (2001)
4. A.Ya. Korolchenko, *Fire and explosion hazard of industrial dust* (Moscow: Himija, 1986) [in Russian]
5. M.I. Neychlelyaev, A.I. Lyubimova, P.M. Petrukhin, *Combating coal dust explosions in mines*, (Moscow: Nedra, 1992) [in Russian]
6. A.Yu. Krainov, V.A. Baimler, Combust., Expl., Shock Waves **38,** 30 (2002)
7. A.Yu. Krainov, V.A. Baimler, Combust., Expl., Shock Waves **38**, 9 (2002)
8. A.Yu., Krainov, K.M. Moiseeva, JPCS **754**, 052003 (2016)
9. D.A. Frank-Kamenetskiy, *Diffusion and Heat Transfer in Chemical Kinetics* (Moscow: Nauka, 1987) [in Russian]
10. R.I. Nigmatulin, *Dynamics of Multiphase Media* (CRC Press; 1990)
11. A.Yu. Krainov, Combust., Expl., Shock Waves **37**, 16 (2001)
12. S.K. Godunov, A.V. Zabrodin, M.Ya. Ivanov, A.N. Kraiko, G.P. Prokopov, *Numerical solution of multi-dimensional problems of gas dynamics* (Moscow: Nauka, 1976) [in Russian]
13. A.N. Kraiko, J. Appl. Math. Mech. **43**, 539 (1979)

An inductive RF discharge in water vapor for atomic-emission spectrometry

Ruslan Medvedev[1,*], and *Dmitriy* Churkin[2]

[1]Lavrentyev Institute of Hydrodynamics SB RAS, Lavrentyev Prosp. 15, 630090, Novosibirsk, Russia
[2]Institute of Laser Physics SB RAS, Lavrentyev Prosp. 13/3, 630090, Novosibirsk, Russia

Abstract. The paper presents the results of experimental measurements of the emission properties of water inductively coupled plasma. The plasma temperature is calculated using the Boltzmann method. It is shown that the temperature decreases with increasing pressure. The plasma resistance, its conductivity, and the thickness of the skin layer are estimated. Analysis shows that the emissivity of water plasma can be optimized while maintaining the power consumption.

1 Introduction

Inductively-coupled plasma (ICP) has a number of distinctive properties, such as purity, high stability, and equilibrium, which makes it attractive for use in applications such as coating, etching, the emission analysis of the chemical composition of substances, purification, and disinfection [1]. Recently, there has been an increased interest in the properties of water vapor as a plasma-forming gas for the listed applications since water vapor is a relatively cheap raw material which dissociates in the discharge to form highly reactive oxygen and OH hydroxyl. The presence of atomic hydrogen in the plasma allows a relatively easy spectrometric analysis of parameters such as the plasma temperature and electron concentration from the hydrogen Balmer series [2].

The possibility of using an inductive discharge in water vapor for atomic emission spectrometry was studied in [3]. In comparison with electrode and diaphragm (capillary) discharges in a liquid, an inductive discharge does not suffer from drawbacks such as the effect of electrodes and the energy loss due to Joule heating of the liquid [4-5]. However, to excite an inductive discharge, it is necessary to reduce pressure in the discharge chamber to $10^{-4} - 10^{-5}$ bar, which leads to difficulties in introducing the sample into the discharge zone and additional requirements for tightness.

Although water ICP has been studied in a number of paper [6-8], appropriate methods for its diagnostics are not available, which makes it difficult to develop systems using water ICP for industrial and analytical applications. Therefore, the study of the energy and spectral characteristics and the search for regularities that allow estimating the parameters of water ICP remains an important problem.

* Corresponding author: ruslan@hydro.nsc.ru

2 Experimental setup

A diagram of the experimental setup is shown in Fig. 1. Inductive discharge was excited in a quartz tube 1 of diameter D = 20–45 mm by a cooper inductor which had N = 2–5 coils. The inductor was supplied with a current with a frequency of 27.12 MHz from a CESAR 2710 high-frequency source. The load power output was $P_s = 600 \pm 50$ W. A vacuum pump was installed on one side of the quartz tube, and a vessel with tap water 2 with a metering valve 3 was placed on the other side. A check valve 4 was installed on the side of the vacuum pump.

Before the RF generator was turned on, air was evacuated from the system to a pressure of ~ 0.1 mbar with the valve 3 closed and the valve 4 open. After this, water vapor was supplied to the system using the metering valve 3. Then the RF generator was turned on, and the discharge was initiated. The discharge spectrum was recorded from the side of the quartz tube through an optical fiber cable 5 using a Solar Systems S-150 spectrometer having a spectral range of λ = 180–1100 nm and a spectral resolution of ~ 0.5 nm. The exposure of spectrum recording was varied depending on the brightness of the recorded lines. The pressure p in the discharge tube was monitored by a vacuum gauge 6 (Testo 552). The discharge burning time was 0.1–0.3 s.

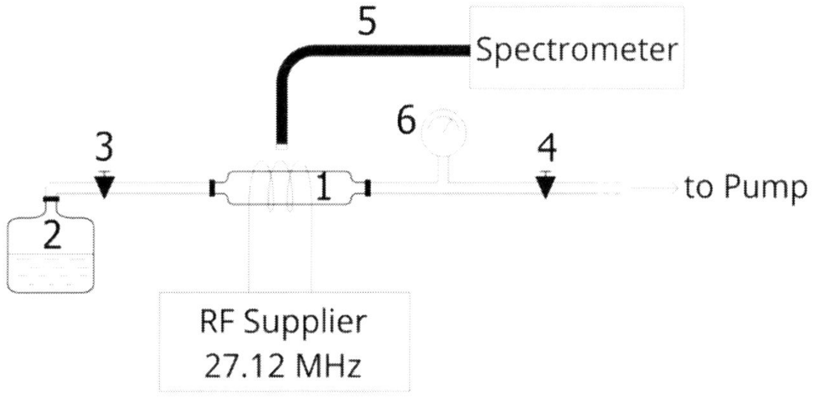

Fig. 1. Diagram of the experimental setup.

Steady discharge burning in water vapor at a power input P_s = 600 W was observed at a pressure p of up to 5 mbar. When the pressure was raised above 10 mbar by increasing the vapor flow with the metering valve 3, the discharge extinguished due to insufficient power.

3 Results and analysis

The circuitry of the setup with the matching circuit is similar to that presented in [9]. Knowing the parallel matching capacitance C, from the circuit parameters we can determine the water plasma resistance by the formula $C = \frac{1}{2\pi f r}\sqrt{\frac{r-R}{R}}$, where R is the resistance of the plasma reduced to the primary winding ($R = R_{pl}N^2$), r is the internal resistance of the generator (in our case, r = 50 Ohm), and f is the generator frequency.

The resistance of water inductively coupled plasma R_{pl} for a tube with D = 45 mm at p = 0.25 mbar was $R_{pl} \approx 4$ Ohm. For D = 20 mm and p = 1 mbar, we have $R_{pl} \approx 1$ Ohm.

Using the expression $R_{pl} = \frac{\pi D}{\sigma l \Delta}$, where σ is the plasma conductivity, l is the width of the plasma loop (approximately equal to the length of the inductor), $\Delta = \frac{1}{\sqrt{\pi f \sigma \mu}}$ is the depth of the skin layer (μ is the magnetic permeability of the plasma) and knowing the resistance of the plasma, we can estimate its integral conductivity and the thickness of the skin layer.

In experiments with $D = 20$ mm and $D = 45$ mm, the plasma conductivity was found to be approximately equal and equal to 10^3 Ohm^{-1}m^{-1} and the depth of the skin layer was 3 mm. These values are close to the values obtained for argon plasma under the same conditions [10].

The temperature of the water ICP was determined from the hydrogen Balmer series using the Boltzmann method [11].

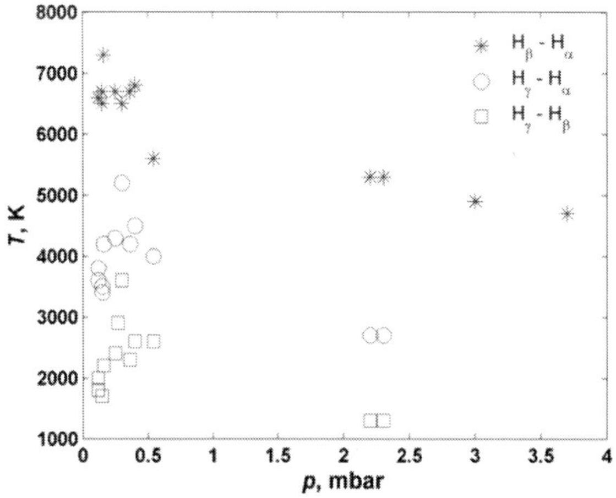

Fig. 2. Plasma temperature, determined from the relative intensities of spectral lines H_α H_β и H_γ, as a function of pressure for $D = 20$ mm, $P_s = 600$ W, and $N = 5$.

Figure 2 shows the plasma temperature as a function of pressure for a tube of diameter $D = 20$ mm, a discharge power $P_s = 600$ W, a number of inductor coils $N = 5$. The temperature is determined from the relative intensities of the first three lines of Balmer hydrogen series ($H_\alpha = 656.3$ nm, $H_\beta = 486.1$ nm, $H_\gamma = 434.1$ nm). It can be seen that the temperatures are not equal: the line H_β has the increased intensity as compared with the case of local thermodynamic equilibrium, so the temperature determined between H_β and H_α is higher and between H_γ and H_β is lower than the actual one. For the tube with $D = 45$ mm the dependency has the same character.

Despite of the impossibility of the accurate temperature determination nevertheless it can be seen that the plasma temperature decreases with increasing the pressure at the same heating power, which is natural since the particle concentration increases in proportion to the pressure.

4 Conclusion

Some characteristics of water ICP, such as resistance, integral conductivity, and temperature were determined.

It is shown that the resistance and conductivity of the plasma are close to those for argon ICP [10]. At an input power of 600 W and a diameter of the discharge tube of 20–45 mm, the maximum temperature of water ICP is 5000–6000 K.

These parameters, obtained at a sufficiently low input power, allow water to be considered as a promising medium for plasma-chemical and analytical applications.

Typical temperatures of argon ICP in emission spectral analysis are 6000–10000 K. To achieve these temperature parameters, it is necessary to increase the power mass density (the power per unit mass of plasma) introduced into the plasma. To do this, it is necessary to increase the power of the HF generator, reduce the volume of the plasma loop, and decrease the pressure.

This work was supported by the Russian Foundation for Basic Research (Grant No. 16-38-60039).

References

1. S.V. Dresvin, *Low-temperature plasma RF and microwave plasma torches* (Nauka, Novosibirsk, 1992) [in Russian].
2. D. Xiao, C. Cheng, J. Shen, Y. Lan, H. Xie, X. Shu, Y. Meng, J. Li, P.K. Chu, Phys. plasmas **21,** 053510 (2014)
3. R.N. Medvedev, D.S. Churkin, I.A. Zarubin, J. Phys.: Conference Series **754,** 102007 (2016)
4. R.N. Medvedev, I.A. Zarubin, A.D. Shokolov, *Advances in Nonequilibrium Processes: Plasma, Combustion, and Atmospheric (collected papers)*, 106 (2014)
5. R.N. Medvedev, I. A. Zarubin, Zavodskaya Laboratoriya. Obrabotka Materialov **79**, 11, 18 (2013) [in Russian]
6. A. Matsutani, H. Ohtsuki, F. Koyama, Journal of Physics: Conference Series **100**, 062022 (2008)
7. Z. Kregar, N. Krstulovic, N. Glavan Vukelic, S. Milosevic, J. Phys. D: Appl. Phys. **42**, 145201 (2009)
8. Y.Q. Tang, C. Trassy, Spectrochimica acta **41B**, 1/2, 143-150 (1986)
9. S. Xu, K. N. Ostrikov, W. Luo, S. Lee, J. Vac. Sci. Technol. A **18**, 2185 (2000)
10. J.T. Gudmundsson, M.A. Lieberman, Plasma Sources Sci. Technol. **6**, 540 (1997)
11. S.J. Hill. *Inductively Coupled Plasma Spectrometry and its Applications* (Oxford: Blackwell Publishing, 2007)

Generation of deterministic eddy structure of the flow as an analogue of the phase transition of the second kind. Development of ideas of academician I. I. Novikov

Olga Mitrofanova*

National Nuclear Research University "MEPHI", 115409, Kashirskoye sh. 31, Moscow, Russia

Abstract. The analysis of the physical processes leading to the structure formation of fluids has been done. The proposed theoretical approach can be used both for the neutral, conductive and magnetic fluids. In the present work the main attention is paid to the vortex and swirl flows of Newtonian fluids. Critical conditions accompanied the occurrence of hydrodynamic instabilities with a structural restructuring of flow field have been identified due to carried out physical experiments on generation of large-scale steady-stable vortex structures in channels of complicated geometry. Examples of topological models of the internal structure of the flow under condition of the phase transition of the second kind without changing of the aggregation state of the environment have been considered. It is shown that for vortex and swirl flows the parameter of helicity associated with the emergence of the internal angular momentum of the fluid and the structuring of the flow depending on the intensity of the local vorticity and orientation of longitudinal helical vortices can be accepted as the order parameter.

Theoretical analysis indicates the overall thermodynamic nature of the crisis phenomena associated with the formation of ordered vortex structures and transition the system to a stable, energetically more favorable state. The thermodynamic approach developed by academician I.I. Novikov is based on the ideas of Gibbs about the stability boundaries and the critical state of matter. General character of the phenomenon "stability crisis" was examined by I.I. Novikov from the thermodynamic point of view in his monograph [1]. Notion critical phenomena in this case refers to different abnomal situations which occur during the movement of liquid or gas. For example: the crisis of flow around, secondary flows of different classes, the crisis of the expiry of the swirl flow (critical flow), the generation of deterministic eddy structure of the flow in the acoustic flows, etc. The phase transition of the second kind corresponds to the case of transition of a thermodynamic system in a continuous manner without decay through a state of zero stability, in which $\left(\partial P/\partial V\right)_T = 0$, with the formation of new stable structures.

* Corresponding author: omitr@yandex.ru

According to [1], the condition for thermodynamic equilibrium of the hydromechanical system is the minimum of the thermodynamic potential – the Gibbs energy Φ at constant pressure P and temperature T. Under this condition, the hydromechanical system is stable against external impact or disturbance.

Given the fact that when the deterministic steady vortex structure of the flow is produced by a moving fluid under the action of mass forces (the nature of which may be varies), and the symmetry of a homogeneous medium is disturbed, in the expression for the thermodynamic potential the Gibbs energy Φ (1) and internal energy U (2) must be entered the term associated with the order parameter (3). This term characterizes the degree of asymmetry of the environment due to the formation of an ordered structure.

$$\Phi = U - TS + pV - \zeta_i \eta_i \tag{1}$$

$$dU = TdS - pdV + \zeta_i d\eta_i \tag{2}$$

$$\zeta_i = (\partial U / \partial \eta_i)_{S,V} \text{ ,} \tag{3}$$

where T – temperature, P – pressure, V – volume, S - entropy, - ζ_i acting on the body due to external force interrelated with the order parameter η_i .

As an illustrative example of the critical restructuring of the flow in Fig.1 it is shown the instability of Taylor.

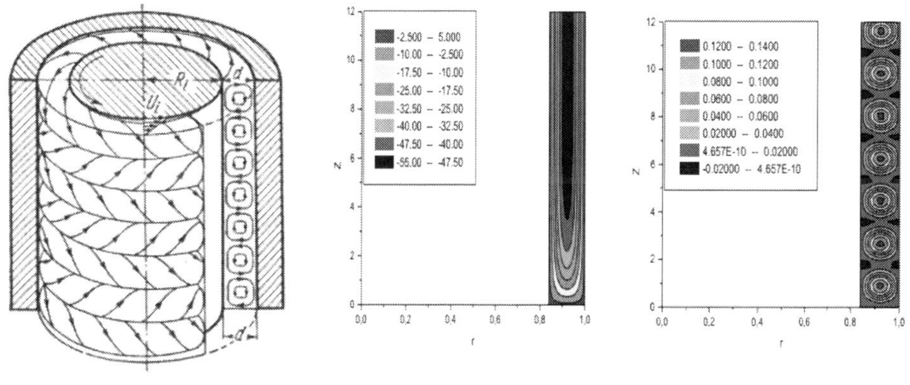

a b c

Fig. 1. The Taylor's vortices: a) the streamlined structure of flow in excess of the critical value of Taylor number, corresponding to the first singular point of the analytic solution μ_1 = 19,65 [3]; b), c) – distributions of tangential velocity in dependence on dimensionless values of the intensity of the flow vorticity K near a singular point : b) K = 19, 64 ($K \to \mu_{1-}$), c) K = 19, 66 ($K \to \mu_{1+}$).

The phenomenon of the crisis of hydrodynamic stability in this case is connected with a sharp restructuring of the flow pattern and the formation of a system of toroidal vortices (Fig. 1 a, c). Thus, as shows the theoretical analysis and comparison with experimental results, the main volume of channel containes non-dissipative vortex structures, obeying the law of a screw flow (Fig. 1 a), and the effect of viscosity is shifted in a narrow region on the boundary of the vortices.

Using the proposed in [1] a thermodynamic approach, which considers the analogy between the phase transition of the second kind and the occurrence of instability, makes it possible to identify the universalism of the development of instabilities typical for nonequilibrium dissipative structures transitions to a new, energetically more favorable forms under changing external conditions.

To identify the most significant physical laws that lead to structural stability of doubly periodic vortex motions with two basic frequencies ω_1 and ω_2, the vortex structures represented in Fig. 2 a, b can be considered as model objects.

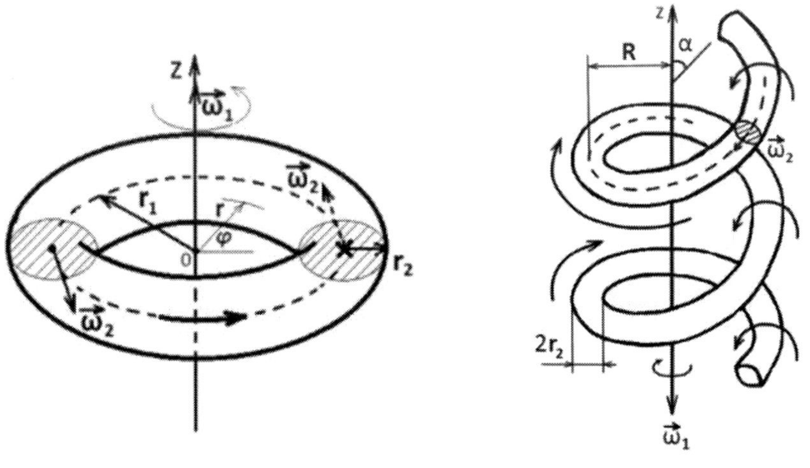

a) b)

Fig. 2. Doubly periodic vortex motion: a) model of a toroidal vortex with double rotation; b) screw-helix structure of the vortex, in this case, the radius r1 corresponds to the radius of curvature of the helix, the twist angle of which around its longitudinal axis z equal to: $r_1 = R / \sin^2 \alpha$.

Since the choice of the order parameter, which can be scalar, vector, and tensor quantity, not limited, it is advisable to use the parameter of helicity H [3] as the order parameter at helical-vortex motion of the fluid:

$$H = \mathbf{u} \cdot \boldsymbol{\omega} \neq 0 \ , \tag{4}$$

where $\mathbf{u}$ is the local flow velocity, $\boldsymbol{\omega} = rot\,\mathbf{u}$ is a vector of vorticity at a given point of the flow.

The condition $\vec{u} \cdot \vec{\omega} \neq 0$ corresponding to appearance of helicity is directly conformed with kinematic condition of a screw motion by Gromeka [4]:

$$rot\,\vec{u} = k\,\vec{u} \ , \tag{5}$$

where k is the vorticity intensity characterizing the ratio of circulation $u_\phi r$ to the stream function (flow rate) $\psi = u_z r^2 / 2$. Physically, this means that the vortex lines are helical lines and coincide with the mechanical trajectories of liquid particles motion [3].

To assess the degree of ordering of the liquid medium during the formation of vortical structures, represented in Fig. 2 a, b, we can obtain the following estimation of the magnitude of helicity in the region of spatial localization of the vortex:

$$H = |\mathbf{u}| \cdot |\boldsymbol{\omega}| = \sqrt{u_1^2 + u_2^2} \cdot \sqrt{\omega_1^2 + \omega_2^2} = \frac{1}{\sqrt{r_1 r_2}} \left(\omega_1^2 r_1^2 + \omega_2^2 r_2^2 \right) = \frac{|\mathbf{u}|}{k}, \qquad (6)$$

u_1, u_2 - components of linear velocity of a screw motion, $|\mathbf{u}| = \sqrt{u_1^2 + u_2^2}$.

According to the theory of self-organized criticality proposed in [5], in non-equilibrium systems can be the processes of self-organization. In addition from the set describing the system variables, you can allocate a small number of control parameters (order parameters) the leading variables that adjusts all other. A characteristic feature of such systems is scale invariance – lack for events or objects their own characteristic scales (sizes, durations, energies, etc.).

Theoretical analysis shows that in this case restructuring the vortex flow structure occurs when changing the intensity of vorticity, which for the considered models (Fig. 2 a, b) is estimated as $k = \frac{1}{\sqrt{r_1 r_2}}$.

In the mathematical solution given in [3] that describes the instability of Taylor (Fig. 1), the existence of bifurcation points with defined values of the parameter k, which may correspond to different values of the radii r_1 and r_2 , indicates scale invariance of the vortex formation [5].

Introduction in consideration of the order parameter gives the possibility to apply the formalism of thermodynamic analysis of phase transitions of the second kind for liquid under condition the formation of stable vortex structure of the flow moving in the field of mass forces.

This work was supported by RFBR grant 16-08-00687 and the program of increasing the competitive ability of National Research Nuclear University MEPhI (agreement with the Ministry of Education and Science of the Russian Federation of August 27, 2013, project no. 02.a03.21.0005).

References

1. I.I. Novikov *Thermodynamics of spinodals and phase transitions* (2000) [in Russian]
2. L.D. Landau, E.M. Lifshits *Theoretical physics: a tutorial. V. Statistical physics. Part I.* – 5-e Izd., (2005)
3. O.V. Mitrofanova *Hydrodynamics and heat transfer of swirl flows in channels of nuclear power installations* (2010) [in Russian
4. I.S. Gromeka, Complete works, Ac. Sci. **3**, 107 (1952)
5. P. Bak, C. Tang, K. Wiesenfeld, *Self-organized criticality: An explanation of the 1/f noise* Phys. Rev. Lett. **59,** 381 (1987)

Studying thermal conductivity of magnetic nanofluids in constant magnetic field

Maxim Pryazhnikov[1,2,*], *Andrey* Minakov[1,2], *Dmitriy* Guzei[1,2], and *Vladimir* Zhigarev[1]

[1]Siberian Federal University, 660041, Krasnoyarsk, Russia
[2]Institute of Thermophysics SB RAS, 630090, Novosibirsk, Russia

Abstract. The thermal conductivity of magnetic nanofluids in a constant magnetic field is investigated. The concentration dependence of the thermal conductivity coefficient of magnetic fluids based on water and iron oxide nanoparticles Fe_3O_4 in a constant magnetic field is obtained. The dependence of the thermal conductivity coefficient of nanofluid on the nanoparticles size in a magnetic field is obtained. The temperature effect on the thermal conductivity of magnetic nanofluid in a constant magnetic field is studied.

1 Introduction

Recently, ferromagnetic nanofluids are actively investigated (carrier fluid + iron oxides nanoparticles). A distinctive feature of ferromagnetic nanofluids is responsiveness to an external magnetic field. In the works of other researchers, the effect of increasing the thermal conductivity was observed. For example, in [1] was a 23 % increase in the thermal conductivity of nanofluid based on kerosene and Fe_3O_4 nanoparticles (10 nm in size and a 7.8 % volume concentration), and 300 % in the absence of a magnetic field. The authors of [2] investigated water-based nanofluids with Fe_3O_4 nanoparticles (10 nm in size and 5 % volume concentration) under the magnetic field action and observed a more than triple increase in thermal conductivity.

2 Experimental apparatus and procedure

Within the framework of our study we measured the ferromagnetic thermal conductivity of nanofluid in a constant magnetic field. Fe_3O_4 nanoparticles were used (nanoparticles powders of the «Plasmotherm», Moscow). Distilled water was base fluid. A standard two-step method was used to prepare nanofluids.

Measurement of thermal conductivity of nanofluids was carried out by the hot-wire method with a Wheatstone bridge circuit. Detailed description of the test bench and testing technique is given in [3]. Wheatstone bridge was used as the basis of the test bench instrumentation to measure an unknown electrical resistance of hot wire. Principle of measuring the thermal conductivity by the hot-wire method is based on a linear dependence

* Corresponding author: arrivent@yandex.ru

between the temperature rise of hot wire and the logarithm of heating time of the wire. A copper wire with a length of 200 mm and a diameter of 75 microns was used in the experiments. The wire was immersed in a vessel of diameter 32 mm containing 200 ml of the fluid under study.

Two permanent neodymium magnets with dimensions 50×30×10 mm were used to create a magnetic field. The magnetic field intensity around the magnet was measured with a milliteslameter TPU-02. The distribution of the magnetic field induction over the magnet is shown in figure 1. The maximum value of the magnetic induction near the magnet is of the order of 0.3T. In experiments, the magnets were located in the immediate vicinity of the measuring cell. Two cases of a magnetic field were considered (see figure 2): a magnetic field created by a single magnet (low magnetic field intensity) and a magnetic field created by two magnets (high magnetic field intensity).

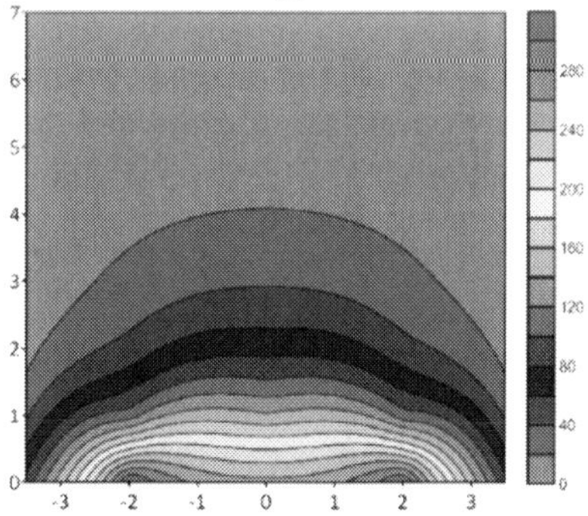

Fig. 1. The magnetic field induction over the magnet (mT).

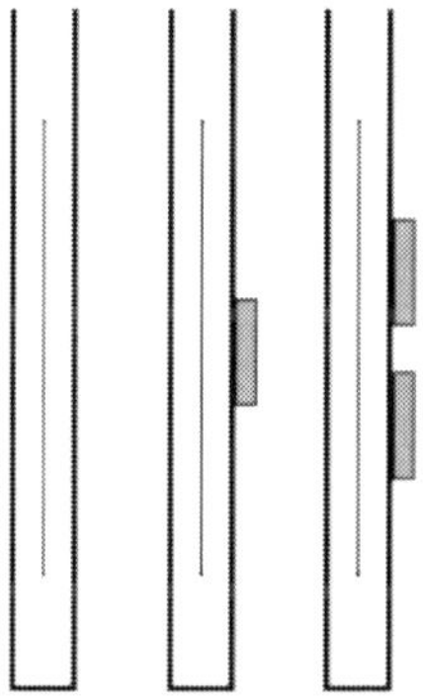

Fig. 2. The arrangement of magnets near the vessel.

3 Results and discussion

The effect of volume nanoparticles density on the thermal conductivity of nanofluid was studied. Fe_3O_4 nanoparticles of 50 nm average size were used. Two volume nanoparticles density in water of 1 and 2% are considered. It is found that the influence of magnetic field intensifies with increase of nanoparticles concentration. As a result of measurements the thermal conductivity of magnetic nanofluid is increased under the influence of magnetic field. For magnetic nanoparticles 2.2 and 2.8 times enhancement in thermal conductivity is achieved for 2 % nanoparticle concentration by the application low and high magnetic field intensity, respectively (see figure 3*a*). The thermal conductivity of magnetic nanofluid intensifies with increase of magnetic field intensity (see figure 3*a*). It should be noted that with increasing magnetic field intensity, an increase and decrease in the thermal conductivity of a magnetic nanofluids can be observed [5, 6].

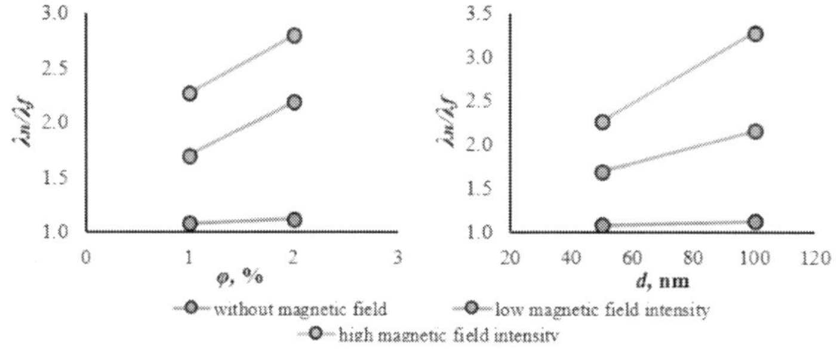

Fig. 3. Dependence of relative thermal conductivity coefficient of nanofluids on nanoparticle concentration (*a*) and on nanoparticle size (*b*).

It is established that the influence of magnetic field intensifies with increase of nanoparticles concentration. The thermal conductivity coefficient intensifies by 5% with an increase the nanoparticles concentration from 1 to 2% without a magnetic field, while increase amounts to 30% under the influence of the magnetic field (see figure 3*a*).

The nanoparticle size effect on the thermal conductivity of nanofluid in a magnetic field was investigated. The study was carried out for Fe_3O_4 nanoparticles. The volume nanoparticles density in water was 1%. Nanoparticles powders of 50 and 100 nm average size were considered. It is found that the thermal conductivity of nanofluid increases with the nanoparticles size, both under the influence of the magnetic field and without it. It was shown for the first time that the thermal conductivity coefficient of nanofluid increases under the influence of a magnetic field with increasing particle size.

The influence of nanoparticle size on the thermal conductivity of nanofluid in magnetic field is studied. The volume nanoparticles density in distilled water was 1%. Figure 3b shows the dependence of the relative thermal conductivity coefficient of the nanofluid (the ratio of the thermal conductivity coefficient of nanofluid to the thermal conductivity coefficient of base fluid) on the nanoparticles size in water.

It is obtained that the thermal conductivity coefficient of nanofluid increases with the nanoparticles size, both under the influence of the magnetic field and without it. Similar behavior was noted earlier for other nanofluids [4]. In absence of magnetic field the addition of nanoparticles with 50 nm average size to water increases the thermal conductivity by 8% in comparison with the thermal conductivity coefficient of water, and nanoparticles of 100 nm size increase by 13%. Increment becomes larger with the application of magnetic field. The thermal conductivity of nanofluid with nanoparticles of size 50 nm at low magnetic field intensity is 70% higher than the thermal conductivity of water, and 11% with 100 nm nanoparticles. In a magnetic field intensity created by two magnets, the growth is 2.27 times for fine nanoparticles (50 nm) and 3.28 times for large nanoparticles (100 nm). Thus, it was shown for the first time that the thermal conductivity coefficient of nanofluid intensifies with an increase particle size under the influence of the magnetic field.

The temperature dependence of the thermal conductivity of nanofluid in a magnetic field is studied. The investigation was carried out for Fe_3O_4 nanoparticles (50 nm). The range of the investigated temperatures is 25-50°C. The dependence of the relative thermal conductivity coefficient of nanofluid on temperature is obtained (see figure 4): the relative thermal conductivity coefficient of a nanofluid without a magnetic field is 1.08 at a temperature of 25°C, it increases linearly with temperature rise and reaches a value of 1.15 at 50°C.

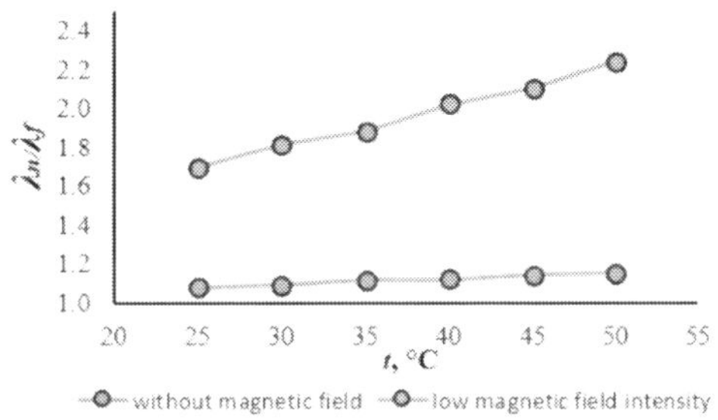

Fig. 4. Dependence of the relative thermal conductivity coefficient of nanofluid on temperature.

The magnetic field effect on temperature dependence of thermal conductivity of magnetic nanofluid is studied. It is shown that the magnetic field effect on thermal conductivity of magnetic nanofluid increases with temperature rise of nanofluid. It should be noted that in [2, 6] it was shown that with increasing temperature, the coefficient of thermal conductivity decreases.

4 Conclusions

The thermal conductivity of magnetic nanofluids in a constant magnetic field is investigated. The concentration dependence of the thermal conductivity coefficient of magnetic fluids based on water and iron oxide nanoparticles Fe_3O_4 in a constant magnetic field is obtained. The dependence of thermal conductivity coefficient of nanofluids on the nanoparticles size in a magnetic field is obtained. As a result of measurements the thermal conductivity of magnetic nanofluid is increased under the influence of magnetic field. The thermal conductivity of magnetic nanofluid intensifies with increase of magnetic field intensity (see figure 3).

The magnetic field effect on temperature dependence of thermal conductivity of magnetic nanofluid is studied. It is shown that the magnetic field effect on thermal conductivity of magnetic nanofluid increases with temperature rise of nanofluid (see figure 4).

The current work is performed at partial support of the projects funded by the Russian Foundation for Basic Research and Krasnoyarsk Regional Fund for Support of Scientific and Scientific-Technical Activities (Contract No. 16-48-243061).

References

1. J. Philip, P.D. Shima, B. Raj, Appl. Phys. Lett. **91**, 203108 (2007).
2. A. Gavili, F. Zabihi, T.D. Isfahani, Sabbaghzadeh, J. Exp. Therm. Fluid Sci. **41**, 94 (2012).
3. A.V. Minakov, M.I. Pryazhnikov, V.Ya. Rudyak, D.V. Guzei, A.S. Lobasov, J. Eng. Phys. Thermophys. **88**, 149 (2015).
4. M. Pryazhnikov, A. Minakov, V. Rudyak, D. Guzei, Int. J. Heat Mass Transfer. **104**, 1275 (2017).
5. C.L. Altan, A. Elkatmis, M. Yüksel, N. Aslan, S. Bucak, J. App. Phys. **110**, 093917 (2011).
6. A. Karimi, M. Goharkhah, M. Ashjaee, M.B. Shafii, Int. J. Thermophys. **36** 2720 (2015).

Modeling of exothermic synthesis of composite with oxide inclusions

Anna Knyazeva[1,2,*], and *Nahum* Travizkii[3]

[1] Tomsk Polytechnic University, 634050 Lenin Av., Tomsk, Russia
[2] Institute of Strength Physics and Material Science, 634055 Akademicheskii pr., 2/4, Tomsk, Russia
[3] University of Erlangen-Nuremberg, D-91058 Martensstr,5, Erlangen, Germany

Abstract. The model is suggested for composite synthesis on the base of metallic matrix with inclusions of aluminum oxide at the conditions close to thermal explosion. Chemical reactions are described with the help of two summary stages. The melting, the dependence of reaction rates on pressure and reaction retardation by the product layer are taken into account. It was shown that during reactions proceeding in any regime the formation of final composition of the matrix does not occur. It agrees with known experimental data.

1 Introduction

Combustion synthesis (the methods of metallothermy and self-propagating high-temperature synthesis) are widely used in various branches of science and technology [1]. The past few years the numerous research techniques of metal matrix composites appear [2], where metallothermy has been used. Among various metal matrix composites very popular have become the composite materials based on light metal alloys. Carbides, oxides, nirides, borides, silicides and fine intermetallides are used as reinforcing inclusions [3]. The principal cause of thermite processes employment is connected with their small energy consumption. However, the difficulties exist in the governing of metallothermic processes with high heat release. The changing in the ratio of the material, adding oxidizers or inert inclusions, external fields of various physical nature etc. are used to optimize the synthesis process. Mathematical modeling serves also to this goal. So, to obtain the composites with minimal porosity, synthesis during isostatic pressing are used. Modeling of this process [4,5] allows studying the conversion regimes. Synthesis together with the extrusion was studied theoretically by authors of [6,7]. However only in [8,9], coupling models were constructed. The influence of inert particles on the combustion and thermal explosion regimes was investigated in [10,11] based on conjugate models. Papers [12,13] includes in the models the detailed reaction mechanisms for investigation of intermetallic composite synthesis on the substrate surface.

* Corresponding author: anna-knyazeva@mail.ru

This paper suggests the thermal explosion model in cylindrical reactor and takes into consideration two summary reactions: aluminum oxide synthesis with the reduction reaction and composite solution formation, correspondingly to experimental data [14].

2 Model formulation

Corresponding to experimental data [14], the green powder mixture placed in cylindrical reactor contains Al, Fe_2O_3; Fe; Cr; Ni in given proportion. The heating of powder mixture can be carried out from the side of lateral surfaces of cylindrical reactor (a) or from the side of plunger, which assists the mixture compaction (b). In a first case, the heating can be assumed as uniform along the reactor height, and hence, the temperature distribution and composition will depend only on radial coordinate. In the approximation of thin layers of the reactor, one-dimensional thermal conductivity equation in cylindrical coordinate system takes a place

$$c\rho \frac{\partial T}{\partial t} = \frac{1}{r}\frac{\partial}{\partial r}\left(\lambda r \frac{\partial T}{\partial r}\right) + W \tag{1}$$

Where the heat capacity c, density ρ and thermal conductivity coefficient depend on the composition and porosity θ, and the function W is the total heat of chemical reactions running in the mixture under the heating. Two summary reactions can be assigned here

$$2Al(L) + Fe_2O_3 \rightarrow Al_2O_3 + 2Fe \tag{2}$$

$$4.9Fe + 3.2Cr + 1.9Ni \rightarrow 10Fe_{0.49}Cr_{0.32}Ni_{0.19} \tag{3}$$

Formal approach permits to write for reaction rates the following expressions

$$\varphi_1 = z_1 C_1^2 C_2 \Phi(C_6 + C_7) exp\left(-\frac{E_1}{RT}\right) \text{ and } \varphi_2 = z_2 C_3^{4.9} C_4^{3.2} C_5^{1.9} \Phi(C_6 + C_7) exp\left(-\frac{E_2}{RT}\right)$$

where C_k are mass concentrations or species; E_1 and E_2 are the activation energies for reactions. z_1, z_2 are pre-exponential factors; R is universal gas constant; the function Φ reflects the reaction retardation due to the reaction product accumulation.

The mass conservation laws are correct for both reactions:

$$\sum_{k=1}^{7} \nu_{ki} m_k = 0, \ i = 1,2,$$

where ν_{ki} are stoichiometric coefficients for k-species in the reaction I; m_k are molar masses. Kinetical equations are written for species

$$\rho \frac{dC_k}{dt} = w_k, \tag{4}$$

where $w_k = \nu_{k1} m_k \varphi_1 + \nu_{k2} m_k \varphi_2$, $k = 1,2,\ldots,7$

The porosity changes in accordance with kinetical equation

$$\frac{d\theta}{dt} = \varphi(\theta, T, P),$$

where P is applied load.

The melting is taken into account through the heat capacity change. Additionally, to diminish the reaction temperature the inert particles could be added to the reaction mixture. It is taken into account trough the heat capacity evaluation also.

When the reaction rates depend on the pressure, usual exponent in Arrhenius equation is changed by the function

$$exp\left(-\frac{E_k - \alpha_k \gamma_k P}{RT}\right),$$

where γ_k is activation volume, α_k is some experimental constant. This law speak that reaction can be activated not only when temperature growths but due to pressure work over activation volume change [15].

The boundary conditions take into account the radiation heating from the source with given temperature and heat losses (or heat accumulation) in the reactor walls.

The problem was solved numerically. Parameters were found in literature and/or were calculated based thermodynamics.

3 Some results

The dynamics of the process depends essentially on reactor size, heating rate, heat exchange conditions and equivalent properties evolution. It was shown numerically that controlling the reactions could be carried out with the help of heating rate and inert particles addition. The first occurs when external source temperature changes. Example of the various reaction regimes are presented in Fig.1,a,b The temperature of external source equals to 1800 K (Figures a) and 2120 K (Figures b).

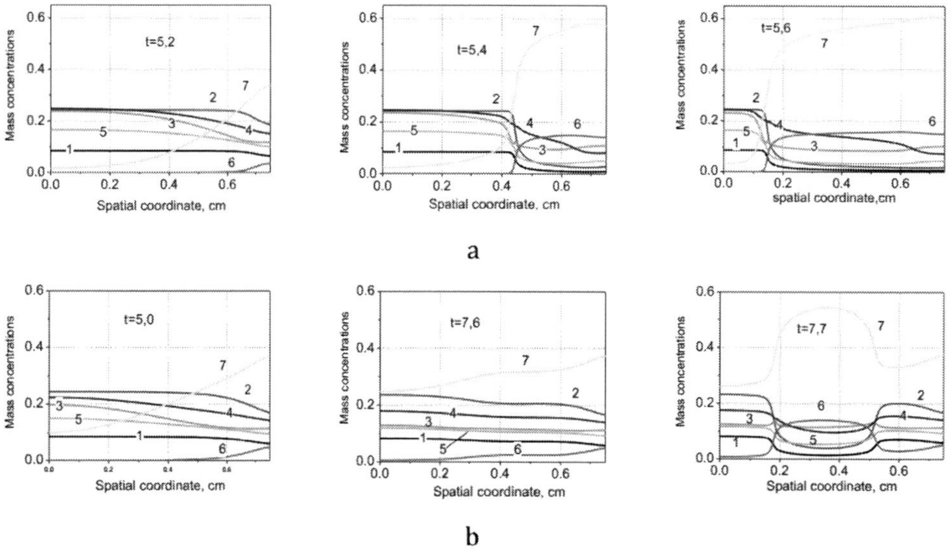

Fig.2. Spatial distribution of reagents and reaction products for different time moments; a) $T_w = 1800$ K, b) $T_W = 2120$ K. 1. - *Al*, 2. - Fe_2O_3; 3. - *Fe*, 4. - *Cr*, 5. - *Ni*, 6. - Al_2O_3, 7. - $Fe_{0.49}Cr_{0.32}Ni_{0.19}$

External heating is broken when temperature of reactor wall reaches 900 K. During a more slow heating, the reaction zone forms which propagates then to reactor center

(Fig.1,a). For high rate heating, the reaction starts near the walls also, however it accelerates then in a volume (Fig.1,b). It was found that for more fast heating, reactions stop near the walls just beginning.

Generally, interrelation between different mechanisms of temperature change in each point (thermal conductivity, chemical heat) accompanying the properties change leads to not uniform composition change. It was observed experimentally [14]. Note, for thick reactor walls, thermal conductivity problem will conjugate. The walls play double function. On the one hand, they need a more heat for the heating, on the other hand, they are additional heat sources when external heating is broken, and the reactions do not completed.

The role of endothermic stage in the thermal explosion was discussed in [16,17] for simple conditions. Suggested here model allows investigate composition change varying the heating conditions. It was detected the first reaction (Fe-reducing) completes often quickly. However matrix formation can continue when reactor is cooling. That agrees with experiment also.

This work was made under support of Russian academic excellence program.

References

1. F.Kh. Urakaev, K.A. Akmalaev, E.S. Orynbekov, B.D. Balgysheva, D.N. Zharlykasimova, Metallurgical and Materials Trans. B **47B**, 2 (2016)
2. E. Fra, A. Janas, A. Kolbus, E. Olejnik, Archives of foundry engineering, **9**, 2 (2009)
3. A. Ibrahim, F. Mohamed, E. Lavernia, J. Materials Sc. **26**, (1997)
4. L.M. Buchatskii, S.I. Khudyaev, G.V. Shkadinskaya, Combust., Explosion and Shock Waves **28**, 1 (1992)
5. V.K. Smolyakov, Combust., Explosion and Shock Waves **34**, 4 (1998)
6. L.S. Stel'makh, N.N. Zhilyaeva, A.S. Stolin, J. Eng. Phys. Thermophys. **63**, 5 (1992)
7. V. E. Ovcharenko, O.V. Lapshin, I.S. Ramazanov, Combust., Explosion and Shock Waves **42**, 3 (2006)
8. A.G. Knyazeva, N.K. Evstigneev, Procedia Computer Sci. **1**, 1. (2010)
9. N.K. Evstigneev, A.G. Knyazeva, J. Appl. Mech. Techn. Phys. **52**, 3 (2011)
10. A.G. Knyazeva, A.A. Chachshina, Combust., Explosion, and Shock Waves **40**, 4 (2004)
11. A.A. Chachshina, A.G. Knyazeva, Chemistry for sustainable development **13**, 2 (2005)
12. S.N. Sorokova, A.G. Knyazeva, Theor. Found. Chem. Eng. **42**, 4 (2008)
13. S.N. Sorokova, A.G. Knyazeva, Theor. Found. Chem. Eng. **44**, 2 (2010)
14. N. Travitzky, P. Kumar, K.H. Sandhage, R. Janssen, N. Claussen, Material Sci. Eng. A **244** (2003)
15. A.G. Knyazeva, Combust., Explosion and Shock Waves **28**, 1 (1992)
16. M.B. Borovikov, U.I. Gol'dshleger, I.A. Burovoi, Combus., Explosion, and Shock Waves **20**, 1 (1984)
17. M.B. Borovikov, U.I. Gol'dshleger, Combust., Explosion, and Shock Waves **17**, 5 (1981)

Thermophysical characteristics of radioactive graphite – water vapor system

Nikolaj Barbin[1,2,3,*], *Anton* Kobelev[1], *Dmitrij* Terent'ev[1], and *Sergej* Alekseev[1,4]

[1] Ural State Fire Service Institute of Emercom of Russia, 620062 Ekaterinburg, Russia
[2] Ural State Agrarian University, 620075 Ekaterinburg, Russia
[3] Ural Federal University named after the first President of Russia B.N. Yeltsin, 620002 Ekaterinburg, Russia
[4] SEC Reliability and Safety of Large Systems of Ural Branch of RAS, 620049 Ekaterinburg, Russia

Abstract. The article considers thermophysical characteristics of radioactive graphite – water vapor system in temperature range 373-3273K. The research was made by thermodynamic modeling method using TERRA software. We determined 4 temperature intervals in which changes of thermophysical characteristics of radioactive graphite – water vapor system occur.

Most of atomic power plants operate according to the scheme where low-enriched uranium is served as a fuel for thermal reactors, graphite is used as a moderator material, and water is used as a reactor coolant. [1].

In Russia and other countries in the world there is a problem of handling with radioactive graphite of uranium-graphite reactive cores after their taking out of service [2].

Radionuclides (^{137}Cs, ^{90}Sr, ^{154}Eu and others) are formed in graphite due to the leak of reactor coolant and transfer of fuel segments.

There are several methods of treatment of radioactive graphite: "Conservation", "Disposal", "Elimination" [3]. First two methods require the construction of deep storages and large financial costs. Method "Elimination" is more effective as the amount of irradiated graphite falls by 90 times [4].

Elimination of irradiated graphite can be performed by the following methods: oxidation by salts, oxidation in molten alkali and carbonates, liquid-phase oxidation in acids, bonding of graphite with metals, oxidation by air, oxygen, carbon dioxide, water vapor [3-6].

The purpose of this research is to study thermophysical characteristics of radioactive graphite – water vapor system in temperature range 373-3273 K.

The tasks of the study are carrying out the experiment with investigated system by thermodynamic modeling method, analysis of thermophysical characteristics (V – specific volume (m^3/kg), S – entropy (kJ/(kg K)), I – total enthalpy (kJ/kg), U – total internal energy (kJ/kg), M – number of moles (mole/kg), Z – mass fraction of condensed phases).

Full-scale experiments do not always give reliable information at high temperatures due to their complexity and measurements errors. So, calculations were made by

* Corresponding author: NMBarbin@mail.ru

thermodynamic modeling using TERRA software which is used for studying systems with complex chemical composition in high temperature conditions.

The software is used to calculate phase composition, thermodynamic and transport properties of arbitrary systems.

The calculation of phase composition and characteristics of equilibrium was made according to reference database for properties of individual substances. The basis of information in the database are thermodynamic, thermophysical and thermochemical properties of individual substances systemized in the Institute of High Temperatures of the Russian Academy of Sciences (database INVATERMO), by national US Bureau of Standards, calculated in Moscow State Technical University using molecular calorimetric and spectrochemical data, published in reference books [7, 8].

Thermodynamic modeling with the use of TERRA software is successfully used in thermal physics [5, 6].

Thermodynamical modeling was carried out at initial pressure of one atmosphere, at initial temperature 373 K and at final temperature 3273 K. Temperature measurement interval – 100 K. Phase composition: gaseous phase – 75% (H_2O – 100% mass), condensed phase – 25% (C ~ 99.98% mass, U ~1.15 10^{-2}% mass, Cl ~1.88 10^{-3}% mass, Ca ~2.69 10^{-4}% mass, Pu ~7.19 10^{-5}% mass, Be ~ 1.19 10^{-5}% mass, Ni ~7.99 10^{-6}% mass, Cs ~3.99 10^{-6}% mass, Sr ~9.99 10^{-6}% mass, Am ~9.99 10^{-6}% mass, Eu ~7.99 10^{-6}% mass.

Figure 1a shows dependence of specific volume on temperature. In temperature range from 373 to 573 K specific volume of the system increases from 1.29 to 2.01 m^3/kg. At temperature 573 K the bend point is observed and in temperature range from 573 to 973 K specific volume increases from 2.01 to 4.98 m^3/kg. At temperature 973 K the bend point is observed and in temperature range 973-2373 K specific volume increases from 4.98 to 12.38 m^3/kg. At temperature 2373 K the bend is observed and in temperature range from 2373 to 3273 K specific volume increases from 12.38 to 19.7 m^3/kg.

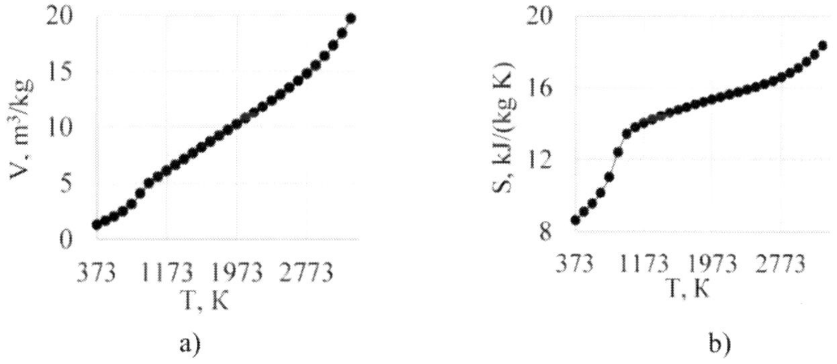

Fig. 1. Dependences of thermophysical characteristics of radioactive graphite – water vapor system on temperature: a) dependence of specific volume on temperature, b) dependence of entropy on temperature.

Figure 1b shows dependence of entropy on temperature. In temperature range from 373 to 573 K entropy increases from 8.63 to 9.56 kJ/(kg K). At temperature 573 K the bend point is observed and in temperature range from 573 to 973 K entropy increases from 9.56 to 13.41 kJ/(kg K). At temperature 973 K the bend point is observed and in temperature range 973-2373 K entropy increases from 13.41 to 15.89 kJ/(kg K). At temperature 2373 K the bend is observed and in temperature range from 2373 to 3273 K entropy increases from 15.89 to 18.35 kJ/(kg K).

Figure 2a shows dependence of total enthalpy on temperature. In temperature range from 373 to 573 K total enthalpy increases from -9873.53 to -9432.54 kJ/kg. At temperature 573 K the bend point is observed and in temperature range from 573 to 973 K total enthalpy increases from -9432.54 to -6368.44 kJ/kg. At temperature 973 K the bend point is observed and in temperature range 973-2373 K total enthalpy increases from -6368.44 to -2482.69 kJ/kg. At temperature 2373 K the bend is observed and total enthalpy increases from -2482.69 to 4704.91 kJ/kg.

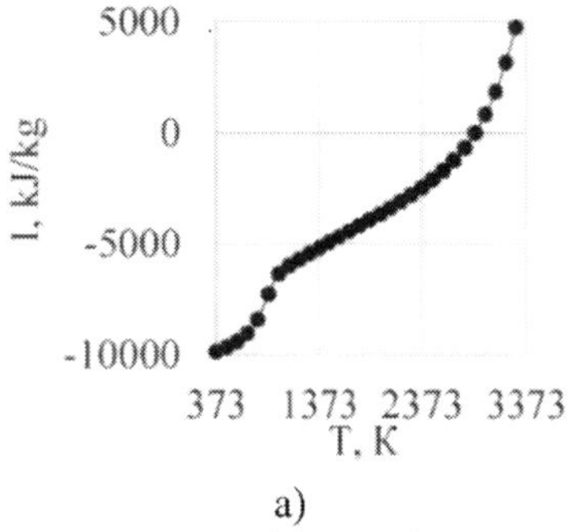

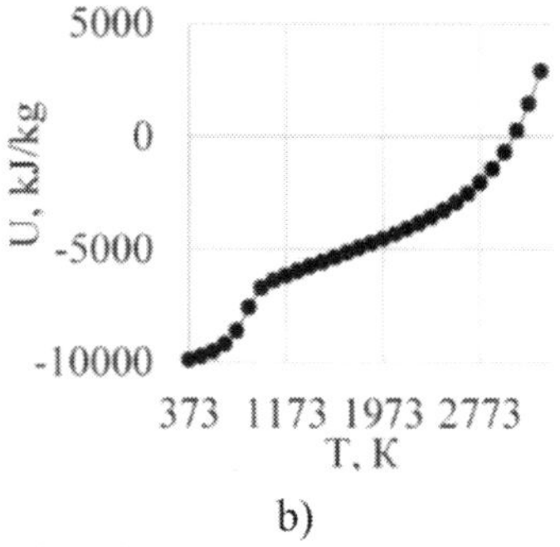

Fig. 2. Dependences of thermophysical characteristics of radioactive graphite – water vapor system on temperature: a) dependence of total enthalpy on temperature, b) dependence of total internal energy on temperature.

Figure 2b shows dependence of total internal energy on temperature. In temperature range from 373 to 573 K total internal energy increases from -9899.44 to -9529.15 kJ/kg. At temperature 573 K the bend point is observed and in temperature range from 573 to 973 K total internal energy increases from -9529.15 to -6713.87 kJ/kg. At temperature 973 K the bend point is observed and in temperature range 973-2373 K total internal energy increases from -6713.87 to -3565.17 kJ/kg. At temperature 2373 K the bend is observed and total internal energy increases from -3565.17 to 2914.22 kJ/kg.

Figure 3a shows dependence of number of component moles on temperature. In temperature range from 373 to 573 K number of component moles in the system decreases from 52.5 to 46.4 mole/kg. At temperature 573 K the bend point is observed and in temperature range from 573 to 973 K number of component moles increases from 46.4 to 61.5 mole/kg. At temperature 973 K the bend point is observed and in temperature range 973-2373 K number of component moles doesn't change and is equal to ~ 62 mole/kg. At temperature 2373 K the bend is observed and number of component moles increases from 62.7 to 72.3 mole/kg.

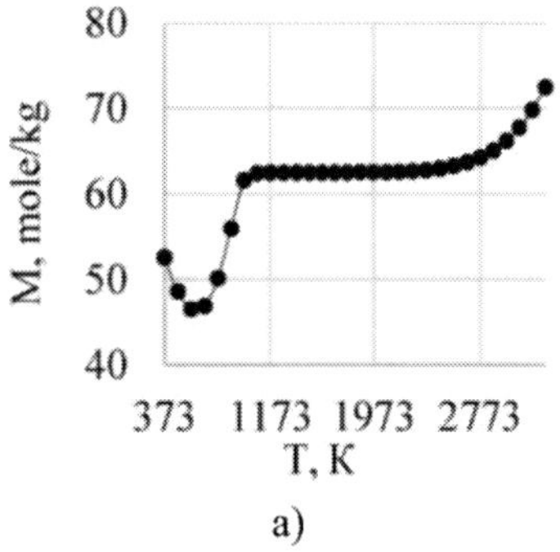

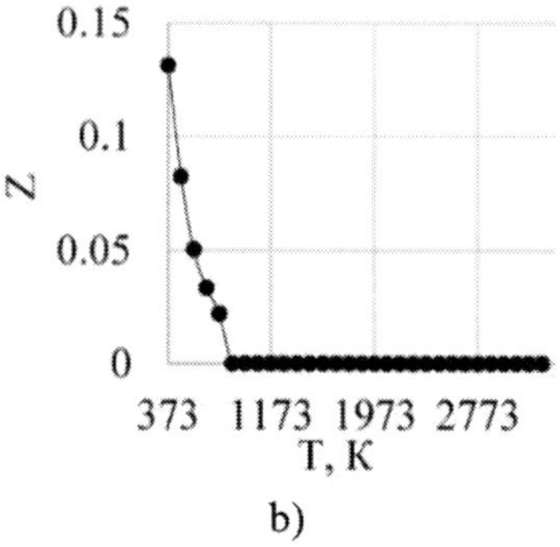

Fig. 3. Dependences of thermophysical characteristics of radioactive graphite – water vapor system on temperature: a) dependence of number of component moles on temperature, b) dependence of mass fraction of condensed phases on temperature.

Figure 3b shows dependence of mass fraction of condensed phases on temperature. In temperature range from 373 to 573 K mass fraction of condensed phases decreases from

0.13 to 0.05. At temperature 573 K the bend point is observed and in temperature range from 573 to 873 K mass fraction of condensed phases decreases from 0.05 to 0.

Condensed C gives the main contribution to thermodynamic properties of the system till its burning temperature (873 K). At temperature > 873 K thermodynamic properties are defined by gaseous phase.

According to the Figure four temperature intervals where changes of thermophysical properties of the system occur are marked.

1 temperature range from 373 to 573 K: changes are connected with the increasing of CH_4 gas content.

2 temperature range from 573 to 973 K: changes are connected with both decreasing of CH_4 gas content and burning of condensed C.

3 temperature range from 973 to 2373 K: changes have a linear behavior and are defined by gas-vapor phase.

4 temperature range > 2373 K: changes are connected with thermal decomposition of gas-vapor phase.

References

1. W.S. Yang, *Advances in reactor concepts: Generation IV reactors* (Purdue University, West Lafayette, 2014)
1. *Nuclear Power Reactors in the World* (International atomic energy agency, Vienna, 2015)
2. *Predisposal Management of Radioactive Waste from Nuclear Power Plants and Research Reactors* (International atomic energy agency, Vienna, 2016)
3. A.M. Grin'ko, V.V. Tokarevskij, *Materials 5th international conference «Cooperation for solving the problem of waste»* (Kharkiv National University of Economics, Kharkiv, 2008)
4. N.M. Barbin, D.I. Terentiev, S.G. Alexeev, T.M. Barbina, J. Radioanal. Nucl. Chem. **307**, 1459 (2016)
5. N.M. Barbin, D.I. Terentiev, S.G. Alexeev, T.M. Barbina, J. Radioanal. Nucl. Chem. **299**, 1747 (2014)
6. V.E. Alemasov, A.F. Dregalin, A.P. Tishin, *Thermodynamic and thermophysical properties of combustion products* (VINITI, Moscow, 1971)
7. L.V. Gurvich, I.V. Vejc, V.L. Medvedev, *Thermodynamic properties of individual substances* (Science, Moscow, 1982)

Experimental study of the effect of air injection on the pressure pulsations in the hydro turbine flow path under different operating conditions

Andrey Minakov[1,2*], *Dmitry* Platonov[1,2], *Alexandra* Maslennikova[1] and *Dmitry* Dekterev[1,2]

[1]Siberian Federal University, 79 Svobodny Ave., 660041 Krasnoyarsk, Russia
[2]Institute of Thermophysics, SB RAS, 1 Lavrentyev Ave., 630090, Novosibirsk, Russia

Abstract. The paper presents an experimental study of oscillatory response in the Francis turbine of hydraulic unit. The experiment was performed on large-scale hydrodynamic test-bench with impeller diameter of 0.3 m. The effect of air injection on the intensity of pressure pulsations was studied at the different regimes in the hydraulic unit. It was revealed that air delivery into the flow path system of the turbine results in almost two-fold reduction of pressure pulsations, but for optimal regimes, there is an increase in pressure pulsations.

1 Introduction

An important task for HPP is regulation of power in the energy system. During load changing hydraulic units repeatedly undergo off-design modes of operation. Under these flow conditions a significant part of swirl, after passing through the water turbine impeller, is remained. The instability of swirling flow may cause the emergency of intense low-frequency hydrodynamic fluctuations that threaten the reliability of turbine design [1-3]. Precessing vortex rope is a serious danger for the hydraulic turbine equipment due to powerful flow pulsations that lead to strong vibrations of hydraulic turbine construction and in the case of resonance can lead to destruction of the equipment. Pressure pulsations generated by the vortex rope precession, may also affect cavitations processes, enhancing cavitations erosion.

Up to date, numerous methods are proposed to stabilize flow in the hydraulic units [4-7]. Each of them has its advantages and disadvantages, and requires further detailed study. In this study, the effect of air delivery on the oscillatory response of the flow was experimentally investigated on the model hydrodynamic stand with Francis turbine.

2 Experimental techniques

The model water turbine with radial-axial impeller is part of the hydropower test-bench at the Laboratory of Hydraulics and Hydraulic Machines of the Siberian Federal University see in Fig. 1. Maximum water head in the test-bench is 3.5 m, the diameter of the impeller

* Corresponding author: tov-andrey@yandex.ru

is D=0.3 m. Hydraulic unit includes a pressure tank, performing the functions of the upper pool, power conduit with the diameter of 400 mm, and penstock valve, equipped with electric drive. Water from the penstock enters the spiral chamber and through the guide vanes flows to the runner. Then water through the draft tube gets into the lower pool with glass walls. The width of the pool was equal to 1.52 meters. Diffusion grid is designed to reduce fluctuations in water level before weir gauge. Weir gauge is used for measuring water flow rate and is fixed by water gauge.

After weir gauge, the water is discharged into the collecting tank and then is diverted through pipe into the pool. From the pool, water is pumped into the pressure tank. The water intake in the collecting tank is performed through the suction cap with mesh. The water flow rate is controlled by gate valve with electric actuator. To record the pressure pulses, the test-bench was equipped with a piezoelectric flow rate sensors (type 014MT) designed to convert rapidly varying and pulsed pressure into an electrical signal. The signal from the sensors came through an amplifier (LE-41) and an analog digital converter (ADC E14-140 l-card) to the computer for further processing.

Fig. 1. Model of water turbine with radial-axial impeller.

3 Experimental results

In the course of experiments we carried out measurements within a wide range of the hydraulic unit operation both with air and without air. The degree of opening of the guide vanes was varied in the experiment, while the pressure remained constant. Then all the necessary parameters, namely, impeller rotation frequency, the flow rate through the hydraulic unit, and generating current and voltage were recorded.

The behaviour of the frequency and the pressure pulsations intensity in the water-conveyance system of the turbine is of great interest for high-head power plants since the precession of a vortex core is a serious hazard to the water turbine equipment. Pressure pulsations were recorded at specified points during five minutes for each of the operation modes.

Figures 2 and 3 below present the experimental results on time variations of pressure pulsation in the draft tube diffuser carried out for two operating modes of the hydrodynamic stand: maximum pressure pulsation regime at opening of 5/15, and near optimum regime at opening of 8/15. From the figures, indicating the point in time corresponding to air injection into the flow path (red line), it is evident that in a regime with maximum pulsations, a significant decrease is observed in pressure pulsations amplitude, while in the regime near to the optimum one, conversely - pressure pulsations increased.

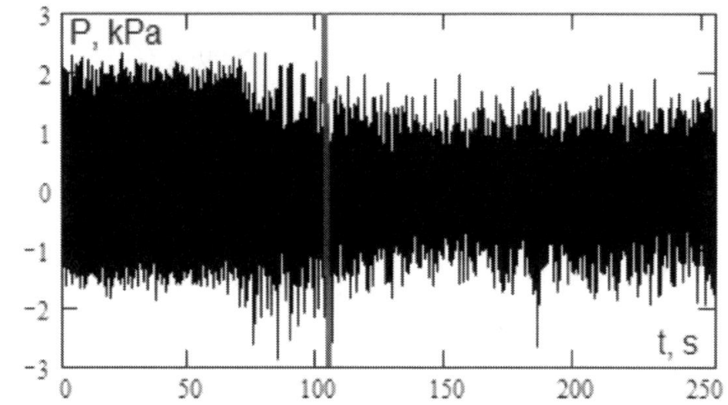

Fig. 2. Time dependence of the pressure pulsation in the draft tube diffuser *(opening 5/15)*.

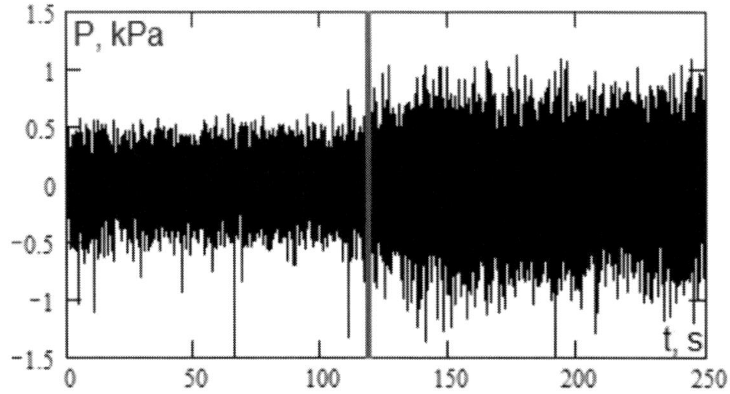

Fig. 3. Time dependence of the pressure pulsation in the draft tube diffuser *(opening 8/15)*.

Figures 4 presents experimental dependence of the pressure pulsation amplitude in the draft tube cone on the opening of the guide vanes. The dependence shows that the air injection allows one to reduce the pressure pulsations in the most dangerous zones by more than 2 times, however, in the region of optimal regimes, the pressure increases. This fact indicates the advisability of air supply only at certain loads on the hydroelectric unit.

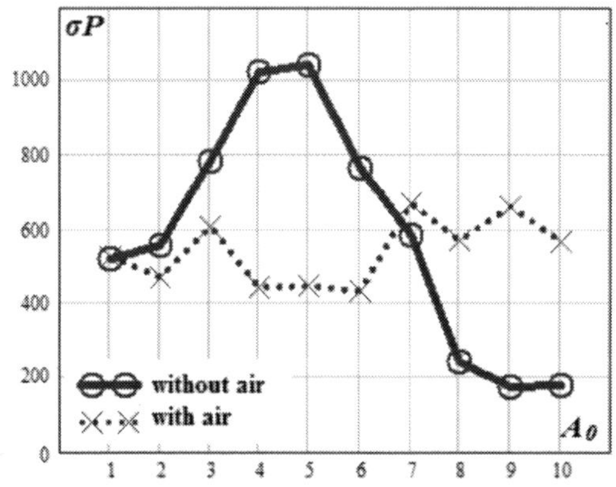

Fig. 4. Dependence of the pressure pulsation amplitude in the draft tube cone on the opening of the guide vanes.

4 Conclusion

The work deals with experimental investigation of pressure pulsations for different flow regimes in the hydraulic unit of the medium-scale hydrodynamic stand with air injection in flow path. The authors obtained correlations between the pressure pulsations and the opening of the guide vanes for constant pressure heads, in the hydraulic turbine taking into account the air phase. Air was supplied into a spiral chamber of the turbine. This method is one of the most promising to reduce loads on the hydraulic unit, caused by the precession of the vortex core in the space behind the runner.

The results of the study have shown that it is possible to significantly reduce the pressure pulsations using air at the maximum pressure pulsations regime. However, this effect needs further investigation. In particular, it is necessary investigating the effect of air injection point.

The current work is performed at partial support of the projects funded by the Russian Foundation for Basic Research and Krasnoyarsk Regional Fund for Support of Scientific and Scientific-Technical Activities (16-41-243081).

References

1. A.V. Minakov, A.V. Sentyabov, D.V. Platonov, A.A. Dekterev, A.V. Zakharov, Comp. & Fluids **111**, 197 (2015)
2. A.V. Minakov, A.V. Sentyabov, D.V. Platonov, A.A. Dekterev, A.V. Zakharov, Int. J. Heat Fluid Flow **53**, 183 (2015)
3. D. Platonov, A. Minakov, D. Dekterev, A. Sentyabov, A. Dekterev, J. Phys. Conf. Ser. **754**, 062007 (2016)
4. H.T. Falvey, U.S. Bureau of Reclamation. (1971)
5. T.L. Wahl, Colorado: Colorado State University, (1990)
6. H. Grein, 10th Symposium of the International Association for Hydraulic Research. (1980)
7. S. Bhan, 5th International Symposium on Hydro Power Fluid Machinery. (1988)

To the question of the interpolation of the phase equilibrium curves for the hydrates of methane and carbon dioxide

Nail Musakaev[1,2*], and *Stanislav* Borodin[1]

[1]Tyumen Branch of Khristianovich Institute of Theoretical and Applied Mechanics of SB RAS, 625026 Tyumen, Russia
[2]Industrial University of Tyumen, 625000 Tyumen, Russia

Abstract. The experimental data base was developed. This database includes 200 points of the phase equilibrium curve of methane hydrate, as well as more than 300 experimental points for carbon dioxide hydrate. This database is based on the work of a large number of researchers. Taking into account the experimental data, empirical formulas are constructed for determining the equilibrium parameters of the methane and carbon dioxide hydrates, as well as the equilibrium curve of "liquid carbon dioxide – gaseous carbon dioxide".

1 Introduction

Natural gas is one of the main energy carriers. For example, in 2014 it accounted for 21.6% of the global electricity production and its share continues growing [1, 2], and the main component of natural gas is methane (77-99%). The largest reserves of methane are concentrated in the methane hydrates. Their total volume ($\approx$ 20000 trillion m^3) is two orders of magnitude higher than the volume of traditional recoverable methane reserves ($\sim$ 250 trillion m^3). Given the increasing demand and the largest amount in comparison with the main types of fossil fuels, the methane produced from gas hydrates is the most promising source of energy.

The most likely cause of global warming is an increase in the concentration of greenhouse gases, in particular CO_2. To combat this phenomenon it is necessary to take measures towards reducing the concentration of carbon dioxide by regularly removing CO_2 from the atmosphere. One of such ways of the carbon dioxide utilization is its disposal in porous reservoirs in the gas hydrate form [3, 4, 5].

To solve the above-described practical problems, a theoretical study is required, which also includes the construction of adequate empirical correlations of the phase equilibrium curves for the methane and carbon dioxide hydrates. The present paper is devoted to the construction of such relations taking into account two refinements. Firstly, the constructed empirical correlations show a good agreement with the experimental data at quadrupole points that is ensured by fixing these points during interpolation (at the remaining points the

* Corresponding author: musakaev@ikz.ru

least squares method is applied). Secondly, they are continuous at quadrupole points that is necessary for the convergence of calculations near such points.

2 The algorithm for constructing empirical formulas

Figure 1 schematically shows the phase diagrams of the hydrates of methane (a) and carbon dioxide (b).

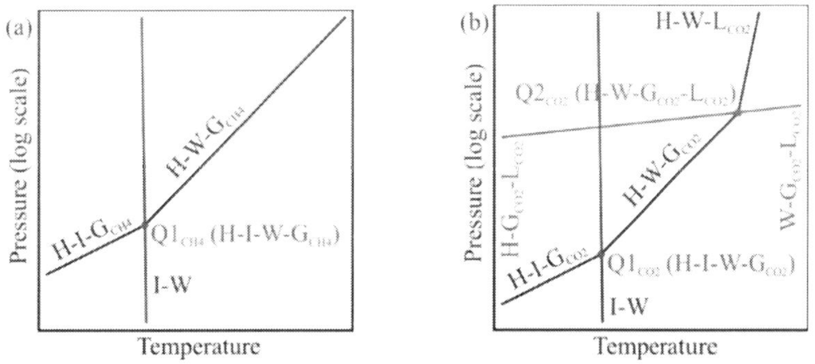

Fig. 1. Schematic representation of the phase equilibrium of methane (a) and carbon dioxide hydrates (b). H – hydrate, I – ice, W – liquid water, G – gas, L – liquid.

The empirical formulas will be constructed in the following form:

$$\ln(p) = A_0 + \frac{A_1}{T} + \frac{A_2}{T^2}, \tag{1}$$

where p is the pressure; T is the temperature; A_0, A_1, A_2 are the coefficients, which are determined from the following system of equations:

$$\begin{cases} A_0 + \dfrac{A_1}{T_Q} + \dfrac{A_2}{T_Q^2} = \ln(p_Q), \\ \displaystyle\sum_{i=1}^{N}\left[A_1\left(\frac{1}{T_i} - \frac{1}{T_Q}\right) + A_2\left(\frac{1}{T_i^2} - \frac{1}{T_Q^2}\right) + \ln(p_Q) - \ln(p_i)\right]^2 \to \min. \end{cases} \tag{2}$$

Here T_i and p_i are the experimental values of the temperature and pressure; T_Q and p_Q are the temperature and pressure in a quadrupole point.

The pressure and temperature values at the quadrupole points will be given based on the data of various authors available from [6]. In this case we take into account that the phase equilibrium temperature I-W is very weakly dependent on the pressure and that this curve must intersect with the phase curves for the hydrates at the points Q1. In the end we get:

$$T(p)_{I\text{-}W} = \text{const} = 273\ \text{K}\ (p \in [0.1, 10]\ \text{MPa}), \tag{3}$$

$$p_{Q1CH4} = 2.6\ \text{MPa}, \quad T_{Q1CH4} = 273\ \text{K}, \tag{4}$$

$$p_{Q1CO2} = 1.2\ \text{MPa}, \quad T_{Q1CO2} = 273\ \text{K}, \tag{5}$$

$$p_{Q2CO2} = 4.5\ \text{MPa}, \quad T_{Q2CO2} = 283\ \text{K}. \tag{6}$$

Using the described above method we find the curves for methane H-I-G_{CH4}, H-W-G_{CH4} and for carbon dioxide H-I-G_{CO2}, H-G_{CO2}-L_{CO2}, W-G_{CO2}-L_{CO2}. To find the curve H-W-G_{CO2} we fix two points $Q1_{CO2}$ and $Q2_{CO2}$, and the coefficients of equation (1) are determined from the following system:

$$\begin{cases} A_0 + \dfrac{A_1}{T_{Q1}} + \dfrac{A_2}{T_{Q1}^2} = \ln(p_{Q1}), \qquad A_0 + \dfrac{A_1}{T_{Q2}} + \dfrac{A_2}{T_{Q2}^2} = \ln(p_{Q2}), \\ A_2 = \dfrac{\sum_{i=1}^{N}\left[\left(\ln\left(\dfrac{p_i}{p_{Q1}}\right) - \dfrac{\ln(p_{Q2}/p_{Q1})}{1/T_{Q2} - 1/T_{Q1}}\left(\dfrac{1}{T_i} - \dfrac{1}{T_{Q1}}\right)\right)\left(\dfrac{1}{T_i} - \dfrac{1}{T_{Q1}}\right)\left(\dfrac{1}{T_i} - \dfrac{1}{T_{Q2}}\right)\right]}{\sum_{i=1}^{N}\left[\left(\dfrac{1}{T_i} - \dfrac{1}{T_{Q1}}\right)\left(\dfrac{1}{T_i} - \dfrac{1}{T_{Q2}}\right)\right]^2}. \end{cases} \tag{7}$$

The empirical formula for the H-W-G_{CO2} curve in the form (1) does not correlate well with the experimental data, so for this curve we will use the following form:

$$p = A_0 + A_1 T + A_2 T^2. \tag{8}$$

3 The empirical formulas and their assessments

Using the described above algorithm, we obtained the equations of the phase equilibrium curves. The coefficients A_0, A_1 and A_2 for these equations are presented in the Table 1. To estimate the accuracy in the Table 1, we also presented the coefficient of determination (R^2), as well as the average (δ_{AV}) and maximum (δ_{MAX}) relative error over all points of the experimental data.

Table 1. The coefficients for empirical formulas and their assessment.

Agent	Curve	Temperature interval, K	Coefficients	R^2	δ_{AV}, %	δ_{MAX}, %
CH_4	H-I-G_{CH4}	[178.2, 273]	$A_0 = \ln(2.6) - A_1/273 - A_2/273^2$; $A_1 = -2768$; $A_2 = 69900$.	0.9984	1.88	4.02
	H-W-G_{CH4}	[273, 302]	$A_0 = \ln(2.6) - A_1/273 - A_2/273^2$; $A_1 = -61987$; $A_2 = 7526200$.	0.9976	2.07	6.95
CO_2	H-I-G_{CO2}	[213.8, 273]	$A_0 = \ln(1.2) - A_1/273 - A_2/273^2$; $A_1 = -13854$; $A_2 = 1351500$.	0.9808	5.22	7.60
	H-W-G_{CO2}	[273, 283]	$A_0 = \ln(1.2) - A_1/273 - A_2/273^2$; $A_1 = \ln(4.5/1.2)/(1/283 - 1/273) - A_2\cdot(1/283 + 1/273)$; $A_2 = 15000000$.	0.9804	3.26	11.46
	H-W-L_{CO2}	[283, 292.7]	$p = A_0 + A_1\cdot T + A_2\cdot T^2$. $A_0 = 4.5 - A_1\cdot 283 - A_2\cdot 283^2$; $A_1 = -815$; $A_2 = 1.45$.	0.9930	13.80	45.45
	H-G_{CO2}-L_{CO2}	[256.5, 283]	$A_0 = \ln(4.5) - A_1/283 - A_2/283^2$; $A_1 = -3478$; $A_2 = 202500$.	0.9995	0.27	1.17
	W-G_{CO2}-L_{CO2}	[283, 304]	$A_0 = \ln(4.5) - A_1/283 - A_2/283^2$; $A_1 = -6376; A_2 = 642700$.	0.9997	0.24	0.63

Fig. 2 shows the experimental data points and calculated phase curves for methane (a) and carbon dioxide (b).

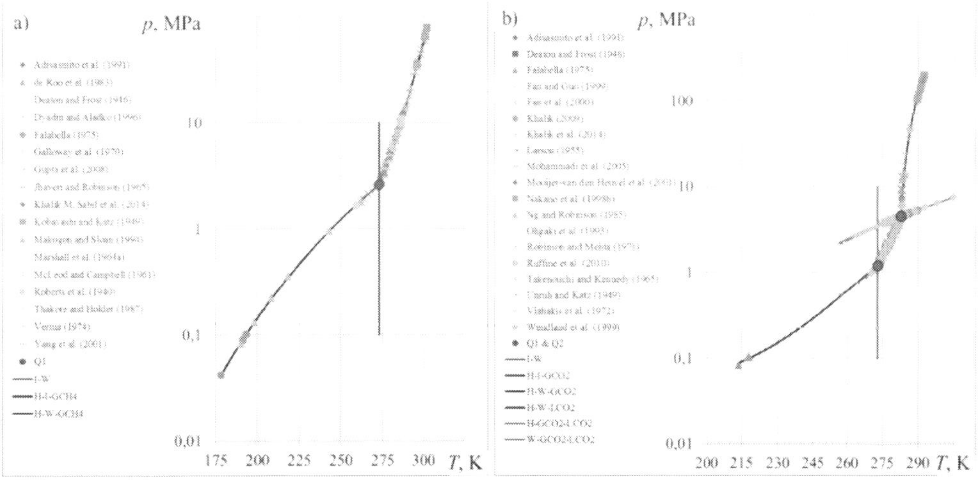

Fig. 2. The experimental data points and calculated curves for methane (a) and carbon dioxide (b).

In this work the empirical formulas for the determination of the equilibrium parameters of the methane and carbon dioxide hydrates, as well as the equilibrium curve of "liquid carbon dioxide – gaseous carbon dioxide", are constructed. The resulting curves intersect at quadrupole points, so that in the calculations there will be no problems with convergence due to discontinuities. At the same time good agreement with the experimental data was reached.

This work was supported by the Russian Foundation for Basic Research (project 17-51-540001).

References

1. Y.F. Makogon, S.A. Holditch, T.Y. Makogon, J. Petroleum Science and Eng. **56,** 54 (2007)
2. Y.F. Makogon, J. Natural Gas Science and Eng. **2**, 49 (2010)
3. G.G. Tsypkin, Fluid Dynamics **49**, 789 (2014)
4. V.Sh. Shagapov, M.K. Khasanov, N.G. Musakaev, Ngoc Hai Duong, Int. J. Heat Mass Transfer **107,** 1030 (2017)
5. G.K. Anderson, J. Chem. Thermodynamics **35,** 1171 (2003)
6. E. Dendy Sloan, Carolyn A. Koh, *Clathrate Hydrates of Natural Gases* (CRC Press, Boca Raton, 2008)

Numerical research of the gas hydrate dissociation to gas and Ice in a reservoir during the gas extraction

Nail Musakaev[1,2*], and *Stanislav* Borodin[1]

[1]Tyumen Branch of Khristianovich Institute of Theoretical and Applied Mechanics of SB RAS, 625026 Tyumen, Russia
[2]Industrial University of Tyumen, 625000 Tyumen, Russia

Abstract. The peculiarities of the gas hydrate decomposition to gas and ice during the gas extraction from the porous medium initially saturated with methane and its hydrate are studied. It is shown that at a negative initial reservoir temperature, the gas hydrate decomposition always occurs with the formation of ice either on the frontal surface or in the extended region. It is established that the extended region of hydrate dissociation is characteristic for large values of the mass flow rate of the gas extraction.

1 Introduction

At present, hydrocarbon gas is one of the most important sources for the chemical and energy industries. At the same time, the largest reserves of methane on our planet are in the form of gas hydrates [1]. For the extraction of natural gas from the gas hydrate deposit it is necessary to decompose the hydrate directly in the reservoir and this can be achieved by the thermal action on the hydrate-saturated reservoir and/or by lowering the pressure at the producing well bottom. If the hydrate decomposition produces gas and ice rather than gas and water, then this allows reducing the energy costs for the gas hydrate deposits development, since the specific heat of the phase transition "hydrate – ice and gas" is significantly lower than that of "hydrate – water and gas" [2]. It is worth noting that the extraction of hydrocarbon raw stock from the gas hydrate deposits is a very complex technological problem that needs thorough and comprehensive theoretical research [1].

The research of the gas hydrate decomposition process in porous structures is presented in [2-5]. In these papers the mathematical models are proposed, and the analytical solutions in plane-parallel and axisymmetric formulations are constructed. It is shown that the gas hydrate decomposition can occurs either at the frontal surface or in the extended region. In this paper the mathematical model of the gas hydrate decomposition in the case of the negative initial reservoir temperature (below 0°C) is constructed. This model takes into account the main features of the process, namely nonisothermal filtering of gas, phase transitions, real gas properties, adiabatic cooling effect, and Joule-Thomson effect.

* Corresponding author: musakaev@ikz.ru

 DOI: 10.1051/matecconf/201711505003

2 Mathematical model

Let the horizontal porous reservoir is initially saturated with methane and its hydrate, whose pressure p_0 and temperature T_0 correspond to the thermodynamic conditions of their existence in the free state:

$$t=0,\ r_w < r < R_k:\quad T=T_0<0\ ^{\circ}\mathrm{C},\ p=p_0,\ S_h=S_{h0},\ S_g=1-S_{h0},$$

where t is the time; r is the radial coordinate; r_w and R_k are the well and reservoir radiuses; S_j ($j = i, h, g$) is the j-th phase saturation in pores (i – ice, h – hydrate, g – gas); S_{h0} is the initial hydrate saturation. The coordinate r is measured from the central well axis.

The gas production occurs with a constant mass flow rate Q (per unit of the well height), and there is no heat flow at the well boundary:

$$t>0,\ r=r_w:\quad r\frac{\partial p^2}{\partial r}=\frac{Qz\mu_g R_g T}{\pi k S_g^3},\quad \frac{\partial T}{\partial r}=0.$$

Here k is the absolute reservoir permeability; μ_g and z are the dynamic viscosity of gas and compressibility factor; R_g is the specific gas constant.

The gas hydrate decomposes due to the lowering of the pressure below the equilibrium pressure, and three characteristic zones can appear in a reservoir (Fig. 1). More specifically, the dissociation of the gas hydrate begins at the boundary between the second and third zones ($r = r_{(d)}$) and ends at the boundary between the first and second zones ($r = r_{(n)}$). The hydrate saturation in the third zone at all points is constant and equal to initial value (S_{h0}).

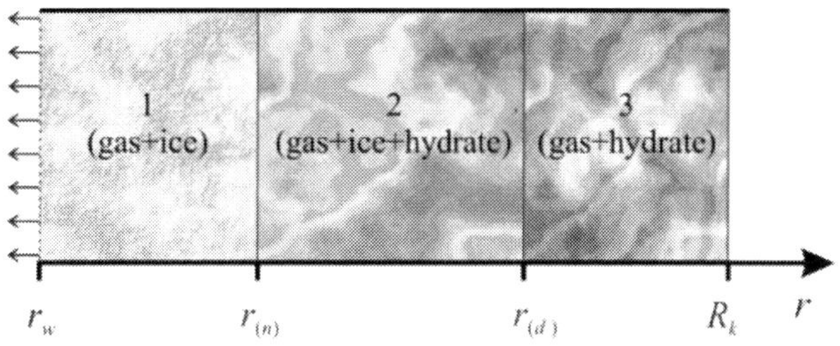

Fig. 1. Scheme of the gas hydrate dissociation into gas and ice occurring in a reservoir.

The conditions on the right boundary for temperature and pressure:

$$t>0,\ r=R_k:\quad \frac{\partial p}{\partial r}=0,\quad \frac{\partial T}{\partial r}=0.$$

The following assumptions are accepted: the porosity of a reservoir m is constant; the skeleton of a porous medium, gas hydrate and ice are incompressible and immovable; the temperatures of the porous medium and its saturants are equal (one-temperature model).

The system of basic equations describing the processes of filtration and heat transfer during the gas hydrate decomposition in a porous medium is based on the laws of conservation of mass and energy, the Darcy law and the equation of the gas state [2, 6, 7]:

$$\frac{\partial}{\partial t}\left(\rho_g S_g+\rho_h S_h G\right)+\frac{1}{r}\frac{\partial}{\partial r}\left(r\rho_g S_g v_g\right)=0,$$

$$\frac{\partial}{\partial t}\left(\rho_i S_i + \rho_h S_h (1-G)\right) = 0,$$

$$mS_g v_g = -\frac{k_g}{\mu_g}\frac{\partial p}{\partial r},$$

$$\rho c\frac{\partial T}{\partial t} = \frac{1}{r}\frac{\partial}{\partial r}\left(r\lambda\frac{\partial T}{\partial r}\right) + \frac{\partial}{\partial t}\left(m\rho_h L_h S_h\right) - \rho_g c_g m S_g v_g\left(\frac{\partial T}{\partial r} + \varepsilon\frac{\partial p}{\partial r}\right) + \rho_g c_g m S_g \eta\frac{\partial p}{\partial t},$$

$$p = z\rho_g R_g T, \quad k_g = kS_g^3, \quad S_g + S_i + S_h = 1,$$

$$\varepsilon = -\frac{1}{\rho_g c_g}\frac{T}{z}\left(\frac{\partial z}{\partial T}\right)_p, \quad \eta = \frac{1}{\rho_g c_g} - \varepsilon .$$

Here ρ_j and c_j ($j = g, i, h$) are the density and specific heat of j-th phase; ρ_{sk}, λ_{sk} and c_{sk} are density, thermal conductivity and specific heat of the porous medium skeleton; v_g is the gas velocity; ε and η are the throttling and adiabatic coefficients; L_h is the heat of hydrate decomposition; ρc and λ are the specific volumetric heat capacity and the thermal conductivity coefficient of the system of "porous medium – saturating substance".

To calculate the gas compressibility factor z we use the Latonov-Gurevich equation:

$$z = \left(0{,}4\cdot\lg\left(T/T_c\right) + 0{,}73\right)^{p/p_c} + 0{,}1\cdot p/p_c .$$

Temperature and pressure in the zone of the hydrate decomposition are interrelated by the phase equilibrium condition [2, 7]:

$$T = T_0 + T_* \ln\left(p/p_{s0}\right),$$

where p_{s0} is the equilibrium pressure corresponding to the initial reservoir temperature T_0; T_* is the empirical parameter which depends on the type of gas hydrate and the phases to which it decomposes.

3 Calculation results

The constructed system of equations was solved numerically using the implicit difference scheme, the tridiagonal matrix algorithm, the simple iteration method, and the original author's method to take into account the phase transition "hydrate – gas and ice".

The calculations were carried out at the following parameters [2, 7, 8]: $p_0 = 2.8$ MPa; $T_0 = 270.15$ K (-3°C); $r_w = 0.1$ m; $R_k = 500$ m; $m = 0.1$; $k = 10^{-14}$ m^2; $S_{h0} = 0.2$; $p_c = 4.6$ MPa; $T_c = 190.56$ K; $T_* = 30$ K; $p_{s0} = 2.54$ MPa; $c_{sk} = 1000$ J/(kg·K); $c_h = 2080$ J/(kg·K); $c_i = 2060$ J/(kg·K); $\lambda_{sk} = 1.5$ W/(m·K); $\lambda_h = 0.45$ W/(m·K); $\lambda_i = 2.2$ W/(m·K); $R_g = 518.3$ J/(kg·K); $\rho_{sk} = 2000$ kg/m^3; $\rho_h = 900$ kg/m^3; $\rho_i = 900$ kg/m^3;$G = 0{,}12$; $L_h = 166$ kJ/kg; the spatial coordinate step $\Delta r = 0.01$ m at $r \in [r_w, 10]$, $\Delta r = 0.1$ m at $r \in (10, 100]$ and $\Delta r = 1$ m at $r \in (100, R_k]$; calculation time $t_{end} = 30$ days; time step $\Delta t = 60$ s; the values c_g, λ_g, μ_g were determined by interpolation of table data for methane.

The testing of the software program was carried out by comparing the results of numerical experiments and calculations carried out taking into account self-similar solutions from [2]. This comparison showed a satisfactory agreement.

Figure 2 shows that depending on the gas mass flow rate Q, three cases are possible. The first case is characterized by the absence of the hydrate decomposition (line 1), in the second case the dissociation of the gas hydrate completely occurs at the frontal boundary (line 2), and the third case is characterized by the extended volume zone of phase transitions (lines 3 and 4).

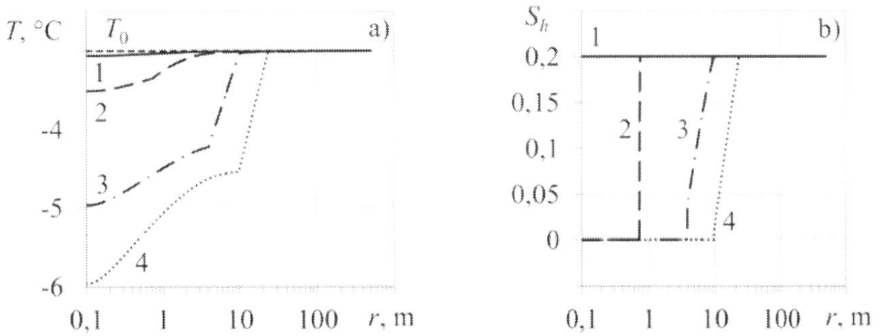

Fig. 2. The distributions of temperature (a) and hydrate saturation (b) by radial coordinate. Lines 1, 2, 3 and 4 correspond to the mass flow rate Q = 100, 300, 500 and 700 kg/(day·m).

Figure 2 shows that the reservoir temperatures is always either equal to or lower the initial reservoir temperature T_0, which is due to absorption of the latent heat of the phase transition and the decrease in gas temperature during filtration due to the adiabatic cooling and Joule-Thomson effects. Consequently, in the case of the negative initial reservoir temperature, the dissociation of gas hydrate will always produce gas and ice. In addition, at large values of the mass flow rate of gas extraction (large values of the pressure gradient) the volume zone of hydrate decomposition arises (lines 3 and 4), and with an increase in the value of Q it becomes impossible to examine the process using the frontal scheme of hydrate decomposition. Also Fig. 2 shows that an increase in the mass flow rate of gas extraction leads to an increase in the extent of the volume zone of hydrate decomposition (second zone) and an increase in the fraction of the dissociated hydrate (lines 3 and 4).

This work was supported by the Grant of the President of the Russian Federation for State Support of Leading Scientific Schools of the Russian Federation (NSh-6987.2016.1).

References

1. Y.F. Makogon, S.A. Holditch, T.Y. Makogon, J. Petroleum Science and Eng. **56** 54 (2007)
2. M.K. Khasanov, N.G. Musakaev, I.K. Gimaltdinov, J. Eng. Phys. Thermophys. **88**, 1052 (2015)
3. G.G. Tsypkin, Doklady Physics **46**, 806 (2001)
4. V.I. Vasil'ev, V.V. Popov, G.G. Tsypkin, Fluid Dynamics **41**, 599 (2006)
5. G.G. Tsypkin, J. Eng. Phys. Thermophys. **74**, 1083 (2001)
6. V.Sh. Shagapov, N.G. Musakaev, M.K. Khasanov, Int. J. Heat Mass Transfer **84,** 1030 (2015)
7. V.Sh. Shagapov, M.K. Khasanov, N.G. Musakaev, Ngoc Hai Duong, Int. J. Heat Mass Transfer **107,** 347 (2017)
8. N.G. Musakaev, S.L. Borodin, Heat Mass Transfer **52**, 981 (2016)

Modeling the multiphase flows in deformable porous media

Lyudmila Perepechko[1,*], *Evgeniy* Romenski[2], *Galina* Reshetova[3], *Sergey* Kireev[3], and *Yury* Perepechko[1,4]

[1]Kutateladze Institute of Thermophysics, SB RAS, 630090 Novosibirsk, Russia
[2]S.L. Sobolev Institute of Mathematics, SB RAS, 630090 Novosibirsk, Russia
[3]Institute of Computational Mathematics and Mathematical Geophysics, SB RAS, 630090 Novosibirsk, Russia
[4]V.S. Sobolev Institute of Geology and Mineralogy, SB RAS, 630090 Novosibirsk, Russia

Abstract. This work proposes the nonlinear model for the flow of mixture of compressible liquids in a porous medium with consideration of finite deformations and thermal effects. Development of this model is based on the method of thermodynamically consistent systems of conservation laws. Numerical analysis of the model is based on the WENO-Runge-Kutta method of the high accuracy. The model is developed to solve the problems arising when studying the different-scale fluid dynamic processes. Evolution of the wave fields in inhomogeneous saturated porous media is considered.

1 Introduction

This study deals with mathematical and numerical modeling of the deformable porous media saturated with a mixture of compressible liquids. The problem of wave process modeling in the multiphase saturated porous media and filtering a mixture of compressible media is relevant for solving a wide range of applied problems in various industries, as well as the problems arising from the analysis of the natural systems behavior, for example, when studying heat and mass transfer in the fluid-magmatic systems. The solution of this problem is especially relevant when creating new technologies for the development of hydrocarbon fields, analysis of dynamic processes in oil wells and borehole formations. The modern methods of hydrocarbon production are characterized by high pressure gradients in the bed and occurrence of significant rock deformations up to their destruction. The non-linear character of the flows under such conditions requires consideration of the shear stresses. The description of such hydrodynamic systems requires a consistent model of the flow of compressible media and deformation of the porous medium, and reliability of results of mathematical and numerical modeling should be provided both by the methods of deriving the coordinated equations of the model and the methods for their solution. In this study, the model of the flow of a liquid mixture of compressible phases through a porous deformable medium is constructed in the framework of the method of thermodynamically

* Corresponding author: ludmila@itp.nsc.ru

consistent system of conservation laws [1]. The equations of motion satisfy the fundamental laws: the laws of conservation of mass, momentum, and energy, invariance relative to Galileo transformations, and fundamentals of thermodynamics. The determining equations are hyperbolic, and this ensures the existence and uniqueness of the solution. An essential property of the derived equations is the possibility of their presentation in the form of conservation laws. These properties of the model equations allow us to apply the effective numerical methods for solving the hyperbolic systems of conservation laws. In this work, we used the algorithm based on the WENO-Runge-Kutta method of high accuracy, successfully used for modeling the two-phase compressible flows and filtering in a saturated porous medium [2-4].

2 Mathematical model

The three-phase medium is a deformable porous medium saturated with a mixture of compressible liquids. The phases are marked with the index "n": where "1" corresponds to the porous matrix, "2" and "3" correspond to saturating liquids, e.g. water and oil. The unit volume of such a medium is characterized by: volume concentrations α_n ($\alpha_1+\alpha_2+\alpha_3=1$, $\phi=1-\alpha_1$ is porosity), mass concentrations c_n ($c_1+c_2+c_3=1$); mass densities and phase velocities ρ_n and u_i^n, and velocities of liquid phases motion relative to the porous matrix $w_i^n=u_i^1-u_i^n$. The three-phase medium is also characterized by density $\rho=\sum_{n=1}^{3}\alpha_n\rho_n$ and velocity $u_i=\sum_{n=1}^{3}c_n u_i^n$ ($c_n=\alpha_n\rho_n/\rho$), tensor of elastic strain gradient F_{ij}, and mass density of entropy s. Inner energy of the three-phase medium $e=e(\rho,\alpha_n,c_n,w_i^n,F_{ij},s)$ is determined as the internal phase energy e^n, averaged by mass concentration, and supplemented with the kinetic energy of relative phase motion:

$$e\left(\rho,\alpha_n,c_n,w_k^n,F_{ij},s\right)=c_1e^1\left(\rho_1,F_{ij},s\right)+c_2e^2\left(\rho_2,s\right)+c_3e^3\left(\rho_3,s\right)+$$
$$+\frac{1}{2}\left(1-c_2\right)c_2w_k^2w_k^2+\frac{1}{2}\left(1-c_3\right)c_3w_k^3w_k^3-c_2c_3w_k^2w_k^3. \qquad (1)$$

Generalized internal energy e depends on total entropy s of the three-phase system due to the assumption of temperature phase equilibrium. Dependence of e^1 on F_{ij} is given by the deviator of Finger tensor g_{ij} [1]. The equation of state is determined by relationships for the pressure in phases $p_n=\rho_n^2e_{\rho_n}^n$ (with $p=\rho^2e_\rho$), shear stresses $\sigma_{ik}=\rho_1F_{kj}e_{F_{ij}}^1$ and temperature $T=e_s$.

Thermodynamically consistent system of conservation laws for the flow of mixture of compressible liquids through a deformable matrix takes the form [3,4]:

$$\frac{\partial\alpha_n\rho_n}{\partial t}+\partial_k\left(\alpha_n\rho_nu_k^n\right)=0, \qquad n=2,3, \qquad (2)$$

$$\frac{\partial\rho\alpha_n}{\partial t}+\partial_k\left(\rho\alpha_nu_k\right)=-\sum_{m=2}^{3}\lambda_{nm}\left(p_1-p_m\right), \qquad n=2,3, \qquad (3)$$

$$\frac{\partial w_k^n}{\partial t}+\partial_k\left(h^1-h^n\right)=e_{kij}u_i\omega_j^n-\sum_{m=2}^{3}\chi_{nm}c_m\left(u_k-u_k^m\right), \quad n=2,3 \qquad (4)$$

$$\frac{\partial\left(\rho u_i\right)}{\partial t}+\partial_k\left(\sum_{n=1}^{3}\alpha_n\rho_nu_i^nu_k^n+\delta_{ik}\sum_{n=1}^{3}\alpha_np_n-\alpha_1\sigma_{ik}\right)=0, \qquad (5)$$

$$\frac{\partial \rho F_{ij}}{\partial t} + \partial_k \left(\rho F_{ij} u_k - \rho F_{kj} u_i \right) = 0, \tag{6}$$

$$\frac{\partial \rho}{\partial t} + \frac{\partial \rho u_k}{\partial x_k} = 0, \tag{7}$$

$$\frac{\partial \rho s}{\partial t} + \partial_k \left(\rho s u_k \right) = \frac{R}{T}. \tag{8}$$

The considered processes of relaxation of volume phase concentrations to the equilibrium value and relaxation of the phase velocities due to interfacial friction determine the form of dissipative function:

$$R = \sum_{n,m=2}^{3} \chi_{nm} \left(u_i - u_i^n \right) \left(u_i - u_i^m \right) + \frac{1}{\rho^2} \sum_{n,m=2}^{3} \lambda_{nm} \left(p_1 - p_n \right) \left(p_1 - p_m \right). \tag{9}$$

Here $h^n = e^n + \frac{1}{2} u_i^n u_i^n + p_n / \rho_n$, $e_{ikj} \omega_j^n = \partial_i w_k^n - \partial_k w_i^n$. Variables ω_j^n satisfy the additional conservation laws [2], which are the compatibility conditions for the solutions of system (2) - (8). This allows us to interpret terms $e_{klj} u_l \omega_j^n$ in (5) as the lower terms. Kinetic coefficients λ_{ij} characterize relaxation of pressures in the phases. The coefficients of interfacial friction χ_{ik} are determined by the Darcy formulas with relative permeabilities k_{nm}.

The equations formulated above ensure the fulfillment of the energy conservation law

$$\frac{\partial}{\partial t} \left(\rho e + \frac{1}{2} \rho u_i u_i \right) + \partial_k \left(\sum_{n=1}^{3} \alpha_n \rho_n u_k^n h^n - u_i \alpha_1 \sigma_{ik} \right) = 0. \tag{10}$$

The model describes the flow of a mixture of compressible liquids through a deformable porous medium in the case of finite deformations with consideration of the thermal phenomena.

3 Modeling results

Propagation of nonlinear wave fields near the boundary of the water-oil mixture and fluid-saturated porous formation in the presence of the oil-water contact near the boundary was investigated as an application to the model. The water-oil contact was set in the calculation region ($5 \times 4\,\mathrm{m}$) at the distance of $1.5\,\mathrm{m}$ to the left of the boundary between the porous matrix and the water-oil mixture (Fig. 1a). The initial volume concentrations in water-oil mixture were $\alpha_{20} = 0.8$ for water and $\alpha_{20} = 0.2$ for oil. The parameters of the porous medium layers were taken as follows: for layers (1) and (2) porosity was $\phi = 0.1$ and permeabilities were $k_{22} = 0.2 \cdot 10^{-12}\,\mathrm{m}^2$, $k_{33} = 0.2 \cdot 10^{-11}\,\mathrm{m}^2$; for layer (2) porosity was $\phi = 0.2$ and permeabilities were $k_{22} = k_{33} = 0.2 \cdot 10^{-10}\,\mathrm{m}^2$. Kinetic coefficients λ_{ij}, characterizing relaxation of pressures in the phases, were taken: $\lambda_{22} = 0.11 \cdot 10^{-6}\,\mathrm{c/m}^2$, $\lambda_{33} = 0.69 \cdot 10^{-6}\,\mathrm{c/m}^2$, and $\lambda_{23} = 0.4 \cdot 10^{-8}\,\mathrm{c/m}^2$. Shear modulus of porous medium were taken $8.9 \cdot 10^9\,\mathrm{Pa}$. The initial values of thermodynamic parameters corresponded to the normal conditions. The initial values of velocities were assumed to be zero.

The stress and pressure fields for the moment of acoustic wave propagation in the region of water-oil contact are shown in Fig. 1. The pressure source of the Ricker type is located in the center of the upper part of the region (3). The location of water-oil contact is marked in Fig. 1 by a dash-dotted line. The boundaries of horizontal layers with different values of porosity and permeability are marked by the dashed lines. Sensitivity of the wave fields to the presence of the water-oil contact and isolated layers in a porous medium is observed. The observed effect is more pronounced for the pressure field.

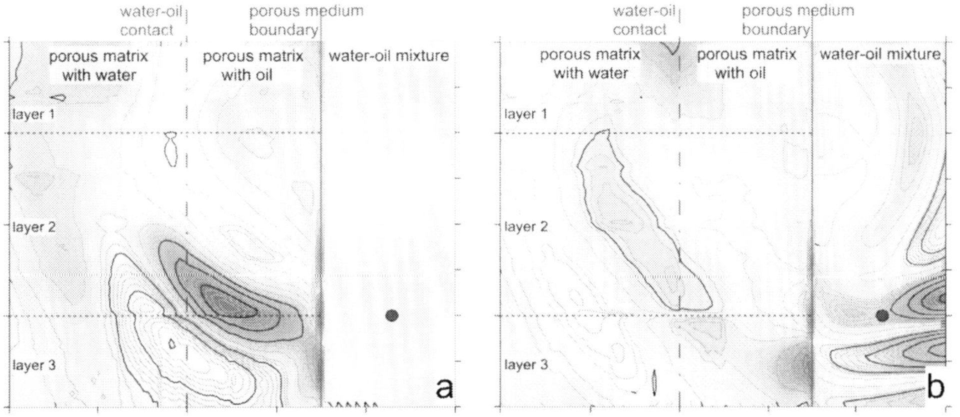

Fig. 1. The fields of stress σ_{xx} (a) and pressure p (b). The pressure source is marked by a circle.

The developed thermodynamically matched mathematical model of multiphase mixtures in the porous media and applied finite-difference algorithm of a high accuracy based on the WENO-Runge-Kutta method make it possible to study the features of the multiphase flow in the deformable porous media in a wide range of model parameters. The model allows us to investigate acoustic wave propagation on a hydrodynamic background, accompanied by dissipative phenomena and temperature variations, and the effect of acoustic action on the character of multiphase flow of fluids in porous media; solve non-stationary problems of non-isothermal multiphase filtration of mixtures of compressible liquids in porous media under the conditions of significant elastoplastic deformations.

The work was supported by the Russian Foundation for Basic Research, grant No. 16-29-15131.

References

1. S. Godunov, E. Romenski, *Elements of Continuum Mechanics and Conservation Laws* (Kluwer Academic/Plenum Publishers, NY, 2003)
2. E. Romenski, D. Drikakis, E.F. Toro, J. Sci. Comput. **42**, 68 (2010)
3. Y.V. Perepechko, E.I. Romenski, G.V. Reshetova, Seismic Technologies **4**, 78 (2014)
4. Yu. Perepechko, E. Romenskiy, G. Reshetova, L. Perepechko, V. Malyshkin, K. Kalgin, S. Kireev, M. Ostapkevich, In book: *Supercomputer Technologies in Science, Education and Industry* 119 (MSU, Moscow, 2013) [in Russian]

Modeling of detonation transition through the field of mixing the reacting and inert gases

Evgeniy Prokhorov[1,*]

[1]Lavrent'ev Institute of Hydrodynamics SB RAS, 630090 Novosibirsk, Russia

Abstract. The non-stationary problem of exciting a plane shock wave by gas detonation in a tube is numerically solved. The case, when the field of mixing the reacting and inert gases filling the closed tube end has finite size, is considered. The influence of mixing field width on the intensity and damping law of excited shock waves is studied. Ignoring energy losses, the problem solution is determined by one dimensionless parameter equal to the ratio of gas mixture volume in the mixing field to the volume of reacting gas located in the tube before the detonation is initiated. By varying this parameter within the range from 0 to 2, the maximal value for the Mach number of the shock wave in inert gas (air) is decreased by about 20%. It is established that decrease pattern of the shock-wave front velocity can be approximately described by the dependence corresponding to the conclusions made from the theory of point explosion for the case of plane adiabatic gas motions.

The non-stationary problem of exciting a plane shock wave by gas detonation in a tube is solved numerically. In contrast to the previous theoretical papers, we consider the case (Fig. 1), where the field of mixing the reacting and inert gases located at the closed end of the tube has finite sizes ($\Delta l \neq 0$).

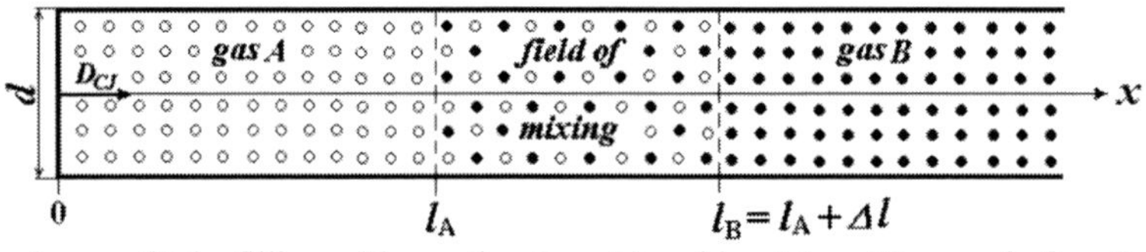

Fig. 1. The scheme of tube filling with reacting (gas A) and inert (gas B) gases before the detonation is initiated.

The mathematical statement of the problem (as in [1]) is based on the approximate model for computation of equilibrium flows of chemically reacting gases describing with high accuracy the variation of molar mass and specific internal energy in the shift of chemical equilibrium of detonation combustion products of hydrocarbons [2]. Moreover, the chemical composition of detonation products is characterized only by relative concentrations of oxygen atoms n_O , carbon atoms n_C ($n_C \leq n_O$), hydrogen atoms n_H ,

* Corresponding author: prokh@hydro.nsc.ru

nitrogen atoms n_N and any other monatomic inert substances n_Z. It is shown that such approach can be also used to describe thermodynamic properties of shock-compressed inert gas.

We performed the computational experiments simulating the process of transforming a detonation wave into a shock wave at the diffuse interface between the reacting gas (acetylene-oxygen mixture: $C_2H_2 + 2.5O_2$) and inert gas (air: $0.21O_2 + 0.78N_2 + 0.01\mathrm{Ar}$). Here it was assumed that in the field of diffusion mixing the concentrations of these gases change linearly. The initial distribution of the parameters of detonation products were given in the form of a simple wave (the Riemann wave) propagating from the rigid wall along the reacting gas with velocity of the Chapman-Jouget detonation D_{CJ}.

We studied the influence of mixing field width Δl on the intensity and the decay pattern of excited shock waves. To keep mass and energy balance, when varying Δl, we assumed that $l = (l_A + l_B)/2 = const$. In this case the volume of reacting gas in the tube before the detonation is always constant, and thus it is easy to estimate its value $l = l_A$ at $\Delta l = 0$ (Fig. 1). Then from a practical point of view the increase of Δl can be interpreted as increasing the diffusion of interface of two contacting gases. Here the coordinate of the left boundary of mixing field $l_A = l - \Delta l/2$ will shift to the closed tube end. Since within the framework of the considered statement of problem $l_A \geq 0$, there is upper limit for the width of mixing field equal to $\Delta l \leq 2l$.

Ignoring energy losses (due to friction and heat removal to the tube walls), we can assume that a characteristic spatial scale of the problem is l. Therefore, at $\Delta l/l = const$ all calculated relationships constructed depending on dimensionless coordinate x/l for different values of l will coincide. Hence, the problem solution has similarity and it is defined by one dimensionless parameter $\Delta l/l$.

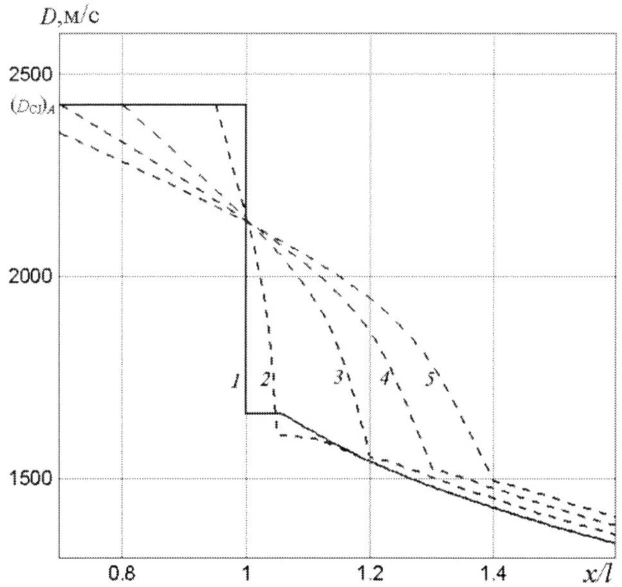

Fig. 2. Wave front velocity D versus its position in the tube. Curve 1 denotes computations at $\Delta l/l = 0$; 2 — $\Delta l/l = 0.1$; 3 — $\Delta l/l = 0.4$; 4 — $\Delta l/l = 0.6$; 5 — $\Delta l/l = 0.8$

Figure 2 demonstrates the dynamics of changing the front velocity D taking place during the process of transforming a detonation wave into a shock wave at different values of mixing field width Δl. The results of computations performed for the case $\Delta l = 0$, where the wave front velocity changes step-wise at the interface between the reacting and inert gases, are also shown. The rapid change of D (curve 1) from the value of the Chapman-Jouget detonation velocity in gas A, i.e., $(D_{CJ})_A$ = 2424 m/s, to the value of $\overline{D}_0$ = 1658 m/s was defined in the solution of the Riemann problem (the problem of discontinuity breakdown).

Interestingly, the proximity of slopes of all curves (see Fig. 2) takes place during the shock wave decay after passing through the mixing zone. At $x > l_B$, the decrease pattern of shock front velocity can be approximately described by the dependence $d(\ln D)/d(\ln x) = -1/2$. This result is in good agreement with the conclusions made from the theory of point explosion for the case of plane adiabatic gas motions.

Curves 2-5 also have a sharp break at $x = l_B$ ($l_B = l + \Delta l/2$), where a detonation wave front transforms into a shock front. At this moment the shock front velocity is maximum. We denote it by $\overline{D}$ ($\overline{D} \le \overline{D}_0$). Then the coordinates of the inflection point in Fig. 2 can be given as $(l_B/l, \overline{D})$. The Mach number $\overline{M} = \overline{D}/(c_0)_B$ for the shock wave propagating along gas B corresponds to this point.

If we draw a line passing through the inflection points of curves 2-5, it must intersect curve 1. With allowance for clarification, it is established that the intersection occurs at $x/l = 1.15$ and corresponds to calculated dependence $D = D(x/l)$ at $\Delta l/l = 0.3$. At $x/l > l_B/l = 1.15$, the curve corresponding to this dependence almost (accurate to 1 %) coincides with curve 1. The above statement is also valid for other calculated curves $D = D(x/l)$ at smaller values of $\Delta l/l$; although for them the inflection point $(l_B/l, \overline{D})$ will be located below curve 1. Here $\overline{D}$ differs slightly from $\overline{D}_0$. For example, for curve 2 this difference is only 3%. It is obvious that curve 2 has a short horizontal segment, as curve 1, which points to the propagation of wave with constant velocity.

This yields a practically significant conclusion that based only on measurements of shock wave front velocity in the tube part at $x/l > 1.15$ in the experiment, it is impossible to answer the question whether the interface between reacting and inert gases was diffused or not up to values of $\Delta l/l = 0.3$ (which may be called critical) before the detonation was initiated.

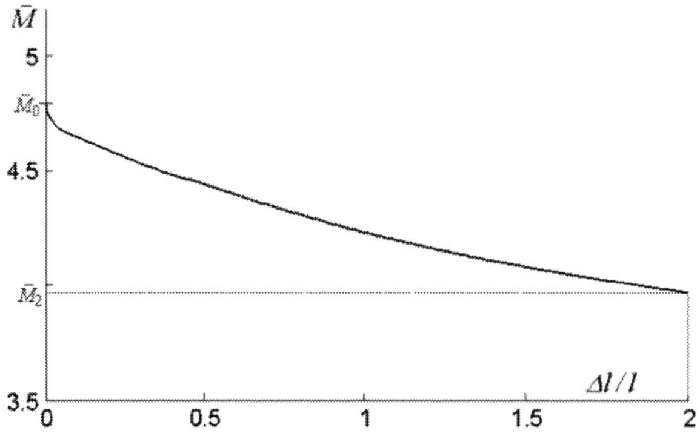

Fig. 3. Maximal Mach number of shock wave $\overline{M}$ excited by detonation in inert gas versus dimensionless parameter $\Delta l/l$.

The dependence shown in Fig. 3 allows estimating the influence of dimensionless mixing zone width $\Delta l/l$, which defines the problem solution, on $\overline{M}$. Thus, the maximal Mach number $\overline{M}$ for a shock wave in gas B has a restriction: $\overline{M}_2 < \overline{M} < \overline{M}_0$, where $\overline{M}_0 = 4.791$ is found by solving the problem of discontinuity breakdown (the Riemann problem) at $\Delta l/l = 0$ and $\overline{M}_2 = 3.973$ is defined at $\Delta l/l = 2$. It follows that when varying parameter $\Delta l/l$, the intensity of shock wave changes slightly ($\overline{M}_0/\overline{M}_2 = 1.206$). Knowing $\overline{M}$ it is easy to calculate gas-dynamic parameters of the shock front, such as density, pressure, temperature, etc., by using known formulae.

Thus, within the framework of the approach simulating the propagation of gas detonation in the medium with variable chemical composition formulated in [1], we conducted numerical solution of the problem of shock-wave excitation in refraction of the Chapman-Jouget detonation wave at diffuse interface between reacting (stoichiometric acetylene-oxygen mixture) and inert (air) gases located at the closed end of the tube. We studied the process of transformation of a detonation wave into a shock wave in the field of mixing the contacting gases, as well as analyzed the effect of mixing field width on the intensity and the decay pattern of excited shock waves.

References

1. E.S. Prokhorov, J. Eng. Phys. Thermophys. **90**, 151 (2017)
2. E.S. Prokhorov, Combust. Explos. Shock Waves **32**, 306 (1996)

Wall shear stress from a single almost spherical and a Taylor bubbles in laminar upward tube flow

Leonid Timkin* and *Roman* Gorelik

Kutateladze Institute of Thermophysics, Pr. Ac. Lavrentev, 1, 630090 Novosibirsk, Russia

Abstract. An experimental electrodiffusional technique with eight double probes is used to detect perturbation the wall shear stress caused by a single bubble in laminar upward tube flow. A small almost spherical and different length Taylor bubbles are considered. The shear stress perturbations by bubbles have a complex structure. It is possible to define three components of perturbation caused by a small bubble. The perturbation by Taylor bubble contains only two components due to the main flow symmetry around the bubble. An unexpectedly long shear stress pulsations zone is registered behind the bubbles.

1 Introduction

Anomalously high wall shear stress in bubbly flow may increase the heat-mass transfer and result in the fast degradation of an industrial apparatus. Numerous works have been devoted to this phenomenon. Many investigations have been dedicated to rising Taylor bubbles, starting from [1, 2]. One of the last review of slug flows focused on the formalization of flow governing laws and the physical mechanisms involved, was given in [3]. In the last few decades, new techniques have been used to deeper understanding of bubble rising. The growing capabilities of data acquisition and processing, along with higher precision of measurement techniques, have resulted in a new understanding of flow features. The rising of single bubble even in laminar upward flow is complex and asymmetric, sometimes including liquid flow reversals. The rising bubbles induce different component of flow perturbations, that may prevail under the various flow conditions, changing the integral flow characteristics. The estimation of different components of bubble induced pertuberation is essential for real flow modeling. This work aims at estimating the single bubble impact on the wall shear stress in laminar flow.

2 Experimental setup

The experimental setup consist of a vertical tube with inner diameter D = 14.8 mm – inner diameter, and 6.0 m in height, with a top and a bottom tank, a pump, a downward tube, and system of thermostabilization. A bubble is generated at the tube inlet in special tank filled

* Corresponding author: timkin@itp.nsc.ru

with liquid. A simple rotated cap is filled with air and overturned to produce a single bubble. Then, the bubble is carried by the flow into the tube. The test section was 5 m distant from the tube inlet. An immersion block with optical windows and a glass tube inside were installed at the top of the test section to take pictures of moving bubbles. A rotameters were used for liquid flow control. A ferri-ferrocyanide electrochemical solution with glycerin addition to increase the viscosity, was used as the test liquid. At a temperature of 20 ± 0.2°C the liquid density was ρ = 1125 kg/m^3, the kinematic viscosity $\nu = 3.5 \cdot 10^{-6}$ m^2/s.

Wall shear stress measurements were performed using an electrodiffusional technique. The eigth double electrodiffusional probes glued into the wall of the test section uniformly over the tube perimeter. Each probe consists of two rectangular platinum electrodes with a thin insulation layer (20 ÷ 30 μm) in between. In experiments with Taylor bubbles, the size of the electrodes was 35 × 210 μm, for small bubbles the size of the electrodes was 40 × 700 μm. The probes were connected to a computer through individual amplifiers and ADC. The detail of probes calibration is fixed in [4]. Photos of bubble types checked in experiments are presented on Fig. 1. The flow characterized by: a superficial liquid velocity V_l, a liquid Reynolds number $Re = D \cdot V_l/\nu$, a bubble's rising velocity V_b, a small bubble diameter d_b, terminal velocity V_∞, Reynolds number $Re_b = d_b \cdot V_\infty/\nu$, and a Taylor bubble length L_b.

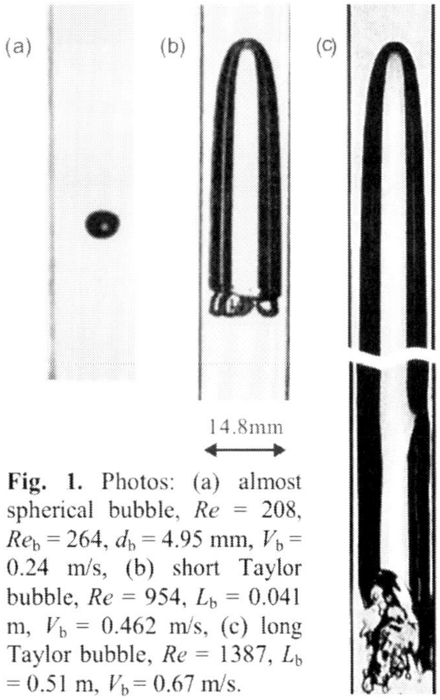

Fig. 1. Photos: (a) almost spherical bubble, Re = 208, Re_b = 264, d_b = 4.95 mm, V_b = 0.24 m/s, (b) short Taylor bubble, Re = 954, L_b = 0.041 m, V_b = 0.462 m/s, (c) long Taylor bubble, Re = 1387, L_b = 0.51 m, V_b = 0.67 m/s.

3 Results

The shear stress perturbations by small bubbles are presented in Fig. 2 (a – c). Prior to the bubble appearance, the shear stress resembles a smooth surface without pulsations, with the shear stress corresponding to single-phase flow. The bubble induces a perturbation of this surface. It is possible to define three components of the perturbation. The first component is a decrease in shear stress, which is caused by an accelerated liquid around the bubble in the direction opposite to the bubble motion. The second component is an intense positive peak. The small bubbles rise out of the tube axis, move closer to one side of the tube wall. In the gap between the bubble and tube wall, the liquid is carried by bubble with high velocity, and generating a peak structure. This peak was previously detected in [5, 6]. The similar structure was obtained in [7, 8]. The third component is a positive pulsations behind the bubble. This component presents a gradual decay of the perturbed shear stress to a single-phase laminar value. In deep laminar flow this component is small, Fig. 2 (a). For higher Reynolds numbers Re = 988 and 1920 the peak component grows and disintegrates into a several peaks, which are distributed over the whole tube perimeter. The peak and pulsation component join together to form a turbulent pulsations zone, completely occupying the tube crossection. At higher liquid Re numbers they are responsible for the anomalously high shear stress in two-phase bubbly flows. There are tested long Taylor bubbles (L_b ~ 0.25 ÷ 0.65 m) and short ones (L_b ~ 0.04 ÷ 0.2 m). Three wall shear stress perturbations caused by Taylor bubbles of length L_b ~ 0.5 m are presented in Fig. 2 (d – f).

 DOI: 10.1051/matecconf/201711505006

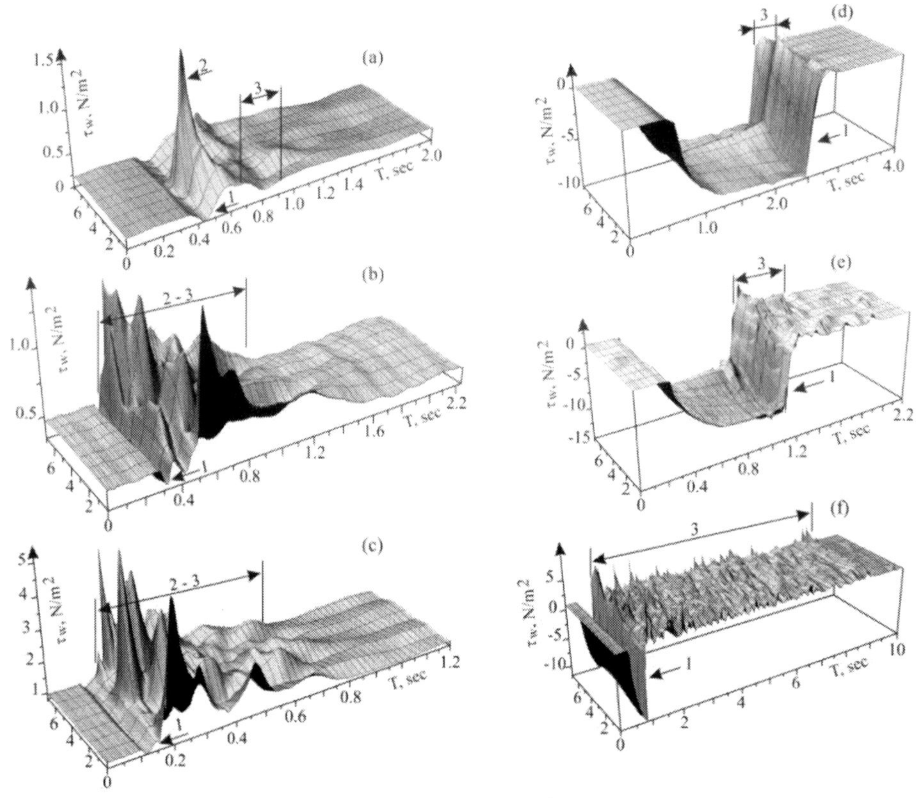

Fig. 2. The perturbations caused by almost similar small bubbles: (a) $Re = 444$, $Re_b = 272$, $d_b = 5.15$ mm, $V_b = 0.33$ m/s; (b) $Re = 988$, $Re_b = 250$, $d_b = 4.63$ mm, $V_b = 0.503$ m/s; (c) $Re = 1920$, $Re_b = 280$, $d_b = 5.36$ mm, $V_b = 0.865$ m/s. The perturbations caused by almost similar Taylor bubbles: (d) $Re = 217$, $L_b = 0.48$ m, $V_b = 0.204$ m/s; (e) $Re = 954$, $L_b = 0.46$ m, $V_b = 0.48$ m/s; (f) $Re = 1844$, $L_b = 0.48$ m, $V_b = 0.87$ m/s. Where: 1 – decreasing component, 2 – peak component, 2 – 3 and 3 – pulsations zone.

Taylor bubble perturbations contain only two components which are the decreasing and the pulsations ones. The peak component is absent due to the main flow symmetry. Again, in front of bubble the shear stress is constant over the tube perimeter. The Taylor bubble pushes the shear stress into the negative values. A downwardly accelerated liquid film in a ring gap between the bubble and the tube wall forms the strong decreasing component. The readings of all probes are almost the same. This proves the fact that the flow around the Taylor bubble is symmetrical. Behind the bubble, the liquid film collides with the liquid entrained by bubble that causes strong mixing. A positive shear stress pulsation appears directly behind the bubble as a pulsations zone, which eventually decays to laminar background flow. In flow with $Re = 217$, Fig. 3 (d), the length of the pulsations zone behind the bubble is about l_p/D ~ 5. For moderate $Re = 954$, Fig. 3 (e), the length increases up to l_p/D ~ 35. For higher Reynolds number, Fig. 3 (f), the length exceeds l_p/D = 400. Previously, in literature a bubble wake was investigated in detail in [2]. The wake length was estimated as $(16 \div 30)D$. Later, according to [9], the length increased up to $50D$.

In the current work l_p considers as the length of the wall shear stress pulsations zone. This length is presented in Fig. 3 (a) for small bubbles, while in Fig. 3 (b) - for Taylor bubbles. The data for Taylor bubbles are divided conditionally in two series for short bubbles ($L_b \sim 0.04 \div 0.2$ m) and long bubbles ($L_b \sim 0.25 \div 0.65$ m). For small bubbles the pulsations zone length – l_p/D increased point-by-point with the liquid Reynolds number. The maximal length is about $70 \div 90D$ that significantly exceeds even the Taylor bubble wake length indicated in the literature. The data for Taylor bubbles length depends on the liquid Reynolds number. Starting from very small values at small Re numbers, the length sharply increased up to 400D for Re ≥ 1500. At higther Re the length for long Taylor bubbles is 1.5 ÷ 2 times bigger than that for short ones.

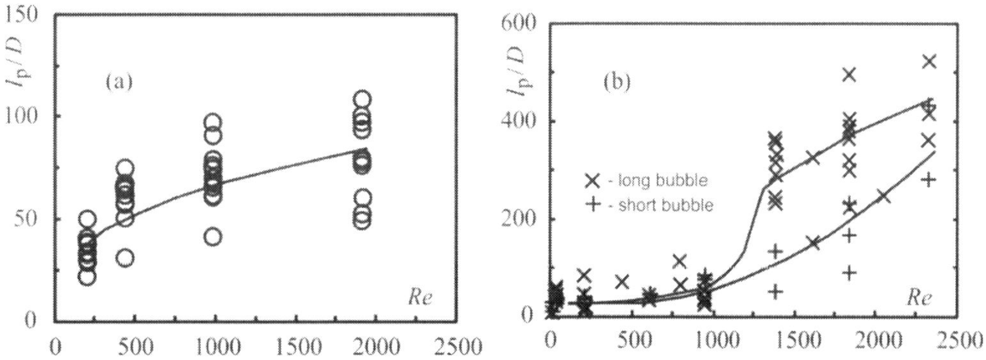

Fig. 3. The pulsations zone length: (a) for almost spherical bubbles with d_b ~ 3 ÷ 6 mm; (b) for short Taylor bubbles with lengths of L_b ~ 0.04 ÷ 0.2 m and long bubbles with with lengths of L_b ~ 0.25 ÷ 0.65 m.

Thus, the length of shear stress pulsations zone significantly depends on the liquid Reynolds number. At high *Re* numbers for small almost spherical bubbles it achieve values of (50 ÷ 90)D and becomes drastically long, exceeding more them 400D for long Taylor bubbles that significantly exceeds the length of bubbles wakes, mentioned in literature.

References

1. D.T. Dumitrescu, Z. Angew. Math. Mech. **23,** 139 (1943)
2. R. Moissis, P. Griffith, J. Heat Transfer, **84**, 29 (1962)
3. A.O. Morgado, J.M. Miranda, J.D.P. Araújo, J.B.L.M. Campos, Int. J. Multiph. Flow **85**, 348 (2016)
4. V.E. Nakoryakov, L.S. Timkin, R.S. Gorelik, Thermophys. Aeromech. **18**, 281 (2011)
5. L.S. Timkin, R.S. Gorelik, P.D. Lobanov, J. Eng. Phys. Thermophys. **78**, 762 (2005)
6. G. Hetsroni, C.F. Li, A. Mosyak, I. Tiselj, Int. J. Multiphase Flow **27**, 1127 (2001)
7. M.N. Descamps, R.V.A. Oliemans, G. Ooms, R.F. Mude, Exp. Fluids **45**, 357 (2008)
8. R. Van Hout, A. Gulitski, D. Barnea, L. Shemer, Int. J. Multiphase Flow **28**, 579 (2002)
9. J.D.P. Araujo, J.M. Miranda, A.M.F.R. Pinto, J.B.L.M. Campos, Int. J. Multiphase Flow **43**, 131 (2012)

A polytropic model of a critical two-phase flow in a bed of spherical particles

Emir Tairov*, and *Polina* Khan

Melentiev Energy System Institute SB RAS, 130, Lermontov St., Irkutsk, 664033, Russia

Abstract. The paper is concerned with a model of isenthalpic flow of vapor-water mixture in a fixed bed of solid particles. The mixture expansion process is considered to be polytropic. Similarly to the known problem of gas dynamics of a granular bed we obtained the relationships for calculation of a critical mass velocity. The results of the calculation based on a theoretical model are compared with the experimental data obtained in the packed beds of steel balls, 2 mm and 4 mm in diameter.

1 Introduction

The first experimental results of the research into a critical flow of a vapor-liquid mixture through the densely-packed beds of spherical particles were presented in a recent paper [1]. Here an important conclusion was drawn on linear dependence between the critical mass velocity and ratio of a particle diameter to bed height $(d/H)^{0.5}$. This dependence, revealed experimentally, coincides with the result obtained with the use of theoretical model of the homogeneous compressible medium motion in a granular bed [2]. An overview study [3] emphasizes that the widely known theoretical models for the determination of a critical flow rate of the two-phase mixture differ from each other in terms of choosing mixture density in a critical cross-section. The analogy between the flow of a one-phase compressible medium and the flow of a two-phase mixture is used to study a critical flow through the nozzles, diaphragms and short pipes in [4]. The authors of [5, 6] applied this analogy to the vapor-liquid mixture flow in the packed beds of solid particles to determine pressure drop. In the present research the isenthalpic expansion of the two-phase mixture is approximated by the equation of polytropic process. The corresponding polytropic coefficient is determined taking account of pressure P, flow quality x, and a phase slip ratio s.

2 Theoretical model

The equations of gas dynamics of a granular bed [2] were used to describe a critical flow of the vapor-liquid mixture through the fixed bed of spherical particles. The governing equations include:
the equation of motion

* Corresponding author: tairov@isem.irk.ru

$$\left[1+\frac{4}{3}(1-m)\right]\rho w\frac{dw}{dy}=-\frac{dP}{dy}-\frac{3}{2}\frac{m(1-m)}{\psi d}\rho w^2\,, \qquad (1)$$

where w – is the velocity of mixture in porous space, m – is the medium porosity, d – is the particle diameter, $\psi = 0.508 + 0.56(1-m)$ – is the relative minimum flow section;
the continuity equation

$$\rho w = \rho_1 w_1\,, \qquad (2)$$

the polytropic equation

$$P/P_1 = (\rho/\rho_1)^n\,. \qquad (3)$$

The integration of the equations with respect to the bed height from 0 to H under the assumption that $d << H$ results in the following relationship for the maximum mass velocity:

$$(\rho w)_{cr} = \left[\frac{2n}{3(n+1)}\frac{d}{H}\frac{\psi}{m(1-m)}P_1\rho_1\right]^{0.5}. \qquad (4)$$

To use equation (4) we need to know the polytropic coefficient n and mixture density ρ_1 at the granular bed inlet. Pressure P_1 and flow quality x_1 at the inlet are considered to be given.

For equilibrium vapor-liquid flow we have

$$\varphi = \left[1+s\frac{\rho''}{\rho'}\frac{1-x}{x}\right]^{-1}, \qquad (5)$$

$$\rho = \rho' - \varphi(\rho' - \rho'') = \rho''\varphi + \rho'(1-\varphi), \qquad (6)$$

where φ – is the void fraction, and $s = w''/w'$ – is the slip ratio. The variables ρ', w' and h' refer to the properties of the saturated liquid, and the variables ρ'', w'' and h'' refer to the properties of the saturated vapor.

The absence of correlation dependences for s as applied to the two-phase flow in granular media creates a certain problem. This makes some authors use in such a case the known relationships obtained for two-phase flows, which, according to these authors, are similar in nature. For example in [6] the slip ratio calculation is based on the recommendation for an adiabatic flow of vapor- water mixture in vertical pipes of a small diameter and ring ducts

$$s = (P_{cr}/P)^{0.38}\,, \qquad (7)$$

where P_{cr} – is the critical pressure. Equation (7) can be applied at the following conditions: $1 < P < 22$ MPa, $400 < \rho w < 3340$ kg/(m^2 s).

In the present research, to close the system of governing equations we obtained an approximating relationship for the slip ratio in critical conditions. The nonlinear regression method based on the experimental data on critical mass velocity $(\rho w)_{cr}$ [1], and equations (4-6) were used to obtain the following expression:

$$s = 1 + \frac{\left(7.0 - 8.0P + 4.0P^2\right)}{\exp\left(\left(0.058 - 0.13P + 0.075P^2\right)/x + \left(2.8 - 3.0P + 1.8P^2\right)x\right)}. \quad (8)$$

Equation (8) was obtained for the critical flow in the granular media under following conditions: $0.6 < P < 1.2$ MPa, $200 < \rho w < 1200$ kg/(m^2 s), $0.02 < x < 0.2$.

Comparison of the slip ratios calculated by equations (7), and (8), is presented in Fig.1. It is seen that for x>0.02 both equations give the values of the slip ratio between 3 and 4. Moreover, equation (8) takes into account the effect of the flow quality.

This research suggests approximating the isenthalpic process by the polytropic one defined by equation (3). The flow quality for isenthalpic process was calculated as follows:

$$x(P) = \frac{h'(P_1)(1 - x_1) + h''(P_1)x_1 - h'(P)}{h''(P) - h'(P)}. \quad (9)$$

Considering the obtained relationship (8), together with equations (3), (5), (6) and (9) we found n for all combinations of the input data, and approximated it as follows (Fig. 2):

$$n(x_1) = 0.42 + 0.45(1 - \exp(-x_1/0.078)), \quad 0.02 < x_1 < 0.20. \quad (10)$$

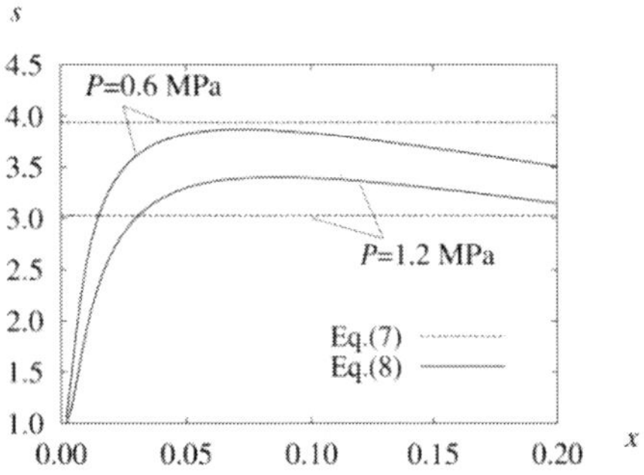

Fig.1. Approximation of the slip ratio.

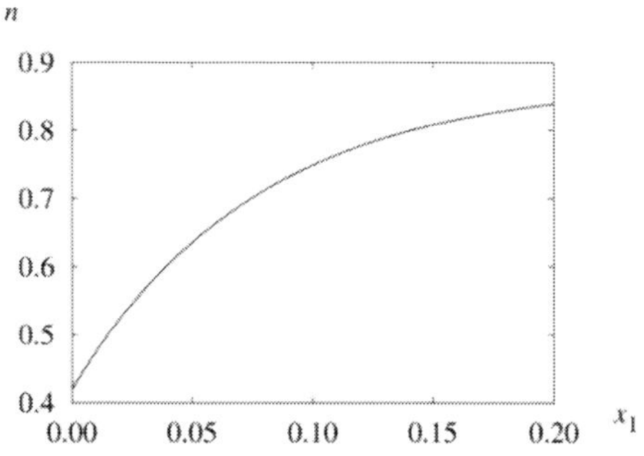

Fig. 2. Approximation of the polytropic coefficient, x1 – is the inlet flow quality.

3 The results of experiments and calculation

A description of an experimental equipment and a technique of the experiments are presented in detail in [1]. The same paper presents a table of results associated with a critical flow of vapor-water mixture through the random packed beds, 250 and 355 mm high. The experiments were carried out at the plant that contained a working segment represented by a vertical section in the form of a pipe with an internal diameter of 39 mm. The pipe had an insert of spherical particles which were steel balls, 2 and 4 mm in diameter. Before the working segment, water passed through the pump, heater and orifice. The high pressure water was throttled through orifice and the thus obtained vapor-liquid mixture was supplied to the working segment inlet. The flow quality x_1 after the orifice was determined by measuring the water temperature and pressure before the orifice, and the pressure of the saturated mixture after the orifice. The pressure applied at the working segment inlet was 0.6, 0.9 and 1.2 MPa. The outlet pressure was near atmospheric one. The flow rate was calculated using the volume method by measuring the time of filling the control balloon placed after the working segment and condenser, where the mixture was changed to liquid and cooled to the room temperature.

In addition to the experiments [1] we obtained new experimental data for shorter packed beds, 50 and 100 mm high, consisting of steel balls, 2 mm in diameter, at a bed inlet pressure $P_1 = 0.6$ MPa (the data for $\left(d/H\right)^{0.5} = 0.14$ and 0.20 in Fig. 3(*a*)).

Figure 3 presents the calculation results for a critical mass velocity and the experimental data. Figure 3(*a*) shows the effect of the ratio of a particle diameter to bed height, while Fig. 3(*b*) illustrates the effect of pressure before the packed bed on $(\rho w)_{cr}$. The difference between calculations and experimental results does not exceed 7% throughout the entire array of available data.

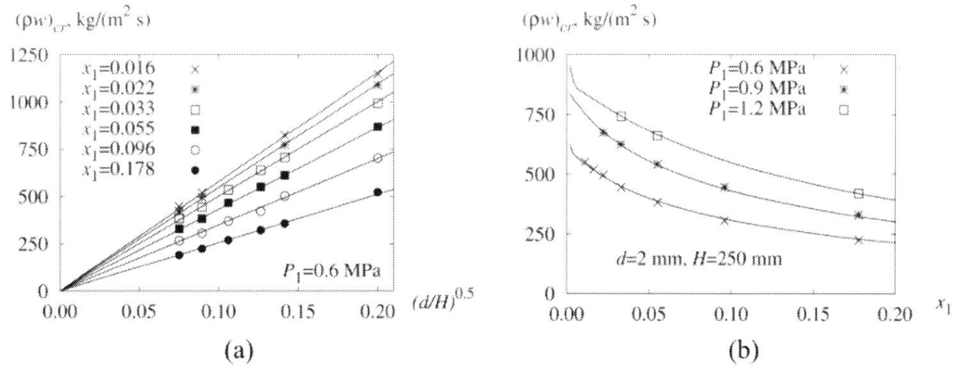

Fig.3. Effect of the ratio of a particle diameter to bed height (a), and the flow quality (b) on the critical mass velocity. Dots are experiment, lines show calculation.

The obtained results make it possible to conclude that the presented model has sufficiently good predictive capabilities for random packed beds of spherical particles within the considered range of operating parameters P_1 and x_1, as well as geometrical parameters *d* and *H*.

The proposed method can find a wider application with the emergence of new data which can be used to specify relationships (8) and (9). The considered approach to the calculation of critical two-phase flows in granular beds can be applied to packed beds of the particles of another geometrical shape different from the spherical one.

The research was performed using the equipment of the multi-access scientific center "High temperature circuit" at Melentiev Energy Systems Institute SB RAS under the support of the Russian Foundation for Basic Research (Project No. 17-08-00709a).

References

1. E.A. Tairov, B.G. Pokusaev, S.M. Bykova, High Temp. **54**, 261 (2016)
2. M.A. Goldstick, *Transfer processes in granular layer* (Novosibirsk, 2005) [in Russian]
3. S. Kim, I. Mudawar, Int. J. Heat Mass Transfer **87**, 497 (2015)
4. R.E. Henry, H.K. Fauske, Trans. ASME J. Heat Transfer **93**, 179 (1971)
5. A.A. Avdeev, R.I. Soziev, High Temp. **46**, 223 (2008)
6. A.A. Avdeev, B.F. Balunov, R.A. Rybin, R.I. Soziev, G.A. Filippov, High Temp. **41**, 377 (2003)

Numerical simulation of heat transfer in falling wavy liquid films on unsteady heat release

Andrey Chernyavskiy[1,*] and *Aleksandr* Pavlenko[1]

[1]Kutateladze Institute of Thermophysics, 1 Acad. Lavrentiev ave., Novosibirsk, 630090, Russia

Abstract. The mathematical model which allows the calculation of the wave surface profile as well as velocity and temperature fields has been presented. The numerical simulation of heat transfer in falling wavy films of liquid nitrogen has been performed. The dependencies of boiling expectation time and total local evaporation time on heat flux density for different inlet Reynolds numbers have been calculated. The regime map which describes the different mechanisms of film decay was obtained by summing up the simulation results. The results of numerical simulation are in satisfactory agreement with the experimental data.

1 Introduction

The liquid flow over a vertical surface is inevitably accompanied by development of wave perturbations on the free film surface which bring considerable contribution to the heat transfer processes and development of crisis phenomena [1–3].

The area of highly efficient heat transfer at the film flow of liquid is limited by the heat flux and depends significantly on the heat release law. When the system reaches the critical heat flux, this leads to decay of the film flow, formation of large-scale dry zones, reduction of heat transfer intensity and as a consequence to a drastic increase in temperature of the heat-releasing surface. All these can cause destruction of the heat-releasing element. The ability to calculate the critical heat fluxes and maximal times of their impact for a given system is required for the design of stably working heat exchangers.

The time of crisis beginning and scenario of film flow decay are characterized by such parameters as time of boiling expectation and time of total local evaporation of the falling liquid film. These values determine maximal times of heat flux influence on the film with certain parameters.

The mathematical model which allows the calculation of the wave surface profile as well as velocity and temperature fields has been presented in the current work. This model allows calculations of the wave characteristics and parameters of liquid film decay at different regime parameters. The results of numerical simulation performed in the framework of the suggested model have been compared with the experimental data obtained for the liquid nitrogen in [2].

* Corresponding author: cherny@ngs.ru

2 Mathematical model

Simulation has been carried out within a 2D calculation area. The abscissa axis is directed along a vertical wall, downstream the liquid film flow. The ordinate axis is orthogonal to the abscissa axis and directed from the wall to the free film surface. The detailed description of the calculation area and the mathematical model has been presented in [4].

2.1 Hydrodynamic model

To describe the processes of liquid flow and evolution of the wave surface profile, in the framework of the current study the 2D hydrodynamic model of Kapitza-Shkadov has been used:

$$\frac{\partial Q}{\partial t}+\frac{6}{5}\frac{\partial}{\partial x}\left(\frac{Q^2}{h}\right)=-\frac{3\nu Q}{h^2}+gh+\frac{\sigma}{\rho}h\frac{\partial^3 h}{\partial x^3}, \tag{1}$$

$$\frac{\partial h}{\partial t}+\frac{\partial Q}{\partial x}=0, \tag{2}$$

$$u=\frac{3Q}{h}\left[\frac{y}{h}-\frac{1}{2}\left(\frac{y}{h}\right)^2\right]=0, \tag{3}$$

$$v=-\int_0^y\frac{\partial u}{\partial x}dy. \tag{4}$$

The method of liquid flow description by means of flow rate characteristic Q allows excluding the dependence on the ordinate and reducing the problem to the pseudo-single-dimensional one. Velocity components in the whole calculation area can be obtained from h and Q by means of expressions (3) and (4). The system of equations (1–4) supplemented with the initial and boundary conditions allows calculation of flow dynamics and evolution of surface profile within the 2D calculation area.

Perturbations at the inlet of the calculation area were set through the addition to the flow rate of the undisturbed film under the condition of keeping the thickness undisturbed:

$$h|_{x=0}=h_0=\left(\frac{3}{4}\frac{\nu^2}{g}\mathrm{Re}_{\mathrm{in}}\right)^{1/3}, \tag{5}$$

$$Q_0=\frac{gh_0^3}{3\nu}=\frac{\nu\,\mathrm{Re}_{\mathrm{in}}}{4}, \tag{6}$$

$$Q|_{x=0}=Q_0(1+RA),\ A(t)\in[-1;1]. \tag{7}$$

Values h_0 and Q_0 set by expressions (5) and (6) are also the initial conditions. In other words, at zero time moment the film is assumed to be waveless and its thickness and flow rate are constant and determined by the conditions of the problem. The additive member in (7) consists of two multipliers: A is a random real number from −1 to 1 and R is the relative perturbations amplitude.

2.2 Heat transfer model

In natural variables the boundaries of the calculation area are not constant because of the unsteady character of the free surface. The following coordinate transformation converting the calculation area into the rectangular one has been used to solve this problem:

$$\eta(t,x,y) = y / h(t,x)\,. \tag{8}$$

After the coordinates transformation the unsteady heat conductivity equation in the liquid film looks like the following:

$$\begin{aligned}&\frac{\partial T}{\partial t}+u\frac{\partial T}{\partial x}+\frac{1}{h}\left[v-\eta\frac{\partial h}{\partial t}-u\eta\frac{\partial h}{\partial x}-2a\frac{\eta}{h}\left(\frac{\partial h}{\partial x}\right)^2+a\eta\frac{\partial^2 h}{\partial x^2}\right]\frac{\partial T}{\partial \eta}=\\&=a\frac{\partial^2 T}{\partial x^2}-2a\frac{\eta}{h}\frac{\partial h}{\partial x}\frac{\partial^2 T}{\partial x\partial \eta}+\frac{a}{h^2}\left[1+\eta^2\left(\frac{\partial h}{\partial x}\right)^2\right]\frac{\partial^2 T}{\partial \eta^2}\end{aligned}, \tag{9}$$

At the inlet of the calculation area and at the free film surface, the temperature was assumed equal to the temperature of liquid saturation. The following boundary condition was used on the heating surface:

$$\frac{\partial T}{\partial \eta}\Big|_{x=0} = -h\frac{q(t,x)}{\lambda}+h\frac{\delta_h C_h \rho_h}{\lambda}\frac{\partial T}{\partial t}\Big|_{\eta=0}\,. \tag{10}$$

To take into account evaporation from the free surface, equation (2) has been modified as follows:

$$\frac{\partial h}{\partial t}+\frac{\partial Q}{\partial x}=\frac{\lambda}{L\rho}\left(\frac{1}{h}\frac{\partial T}{\partial \eta}\Big|_{\eta=1}-\frac{\partial h}{\partial x}\frac{\partial T}{\partial x}\Big|_{\eta=1}\right). \tag{11}$$

3 Results and conclusions

The boiling expectation time was determined as the minimum time of reaching the given overheat by the heating surface. In the framework of the present work three threshold values of overheating for the liquid nitrogen under conditions of atmospheric pressure were used: two theoretically calculated values, $\Delta T_{b.e} = 26$ K, corresponding to 95% and $\Delta T_{b.e} = 32$ K, corresponding to 100% of probability of boiling under the conditions of homogeneous nucleation [5], and one approximation for experimentally retrieved dependence of overheat on heat flux density obtained for boiling-up under the condition of heterogeneous nucleation:

$$\Delta T_{b.e}(q) = 32 - 21.1585\cdot 0.93^{q},\ \ q[W/cm^2]\,. \tag{12}$$

The calculated boiling expectation times (Fig. 1) are in satisfactory agreement with the experimental data [2].

The regime map which describes the different mechanisms of film decay was obtained by summing up the results of numerical simulation (Fig. 2). Calculated curves 1, 2 and 3 divide the area of the investigated Reynolds numbers and heat flux densities into the set of regions. Region I is characterized by the absence of crisis; region II describes the set of the problem parameters, where liquid film decay will be performed by total local evaporation without liquid boiling-up; region III describes the set of parameters, where the liquid film decay will be performed by the film boiling-up. Calculated critical heat flux densities of the total local evaporation 3 agree satisfactorily with the values calculated for steady heat release conditions by the formula, which has been presented in [6] - curve 4.

According to the presented regime map, it is possible to make a conclusion that boiling-up becomes the prevailing decay mechanism independent of the heat flux density when the inlet Reynolds number Re_{in} exceeds 130 – 220 depending on liquid and heater parameters.

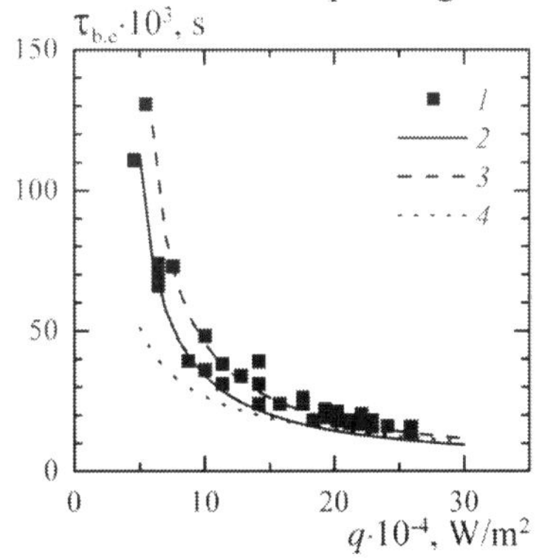

Fig. 1. Comparison between experimental and calculated boiling expectation times, Re_{in} = 230: 1 – Experiment [2]; 2 – $\Delta T_{b.e}$ = 26 K; 3 – $\Delta T_{b.e}$ = 32 K; 4 – $\Delta T_{b.e} = \Delta T_{b.e}(q)$.

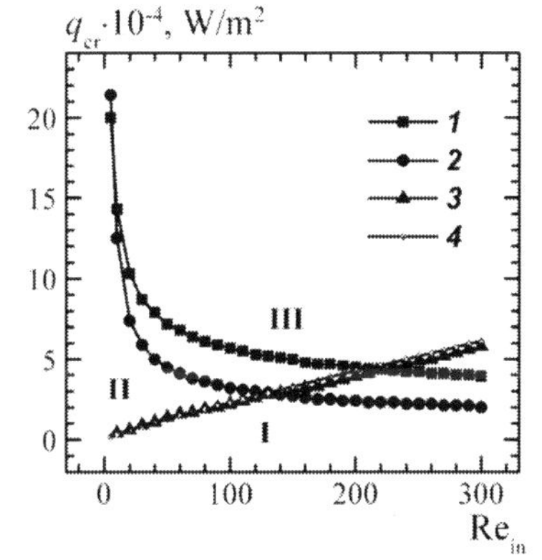

Fig. 2. Regime map. 1,2 – Boiling-up, $\Delta T_{b.e}$ = 26 K, $\Delta T_{b.e} = \Delta T_{b.e}(q)$ respectively; 3 – Full local evaporation; 4 – Full local evaporation on steady heat release [6].

The study was performed at Kutateladze Institute of Thermophysics SB RAS and was financially supported by the grant of the Russian Science Foundation (project No. 14-49-00010).

References

1. E.G. Vorontsov, Theor. Found. Chem. Eng. **33** 117 (1999)
2. A.N. Pavlenko, A.S. Surtaev, A.M. Matsekh, High Temp. **45**, 826 (2007)
3. C.N. Markides, R. Mathie, A. Charogiannis, Int. J. Heat Mass Transfer **93**, 872 (2016)
4. A.N. Chernyavskiy, A.N. Pavlenko, Int. J. Heat Mass Transfer **105**, 648 (2017)
5. V.P. Skripov et al., *Thermophysical Properties of Liquids in the Metastable (Superheated) State* (Gordon and Breach Science Publishers, New York, 1988)
6. A.N. Pavlenko, V.V. Lel', Rus. J. Eng. Therm., **7**, 177 (1997)

Direct numerical simulation of two-phase gas dynamic flows with phase transition for water and for liquid sodium

Anna Aksenova[1], *Vladimir* Chudanov[1*], *Alexey* Leonov[2] and *Artem* Makarevich[1]

[1]IBRAE, Nuclear Safety Institute, 115191, Moscow, B. Tulskaya 52, Russia
[2]MEPhi, National Research Nuclear University, 115409, Moscow, Kashirskoe highway 31, Russia

Abstract. In this paper the results of the calculation for one-dimensional problems with water, as well as, a two-dimensional calculation simulating an experimental study of heat transfer and resistance in the vertical pipe with boiling sodium subject to forced lift conditions are presented.

Introduction

There are two classes of methods to simulate two-phase flows with interface (boundary) between phases: 1) the interface is considered as a sharp break; 2) the interface is considered as a diffusion region, like contact discontinuities in the gas dynamics. Models that allow describing the diffusion regions are divided into nonequilibrium (there is no equality of phase pressures and phase velocities) [1] and equilibrium (phases are in mechanical equilibrium) [2]. Numerical approximation for equilibrium models proves to be simpler, but in this case a number of problems arises [3]: 1) the nonmonotonic behavior of a two-phase mixture sound velocity as a function of the volume fraction, that can lead to errors in the calculation of the passage of waves across the interface; 2) non-conservative form of the equation for the volume fraction. Even using of specially obtained relations on the front of the shock wave [4], it is not always possible to achieve a coincidence between the numerical and analytical solutions, since the averaging of non-conservative variables has no physical meaning. This problem was overcome in [3], using the relaxation algorithm instead of the averaging procedure. A more robust approach involves the model that is nonequilibrium only by pressure [5]. In this model dependence of the sound speed on the volume fraction turns out to be monotonous and equal to the "frozen" mixture sound speed. Such imaginary complication of the model leads to a simplification of the numerical algorithm [6]. The involving of the surface tension forces in such two-phase model is realized using the Hamiltonian principle of least action. As a result, in the equations for the momentum and total energy of the two-phase mixture the additional tensor components appear. These terms contain the second spatial derivatives of the mass fraction of the liquid phase [7]. In this paper, the non-equilibrium pressure model is used to calculate the two-phase flows. This model takes into account the surface tension forces using modified relations at the shock front [8]. To simulate phase transition the equilibrium model [9] is

* Corresponding author: chud@ibrae.ac.ru

used. In this model heat and mass transfer processes set the two-phase system into a state with the same values of pressures, temperatures, and Gibbs energies. As closure relations the equations of state "stiffened" ($e=\dfrac{p+\gamma\pi}{\rho(\gamma-1)}+q$) or "Noble-Abel-Stiffened" ($e=\dfrac{(p+\gamma\pi)\times(1-\rho b)}{\rho(\gamma-1)}+q$) are used. The "Noble-Abel-Stiffened" system, in addition to the attractive forces, allows taking into account the molecular repulsive forces [10], which can significantly improve the accuracy of direct numerical simulation. The constant parameters for the specified equation of state are found from the experimental saturation curves for the enthalpies, densities and pressure in the fixed range of the temperatures [11, 12]. Calculations related to the influence of capillary effects in two-phase flows are presented in [13]. In this paper the results of the calculation for one-dimensional problems with water, as well as, a two-dimensional calculation simulating an experimental study of heat transfer and resistance in the vertical pipe with boiling sodium subject to forced lift conditions are presented. In conclusion, the shortcomings of the presented technique for the direct numerical simulation of mass transfer and variants of it modification are indicated.

1 Cavitation in the water

The calculation of cavitation in the water was performed for the "stiffened" EOS with the coefficients presented in [14]. The tube with water of density 1150 kg/m^3 at atmospheric pressure is 1 meter long and the initial discontinuity is located at 0.5 meter. The liquid is assumed to contain a uniformly distributed small amount of vapor, 10^{-2} in the whole domain. The temperature of both phases is the same. The speed of the two-phase medium in the left part of the tube is equal -2 m/s, and the velocity in the right-hand part of the pipe is equal to 2 m/s. In this calculation, an explicit interface between phases is initially absent, cavitation occurs in a two-phase mixture with a small fraction of the gas phase. The results of the cavitation calculation for the "stiffened" EOS (Table 1) using a grid of 4096 cells at time of 3.2 msec are presented in Figure 1. The profiles of pressure (a) and speed (b) are shown. Each graph is obtained using the model described in this paper with taking to account of the mass transfer and without of one. The results of the presented calculations qualitatively coincide with the dependences obtained in [14] for the same problem.

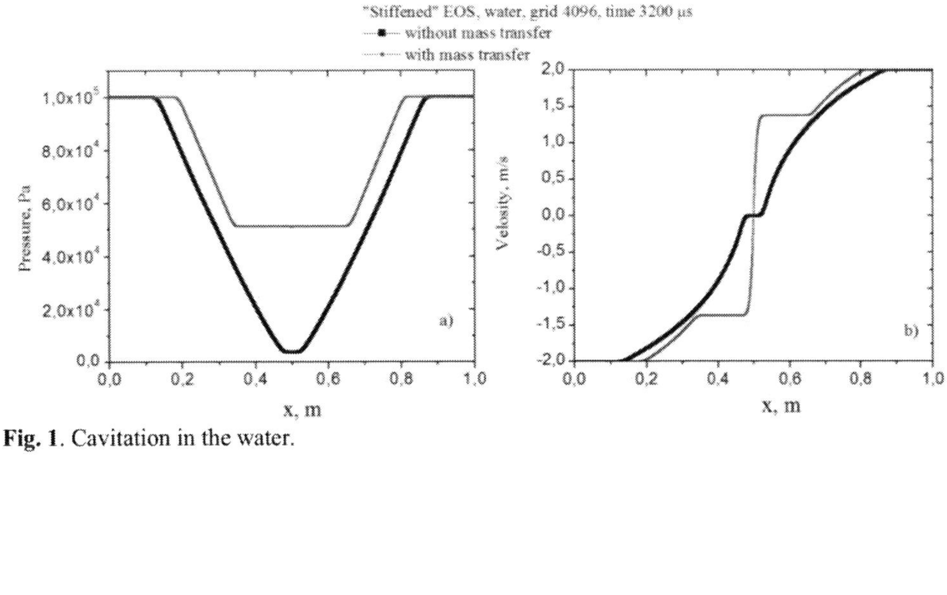

Fig. 1. Cavitation in the water.

Table 1. "Stiffened" EOS coefficients for water [14].

Phase	γ	π, Pa	q, J/kg	q', J/kg/K	C_V, J/kg/K
Liquid	2.35	1×10^9	-1167×10^3	0	1816
Gas	1.43	0	2030×10^3	-23400	1040

2 Simulation of 2D heat and mass transfer in a vertical channel with boiling sodium subject to forced lift conditions

The vertical dimension of the working region of a flat channel with transverse dimension of 5 mm is 70 cm. The heat flux through the side walls was assumed to be ~ 0.57 MW /m^2. The initial volume fraction of sodium vapor in the whole pipe is 10^{-5}, and the initial temperature of two-phase sodium mixture was assumed equal to 1153 K everywhere, which is 3 K less than the boiling point at atmospheric pressure. The distribution of the initial vertical velocity corresponds to a mass flow of 166 kg/m^2/s. At the lower boundary, constant values of temperature and volume fraction are maintained, equal to the initial values. A constant mass flow is also maintained at the lower boundary. At the upper boundary, atmospheric pressure is fixed. The calculation was carried out using the "Stiffened" EOS with the parameters shown in Table 2.

Table 2. "Stiffened" EOS coefficients for sodium.

Phase	γ	π, Pa	q, J/kg	q', J/kg/K	C_V, J/kg/K
Liquid	1.48	3.67×10^8	-265030	0	864
Gas	1.49	0	2086460	-4149	1690

A vertical profile of the pressure averaged over the cross section at time instant 0.22 s is shown if Figure 2.

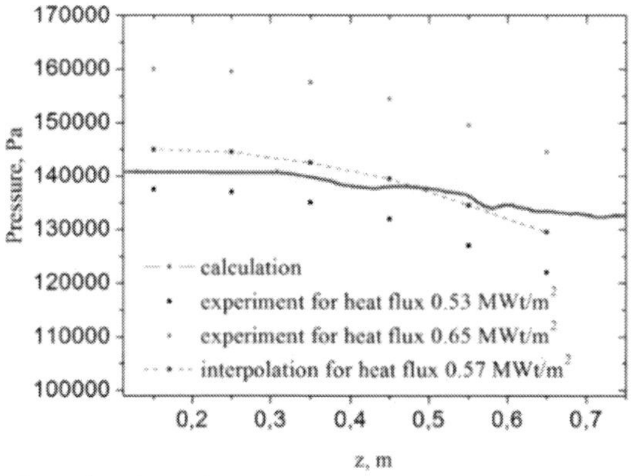

Fig. 2. Pressure distribution along the channel

The points correspond to the experimental values of the pressure for two different values of the heat fluxes [16]. The dashed curve corresponds to the interpolation of the experimental values for the heat flux of 0.57 MW/m^2 assuming a linear dependence of the pressure values on the heat flux. A good qualitative agreement of the calculated and experimental values is obtained. The difference in the magnitude of the pressure difference over the working region of the pipe can be related to the "flat" approximation of the model calculation, which leads to different values of the ratio of the transverse length of the heating section to the enclosed area.

Conclusion

The presented numerical method was constructed for 2D calculations of two phase mixture flows combining the noequilibrium pressure model approach [5] with conservative inclusion of surface tension forces using the Hamiltonian principle of least action [7] and with phase transition approach splitting heat and mass exchanges processes [9]. This numerical method is explicit and realized in parallel programming mode and allows one to obtain qualitative agreement with the experimental data. However, it should be noted that the "stiffness" of the non-linear system of equations in the equilibrium model provides not always an accurate simulation of mass transfer [14]. In such cases, the assumption of complete evaporation or condensation in the considered grid cell is applied [17]. Further development of a numerical technique for direct simulation of mass transfer in two-phase flows is associated with the use of a simplified relaxation solver, which ensures convergence to the same solutions as the conventional solver, but is much simpler and robust, since the equilibrium mass fraction is linearized after evaporation (or condensation) on the value of the equilibrium pressure [18]. It is necessary to mention that DNS is very sensitive tool for grid changing. The special study will be fulfilled for 2D computations using DNS and grid for this method.

References

1. R. Abgrall, R.Saurel, J. Comput. Phys. **186**, 361 (2003)
2. R. Saurel, F. Petitpas, R. Abgrall, Fluid Mech. **607**, 313 (2008)
3. F. Petitpas, E. Franquet, R. Saurel, O. Le Metayer, J. Comp. Phys. **225**, 2214 (2007)
4. R. Saurel, O. Le Metayer, J. Massoni, S. Gavrilyuk, Shock Waves **16,** 209 (2007)
5. R. Saurel, F. Petiptas, R.A. Berry, J. Comp. Phys. **228**, 1678 (2009)
6. A. Leonov, V. Chudanov, A. Aksenova, Proc. IBRAE RAS, Issue 14, Moscow, Nauka, (2013)
7. R. Berry, R. Saurel, F. Petitpas, E. Daniel, O. Le Metayer, S. Gavrilyuk, N. Dovetta, R. Martineau, INL Report, Contract DE-AC07-05ID14517, October (2008).
8. A. Aksenova, A. Leonov, A. Makarevich, V. Chudanov, Proc. IBRAE RAS, Issue 16, Moscow (2016)
9. A. Zein, M. Hantke, G. Warnecke, J. Comput. Phys. **229**, 2964 (2010)
10. R. Saurel, P. Boivinc, O. Le M´etayer, Computers & Fluids (to be published), 2016
11. O. Le Metayer, J. Massoni, R. Saurel, Int. J. Therm. Sci. **43**, 265 (2004)
12. O. Le Métayer, R. Saurel, Phys. Fluids **28**, 4 (2016)
13. V. Chudanov, A. Aksenova, A. Leonov, Int. J. Math. Mod. Meth. App. Sci., **8**, 248 (2014)
14. M. Pelanti, K.M. Shyue, J. Comput. Phys. **259**, 331 (2014)
15. J. Fink, L. Leibowitz, *Thermodynamic and transport properties of sodium liquid and vapor*, ANL 9700, South Cass Avenue Argonne, Illinois, U.S.A., 60439 (1995)
16. Yu. Zeigarnik, V. Litvinov, High Temp. **5**, 1116 (1977)
17. G. Faccanoni, S. Kokh, G. Allaire, *Modelling and simulation of liquid-vapor phase transition in compressible flows based on thermodynamical equilibrium*, ESAIM: M2AN **46**, 1029 (2012)
18. A. Chiapolino, P. Boivin, R. Saurel, Int. J. Num. Meth. Fluids **83**, 583 (2017)

Investigating characteristics of liquid hydrocarbon spraying by a steam jet

Igor Anufriev, *Evgeniy* Kopyev, and *Evgeniy* Shadrin*

Kutateladze Institute of Thermophysics SB RAS, Novosibirsk, Russia

Abstract. Using the shadow photography method, the dispersed composition of a gas-droplet flow has been investigated at diesel fuel spraying by a superheated steam jet. The distributions of droplet sizes have been obtained in a wide range of parameters (steam and fuel consumption). Dependencies of droplets size on steam consumption have been found. The optimum values of steam and fuel consumption, at which the smallest prevailing droplet size (below 30 μm) is reached, have been found.

Jet streams created by injectors are currently used in various fields of science and technology: cooling and fire extinguishing systems, medical inhalers, internal combustion engines, boiler equipment and many others. To improve the operation of devices with spraying nozzles, a detailed study of characteristics of the generated gas-droplet flow is required. Obtaining information on the velocity distribution and droplet sizes in the investigated jet stream allows justifying scientifically the selection of optimal regime parameters of the device. In particular, it is important when designing effective burners, used for utilization of substandard hydrocarbon fuels and industrial waste. Studies of the combustion of such liquid hydrocarbons, carried out by the authors earlier [1-3], have shown that the process of their ignition and combustion is greatly intensified when a superheated steam jet is fed into the combustion zone. The environmental characteristics of the process are improved as well. However, the use of evaporative-type burner devices does not allow achieving high productivity (power).

In this paper, a promising method of spraying a liquid hydrocarbon fuel with a steam jet [4] is experimentally studied. A distinctive feature of this method is that the fuel and spraying medium (carrier phase – steam), are not pre-mixed: the steam is supplied from the nozzle in the form of a jet, on which liquid fuel flows, thereby forming a fine gas-droplet flow. In practice this is an important advantage because of no contact between the fuel and the nozzle, and, hence, no coking of its surfaces and subsequent failures in the burner operation.

Using the shadow photography method [5] the dispersed composition of the gas-droplet flow was investigated at diesel fuel spraying by a superheated steam jet. The experimental setup is shown in Figure 1. This method is based on registering a shadow photograph of an object having a refractive index, different from ambient medium. At that, behind the investigated object (relative to the camera) there is a diffuse light source with a uniform

*Corresponding author: evgen_zavita@mail.ru

spatial distribution of intensity. The focusing plane of the camera lens is close to the object of investigation to obtain the greatest sharpness of the shadow photo. Digital analysis of the shadow image allows identifying the boundary and position of the object. This method allows measuring the droplet sizes within the range from 10 to 1000 μm.

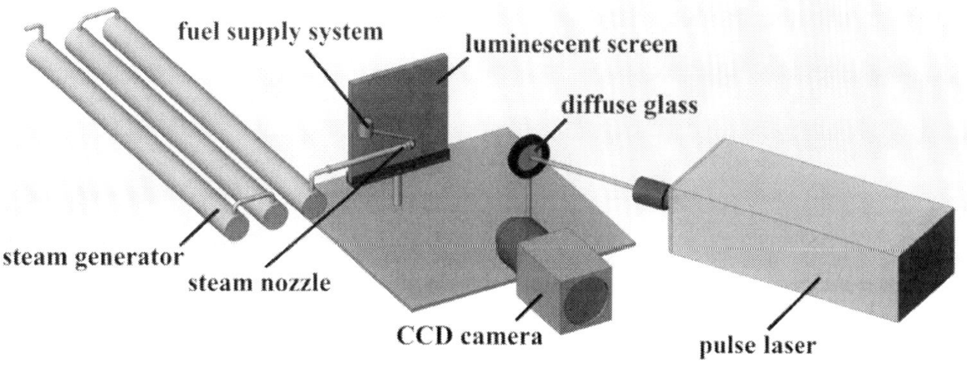

Fig.1. Scheme of experimental setup for studying characteristics of liquid fuel sprayed with a steam jet.

In the experiments, the CCD camera Videoscan 4021 (with a resolution of 2048 × 2048 pixels, a shooting frequency up to 1.25 Hz, and an exposure time of 28 ms) and the Tamron SP AF macro lens with a focal length of 180 mm were used to reduce the size of the measuring area to 10 × 10 mm with good spatial resolution (magnification 1.5:1). A background screen with a rhodamine-based luminescent coating preliminarily illuminated by a defocused beam of a Nd:YAG QuantelEVG pulse laser (wavelength 532 nm, pulse energy up to 145 mJ, and pulse duration 10 ns) was used as a light source. A threshold light filter (560 nm) whose bandwidth corresponds to the wavelength of light re-emitted by rhodamine was used to increase the shadow photo contrast. A coordinate-moving device with a positioning accuracy of 0.1 mm was used to move the camera along the axis of the nozzle that enabled examining of the flow at different distances from the nozzle.

The measurements were carried out at different distances from the nozzle and at different operating parameters: the steam flowrate was varied within 100-600 g/h, and the fuel consumption ranged to 300-600 g/h. The temperature of the superheated steam at the outlet from the steam generator was 150-550°C, and a pressure inside the steam generator was up to 10 atm.

For digital processing of the obtained shadow images the "Bubbles Identification" algorithm, implemented in the ActualFlow software [6], was used. It includes high-pass filtering algorithm to identify the boundaries of images of registered objects, algorithm for binarization by a threshold value, and algorithm for determining the position and the diameter of spherical droplets. Figure 2 shows the image processed using the specified algorithm.

Image processing has resulted in the data on the dispersed composition of fuel droplets under various spraying regimes. Figure 3 shows the characteristic dependences of the droplet size on steam flowrate and distance from the nozzle. The analysis of the results shows (Fig. 3-a) that the droplets size depends on steam flowrate: the share of small droplets in the flow (10-20 μm) increases with the steam consumption (vapor velocity at the outlet from the nozzle). At the same time, the droplets size is almost independent of the distance from the nozzle (Fig. 3-b), which indicates the efficiency of fuel spraying by the proposed method. There is a slight decrease in the size of the droplets as they move away from the nozzle, which, apparently, is associated with evaporation. At steam consumption of 600 g/h and fuel consumption of 300 g/h, the prevailing droplets size in the flow is 10-30 μm. An increase in fuel consumption or a decrease in steam consumption led to an increase in the characteristic droplets size.

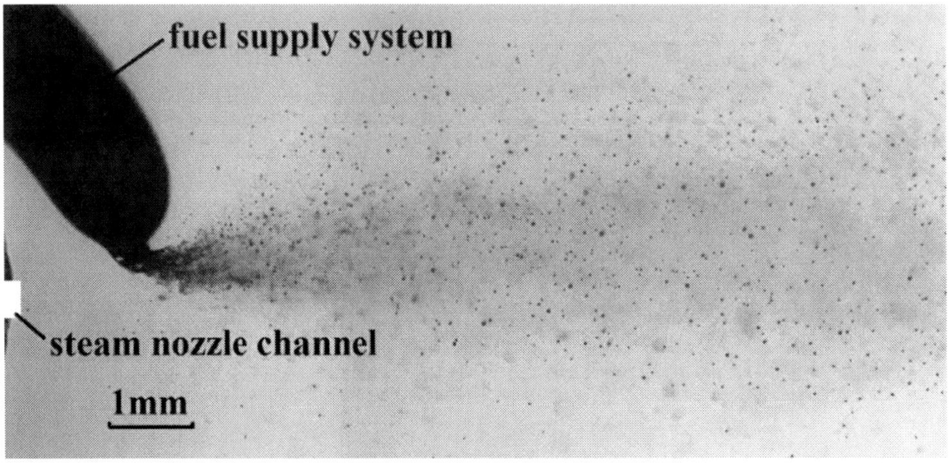

Fig.2. Result of processing of the shadow image of the fuel, sprayed with a steam jet in the air (the identified particles are indicated by marks).

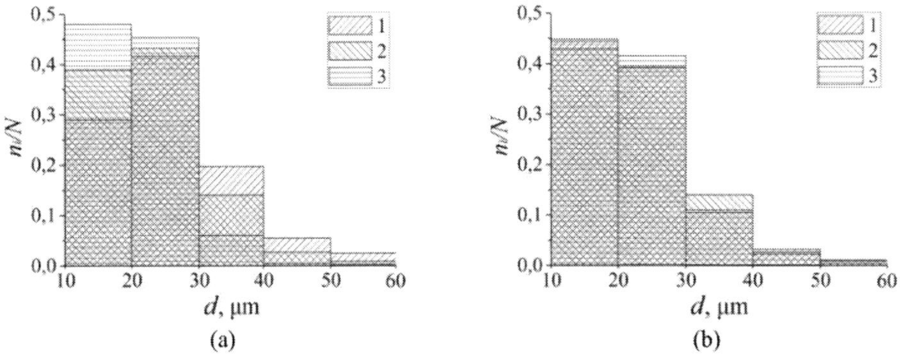

Fig.3. Dispersed composition of fuel droplets, formed by spraying diesel fuel with a superheated steam jet in the atmosphere: (a) for different steam consumption (1 – 100 g/h, 2 – 300 g/h, 3 – 600 g/h), with a fuel consumption of 300 g/h, at a distance of 10 mm from the nozzle; (b) at different distances from the nozzle (1 – 0-10 mm, 2 – 10-20 mm, 3 – 20-30 mm), at a steam consumption of 300 g/h and a fuel consumption of 300 g/h. (n_i is the number of droplets with dimensions from a certain range, and N is the total number of droplets identified by the algorithm, d – is the size of droplet).

As a result, the characteristics of the liquid hydrocarbon fuel spraying by a superheated steam jet have been studied using the method of shadow photography. The optimal values for steam and fuel consumption have been found, which are 600 and 300 g/h for steam and fuel, respectively. At that, the smallest prevailing droplets size (below 30 μm) in the flow is achieved. This dispersion result ensures efficient fuel combustion. At the same time, the effect of superheated steam temperature on the characteristics of spraying is not noticeable.

The research was carried out with the financial support of the Russian Science Foundation (project No. 17-79-20209)

References

1. S.V. Alekseenko, I.S. Anufriev, M.S. Vigriyanov, E.P. Kopyev, O.V. Sharypov, Combustion, Explosion, and Shock Waves, **52,** 286 (2016)
2. S.V. Alekseenko, I.S. Anufriev, M.S. Vigriyanov, V.M. Dulin, E.P. Kopyev, O.V. Sharypov, Thermophysics and Aeromechanics, **21**, 393 (2014)
3. I.S. Anufriev, E.P. Kopyev, E.L. Loboda, Proc. SPIE, **9292**, (2014)
4. M.S. Vigriyanov, S.V. Alekseenko, I.S. Anufriev, O.V. Sharypov, RF Patent No. 2579298 (2016) [in Russian]
5. G.V. Kuznetsov, P.A. Strizhak, R.S. Volkov, O.V. Vysokomornaya, Int. J. Therm. Sci., **108**, 218 (2016)
6. User's Guide: Identify images of circular objects by delineating borders on images obtained by shadow photography and PFBI methods, (2012) [in Russian]

An experimental investigation of flow boiling heat transfer of R-141b and R-1234yf in microchannels

Alisher Shamirzaev*

Kutateladze Institute of Thermophysics SB RAS, 630090 Novosibirsk, Russia

Abstract. This study presents experimental results of flow boiling heat transfer of refrigerants R-141b and R-1234yf in a horizontal microchannel heat sink. The copper microchannel heat sink contains 21 microchannels with 335x930 μm cross-section. Distribution of local heat transfer coefficients along the length and width of the microchannel plate were measured in the range of external heat fluxes from 50 to 550 kW/m^2. Finally, comparisons with predictions according to the model of flow boiling heat transfer are reported for the data sets.

1 Introduction

Studies of the processes of heat and mass transfer in microsystems with phase transformations are the fast-developing field of science. This is due to the growth of technological applications, which require the transfer of large heat and mass flows in a limited volume, including compact steam generators/condensers of power and cryogenic devices, microprocessor cooling systems [1]. At channel size reduction, a wide variety of the phenomena arise which are not typical for conventional tubes. That confirmed by experimental data from [2–6] and other papers. In despite of a large number of published methods for predicting heat transfer in microchannels, the application of these methods requires verification, and the determination of heat transfer for various heat transfer modes in boiling in microchannels, as noted in [7], remains actual. The objective of this study is to experimentally establish how the local flow boiling heat transfer coefficient correlates with the heat flux, mass flux and vapor quality.

2 Experimental equipment and methods

Figure 1 shows the setup for investigation of refrigerants flow boiling heat transfer in the microchannel heat sink. Liquid refrigerant is supplied from the cooler through the filter and flow controller Bronkhorst HI-TECH to the temperature control system via the pump. Then the refrigerant goes through the pre-evaporator to achieve the flow with desired vapor quality. After passing through to the microchannel heat sink it goes to the condenser.

* Corresponding author: alisham@ngs.ru

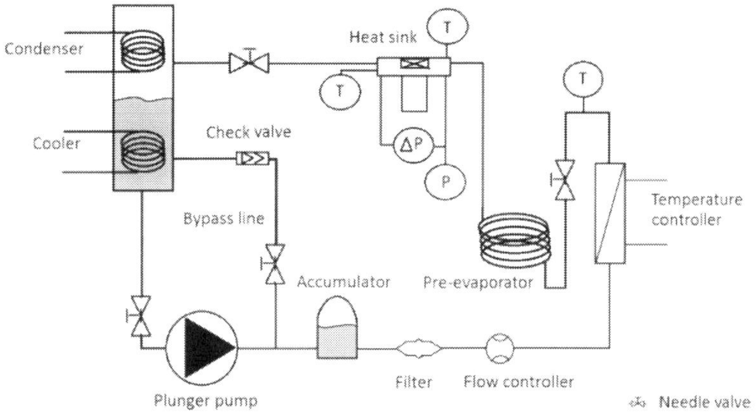

Fig. 1. Schematic of the experimental setup.

Twenty-one microchannels were made by precise milling of 335 μm wide and 930 μm deep micro-slots into the 20x40 mm top surface of the oxygen-free copper plate of the microchannel heat sink. The distance between channels equals to 650 μm (fin thickness) at total plate thickness of 2.5 mm. Wall roughness Ra was estimated at 1 μm, using surface measurement by an optic microscope. The microchannel plate and heating block were mounted in the stainless steel container with two sections, divided by the partition wall for the local heat flux measurements, see fig. 2. The thickness of the partition wall equaled 2.51 mm. Tin gaskets with 0.5 mm thickness and thermocouples inside were placed between the microchannel plate and partition wall as far as between the heating block and the partition wall. Gaskets were melted during the assembly process to reduce the contact thermal resistance. After mounting, the microchannel plate was closed by the copper cover and soldered.

Heat was supplied to the test section by means of the copper block with mounted heat cartridges, which can supply the heat flux up to 550 kW/m^2. During the test, temperatures of tin gaskets were measured in four cross-sections along the heat sink at distances of 5 mm, 15 mm, 25 mm, and 35 mm from the microchannels inlet, see fig.2. The numerical calculations were performed to obtain actual temperatures of the internal channel wall and temperature difference on the partition wall. Due to small fin height and high thermal conductivity of copper the temperature of the internal channel wall was almost uniform in the cross-section.

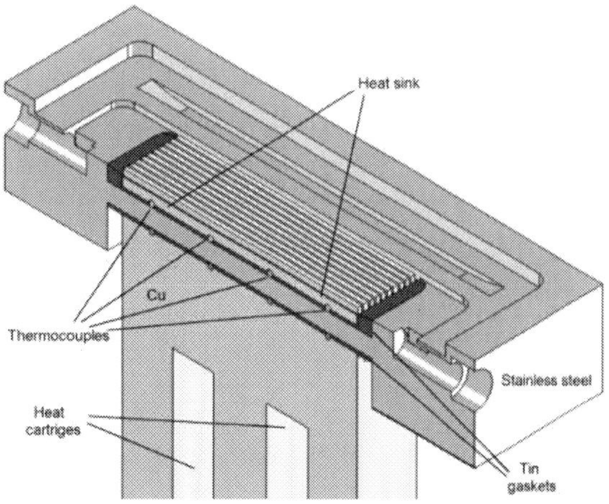

Fig. 2. Schematic of the test section.

In each cross-section, two thermocouples were placed along the width of the microchannel plate. It allowed controlling the even flow distribution in the microchannel plate. To manage flow maldistribution the inlet and outlet ports were connected to the microchannel plate through inlet and outlet constrictions, see fig. 2. Pressures and temperatures in the inlet and outlet restrictions were measured using the pressure probes and insulated thermocouples. The heat losses from the test section were calibrated and did not exceed 0.19 W/K. The liquid flow rate was measured by the flow controller with accuracy of 0.02 g/s.

The heat transfer coefficient was calculated using internal wall temperatures $T_{w,i}$:

$$h_i = q_{w,i}/(T_{w,i} - T_{sat,i}) \tag{1}$$

where $T_{sat,i}$ is the saturation temperature related to the i-th thermocouple and $q_{w,i}$ is the local heat flux from the inner wall to the flow. The saturation temperature related to the i-th thermocouple was determined using linear approximation of the measured input and output pressure and the dependence of the saturation temperature on pressure for the tested refrigerant. Local heat flux was determined from temperature difference $\Delta T_{w,i}$ through the partition wall as follows:

$$q^*_{w,i} = \left(\frac{\delta_{st}}{\lambda_{st}} + \frac{\delta_{tin}}{\lambda_{tin}}\right)^{-1} \Delta T_{w,i},\quad q_{w,i} = q^*_{w,i}\frac{A_{hs}}{A_{in}} \tag{2}$$

Here, $q^*_{w,i}$ is the external local heat flux, λ_{st} and λ_{tin} are thermal conductivities of stainless steel and tin, respectively, δ_{st} and δ_{tin} are thicknesses of the partition wall and tin gasket, A_{hs} is the external area of heat sink, and A_{in} is the inner area of heat sink. The vapour quality at the heat sink inlet was determined considering heat production in the pre-evaporator Q_{coil} and liquid temperature at the pre-evaporator inlet T_0 as follows:

$$x_0 = [Q_{coil} - Q_{lost,coil} - m \cdot C_{p,liq}(T_{sat,in} - T_0)]/(mh_{fg}) \tag{3}$$

where m is the mass flow rate, h_{fg} is the latent heat of vaporization, and $C_{p,liq}$ is the specific heat of liquid. The vapor quality along the heat sink was calculated as follows:

$$x_{L_i} = x_0 + \left(\int_0^{L_i} W\langle q_w\rangle dL - Q_{lost,hs} l_i / L_{hs}\right) \Big/ (m \cdot h_{fg}), \tag{4}$$

where L_{hs} is the heat sink length, $Q_{lost,hs}$ is the heat loss, W is the heat sink width, $\langle q_w \rangle$ is the cross section averaged local heat flux supplied to heat sink, and x_0 is vapor quality at the heat sink inlet.

We have verified the measurement procedure using tests on measuring the heat transfer coefficient distribution along the heat sink length at subcooled refrigerant flowing. The data from single-phase experiments, taking into account the initial section, were in good agreement with the data from [8].

3 Flow boiling data

For both investigated refrigerants, the dependences of heat fluxes on wall overheating for fixed pressure and mass flow have been obtained. The comparisons of experimental data with pool boiling correlation and prediction by [9] are presented on fig. 3. In all experimental data from fig. 3 the initial vapor quality was changed for the local vapor quality in the point of measurement

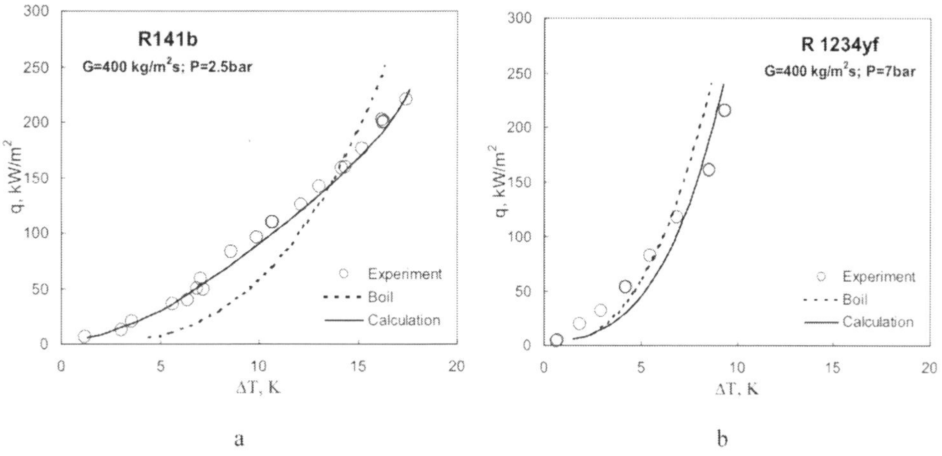

Fig. 3. Dependence of heat flux from overheating: (a) for R141b; (b) for R1234yf.

to be in range from 0.15 to 0.3. According to [10] Cooper correlation [11] was used to calculate R141b pool boiling (dotted lines fig. 3a). For calculation of R1234yf pool boiling the correlation [6] was used (dotted lines fig. 3b). The corresponding correlations for pool boiling were used to calculate flow boiling heat transfer with [3] (solid line fig.3). At the considered conditions, the measured flow boiling heat transfer coefficients exceeded pool boiling prediction when local heat flux was less than 100 kW/m^2 and were in good agreement with prediction by [3] in all investigated range of heat fluxes.

This work was supported by RFBR grant 15-08-07506-a.

References

1. V. Kuznetsov, J. Phys.: Conf. Ser. **754**, 032012 (2016)
2. V.V. Kuznetsov, A.S. Shamirzaev, Heat Transfer Eng. **28**, 738 (2007)
3. V.V. Kuznetsov, A.S. Shamirzaev, Micrograviry Sci. Tech. **21**, 305 (2009)
4. V.V. Kuznetsov, A.S. Shamirzaev, I.A. Kozulin, S.P. Kozlov, Heat Transfer Engineering **34**, 235 (2013)
5. V. V. Kuznetsov, S. V. Dimov, P. A. Houghton, A. S. Shamirzaev, S. Sunder, *Proc. 1st Int. Conf. on Microchannels and Minichannels. ASME*, 683 (2003)
6. V.V. Kuznetsov, et al. J. of Phys.: Conf. Series **395**, 012093 (2012)
7. S.S. Bertsch, E.A Groll, S.V. Garimella, Int. J. Heat Mass Transfer **52,** 2110 (2009)
8. R.J. Phillips, *Forced-Convection, Liquid-Cooled, Microchannel Heat Sinks*, / Technical rept. Massachusetts Inst. of Tech., Lexington Lincoln Lab, (1988)
9. V.V. Kuznetsov, A.S. Shamirzaev, Heat Transfer Eng. **37**, 1105 (2016)
10. V. Trisaksri, S. Wongwises, Int. J. of Heat and Mass Transfer **52**, 1582 (2009)
11. M.G. Cooper, *1st UK Natl. Heat Transfer Conf.* (I. Chem. E. Symp. series No. 86) **2** 785 (1984)
12. D. Jung, H. Lee, D. Bae, and S. Oho, Int. J. Refrig. **27**, 409 (2004)

Investigation of the effect of passive vortex inserts of different geometrical shapes on the vibrations reduction efficiency in pipelines with two-phase flow

Vladimir I. Velkin*, *Sergey E.* Shcheklein, *Hossain* Ismail, *A.* Nikitin, and *G.*Chikansev

Ural Federal University named after the first President of Russia B. N. Yeltsin
19 Mira str., Yekaterinburg 620002, Russia

Abstract. This paper describes the experimental test bench used in coolant flow research with application of adjusted optical laser system. It is based on the technology of pulsed particle visualization of micro tracers, i.e. Particle Image Velocimetry (PIV system) that allows for the analysis of the microstructure of the hydrodynamic liquid flow passed through the vortex inserts. Optical measurement (PIV-method) of hydro- and aerodynamic characteristics of the coolant flow allowed obtaining the velocity distribution in straight and various curved sections of the pipeline in order to reliably define the vortex formation condition and the occurrence of the pressure gradient on the outer and inner walls of the curved sections. This gives us the opportunity to verify the calculation models describing the problem of passive control (decrease) of the displacement of pipelines of power equipment, and thus to improve the reliability and safe operation of the system in general.

1 Introduction

Reliable operation of power equipment of thermal and nuclear power plants is connected with the solution to the problem of reducing the vibrations of pipelines transporting two-phase flows. Significant levels of vibration, particularly in pipelines with two-phase flow condensation, lead to a sharp increase in the cost of their maintenance, reduction of reliability of the energy system, and in critical cases causes the depressurization of the coolant flow paths due to fatigue failure of the pipeline metal [1]. One of the directions in solving the problem of reducing vibration in pipelines is carrying out research towards optimizing the geometry of passive devices that affect the hydrodynamics of a two-phase flow [2]. The two-phase flow in the pipelines of the power equipment is realized in the form of various hydrodynamic structures and manifests itself in the form of pulsations of the main parameters of the stream. To reduce or eliminate this phenomenon, various kinds of active and passive devices are used that change the hydrodynamic flow structure [3].

* Corresponding author: v.i.velkin@urfu.ru

2 Experimental test bench

For this research purpose an experimental test bench was designed that enabled carrying out forced circulation of the mixture consisting of coolant and injected air to form various modes of two-phase flow (Fig. 1).

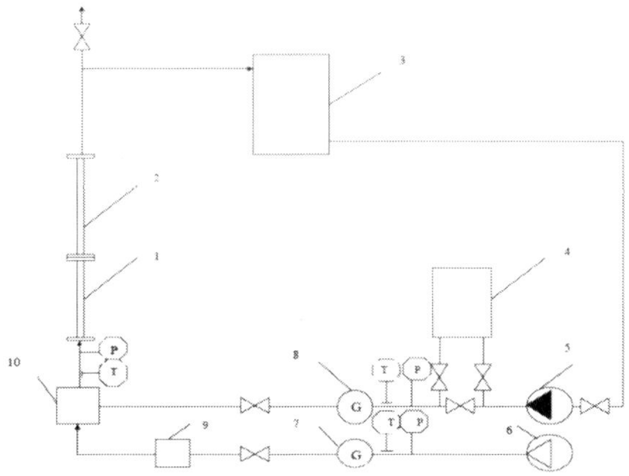

Fig.1. The schematic diagram of the experimental test bench 1 - vortex insertion; 2 - test section of the channel; 3 - water tank; 4 - control panel; 5 – pump; 6 - compressor; 7 – air flowmeter; 8 – water flowmeter;9 - damper; 10- blending unit.

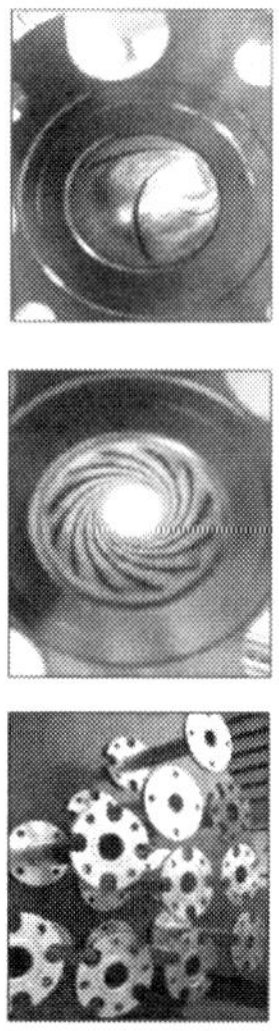

Fig.2. Photos of vortex inserts: a) with three grooves; b) with twelve grooves; c) - a set of investigated samples of vortex inserts with different internalsurface geometries.

Effectiveness analysis of various kinds of vortex inserts relating to passive means of reducing the vibration of pipelines has shown that the use of inserts with coiled surface grooves is quite promising and has a certain advantage over the other devices [4]. In particular, they have less hydraulic resistance and can be installed in series along the length of the pipeline before entering the turn. Figure 2 shows vortex inserts with different internal surface geometries.

Investigations of the twisting effect in two-phase flow with benefit of vortex inserts with involute grooves were carried out on a hydrodynamic test bench with an air-water flow circulating in a closed loop of 45mm-diameter pipe with transparent Plexiglas sections for flow visualization [5].

3 Comparison of vortex inserts

We take the ratio between the kinetic energy of the swirling part of the flow ...J the energy of the translational (axial) flow at the outlet of the swirler as the energy characteristic of the twist intensity.

$$\varepsilon = (tg\,\varphi)^2 nf/F_0; \quad tg\,\varphi = U_\varphi/U_0 \tag{1}$$

Where n – is the number of swirl grooves; f and F_0 – are the values of the total cross-sectional area of the grooves and the cylindrical channel; $U\varphi$ and U_0 – are the tangential and axial flow velocities.

According to the Darcy-Weisbach equation for a single-phase flow the hydraulic characteristics of the vortex inserts can be figured out from the pressure drop measurements:

$$\xi = 2\,(d\,/\,L)\,(\Delta p/\rho U_o^{\,2}) \qquad (2)$$

where ξ – is the coefficient of hydraulic resistance (CHR) of the insert; d and L – are the hydraulic diameter and length of the insert.

4 Study of the vortex inserts effectiveness using the PIV method

Theoretical and experimental data obtained during the experiments on the test bench need to be analyzed with the help of more sophisticated methods and tools for studying micro flows. To verify the results of mathematical modeling within the framework of this task, a digital image velocimetry (PIV) method was applied. It makes it possible to measure the velocity fields in the channels within the submillimeter range with a resolution of the vector field up to 2-5 μm / vector in the cross section just a few μm thick. The determination of the displacement is based on the application of correlation methods to the tracer pictures using regular partitioning into elementary regions.

For the experiments, we used the "Polis" measuring system, which made it possible to measure the spatial distributions of two and three components of the instantaneous flow velocity, to calculate the spatial derivatives of the velocity, and to calculate a whole set of mixed statistical moments (up to the fourth order) [6]. The scheme of the used flow-measuring structure is shown in Fig. 3.

Recording of the reflected optical pulses is carried out by the cross-correlation camera, which is designed for double adjustment of trace patterns in the stream and can work both in the two-frame and single-frame modes. The test bench is equipped with a tracer device (High powered Fog generator Martin Magnum 1800). The facility allows the study of flow regimes at Reynolds number 10^4-10^6 that ensures the applicability of the results obtained for the analysis of aerodynamics of real-scale installations [7, 8]. To measure the velocity fields in the concerned model sections we applied the PIV-system "Polis" developed at the Institute of Thermophysics SB RAS [9,10]. Figure 4 shows the photo visualization of the fluid flow swirling (f = 0) after transition through the vortex insert.

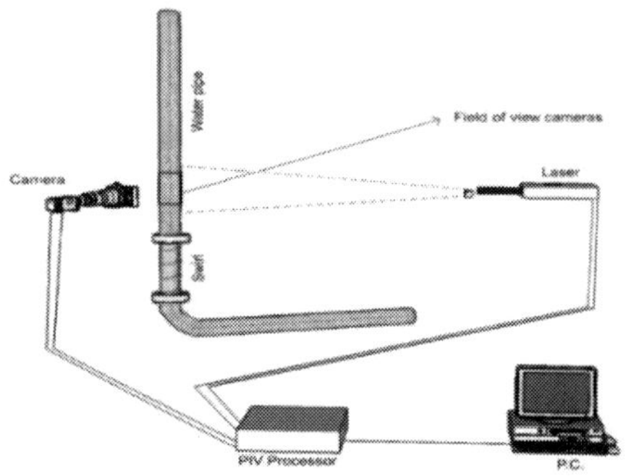

Fig.3. The scheme of the flow-measuring coolant flow after transition of the structure "Polis" as part of a laser installation, receiving chamber, synchronizer, and PC-image processing unit.

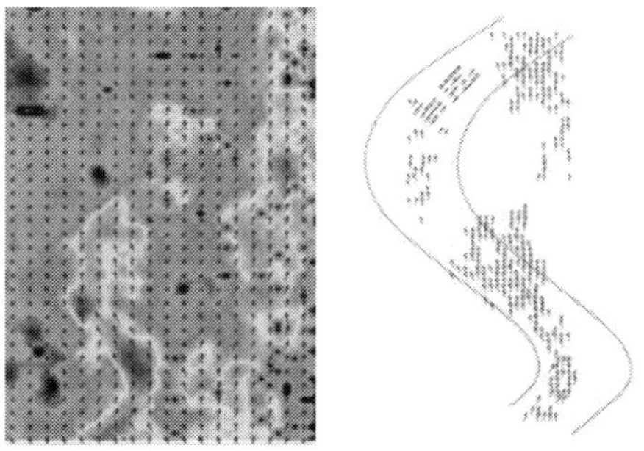

a) b)

Fig.4. a) Color photo visualization of the fluid flow swirling in the absence of gas after transition through the vortex insert; b) the tracer of the velocity vector of the swirl insert at the section L = 0-15 cm.

The different colors of the vectors in Fig. 6a correspond to different velocities. After a small image processing using the application program, we get a picture of the velocity vectors of a swirling fluid flow (Fig. 6b). The presented flow vector of the flow velocity vectors on the straight section of the pipeline demonstrates the distribution of the velocity vectors after leaving the swirl insert and shows the dispersion of the pressure gradient of the aqueous medium from the outer wall throughout the pipeline.

Conclusions

1. We designed the vortex insertion devices for passive control of pipeline vibrations which allow reducing the vibration displacement level of pipelines with two-phase flows by factor of 1.5-2 at a relatively low hydraulic resistance.
2. To verify the results of the simulation, an experimental test bench was designed and an optical laser system based on the pulsed imaging technology of micro trackers particles (PIV-system) was adjusted for measurements.
3. The system of optical measurement of hydrodynamic characteristics of the coolant flow (PIV-method) will allow obtaining velocity distributions in straight and various curved sections of the pipeline, as well as solving problems of passive control (reduction) of vibration displacement of pipelines of power equipment, and increasing reliability and safety of operation in general.

References

1. V.I. Baranenko, S.G. Oleinik, O.A. Belyakov, R.S. Istomin, A.V. Kumov, *Fourth International Scientific and Technical Conference" Ensuring the Safety of NPPs with VVR*, Podolsk, FSUE OKB" GIDROPRESS. (2005) [in Russian]
2. E.D. Fedorovich, B.S. Fokin, A.F. Axelrod & others, M: Energoatomizdat (1989). 168
3. M.Ya. Belenky., M.A Gotovsky., B.S. Fokin, Teploenergetika. **3**, 41 (1996)
4. V.I. Kasina, A.I. Leontiev, Heat-power Engineering **3**, 40 (2005)
5. V.I. Velkin, D.S.Komoza, A.Yu. Krutikov, V.V.Khnykina, News of Higher Educational Establishments "Nuclear Energy", **1**, 4 (2009) [in Russian]
6. E.K. Akhmetbekov, A.V. Bilsky, D.M. Markovich, A.A. Maslov, P.A. Polivanov, I.S. Tsyryulnikov, M.I. Yaroslavtsev, Thermophys. Aeromech. **16**, 343 (2009)
7. A.A. Maslov, S.G. Mironov, Izv. RAS. FGM. **6**, 155 (1996)
8. F. Scarano. *Overwiew of PIV in supersonic flows. Particle Image Velocimetry* / Ed. By A. Schroeder, C.E. Willert. (Springer-Verlag, Berlin, Heidelberg, 2008)
9. M.P. Tokarev, D.M. Markovich, A.V. Bilsky, Computational technologies. **12**, 109 (2007)
10. S.V. Alekseenko, A.V. Bilsky, D.M. Markovich, Devices and experimental technique. **47**, 703 (2004)

Regimes with periodical pressure pulsation in Francis draft tube

Ivan Litvinov*, *Evgeny* Gorelikov, and *Sergey* Shtork

Kutateladze Institute of Thermophysics, SB RAS, Novosibirsk, Russia

Abstract. The paper reports on results of the experimental study of flow and pressure pulsations in a laboratory model of the draft tube of Francis-99 hydro turbine over a broad range of operating regimes. A novelty is the rapid prototyping and 3-d printing of the inflow swirl generators that makes it possible to acquire rapidly a large amount of experimental data for a variety of designs and operating conditions. The frequency characteristics of flow and the phase-average pressure distribution in draft tube cone have been measured with the aid of special acoustic sensors.

1 Introduction

Expansion of the range of sustainable and reliable operating regimes of the hydro turbine, including those at nonoptimal conditions is an important task of the hydro power industry. Often at nonoptimal operation, especially for the Francis turbine, the flow entering into the draft tube (DT) keeps significant residual swirling intensity. Some peculiar phenomena of vortex instability have appeared in the decelerating swirl flow, which are named as vortex breakdown and precessing vortex core (PVC) or precessing vortex "rope". The effects associated with the transformation of an axisymmetric swirl flow in one or more precessing vortexes causes dangerous periodic vibrations on the draft tube walls [1, 2].

In present paper we report on approach to studying regimes with vortex instability and high level of pressure pulsations using «rapid» experimental methods and the air model of hydro turbine draft tube. During the experiments pressure pulsations on draft tube walls have been studied with acoustic sensors for different regimes (about 900 regimes) simulating different loadings of the turbine generator. Regimes with high level of pressure pulsations and PVC formation were identified. Further detailed studies provide new information on unsteady vortex structures under nonoptimal regimes in hydro turbine draft tube.

2 Experimental setup

Figure 1 shows sketch of the test section which includes a scale-down geometrical model of the Francis-99 draft tube [3] with an inlet diameter D=100 mm and a special device for generating the inflow into the conical part of the draft tube. Air at atmospheric pressure was

* Corresponding author: litvinov@itp.nsc.ru

used as the working medium [4]. This allows reducing significantly the cost and simplify the fabrication of the working channel based e.g. on the available plastic materials. To generate the required flow distribution at the inlet to the draft tube, a pair of swirlers was used: a stationary swirler, acting as guide vanes and a rotary one is an analog of the turbine runner, as in [5, 6]. However, in contrast to those studies, which used a freely rotating impeller, here the rotor speed is controlled by the servomotor at a specified frequency. The geometry of the swirlers was determined by numerical simulation, targeted to reproduce best the velocity distributions at the draft tube inlet for different operating modes of a radial-axial Francis hydro turbine [7, 8]. On the basis of the calculated geometry, the physical models of the swirler pairs were fabricated using 3-d printing. The experimental setup included a computer-aided system for controlling the airflow and rotation frequency of the rotor. During the experiments, the pressure pulsation levels at the walls of the draft tube model were recorded with high-sensitivity microphones. To register distribution of pressure in draft tube cone, phase-averaged technique was applied; the special probes in the form of a thin tube of 2-mm diameter were connected to microphones as described in [9].

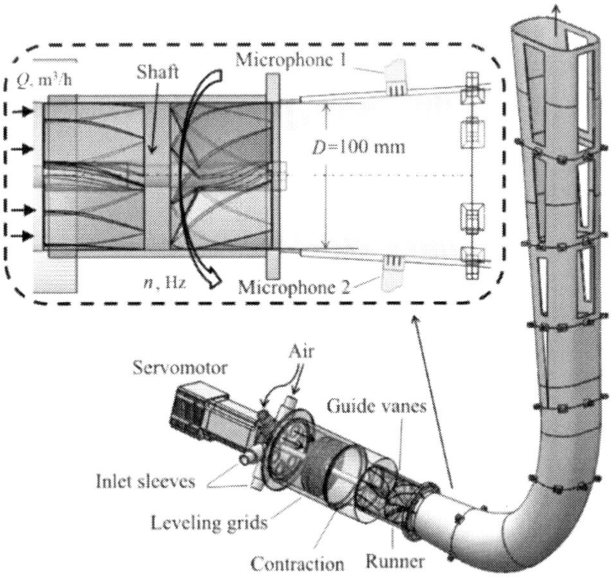

Fig. 1. Schematic of the experimental setup.

3 Results

Figure 2 presents sample results for the pair of swirlers, designed for optimal operating conditions (the best efficiency point - BEP), corresponding to the volumetric flow rate Q_c=48.5 l/s and the runner rotation speed n_c = 40.5 Hz. The plot on the left illustrates the pressure pulsation levels for almost 900 operation regimes, shown by spectral peaks obtained from the differential signals of two microphones positioned as shown in Fig.1. This method allows extracting the flow instability mode associated with the precessing vortex core (PVC) [10]. It can be seen that a departure from the BEP, for example by decreasing the flow rate below Q_c at a fixed rotor speed $n=n_c$ (partial load) leads to an increase in the pulsation level. The elevated flow pulsations are caused by the PVC, precessing with a dominant frequency, which is clearly visible in the spectrum of the acoustic signal, Fig. 2(c) right. Starting from the flow rate Q=0.4Q_c the peak with dominant frequency emerges; further at rate of 0.5Q_c the peak amplitude reaches a maximum. When

DOI: 10.1051/matecconf/201711505013

approaching the design condition (Q_c) the PVC frequency disappears from the spectrum completely. The obtained amplitude (normalized by bulk velocity U_0 in the cone inlet) of the dominant frequency in pressure spectra for each value of Q/Q_c are shown in Fig. 2b. From the data, it is seen that a local extremum arises (Q=0.5Q_c). This extremum occurs due to the formation of the precessing vortex core in the draft tube cone.

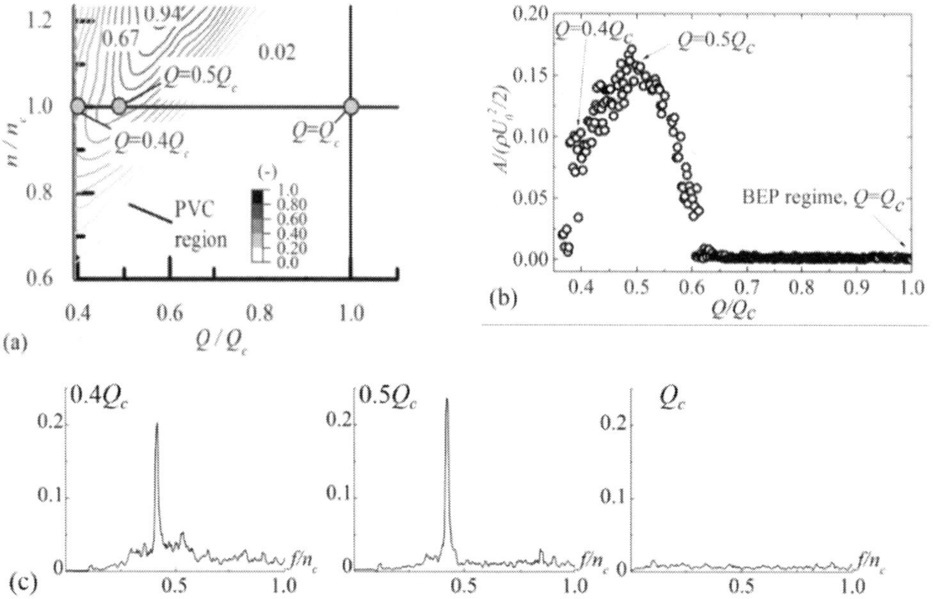

Fig. 2. Pressure pulsation level (arbitrary units) vs. regime parameters: Q is the air flow; n- the runner rotation speed (a); pressure pulsation distribution along the line n=n_c (b); typical spectra: Q=0.4Q_c, 0.5 Q_c, Q_c (BEP), vertical axes in arbitrary units.

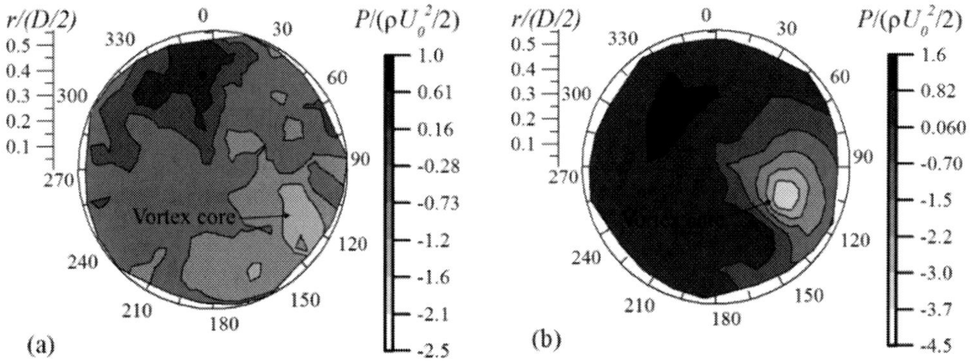

Fig. 3. Experimental phase-averaged pressure field in DT cross section (in the center of cone length): at 0.4 Q_c (a), at 0.5Q_c (b).

For better understanding a structure of the precessing vortex, we performed phase-averaged experiment using two acoustic sensors [9] for two regimes Q=0.4Q_c and Q=0.5Q_c (Fig. 3). It is seen that in the cross section of the draft tube the pressure field has more than one minima for Q=0.4Q_c. These minima correspond to the centers of two mild vortex structures. When the flow rate grows up to Q=0.5Q_c one powerful concentrated precessing vortex core is formed in the draft tube cone (Fig.3b). In the vortex core center (region with minimal pressure) the pressure depression doubled. Consequently, the intensity of pressure pulsations in the draft tube cone reaches the maximum value (Fig.2).

With further increase in flow rate, the flow swirl decreases, and a concentrated vortex after the impeller loses its intensity. The radius of vortex precession after the impeller decreases, and the intensity of pressure pulsation decreases as well. In the BEP regime (Q_c) the concentrated vortex after the impeller breaks down and phase-averaged technique is not able to catch them.

4 Conclusions

The paper illustrates the usefulness of the rapid prototyping of physical models for the experimental study of flow unsteadiness and emergence of the unsteady vortex structures for a variety of swirler designs and a range of operating conditions. Despite the neglect of cavitation and possible scaling effects, such results can provide insight into the effect of various design and operating parameters on the flow physics, as well as serve for verification of the numerical simulations.

The pressure measurements with sophisticated acoustic sensors confirmed that the regimes with PVC formed in draft tube cone of the hydro turbine at part load regime (0.5 Q_c). Using phase-averaged technique new information on the precessing vortex features in draft tube has been found.

This work was funded by the Russian Science Foundation grant No. 14-19-00487.

References

1. H.T. Falvey, *Draft Tube Surges* (Bureau of Reclamation, 1971)
2. R.-K. Zhang, Q.-D. Cai, J.-Z. Wu, Y.-L. Wu, S.-H. Liu, L. Zhang, Mod. Phys. Lett. B **19**, 1527 (2005)
3. M. Cervantes, C.H. Trivedi, O.-G. Dahlhaug, T. Nielsen, J. Phys. Conf. Ser. **579**, 011001 (2015)
4. M. Nishi, K. Yoshida, M. Yano, M. Okamoto, K. Miyagawa, S. Liu, IAHR Proceedings, Romania (2007)
5. R. F. Susan-Resiga, S. Muntean, C. Tanasa, A. Bosioc, GRoWTH Proceeding, (2008)
6. S. Skripkin, M. Tsoy, S. Shtork, K. Hanjalić, J. Hydraul. Res. **54**, 450-460 (2016)
7. V. Sonin, A. Ustimenko, P. Kuibin, I. Litvinov, S. Shtork, IOP Conf. Ser. **49**, 082020 (2016).
8. S. Shtork, I. Litvinov, A. Mitryakov, K. Hanjalic, EPJ Conf. Ser. **143**, 02103 (2017)
9. I. Litvinov, S. Shtork, P. Kuibin, S. Alekseenko, K. Hanjalic, Int. J. Heat Fluid Flow **42**, 251 (2013)
10. E. Fernandes, M. Heitor, S. Shtork, Exp. Fluids **40**, 267 (2006)

Features of reducing the turbulent friction of a liquid on the channel wall by gas-saturation

Aleksei Evseev*, *Leonid* Maltsev

Kutateladze Institute of Thermophysics, Novosibirsk, 630090, Russia

Abstract. The report presents the results of an experimental study of the efficiency of reducing the local friction at gas saturation of the turbulent boundary layer (TBL) in the input section of the channel at different gravitational orientation of the wall, and its dependence on the structure of gas-liquid flow. Profiles of gas concentration have a peak near the wall, which increases with the gas flow increase. The growth of concentration in the near-wall zone leads to rapid coalescence of bubbles, as a result of which the flow in TBL transits to the film-bubble regime with increasing the buoyancy effect of the gas phase, especially at low flow rates. It is shown that the key parameter of friction reduction by gas saturation is the gas phase concentration in the inner region of the boundary layer, whose magnitude is determined by the gas flow rate, the flow velocity, the distance downstream behind the gas generator, and the gravitational orientation of the wall.

1 Introduction

The technology for reducing turbulent friction using microbubbles has received increasing attention due to such advantages as high efficiency (80%), environmental friendliness, simplicity of design and low costs.

The state-of-the-art in research and the latest achievements in the reduction of friction using gas saturation of the turbulent flow are given in the reviews [1, 2].

The aim of the research is to measure shear stresses on the top and bottom channel walls at introducing gas into the flow through a porous wall, and to determine the dependence of friction reduction on the profile of gas concentration in the range of numbers $Re_x = (0.23\text{-}1.5)\times10^7$ and specific gas flow rates up to $q = 0.05$ m^2/s

2 Experimental setup and experimental technique

Studies on the effect of gas saturation on TBL and friction at the wall were carried out in the input section of the channel with section of 100 mm (height)×300 mm (width) and a length of 2200 mm. Water from the receiving tank was directed into the channel by a centrifugal pump through a honeycomb and a confuser (1:10). The gas bubbles were removed from the flow to an open receiving tank. The gas was injected into the flow

* Corresponding author: evseev@itp.nsc.ru

through a porous (20 μm) inset (140×172 mm), which was mounted flush with the channel wall at a distance 429/615 mm from the confusor in an open or closed working part. The gas flow was varied in the experiments up to 7 l/s and was measured with a standard orifice plate with an accuracy of ±3%. The local friction on the channel wall with accuracy of 10% was measured by sensors of the type "floating wall" (diameter 23 mm). Sensors (P1-P4)/(P5-P9) were mounted flush on the top/bottom channel wall at distances X*=93, 168, 243 and 488 mm downstream over the porous inset. Measurements of the local gas concentration (φ) were carried out using a fiber optic sensor. The diameter of the sensor was 0.15 mm. The measurement error was ±10%. First, we measured the local friction on the bottom wall of the channel behind the gas generator (X*=93, 168, 243 and 488 mm) with gas saturation of TBL at three flow velocities U=3.6, 6.3-6.5 and 10.2 m/s. Then, instead of the block of local friction the unit with a fiber sensor was mounted, and the gas concentration profile was measured in the sections of the boundary layer at X*=50, 200 and 445 mm. Similarly, measurements were carried out in a hydrodynamic tunnel on the upper wall of the working part. Experimental methodology is presented in [3].

3 Experimental results

Fig. 1 a, b show the dependence of the drag reduction ($DR=1-C_f/C_{f0}$) on the specific gas flow rate (q=Q/b) on the top (Fig. 1a) and bottom channel walls (Fig. 1b), where b = 140 mm is the width of the gas generator. For comparison, the graph shows data [4]. By reducing friction in TBL using gas saturation it is proposed to consider three modes depending on the specific gas flow rate (q): bubble regime (DR up to ~20%); transitional film-bubble mode (DR~ (20 -80)%); film mode (ALDR), which is characterized by a thin air layer (film) between the surface and the liquid flow with friction reduction of 90 % ± 9 %. The reduction in local friction increases with increasing q, but the steepness of the data varies at different X* and flow velocities on the upper and lower walls. At a flow rate U=10.2 m/s the dependence of the effect on q for the upper and lower walls is almost identical. At q = const the reduction of the flow rate leads to a larger reduction of friction. Near the generator q has its optimum (bottom wall), when the DR effect is maximum.

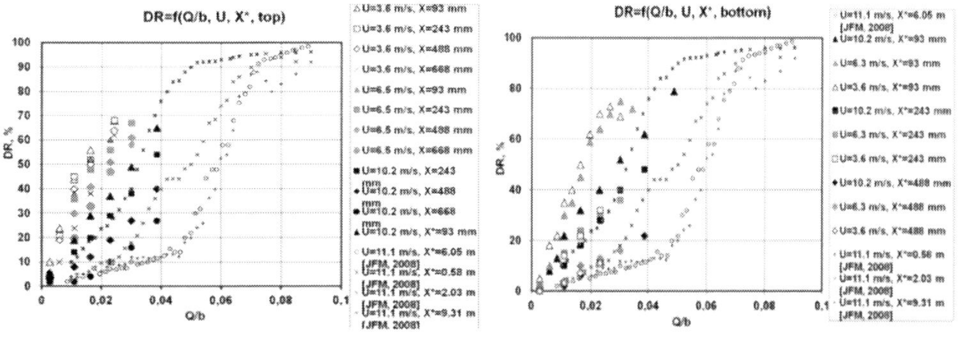

Fig. 1 a, b. Dependence of drag reduction at top (a) and bottom (б) channel walls on q=Q/b.

Fig. 2 a, b show the dimensionless concentration profiles of the gas phase in the bottom (a) and top (b) walls of the channel at X* = 445 mm for three flow velocities and three gas flow rates. Here δ is the thickness of the boundary layer without gas injection. For maximum flow rate (U = 10.2 m/s) and Q = 2.3, l/s the concentration peak was reduced to φ_m =22% / φ_m =16%, and moved away from the wall at a distance of 3 and 4 mm (top/bottom). But for U = 3.6 m/s and Q = 2.3, l/s the peak (φ_m =65%) at the upper wall remained at a distance Y≈1 mm, which is close to the concentration at X*= 50 mm (70%). When increasing the gas flow rate there is a transition from bubble to film-bubble regime of the mixture flow, resulting in deformation of the concentration profile (Fig. 2b) due to the effect of buoyancy.

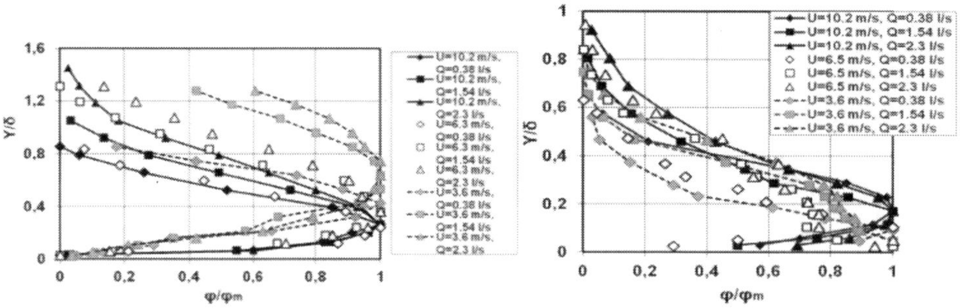

Fig. 2. a, b. The dimensionless concentration profiles of gas phase at the lower (a) and top (b) walls of the channel at X* = 445 mm for three flow velocities and three gas flow rates.

Mechanisms of friction reduction at gas saturation of TBL were considered in many theoretical works, an overview of which can be found in [5]. Fig. 6 shows the results of measurements in which the relative reduction in friction is a function of φ*, that is, the average gas concentration in the near-wall zone of the boundary layer ($Y^+<250$). Here, the straight line 1 corresponds to a change in the mixture density in the near-wall zone of the boundary layer, and the curve 2 corresponds to a theoretical dependence, taking into account changes in the mixture density, increase in its effective viscosity and modification of turbulence [5]. All known data on friction reduction at gas saturation are generalized by the dependence on gas concentration φ* in the near-wall zone of TBL, which is a key parameter. This confirms the hypothesis that the principal features of the mechanism of friction reduction at gas saturation by bubbles ($18<d^+<200$-250) is the decrease in mixture density near the wall, increase in its effective viscosity, as well as modification of turbulence at the interaction of bubbles with vortex sheath in the inner region of the boundary layer.

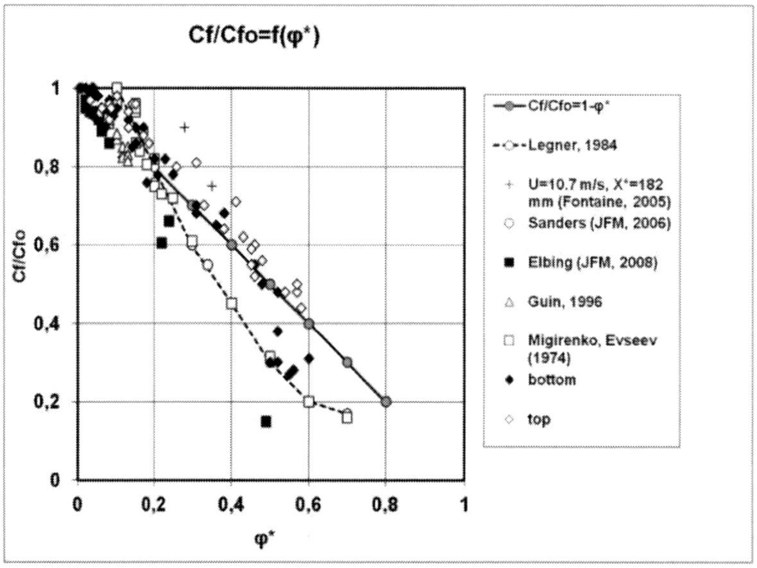

Fig. 3. The dependence of the effect of friction reduction at gas saturation on the average gas concentration **φ*** in the inner region of the turbulent boundary layer $Y^+<250$).

4 Conclusion

The reduction of friction on the upper and lower walls of the channel (inlet section) arising at gas introduction through the porous coating was investigated in the range of Reynolds numbers $Re_x = (0.23\text{-}1.5)10^7$. The friction at the wall was measured by sensors of the "floating" wall, and the gas concentration in TBL was measured by the fiber-optic sensor.
With the growth of Q the maximum effect of reducing friction on the walls amounted to 70-80%. But for the upper wall, the effect remained at a high level downstream, and on the bottom it was lost.
The concentration profiles had a peak near the wall, which increased with increasing gas flow rate up to ~ 60±10%. The growth of concentration in the near-wall area led to intensive coalescence of bubbles, which resulted in the transition of the flow in the boundary layer to the film-bubble regime with increasing buoyancy effect of the gas phase, especially at low flow rates. Downstream, the outflow of air bubbles from the wall area led to a rapid loss of effect on the lower wall, and at the upper wall the effect remained at a high level. Therefore, the key parameter to reduce friction with gas saturation is the concentration of the gas phase in the buffer zone of TBL, the value of which is determined by the gas flow rate, the fluid velocity, the distance downstream behind the gas generator and the gravitational orientation of the wall.

The work was performed with financial support of the Russian Science Foundation (Grant No. 14-19-01685, Leading scientist Markovich D.M.) provided through IT SB RAS.

References

1. L.S. Ceccio, Annu. Rev. Fluid Mech. **42,** 89 (2010)
2. A. Hashim, O.B. Yaakob, K.K. Koh, N. Ismail, Y.M. Ahmed, Jurnal Teknologi (Sciences & Engineering). **74,** 5 39 (2015)
3. A.R. Evseev, L.I. Maltsev, J. Int. Scientific Publications: Materials, Method & Technologies. **6**. Part 3. (2012)
4. B.R. Elbing, E.S. Winkel, K.A. Lay, S.L. Ceccio, D.R. Dowling, M. Perlin, J. Fluid Mech. **612,** 201 (2008)
5. H.H. Legner, Phys. Fluids. **27,** 356 (1984)

Experimental study of gas-liquid flow local characteristics in rectangular microchannel

German Bartkus[1,2,*]

[1] SB RAS, Kutateladze Institute of Thermophysics, 630090 Ac. Lavrentieva ave. 1, Novosibirsk, Russia
[2] Novosibirsk State University, 630090 st. Pirogova 1, Novosibirsk, Russia

Abstract. Using high-speed video recording and the method of dual laser scanning the gas-liquid flow was investigated in a rectangular microchannel with an aspect ratio of 0.74 (cross section 269×362 μm). The T-mixer was used at the channel's inlet for the two-phase flow formation. The peculiarity of this work is using a number of liquids (ethanol, distilled water, 40% aqueous ethanol) with different physical properties, including surface tension, viscosity, and density, with nitrogen. Experiments were carried out for the vertically upward and horizontal flow. Using laser scanning method the maps of flow patterns were obtained for all mixtures.

1 Introduction

Significant attention is paid now to investigations of gas-liquid and liquid-liquid flows in microchannels. This fact is connected with using of microchannels in high-intensity cooling systems, based on microchannel heat exchangers, in chemical reactors, and biological chips for rapid blood test [1]. This shows that study of local characteristics is urgent because these characteristics influence on the intensity of heat and mass exchange processes in various technological and natural processes [2-4]. The hydrodynamics of gas-liquid and liquid-liquid flows in microchannels was studied in a number of works in which the main flow patterns were defined, e.g. [3, 5, 6]. In paper [7] local characteristics of the flows (such as liquid distribution in microchannel) were investigated. These characteristics are used for creating heat and mass transfer models such as for example in [8, 9].The present study aims at conducting the experiments of gas-liquid flow using liquids with different physical properties and generalizing flow pattern maps for vertical and horizontal flows.

2 Experimental equipment and methods

The experimental setup for studying the upward vertical and horizontal gas-liquid flow in a rectangular microchannel is shown in Fig. 1 (a). Nitrogen, which is fed from a high-pressure tank (11) via the control valve (10) and gas flow controller (4) to the test section, is used as working gas. Water flows from the tank (7) with filter (filter rating 1μm) through a fluid flow controller (3) and further into the T-shaped mixer at the inlet of the test section,

* Corresponding author: germanbartkus@gmail.com

i.e. microchannel (2). Connectors (8, 9) allow removing the tank and pour liquid very rapidly. Gas and liquid flow controllers (3, 4) by Bronkhorst are powered by power supplies (5, 6) of the same company. Test section is a microchannel, assembled from glued glass and aluminum plates with length L = 30 cm. The hydraulic diameter is much smaller than the capillary constant. From the exit of test section, the gas-liquid mixture is moved to the liquid tank (1), where the gas escapes to the atmosphere.

The study of gas-liquid flow patterns was conducted using the method of dual laser scanning, see Fig. 1 (b). Two lasers (12, 13) are positioned so that the laser beams (15) illuminate one side of the rectangular channel. A lens (14) is used to change the diameter of the laser spot. The lasers are powered using voltage power source. The intensity of transmitted light is measured by photodiodes (16) arranged at the opposite side of the channel. The electric voltage from the photodiodes is analyzed using high-speed ADC and processed on the computer over the period from one to two minutes. This method is very useful for determining statistical characteristics of two-phase flow, e.g. [7]. Flow patterns are also registered using the high-speed camera Optronis CX600x2 (17). The first laser is located at a distance of 6 cm from the inlet, while camera – at 17.7 cm from the inlet, and second laser – at 22.4 cm for observing the stable gas-liquid flow. Physical properties of used liquids are shown in Table 1.

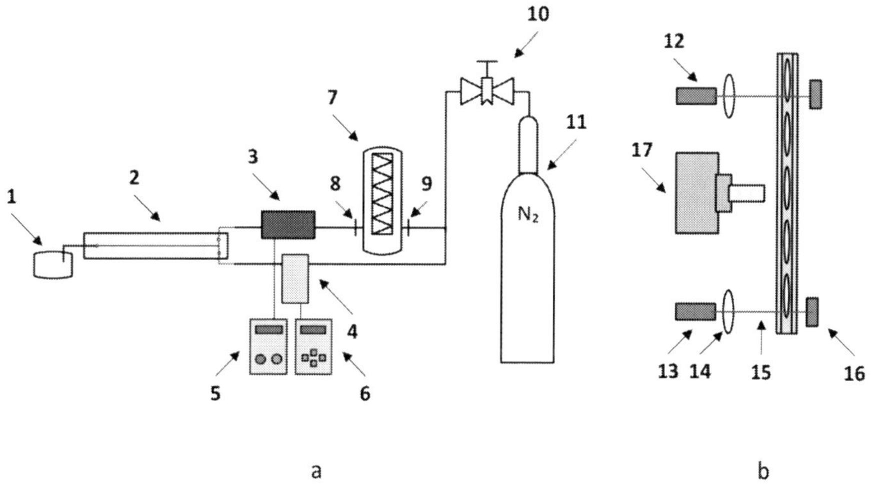

Fig. 1. Experimental setup (a), scheme of the dual laser scanning method (b).

Table 1. Properties of used liquids

	Physical properties		
Liquid	**Density, ρ [kg/m³]**	**Viscosity, η [mPa·s]**	**Surface tension, σ [H/m]**
Distilled water	998.2	1.004	0.073
Ethanol	789	1.190	0.0223
40% aqueous ethanol solution	935.2	2.910	0.03

3 Two-phase gas-liquid flow in microchannels

The high-speed video and laser scanning method show that the periodic flow with elongated gas bubbles (Taylor bubbles), the transition flow, and the annular flow are the main flow regimes in microchannels with the gap substantially less than the capillary constant.

(a) (b) (c)

Fig. 2. Visualization of ethanol-nitrogen flow in horizontal microchannel 269×362 μm: a – photo of elongated bubble flow, J_{lig}=0.066 m/s J_{gas}=0.072 m/s, b – transition flow, J_{lig}=0.066 m/s J_{gas}=1.14 m/s, c – annular flow, J_{lig}=0.054 m/s J_{gas}=3.82 m/s.

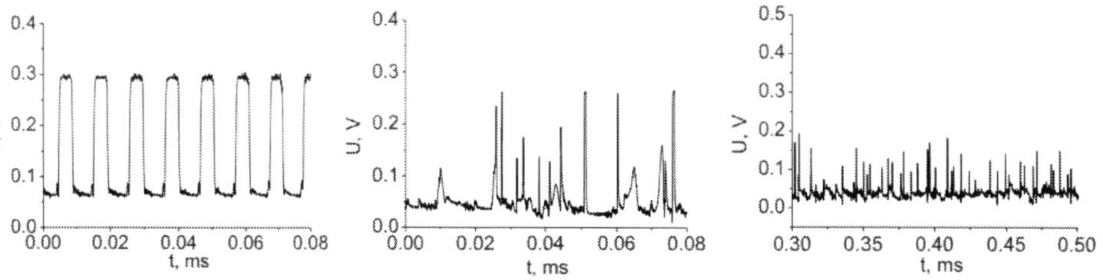

(a) (b) (c)

Fig. 3. Horizontal flow in microchannel of 269x362 μm (ethanol-nitrogen). The dynamics of voltage signal on the photodiode for elongated-bubble flow at superficial velocities of liquid and gas J_{liq}=0.072 m/s, J_{gas}=0.147 m/s (a), transition flow – J_{liq}= 0.072 m/s, J_{gas}= 1.399 m/s (b), annular flow – J_{liq}= 0.036 m/s, J_{gas}= 3.62 m/s (c).

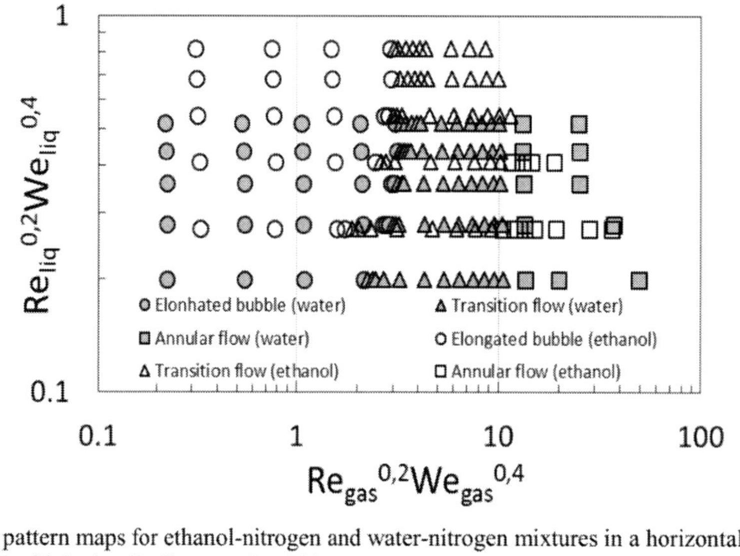

Fig. 4. Flow pattern maps for ethanol-nitrogen and water-nitrogen mixtures in a horizontal microchannel with hydraulic diameter D_h=309 μm.

The photograph and voltage signal from photodiode for three regimes are presented in Figs. 2 and 3 for horizontal ethanol-nitrogen experiment in 269x362 μm microchannel.

4 Flow pattern maps

Flow pattern maps for ethanol-nitrogen and water-nitrogen flows in horizontal microchannel were constructed in the coordinates of the dimensionless complex $Re^{0.2}We^{0.4}$ proposed by Waelchli [9], and representing the combination of Re and We numbers. Formulas for *Re* (1) and *We* (2) numbers are shown below:

$$Re_{liq,gas} = \frac{J_{liq,gas} \cdot D_h}{\nu_{liq,gas}} \quad (1)$$

$$We_{liq,gas} = \frac{\Delta\rho \cdot J_{liq,gas}^{2} \cdot D_h}{\sigma}. \quad (2)$$

The maps are presented in Fig. 4. Offered coordinates give a good generalization of the experimental data for boundaries of regimes for liquids with different properties.

5 Conclusion

We carried out the experiments, aimed at studying the influence of liquid physical properties (surface tension, viscosity, density) and the orientation of the microchannel in space (horizontal, vertical) on the flow regimes and the regions of their existence. Ethanol, distilled water, and 40% aqueous ethanol solution were used as the liquid phase, while nitrogen was taken as the gas phase. It was established that for the present microchannel with a hydraulic diameter substantially less than the capillary constant, three basic flow regimes are formed: flow with elongated gas bubbles, the transition, and annular flow. The effect of the different orientation of the microchannel (horizontal, vertical) on the boundaries of flow regimes is negligible that shows the dominance of capillary forces. Based on the visualization and laser scanning method, flow pattern maps were constructed for 2 types of mixtures using the dimensionless complex $Re^{0.2}We^{0.4}$, proposed by Waelchli [9]. The dimensionless complex makes it possible to take into account the physical properties of the liquid and gives a good generalization of the experimental data.

This work was performed by a grant of the Russian Science Foundation (project No. 16-19-10519).

References

1. V. Kuznetsov, J. Phys.: Conf. Ser. **754**, 032012 (2016)
2. V. Kuznetsov, S. Dimov, P. Houghton, *1st Int. Conf. on Microchannels and Minichannels. ASME, 683* (2003)
3. V. Kuznetsov, I. Kozulin, A. Shamirzaev, J. Phys.: Conf. Ser. **395**, 012093 (2012)
4. V. Kuznetsov, A. Shamirzaev, I. Kozulin, Heat Tran. Eng. **34**, 235 (2013)
5. A. Kovalev, A. Yagodnitsyna, A. Bilsky, Chem. Eng. J. **303**, 547 (2016)
6. G. Bartkus, V. Kuznetsov, MATEC Web of Conferences **84**, 00005 (2016)
7. I. Kozulin, V. Kuznetsov, J. Appl. Mech. Tech. Phys. **52**, 956 (2011)
8. V. Kuznetsov, A. Shamirzaev, Heat Tran. Eng. **37**, 1105 (2016)
9. S. Waelchli, P. Rudolf von Rohr, Int. J. Multiphase Flow **32**, 791 (2006)

MATEC Web of Conferences 115, 05016 (2017) DOI: 10.1051/matecconf/201711505016
STS-33

The effect of outflowing water coolant with supercritical parameters on a barrier

Maksim Alekseev[1,*], *Ivan* Vozhakov[1,2], *Sergey* Lezhnin[1,2], and *Nikolay* Pribaturin[1]

[1]Kutateladze Institute of Thermophysics SB RAS, Novosibirsk, Russia
[2]Novosibirsk State University, Novosibirsk, Russia

Abstract. The outflow of supercritical coolant with different initial parameters and its impact on the barrier have been numerically simulated. Spatial and axial distributions of pressure and steam quality are presented. The force acting on the barrier at different parameters of the outflow has been calculated.

1 Introduction

In connection with increasing intensity in modern power plants, working pressures and temperatures may approach or even exceed critical thermodynamic parameters. It is extremely important to be able to predict scenarios of dynamic processes at sudden depressurization of the work section. In particular, of great interest is to determine impacts on various obstacles at low (wave) and large times of the energy heat carrier outflow. As it was shown in [1], under non-stationary outflow of the superheated coolant into the atmosphere, an unsteady compression wave, which may damage other elements of the power plant, is generated into the environment. There are many experimental studies of the stationary jet outflow of superheated steam and boiling coolant on the barrier under high pressure. For example, in [2] the impact of the steam jet on the obstacle and the pressure distribution over the radius on the obstacle were measured. The similarity of the stationary outflow of superheated steam and the one of the gas jet was shown. Experimental study of the stationary outflow of the boiling coolant from the pipe end on the barrier in an industrial scale was carried out in [3].

2 Governing equations

The present study considers the axisymmetric problem of the water coolant outflow from a high pressure vessel with the aim to obtain the characteristics of wave impact on the outer barrier and the outflow parameters. The problem is solved by numerical simulation of the system of Euler equations, the equation of continuity and energy equations for a two-phase homogeneous mixture. For calculations based on computing complex LCPFCT [4] a software code was developed for solving a wide class of problems, arising during depressurization of vessels or piping with superheated water coolant. To close the system of

* Corresponding author: alekseev@itp.nsc.ru

equations it is required to set the equation of state of thermodynamically equilibrium two-phase mixture in the form of

$$p = F(\rho, E) \tag{1}$$

Here ρ is the density, p is the pressure, and E is the specific internal energy.
Density, internal energy, other thermodynamic parameters and potentials of the mixture in equilibrium depend on pairs of variables, e.g. X and p [5].

$$\frac{1}{\rho} = (1 - X)\frac{1}{\rho_1(p)} + X\frac{1}{\rho_2(p)} \tag{2}$$

$$H = (1 - X)H_1(p) + XH_2(p),\ E = (1 - X)E_1(p) + XE_2(p) = H - p/\rho \tag{3}$$

Here H is the specific enthalpy of the mixture, X is the steam quality, and L is the latent heat of vaporization. Index 1 refers to liquid, and 2 refers to vapor.
Using the expression (3) and the equality $L = H_2 - H_1$, it is possible to obtain

$$E = (1 - X)E_1 + XE_2 = H - p/\rho = H_1 + XL - p/\rho \tag{4}$$

Since from (2) follows

$$X(p,\rho) = \frac{1/\rho - 1/\rho_1(p)}{1/\rho_2(p) - 1/\rho_1(p)}, \tag{5}$$

then, from (4) we obtain the equation of state of the two-phase mixture in the form of relationship between pressure, density and specific internal energy:

$$E(p,\rho) = H_1(p) - \frac{L(p)/\rho_1 p)}{1/\rho_2(p) - 1/\rho_1(p)} + \left[\frac{L(p)}{1/\rho_2(p) - 1/\rho_1(p)} - p\right]\cdot\frac{1}{\rho} \tag{6}$$

From equation (6) the dependence (1) was numerically tabulated in advance to further use in the numerical simulation. Calculation of properties of steam and water in sub- and supercritical states was performed using the software package TTSE [6].

3 Modeling results and discussion

In the present work, the outflow of water with different initial supercritical parameters on the barrier was numerically simulated. On the PV-diagram with initial P_0=23 MPa and T_0=370 ^{0}C, the adiabatic line passes to the left of the critical point and crosses the line of boiling, i.e. enters the two-phase region with void fraction equal to null [7]. In the case when the initial parameters take values P_0=23 MPa and T_0=380 ^{0}C, the adiabatic line passes to the right of the critical point and crosses the line of condensation, i.e. enters the two-phase region with void fraction equal to unity.

Figure 1 presents spatial distributions of normalized pressure gradient and steam quality. In the initial moment of time at depressurization, the generation of compression waves occurs. The compression wave moves towards the obstacle and is reflected from it. Behind the compression wave, forms the area of the two-phase mixture outflow, consisting of the Mach disk and oblique shock wave. The shape of the Mach disk and the lateral suspension surges differ for the cases of evaporation and condensation. At vapor outflow with condensation, the form of the Mach disk and the lateral suspension surges are similar to the form, arising at a gas-dynamic outflow. At liquid outflow with evaporation, the shape of the lateral suspension surges with the Mach disk approaches hemispherical.

Figures 2, 3 show axial profiles of pressure and steam quality at various points in time. The vertical solid line shows the coordinate of the target, and the vertical dotted line shows the coordinate of the channel end. The amplitudes of compression waves at propagation and reflection from the obstacle are similar for the process of evaporation and condensation.

 DOI: 10.1051/matecconf/201711505016

Together with the compression wave, the rarefaction wave moves in the interior of the channel. From the drawings it is seen that the propagation velocity of the condensation

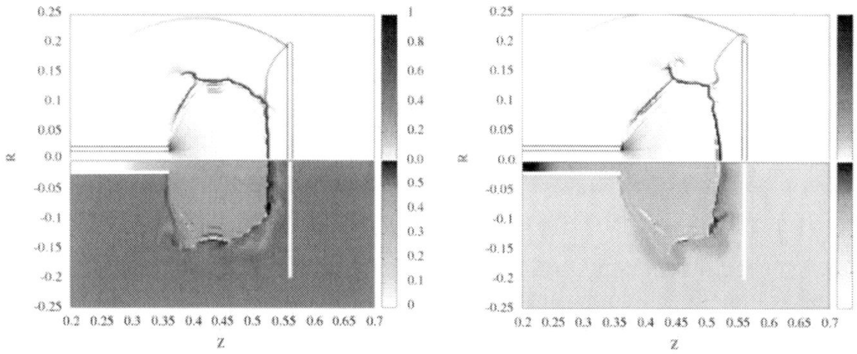

Fig. 1. Spatial distribution of normalized pressure gradient (top) and steam quality (bottom) at time t=0.5 ms for initial conditions P_0=23 МПа, T_0=370 ^{0}C (left) and T_0=380 ^{0}C (right).

front is higher than that of the evaporation front inside the channel. The main change of steam quality in case of condensation occurs inside the channel, and for the case of evaporation, the phase transition continues outside of the channel as well. In addition, the case of condensation is characterized by a sharp pressure drop at the end channel, while for the case with evaporation, this pressure drop is not observed.

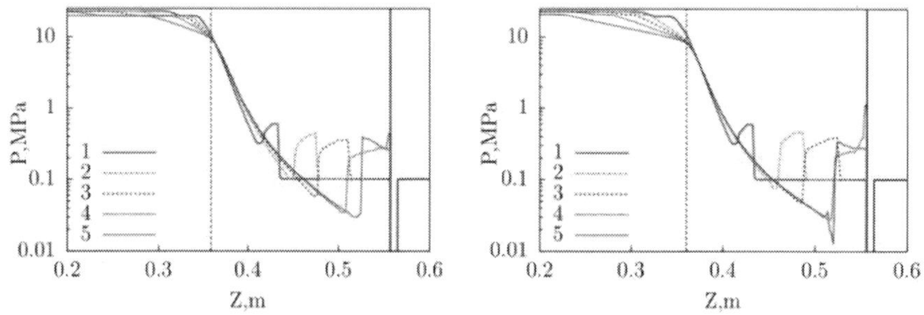

Fig. 2. Axial pressure distribution in different moments of time (1 – 0.09 ms, 2 – 0.15 ms, 3 – 0.21 ms, 4 – 0.3 ms, and 5 – 0.51 ms) for initial conditions P_0=23 МПа, T_0=370 ^{0}C (left) and T_0=380 ^{0}C (right). Solid vertical line is the coordinate of the target, and the dotted vertical line is the coordinate of the channel end.

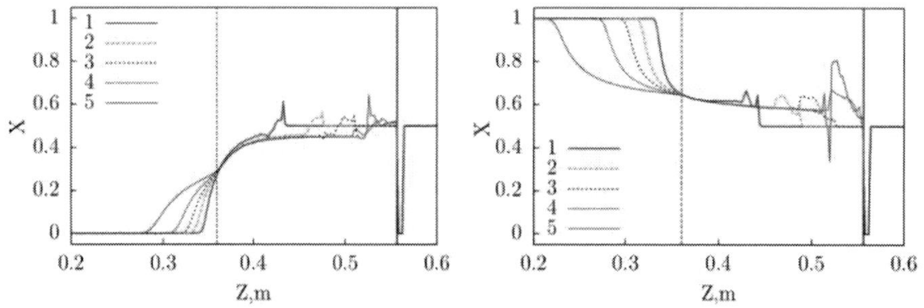

Fig. 3. Axial distribution of steam quality at different points in time (1 – 0.09 ms, 2 – 0.15 ms, 3 – 0.21 ms, 4 – 0.3 ms, and 5 – 0.51 ms) for initial conditions P_0=23 MPa, T_0=370 ^{0}C (left) and T_0=380 ^{0}C (right). Solid vertical line is the coordinate of the target, and the dotted vertical line is the coordinate of the channel end.

Figure 4 shows the time dependence of the force, acting on the barrier and appearing during the water coolant outflow. The differences between the cases of evaporation and condensation are due to different velocities of perturbation waves' propagation in the external space (time shift). In addition, it is seen that the maximum value of the force acting on the barrier is reached at time moments 0.45 ms (T_0=380 ^{0}C) and 0.5 ms (T_0=370 ^{0}C), that is followed by fall.

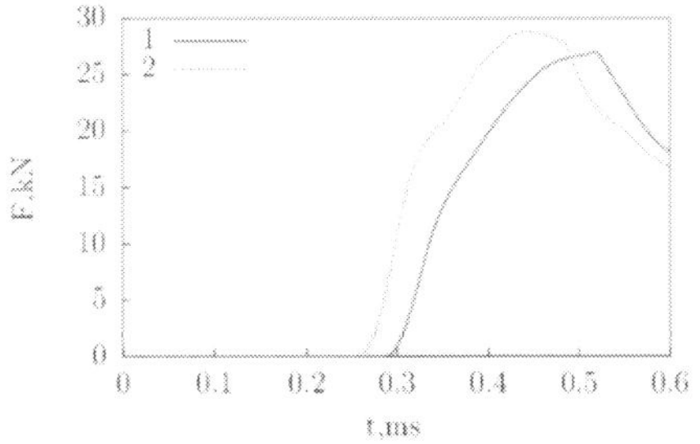

Fig. 4. Time dependence of the force acting on the barrier at outflow of water in supercritical state. 1– P_0=23 MPa, T_0=370 ^{0}C, 2 – P_0=23 MPa, T_0=380 ^{0}C.

4 Conclusion

The outflow of water with initial supercritical parameters has been numerically simulated using the equilibrium model of phase transition. It is shown that, depending on initial parameters, there are various options for the outflow development. The options are associated with the presence of the "opposite" phase transitions: evaporation or condensation. At steam outflow with condensation, the form of the Mach disk and the lateral suspension surges are similar to the form, arising at gas-dynamic outflow. At the liquid outflow with evaporation, the shape of the lateral suspension surges with a Mach disk approaches hemispherical. The results of the calculation of the spatial and axial distribution of pressure and steam quality are presented. It is noted that in the case of condensation, phase transition occurs mainly within the channel, and in the case of evaporation, it takes place both inside and outside of the channel. Forces acting on the barrier are calculated at different initial parameters.

The work was financially supported by the Russian Science Foundation (grant No14-29-00093).

References

1. G.V. Gofman, A.E. Kroshilin, B.I. Nigmatulin, High Temp. **19** 897 (1981)
2. F. Masuda, Nuclear Eng. Design **67** 273 (1982)
3. T. Isozaki, S. Miyazono, Nuclear Eng. Design **96** 1 (1986)
4. J.P. Boris, A.M. Landsberg, E.S. Oran, J.H. Garder, Naval Research Lab Washington DC (1976)
5. M.D. Weisman, *Thermodynamics of Vapor-Liquid Flows* (1967)
6. Guideline on the TTSE method for calculation of thermodynamic properties of water and steam applied to IAPWS-95 (2003)
7. I.S. Vozhakov, M.V. Alekseev, S.I. Lezhnin, N.A. Pribaturin, J. Phys.: Conference Series **754** 032021 (2016)

Near-wall liquid film outflow of water-ethanol mixture from cylindrical channel into vacuum

Yuri Vyazov, *Pavel* Votinov, and *Igor* Yarygin*

Kutateladze Institute of Thermophysics SB RAS, Rarefied Gases Laboratory, 630090, Lavrentieva ave. 1, Novosibirsk, Russia

Abstract. The problem of joint near-wall liquid film ejection with gas flow from a cylindrical channel into vacuum is studied experimentally. A new technique for measuring thickness and velocity of the liquid film inside the channel with the help of capacity-type probes is presented. It is shown that co-current gas flow has strong impact on the near-wall film, leading to intense wave formation, detachment of droplets from the film surface, and formation of dry spots.

1 Introduction

Expansion of gases, liquids and gas-liquid mixes into vacuum is of great interest for a number of scientific and technological applications. Recently a lot of attention was attracted by the jet expansion of gases from nozzles of various geometry (supersonic, transonic, axisymmetric, flat, etc.). It was established that so-called underexpanded flow is formed behind exit cross-section of a nozzle under gas expansion into vacuum. This flow includes area of free supersonic expansion (the core of the jet), as well as oblique shock waves and Mach disk [1]. The flow conditions in a jet and a set of physical processes accompanying supersonic gas expansion are mainly characterized by nozzle geometry, gas composition, its initial parameters inside the nozzle (stagnation temperature and pressure) as well as conditions in the vacuum chamber. One of the special characteristics of gas jet expansion into vacuum is Prandtl-Mayer expansion fan at the exit edge of the nozzle, in which gas turns at angles over 90° relative to jet axis. Such flows are called backflows, and they form a peripheral part of the jet.

However the number of studies published concerning ejection of liquids and gas-liquid flows into vacuum is very limited and most of them are devoted to space applications, such as ejection of liquids from spacecrafts during purging of propellant refueling systems, operation of orientation thrusters in which fuel film is used for cooling walls of combustion chamber and a nozzle, etc [2,3]. As saturated vapor pressure of liquids used is usually several orders of magnitude higher than pressure in surrounding space, liquid becomes "instantly overheated" being ejected into vacuum. This leads to liquid film boiling, its disintegration into droplets, and phase transitions on the surface and inside droplets. Also the co-current flow of gas has significant effect on behavior of a wall liquid film both when moving inside the channel and under ejection into vacuum. In experimental studies [4] it

* Corresponding author : yarygin@itp.nsc.ru

was shown that at the exit edge of the nozzle the wall film of liquid not only breaks up into droplets, but also emerges on the external surface of the channel, moving on it in the opposite direction, even against gravitation force. Under ejection into atmosphere the revealed effect is absent, and the ordinary gas-droplet flow takes place behind the exit cross-section of the channel. Another important feature of wall liquid film behavior inside the channel noted in [5] is presence of high-velocity co-current gas flow. Gas velocity under expansion from the cylindrical channel into vacuum reaches sonic velocity at the exit cross-section. This leads to rather high (over 10) values of Weber number We of co-current gas flow. According to current understanding [6] detachment of droplets from the film surface and their carrying away by co-current gas flow takes place at Weber numbers We > 2. Experimental studies [5] have shown that amount of liquid carried away by gas flow can reach significant values (tens of percent of the liquid initial flow rate). Note that only ethanol was used as working liquid in studies [4,5]. Current paper is devoted to experimental study of water-ethanol mixtures (with concentration from 0 to 100%) ejection with co-current gas flow from cylindrical channel into rarefied space. Some aspects of the problem concerning behavior of near-wall films of binary mixtures at the exit edge were presented in [7], and special attention in this paper is paid to interaction of the liquid film with co-current flow inside the channel.

2 Experimental setup and measurement techniques

The main goal of our experiments was to perform measurements of wall liquid film local parameters (thickness and velocity). Current methods of liquid film diagnostics can be separated into two classes: contact (probe) and non-contact (optical). In our study the probe method was chosen, namely the method using capacity-type probes [8]. The scheme of test section is shown in Fig. 1.

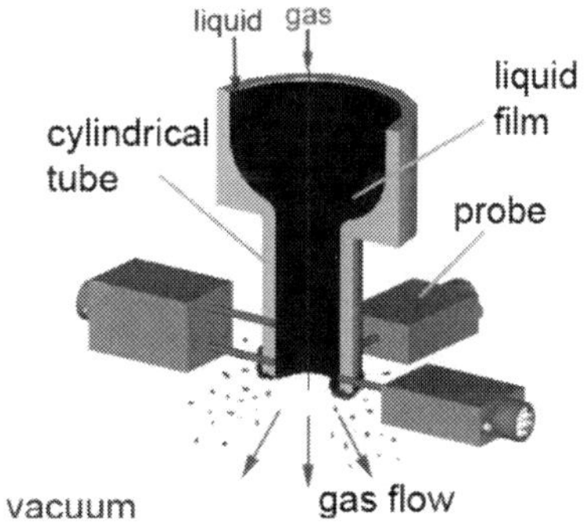

Fig. 1. Scheme of test section.

The key idea of capacity-type probe diagnostics is to measure liquid layer capacity over the probe and to convert probe indications into film thickness with the help of special calibration technique. This method has good both spatial (about 1 mm) and temporal (about 1 ms) resolutions. Two sequentially located probes *3* were used for wall film velocity measurement. Four probes *2* located through 90° were used for thickness measurement. Experiments were carried out on the VIKING vacuum gas-dynamic setup at Kutateladze Institute of Thermophysics [9]. The large volume of the working chamber (150 m^3) allowed operation in pulse mode with high flow rates of gas and liquid. Cylindrical channel (nozzle) made of brass with 5 mm diameter, 20 mm length and 1 mm wall thickness was mounted vertically in the vacuum chamber, with exit part facing downwards. The liquid entered through a circular gap of 0.1 mm width into the stagnation chamber of the nozzle and moved down the nozzle walls as a film. Simultaneously, gas was blown through the nozzle.

DOI: 10.1051/matecconf/201711505017

Purified air was used as working gas, and water-ethanol mixture with volumetric concentration of 0, 50 and 100% was used as working liquid. It should be mentioned that viscosity of water-ethanol mixture can exceed viscosity of pure components by several factors depending on concentration and temperature. Initially gas and liquid were at room temperature. Experiments were carried out in pulse modes. Usually the pulse length was 5 seconds, and the pressure in vacuum chamber increased from initial 1 Pa up to less than 1.1 Pa. Mass flow rate of gas and liquid varied from 0.5 to 20 g/s and from 0.6 to 2 g/s, respectively.

3 Results

An example of relationship between liquid film thickness and time under expansion with co-current gas flow from the cylindrical channel into vacuum is shown in Fig. 2 for ethanol (a) and water (b).

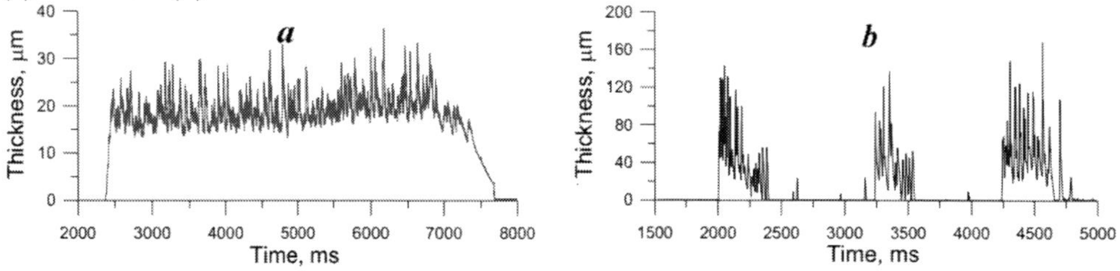

Fig. 2. Liquid film thickness. $Re_{gas} = 4.7 \cdot 10^4$. a – ethanol film, b – water film.

From Fig. 2a one can see timing of liquid film arrival to the probe. Further liquid film with thickness of several tens of microns is registered on the probe during feeding time. This film has wave structure with rather high amplitude and frequency. Then rapid drop of film thickness happens after liquid supply is cut off. Waves on the film surface have steep front and flat rolloff. We believe that wave characteristics mainly depend on parameters of co-current gas flow (its Reynolds number). Liquid film behavior changes under ejection of water. Under certain condition we observed multiple arrivals of film to the probe and decrease of film thickness to zero. At the same time average film thickness was rather high. Such behavior of the film seems to be caused by formation of dry spots on the channel surface and film flow in the form of separate streamlets – rivulets. This is probably due to high value of water surface tension coefficient and low wettability of the surface.

Experimental data on measurements of liquid film thickness under motion with high-velocity co-current gas flow inside the channel are shown in Fig. 3 for ethanol, water and their mixture.

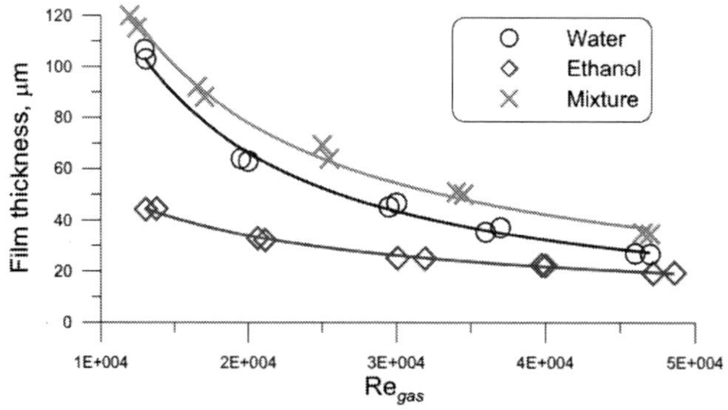

Fig. 3. Dependence of liquid film thickness on Reynolds number of gas flow.

One might expect that the film thickness for water, which has the lowest viscosity among working liquids, should be the lowest among all the working liquids used at identical Reynolds numbers of the film. At the same time for water-ethanol mixture with concentration of 50% the thickness should be the highest. Film thickness of mixture was indeed the highest, however film thickness for water was much higher, than that for ethanol. It is caused by formation of dry spots and the rivulet flow mode of water film. Note presence of droplets detachment from liquid film surface in our experiments that depends on Weber number of co-current gas flow. Since Weber number is in inverse proportion to coefficient of surface tension of liquid, the highest Weber numbers are achieved for film flows of ethanol and water-ethanol mixture. Under conditions of our experiments Weber numbers exceeded 100 that lead to very intensive (tens of percent from the initial flow rate of liquid) detachment of droplets from the film surface along the channel and decrease of film thickness at the channel exit cross-section.

4 Conclusion

A series of experimental studies on liquid film interaction with high-velocity gas flow under expansion from a cylindrical channel into vacuum was carried out. These experiments allowed measuring average thickness and velocity of near-wall liquid film, and establishing special features of film interaction with gas flow (wave formation, detachment of droplets from film surface by gas flow). It was found that under expansion into the vacuum physical properties of liquids and parameters of co-current gas flow determine film behavior inside the channel. It was shown that surface tension coefficient of liquid essentially affect rupture of thin films, formation of dry spots on the channel surface and transition to a rivulet flow mode. Our results expand existing knowledge on the features of liquids ejection from channels into the vacuum in the form of thin films, and could be relevant for verifying two-phase flow models.

The study was supported by the Russian Foundation for Basic Research, research project No. 16-38-00406.

References

1. S. Crist, P. Sherman, D. Glass, AIAA J. **4**, 68 (1966)
2. H. Trinks, I. Kaelsch, AIAA Paper No. 87-1603 (1987)
3. V. Yarygin, Yu. Gerasimov, A. Krylov, V. Prikhodko, I. Yarygin, Microgravity Sci. Technol. **23**, 15 (2011)
4. V.G. Prikhodko, S.F. Chekmarev, V.N. Yarygin, I.V. Yarygin, Doklady Phys. **49**, 119 (2004)
5. V. Prikhodko, I. Yarygin and V. Yarygin, Microgravity Sci. Technol. **21**, 277 (2009)
6. I.I. Gogonin, J. Eng. Phys. Thermoph. **77**, 454 (2004)
7. Yu. Vyazov, V. Prikhodko, I. Yarygin, V. Yarygin, MATEC Web of Conferences **84**, 00040 (2016)
8. A.F. Serov, S.V. Kotov, A.D. Nazarov, A.N. Pavlenko, N.I. Pecherkin, V.Yu. Chehovich, Instrum. Exp. Tech. **40**, 136 (1997)
9. V.G. Prikhodko, G.A. Khramov, V.N. Yarygin, Instrum. Exp. Tech. **39**, 309 (1996)

Heat transfer of bubbly flow on inner wall of annular channel

Pavel Lobanov[1,*], *Aleksandr* Kurdyumov[1], and *Aleksandr* Svetonosov[2]

[1]Kutateladze Institute of Thermophysics, Novosibirsk, 630090, Acad. Lavrentiev ave., 1, Russia
[2]Novosibirsk State University, Novosibirsk, 630090, Pirogova, 2, Russia

Abstract. Experimental investigations of heat transfer from the heated wall to the two-phase bubbly flow were performed in vertical annular channel using air-water system. The IR-thermography and miniature temperature sensors were used to measure heat transfer coefficients. The influence of bubbles on heat transfer is shown in comparison with the case of single phase flow. The presence of bubbles in the flow leads to heat transfer intensification in the annular channel even for low void fractions.

1 Introduction

An important problem of creating the energy-efficient equipment is the study of passive heat transfer intensifiers. One of the simplest examples of such equipment is a pipe-in-pipe heat exchanger in which heat is transferred between the flows in the inner and outer pipes and the medium around the outer pipe.

Often the temperature of the medium in such heat exchangers exceeds the critical one that causes boiling of the coolant. This leads to the appearance of vapor bubbles in the flow. Despite a rather large number of studies, a number of questions are still open even for a single-phase flow. Studies of the local structure of the two-phase gas-liquid flow, and especially heat transfer in such flows, are very limited in the literature [1-6].

Therefore, studying the local structure of bubble gas-liquid flow in annular channels as well as the effects of gas bubble addition on heat transfer is a relevant task.

2 Experimental setup

The experimental setup was a closed circuit in a liquid and an open circuit in the gas phase. The test liquid, i.e. distilled water, was pumped by the centrifugal pump to the test section of the main tank. The temperature of the liquid was controlled at the level of 25 ± 0.2 ° C. The flow rate was monitored using an ultrasonic flowmeter.

The section, simulating the pipe-in-pipe heat exchanger, was an axisymmetric annular channel (Fig. 1 a). To center the inner tube we used the distancing elements installed to avoid flow disturbance in the measurement area. The hydraulic diameter of the annular

* Corresponding author: lobanov@itp.nsc.ru

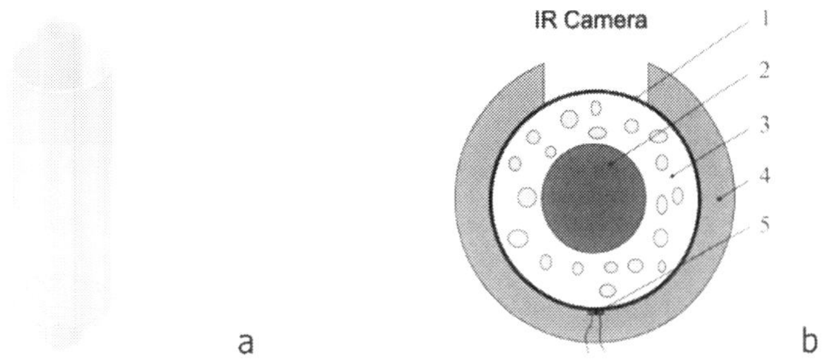

Fig. 1. Schematic of the test section: a – annular channel; b - cross section of heat transfer measurement unit: 1 - thin stainless steel wall; 2 - inner pipe; 3 - two-phase flow; 4 - thermal insulation; 5 - RTDs.

channel (D_h= D - d) was equal to 22 mm. The length of the flow stabilization area before the measurement region was 80 Dh.

The gas flow rate was determined using Bronkhorst (up to 1 l/min) and Aalborg (up to 5 l/min) gas flow controllers. The measurement error of liquid and gas flow rates was ±2 % of the measured value.

Generation of bubbles was carried out by feeding air into the liquid flow through the capillaries of 50 mm length and an internal diameter of 0.7 mm, uniformly distributed in the channel cross-section. Twenty four capillaries were used for the annular channel. Bubbles' size measurements were carried out using video shooting with shadow illumination and subsequent image processing. To reduce optical distortion a box with a square cross-section, filled with immersion liquid, was used.

For experimental studies of heat transfer from the heated wall to the bubble flow, we used the test section, made of thin-walled stainless steel [7] (Fig. 1 b). The inner diameter of the test section was 42 mm, the wall thickness was 0.5 mm, and the heating area length was 500 mm. The section was heated using electric current. Current was supplied using a laboratory power source with power up to 2 kW, used for power control. The source was connected to a step-down transformer, allowing for supply of the alternating current within the range of 300-400 A to the test section at an operating voltage of 2-2.4 V. Current strength was measured using a measuring transformer, and the voltage in the work area was measured with a precision multimer with a measuring error of 0.02 V. Behind the measuring area there was a pipe section 42 mm in diameter and a length of 1 m to avoid the output effects.

The infrared thermography was used as a main measuring method. To increase emissivity in thermographic measurements, the external surface of the working area was painted black. For the measurements we used a thermal imager Fluke Ti32. For comparison, 3 thermal resistors Pt1000 were installed on the opposite side, at distances z = 68, 230 and 405 mm on the surface of the work area. Thermal resistors were pressed against the pipe surface through a thin layer of thermal paste. Around the heat resistors there was insulation to reduce heat loss to the environment (Fig. 1 b). The results obtained by different methods are in good agreement. At the input and output of the work area the control resistance thermometers were set to assess the degree of heating of the working fluid while passing through this section.

In the experiments on heat transfer distribution, voltage was supplied to the work area. After the time needed for the flow stabilization, 20 thermal images were recorded. At the same time, temperatures at the work area input and output, as well as the signals of the temperature sensors, mounted on the surface of the heated wall, were recorded. Given the axial symmetry of the flow, thermal images were processed to vertically average regions, corresponding to the entire height of the work area and the distance from the axis of

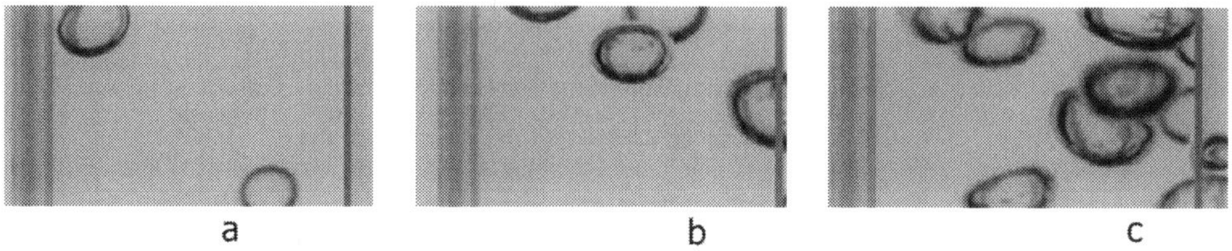

Fig. 2. Photographs of bubbles in annular gap. Re = 7500. Blue lines - outer wall of the annular channel, red lines - inner one: a - β=0.007; b - β - 0.02; c - β = 0.033

symmetry of ±20% from the diameter of the heated tube. For all modes the wall temperature was determined. Using the known heat flux on the wall, temperatures at the input and output of the work area and the field of temperature distribution in the work area, the heat transfer coefficient h and the Nusselt number - Nu were calculated that was the main result of our experimental studies.

In the experiments, the repeated measurements of the temperature of the work area at the wall without heating were performed (over 100 measurements). The temperature of the liquid according to the temperature sensors, installed in the tank, as well as before and after the measurement section, was 25±0.1 ° C, and the ambient temperature was ~27.5 °C. At that, the deviation of the temperature of the channel wall, determined by means of infrared thermography, was within the range t_W = 25±0.15 °C. Thus, we can assume that the error of the wall temperature measured with a thermal imager in our case is ±0.15 °C. The measured temperature difference of the liquid flow and the wall depending on the flow regime and position of the measured points was within the range 4–5 ° C that leads to the uncertainty in determining the difference of wall temperature from ±3% to ±4%.

3 Results of experiments

The measurements were carried out at Reynolds numbers of the liquid phase Re = ($u*D_h/v$) - 5000-10000, the gas content was up to 3.3%. Photographs of bubbly flow in the annular channel for different flow conditions are shown in Fig. 2. The outer and inner walls of the channel are shown by colour lines. Equivalent diameter of bubbles in our experiments was within the range 2-5 mm. Small bubbles were observed to tend migrating to the channel walls, while large bubbles moved in the whole volume of the test section.

Thermograms, obtained using IR camera are shown in Fig. 3. One can see that incensement of gas flow rate leads to decrease of the wall temperature. Further processing of presented thermal images and signals of temperature sensors allows us to obtain heat transfer characteristics.

Relative heat transfer intensification at different gas void fractions and Re is shown in

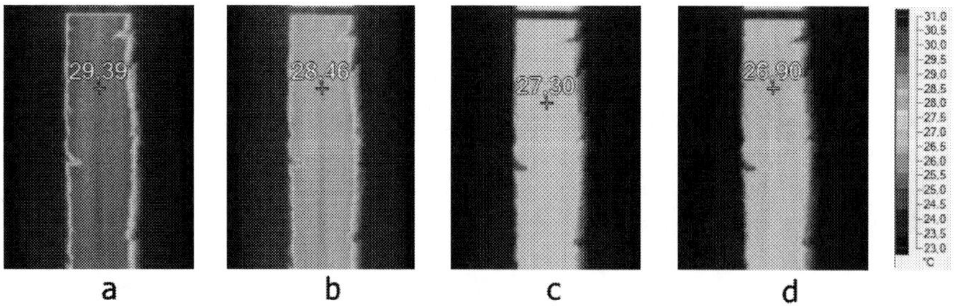

Fig. 3. IR images of the outer wall of the annulus:
a - single phase flow; b - β=0.007; c - β - 0.02; d - β = 0.033

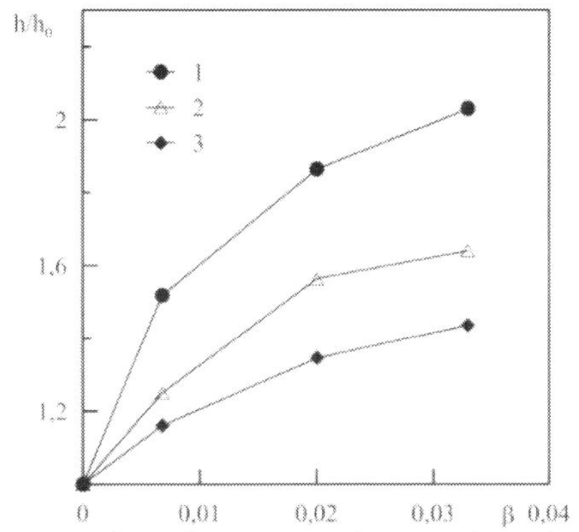

Fig. 4. Heat transfer enhancement in two phase bubbly flow:
1 - Re = 5000; 2 - Re = 7500; 3 - Re = 10000.

Fig. 4. Adding gas bubbles to the liquid flow results in heat transfer enhancement. The increase in gas flow rate leads to an increase in difference for the cases of single-phase and two-phase flows. For Re=5000 and β=3% heat transfer enhances almost twice, while for Re=10000 β=3% the ratio is approximately 1.4.

Similar results were previously obtained in the IT SB RAS, at investigating the wall shear stress and the local flow structure in bubble flows at comparable Reynolds numbers [5]. It was shown that the intensification of heat transfer occurs due to transformation of velocity profiles of two-phase flows that leads to a substantial increase in the fluid velocity gradient in the near-wall region.

Conclusions

The study of heat transfer from the heated wall to single- and two-phase flows has been carried out. It is shown that the addition of gas bubbles to the flow leads to the heat transfer intensification. At that, the significance of this effect decreases with increasing Reynolds number of liquid that was reported earlier by other authors. The effects of heat transfer increase due to appearing bubbles should be taken into account in the calculations when adding a dispersed phase in the flow or in the beginning of boiling.

We can conclude that at small Reynolds numbers in the liquid phase, the addition of the dispersed phase is an effective method of heat transfer intensification. However, with increasing Re the effect of bubbles on heat transfer decreases.

This work was supported by the Russian Foundation for Basic Research (project No. 15-38-21040).

References

1. B. Ozar, J. J. Jeong, A. Dixit, J. E. Juliá, T. Hibiki, M. Ishii, Chem. Eng. Sci. **63,** 3998 (2008)
2. T. Hibiki, R. Situ, Y. Mi, M. Ishii, Int. J. Heat Mass Transfer. **46**, 427 (2003a)
3. T. Hibiki, R. Situ, Y. Mi, M. Ishii, Int. J. Heat Mass Transfer. **46**, 1479 (2003b)
4. R. Situ, T. Hibiki, X. Sun, Y. Mi, M. Ishii, Int. J. Heat Mass Transfer. **47**, 5351 (2004)
5. O.N. Kashinskii, A.S. Kurdyumov P.D. Lobanov, Thermophys Aeromech. **18** 37 (2011)

6. P.D. Lobanov, M.V. Timoshevskiy, J. Phys.: Conf. Ser. **754**, 032013 (2016)
7. O.N. Kashinsky, V.V. Randin, A.V. Chinak, Thermophys. Aeromech. **20**, 391 (2013)

The distribution of the gas phase in the bubble flow in the vertical assembly of 3x3 rods with gas injection from a single source

Maksim Vorobyev, *Oleg* Kashinsky, and *Pavel* Lobanov*

Kutateladze Institute of Thermophysics, Novosibirsk, 630090, Acad. Lavrentiev ave., 1, Russia

Abstract. The work is devoted to an experimental study of the distribution of gas along the channel cross-section in the form of an assembly of vertical rods 3x3 in a square arrangement. The gas was introduced into the channel from the corner rod through a single capillary. Upstream, a local gas content sensor was attached to the the central fuel element simulator. The position of the sensor was changed by rotating the rod around its axis. The experiment was carried out at Reynolds numbers Re = 4500 - 9000 and flow rate gas contents of 2 - 4%.

1 Introduction

In fuel assemblies of nuclear power plants, there is a possibility of boiling of the coolant on the surface of fuel elements. As a result of this process, a two-phase vapor-liquid flow occurs. In the literature there is a large number of works devoted to the investigation of gas and vapor-liquid flows in pipes. In particular, it is shown that with dispersed phase occurring the local structure of the flows is reconstructed, also the key thermohydraulic parameters, such as hydraulic resistance and heat transfer coefficients, can change significantly [1 - 6]. Due to the above, an actual task is to study the regularities of distribution of the disperse phase during the movement of two-phase flows in assemblies of rods and its effect on the thermal-hydraulic flow characteristics. The task of this work was to study the distribution of gas at a certain distance from a single source of the gas phase in the assembly of vertical rods 3X3 in a square arrangement.

2 Experimental setup and experimental technique

The experimental setup (Fig. 1) was a closed pipe filled with a fluid. Using the centrifugal pump (1), the liquid (distilled water with the addition of sodium carbonate) from the tank (2) was fed into a working channel of square section (3), which was an assembly of nine vertically oriented rods 9.1 mm in diameter. The geometry of the assembly model was fixed with using of spacing grid simulating the flow in the fuel assemblies of the reactor plants. A stainless steel capillary (4) was soldered into the one of the corner rods, through which gas (atmospheric air) using was supplied into the fluid flow. Gas flow rate was regulated by

* Corresponding author: lobanov@itp.nsc.ru

Bronkhorst controller. The capillary was bent and directed along the fluid flow. The rod supplying gas was oriented in such a manner that the point of introduction of the gas into the flow was turned towards the centre of the channel. To the central simulator of fuel element was attached a sensor of local gas content (5), which is a platinum wire welded into a glass capillary. This detector was used in the mode of the conductivity sensor. An alternating voltage of 100 kHz was applied to it. Its current increased and straightened. The low-frequency envelope of this signal had two levels corresponding to the detection of the sensor element in the liquid and gas phases. The signal of this sensor was used to determine the local gas content. The position of the sensor was changed by rotating the rod around its axis, which made it possible to obtain a diagram of the distribution of the local gas content along the circumference, the pitch was 15 °. The flow rate was measured using a calibrated flow diaphragm (6). The pressure drop across the diaphragm was measured using a differential pressure gauge Metran-100 DD. The gas-liquid mixture from the outlet of the channel through a flexible tube was fed into the separation tank (7), where gas and liquid were separated, and then the liquid was merged into the main tank through a return pipe.

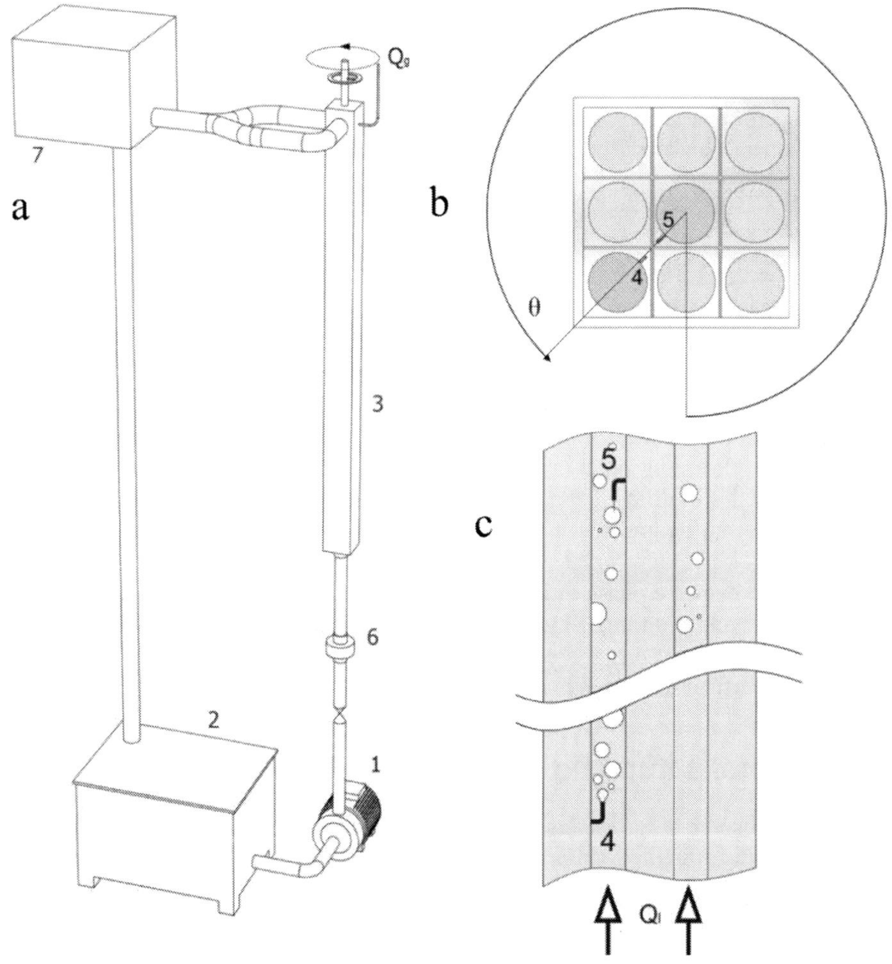

Fig. 1. (a) The scheme of the experimental setup; (b) Working channel top view; (c) Working channel side view.

3 Experimental results

The experiment was carried out at Reynolds numbers Re = 4500 - 9000 and gas flow rates from 500-1000 ml / min, which corresponded to gas content from 2% to 4%. The hydraulic diameter, included in the Reynolds number, was defined as the quadruple ratio of the channel area to its perimeter.

Distributions of local gas content around the central fuel element were obtained at different Reynolds numbers and gas consumption rates. The measuring section was located above the spacing grid. In Fig. 2a the dependences for Re = 4500 - 9000 and gas flow rate of 1000ml/min are presented. In Fig. 2 b the results are presented for a constant liquid flow rate (Re = 4500), but at different gas flow rates.

One can see that there is a significant uneven distribution of the local gas content. The presence of the spacing grid leads to an increased diffusion of the gas phase along the channel cross-section.

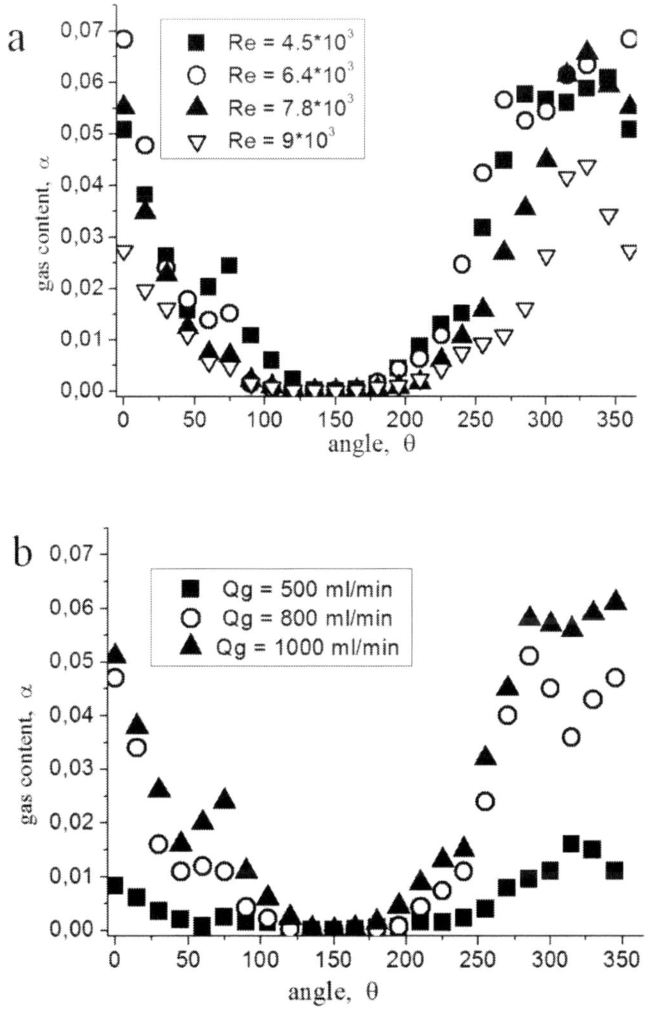

Fig. 2. Dependence of the local gas content on the angle of rotation of the sensor; (a) different Reynolds numbers, Q_g = 1000 ml/min; (b) different gas flow rates, Re = 4500.

References

1. V.E. Nakoryakov, O.N. Kashinsky, A.P. Burdukov, V.P. Odnoral. Int. J. Multiphase Flow **7**, 63 (1981)
2. S.K. Wang, S.J. Lee, O.S. Jones Jr., R.T. Lahey Jr., Int. J. Multiphase Flow **13**, 327 (1987)
3. M.A. Pakhomov, V.I. Terekhov, Int. J. Heat Mass Transfer **101**, 1251 (2016)
4. D. Métrailler, S. Reboux, D. Lakehal, Nuclear Eng. Design (2017) (to be published)
5. J.P. Schlegel, S. Sharma, R.M. Cuenca, T. Hibiki, M. Ishii, Int. J. Heat Fluid Flow **47**, 42 (2014)
6. S. Dabiri, G. Tryggvason, Chem. Eng. Sci. **122**, 106 (2015)

Generation and development of surface waves on the interface boundary of viscous fluid oscillating drop

Leonid Tonkov and *Alena* Chernova*

Institute of Mechanics UB RAS, 426067, 34 T. Baramzinoy Str., Izhevsk, Russia

Abstract. Numerical experimental methods for surface waves' generation and development processes associated with oscillations of small liquid droplet on a solid surface vibrating at a frequency close to 1 kHz are used. Internal flows formed inside the droplet are characterized by vivid three-dimensionality, complex topology, are determined by the contact angle and surface waves propagation. Interaction processes are analysed. The achieved results are compared with the experimental ones.

Introduction

When small liquid droplet oscillates in heave we can observe surface ripple waves excitement. Its dynamics depends on viscosity, surface tension and liquid density as well as external action characteristics. Let's consider preliminary stage of the phenomenon with small amplitudes that endure liquid heave oscillations on its surface to initiate capillary – gravity waves. The present thesis deals with a numerical research on internal flows in a small amount of liquid (a droplet) placed on a low vibrating solid surface and its influence on the generation and development of surface waves.

We can define some articles [1-3] in experimental hydrodynamics that study oscillations of liquid with ultra-small volume (5 μl) placed on a solid substrate vibrating at low frequencies (up to 800 Hz) and with low amplitudes (up to 22 μm). Data on droplet free surface depending on the vibrating surface location are provided, and droplet shapes and internal capillary flows corresponding to various surface vibration modes are also considered.

Current analytical works are dedicated to the droplet free surface profile location matters [4, 5] and the dynamic process is described only in works [6, 7]. Thus only work [7] studies problems concerning surface waves distribution on a free surface of an ideal liquid droplet under axisymmetric statement.

Current numerical works studying the droplet dynamics on vibrating surfaces are very few [8, 10]. The work [10] numerically studies the oscillation and spraying process of a single liquid droplet from the solid rod. In this study the physical problem statement has several rather strong assumptions: the initial droplet configuration was defined as half sphere, in addition to this the possible movement of a contact line was not considered; in

* Corresponding author: alicaaa@gmail.com

hydrodynamic equations, cycling acceleration of a substrate is added to the gravity acceleration. Also the work [10] describes only solutions for the two dimensional problem.

Experimental research proves that internal flows in a droplet are characterized as vividly 3 dimensional. The volume of fluid (VOF) method was applied in this work.

1 Problem statement

We consider the problem of liquid droplet motion caused by vertical movements of a solid substrate. Let the area $\Omega \in R^3$ be filled with biphasic medium and $\Omega = \Omega_1 \Omega_2$ with subarea Ω_2 containing a gas, and Ω_1 containing a liquid; Γ_0 is the phase boundary, and Γ_1 is the atmospheric boundary of computational region. At the initial instant $t = 0$ the substrate starts to oscillate in heave at a given frequency (f=635 Hz) and amplitude (a=3.5 μm) corresponding to the 8th mode [2] of shape oscillation. A liquid drop placed on the substrate also starts to oscillate in heave.

The experimental apparatus and procedure is described in article [2]. Water droplet of 5μl volume was placed in the substrate center (fig. 1), thus the wetting angle was equal to $\Theta = 115 \pm 1\,°$, and the contact diameter and droplet height, 2.02 mm and 1.52 mm, respectively.

2 Methods and algorithms

The VOF method [11] describes the droplet dynamics as a system of two none-mixable uncompressible viscous fluids; the motion of each is described by Navier-Stokes equations, a continuity equation and additional nonstationary volume fraction transport equation [12]. The study [14] shows that free surface dynamics requires accounting for the wetting angle dynamic change in a substrate/liquid/gas triple point for adequate solution. The contact angle description will be performed according to the model

$$\Theta = \Theta_0 + (\Theta_A - \Theta_R)\,\mathrm{th}\left(\frac{u_W}{u_\Theta}\right),$$

thereby in initial conservation equation system, written according to the VOF method, only yhr motion equation will be changed

$$\frac{\partial \rho U}{\partial t} + \nabla \cdot (\rho U U) = -\nabla p + \nabla \cdot \tau - \sigma K \nabla \alpha \delta(\alpha)(1 - \cos\Theta) + \rho g,$$

where K – surface curvature Γ_0, σ – surface tension coefficient, Θ – wetting angle, α – scalar-valued function [12], aiming as liquid volume fraction.

Boundary conditions are defined in a following way: - on a border Γ_1: $\partial u/\partial n = 0$, $p = p_0$, where n – outward normal; - on liquid/substrate border $\partial p/\partial n = 0$, $u_x = u_z = 0$, $u_y = u_{\Omega 0}$ where $u_{\Omega 0}$ – velocity of substrate motion.

Studies [13, 14] describe the detailed process of this equation system solution. The cartesian grid consists of 364,140 hexagons and has axial and radial refinement. Studis [13, 14] study the liquid droplet oscillation process on a substrate vibrating at low frequencies corresponding to low (second and fourth) oscillation modes. Let's consider that solution of the liquid oscillation at frequency corresponding to high (6 and 8) oscillation modes require correction of Θ_R and Θ_A, in respect to similar constants determined for low modes of liquid oscillations.

3 Numerical results

This study investigates problem on oscillation at frequency 635 Hz of a water drop 5 μl in volume on a solid substrate, that is experimentally described [6]. Mathematical simulation of liquid motion under oscillations was performed using volume-of-fluid method in a cell in two stages: first – the solution of liquid droplet stationary problem [14]; second – the liquid droplet dynamics simulation with vibrating substrate acting on it.

Calculated droplet vertical cross-section profiles (fig. 1, a) fit the experimental ones (fig.1, b) – 4 circle brake lines of free surface were detected [2] that proves the mathematical model and results of computational procedure.

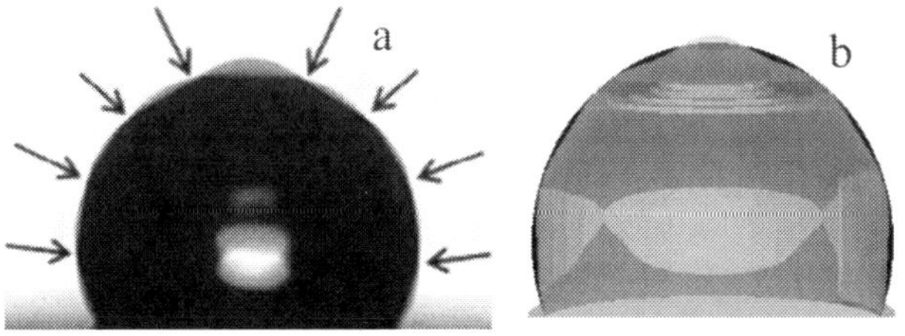

Fig. 1. Form of a droplet obtained in the result of: a) experiment [2]; b) present calculation

Fig. 2 describes the velocity field and free surface deformation. Fig. 2 shows that interaction of the diverse flows leads to complex topological as well as whirl structures generation in a droplet. From the point of free surface the described process is followed by generation and development (motion) of surface waves that are localized by topological peculiarities of liquid flow. We need to point out that direction of velocity vectors in a liquid droplet is significantly determined by the substrate motion [14].

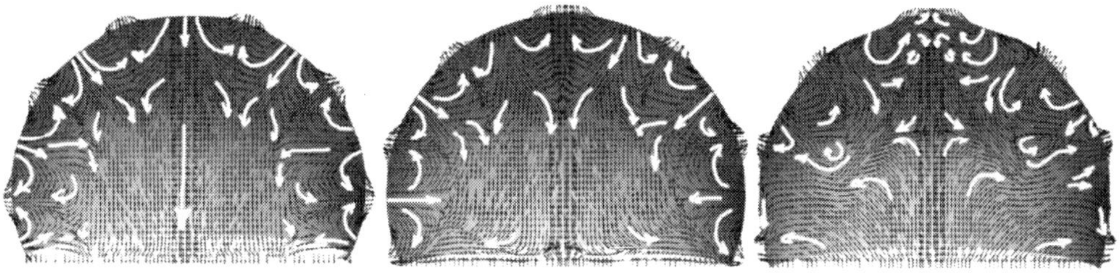

7.25*t* 7.75*t* 8*t*

Fig. 2. Internal droplet flows and free surface deformation, $t = \mathrm{t} / T$

Thus, we can suggest that surface waves' generation and development is connected with local topological peculiarities of internal flows, their transformation and mutual interaction.

This work was partially supported by RFBR according to the research project №16-38-00127-mol_a and 16-41-180276 r_a.

References

1. Y.S. Shin, H.Ch. Lim, Eur. Phys. J.E. **37**, 1 (2014)
2. H. Kim, H.Ch. Lim, J. Phys. Chem. B. **119**, 6740 (2015)

3. Ch.S. Park, H. Kim, H.Ch. Lim, Exp. Thermal Fluid Sci. **78**, 112 (2016)
4. V.Z. Kanchukaev, Pisma v ZhPh. 30(2), 12 (2004)
5. D.V. Lyubimov, T.P. Lyubimova, S.V. Shklyaev, Fluid Dynamics **39**, 851 (2004)
6. A.O. Ivancov, Bulletin of Perm University. Series: Physics. **3,** 16 (2012)
7. D.V. Lyubimov, T.P. Lyubimova, S.V. Shklyaev, Phys. Fluids **18**, 012101 (2006)
8. Y. Li, A. Umemura, Int. J. Multiphase Flow **60**, 64 (2014)
9. G.B. Foote, J. Comp. Phys. **11,** 507 (1973)
10. A.J. James, M. Smith, A. Glezer J. Fluid Mech. **476**, 29 (2003)
11. V.A. Kalinichenko, S.V. Nesterov, A.N. So, Fluid Dynamics **50**, 535 (2015)
12. C.W. Hirt, B.D. Nichols, J. Comp. Phys. **39**, 201 (1981)
13. A.A. Chernova, S.P. Kopysov, L.E. Tonkov, IOP Conf. Ser.: Mater. Sci. Eng. **158**, 012026 (2016)
14. I.M. Kuzmin, A.S. Sarmakeeva, A.A. Chernova, Chemical Physics and Mesoscopy. **18**, 515 (2016)

Experimental investigation of rotation resistance moment energy spectra in multicylindrical circular Couette system with independently rotating cylinders

Anatoly Serov[1,2*], *Aleksandr* Nazarov[1,3], and *Valery* Mamonov[1]

[1]Kutateladze Institute of Thermophysics, Siberian Branch of the RAS, Novosibirsk, Russia
[2]Novosibirsk state technical University, Novosibirsk, Russia
[3]Novosibirsk state University, Novosibirsk, Russia

Abstract. The torque of the rotational resistance in the Ku-Etta multi-cylinder system rotating in the direction towards each other is measured. The experiments were carried out for three values of the kinematic viscosity of the working fluid that fills the multicylinder system: water at a temperature of 24 °C (viscosity 0.9 cSt), an aqueous solution of glycerol at 20 ° C and 41 °C (2.5 cSt and 5.2 cSt). An attempt is made to investigate the features of a viscous flow in the multicolor Couette flow system from the analysis of the energy spectra of the moment of resistance to rotation of cylinders.

Introduction

Searching for the most cost-effective heat resources for liquid heating led us to the usage of dissipative effects, characterizing an ability of a liquid to interact with solid surfaces. Use of mechanical power of wind rotors with vertical axis for driving liquid heat generators takes special place. In particular, it is of interest to consider a device, where the heat is released in a liquid volume placed in the narrow annular gap between the coaxial rotating towards each other cylinders, as a such heat generator [1, 2].

A large number of experimental and theoretical researches has been published concerning the flow between concentric rotating cylinders (circular Couette flow) since the first investigations of Mallock (1888, 1896) and Couette (1890) were conducted. The authors [3] demonstrated a detailed map of current regimes in circular gap between rotating towards each other cylinders as Reynolds numbers' functions for inner and outer cylinders (Re_i and Re_0 respectively) based on proper research. Optical method of polymer flakes distribution in water suspension registration applied by authors [3] made it possible to describe actions and energy spectra of complex structural formations appearing in the circular gap in a wide range of Reynolds numbers' change.

* Corresponding author: serov@itp.nsc.ru

Results

The article which continues to [2] deals with an attempt to study the features of a viscous liquid flow in a multi-ring Couette system with rotating towards each other coaxial cylinders analyzing rotation resistance moment energy spectra for these cylinders. An experimental setup sceme and a photograph of a main part – heat generator with milti-cilyndrical rotor system – are shown on the figure 1. Role of the wind rotors in the sceme is performed with low-speed electric motors with variable rotating speed. The setup allows to study the effect of used liquid parameters and «rotor» rotation rate on their resistance to counter-rotation depending on used liquid flow structure in multi-cylindrical «rotor» system gaps. The measurement of total «rotor» rotation resistance moment was conducted as follows: lower drive stator was slowed down, upper drive stator was suspended on a freely rotating suspension axis. Total opposite «rotor» rotation resistance moment was transferred through lever (L = 195 mm) to dynamometer tesometric sensor 4 (figure 1). Dynamometer, thermometer and two-channel tachometer data were transferred through microprocessor data acquisition and pre-processing unit to computer for final processing and storage.

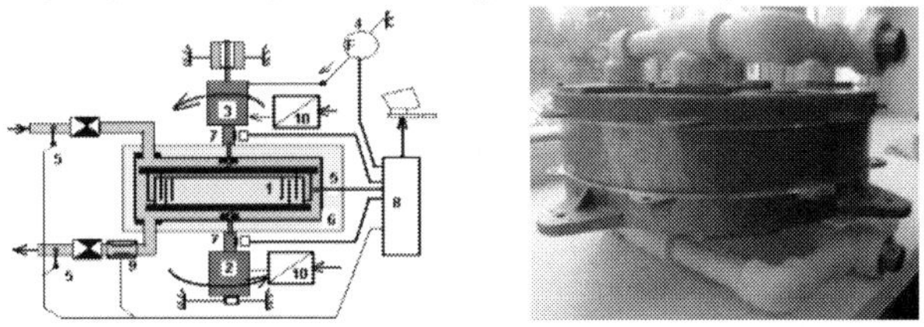

a. b.

Fig. 1. Scheme of the experimental setup (a) and photograph (b) of the block of the heat source. 1 - "the rotors"; 2, 3 - drive; 4 - digital dynamometer; 5 - temperature sensor; 6 - insulation; 7 - tachometer; 8 - microprocessor unit for data processing; 9 - flowmeter; 10 - control unit rotation speed of the drive.

Based on measurement of total «rotor» rotation resistance moment the capacity released from heat generator was determined with the formulae: $N = M\Omega$, where N [W] – power, M [$N{\cdot}m$] – total moment measured, Ω [1/s] – rotational rate equal to arithmetical mean value of rotational speeds of upper and lower «rotors» towards to fixed coordinate system. Heat power released with heat generator was also determined from these calculations. A comparison of heat generator released power measurement results resulting from measurement of «rotor» rotation resistance moment and from heat measurements is shown on the figure 2. Here $Re = \Omega R\delta/\nu$ – Reynolds' number for multi-cylindrical system, R – mean system cylinder radius, δ – mean gap between system cylinders, ν – used liquid kinematic viscosity. Power results measured with two independent methods coincide with the high degree of accuracy, which confirms the reliability of the method used for heat generator released power measurement.

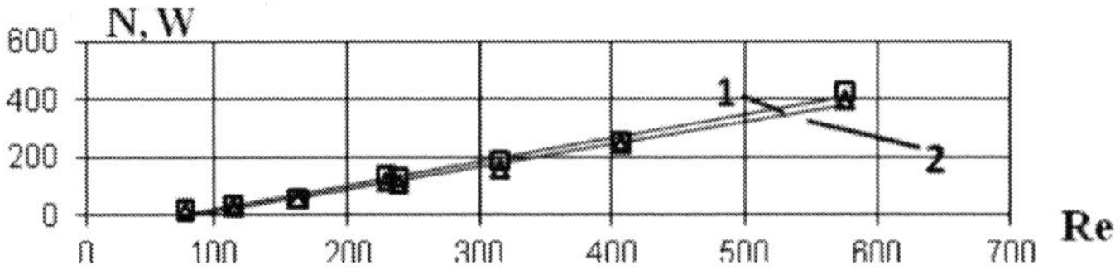

1 - heat power, 2- mechanical power

Fig. 2. Comparison of heat and mechanical power emitted by the heat source.

The article deals with spectra investigation in the range of small Taylor numbers ($Ta = Re\sqrt{\delta / R}$) Ta = (1 - 400), where the impact of appearing wave vortical flow on rotation resistance is observed, which greatly determines heat processes power in gaps.

There is an area f_2 = (0 – 75) Hz of the main energy with intensity (~ 50dB) and two frequency bands $f_{3,4}$=(160 – 240; 235 – 360) Hz with energy (~ 15dB). It should be noted, that spectral analysis was conducted using ten average spectra for each rotation speed. For more detailed studying these spectra were divided to 5 frequency ranges (figure 3).

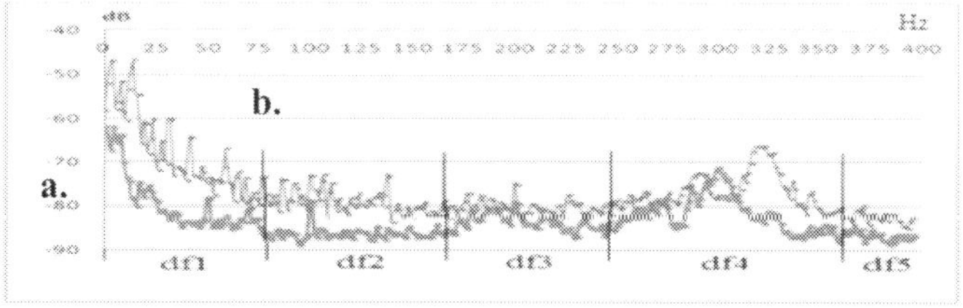

Fig. 3. Mean energy spectra at opposite «rotor» rotation regime for two rotation rates. a - Ω = 15.5, 1/s; b - Ω = 28.1, 1/s.

Data concerning power input for individual harmonics of the whirl structures to total energy for marked frequency areas are presented in table 1 based on spectral analysis (figure 3). The data show that the main input comes from low-frequency range pulsation.

Table 1. Data power individual harmonics.

Frequency band, Hz		%, harmonics, Ω = 28.1, 1/s	%, harmonics, Ω = 15.5, 1/s
df1	0.3 - 75	95.32	68.75
df2	75 - 165	1.01	3.77
df3	165 - 240	0.52	5.84
df4	240 - 360	3.03	20.49
df5	360 - 395	0.14	1.28

Subrange *df1* (0 – 75 Hz) is different due to the fact that it gives the greatest input to the released heat powery. These data coincide well with the data published in [3] concerning low-frequency pulsation wave spiral appearance (INT) for Reynolds numbers from 20 up to 400. On the figure 4 there are data describing energy input distribution in low-frequency (0 – 75 Hz) and in high-frequency spectral areas (165 – 355 Hz). The maximum peak capacity values for rotation rate Ω = 28.1 1/s in low-frequency area (figure 4, peaks 1 and 3) are 80 times higher than values in high-frequency area (figure 4, peak 9). The peak power values for rotation rate Ω = 15.5 1/s are distributed in the other way: low-frequency values (figure 4, peaks 5, 6) are only 10 times higher than similar values in high-frequency area (figure 4, peaks 7, 8).

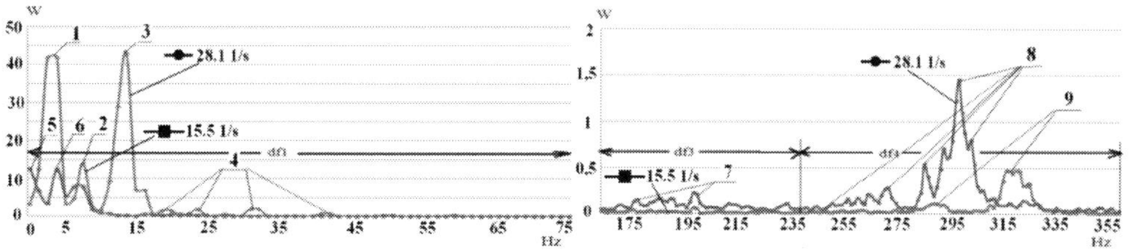

Fig. 4. Frequency power spectrum of the harmonics in the frequency band F = (0.3 - 395) Hz. Stripe *df1* (2.7 - 75) Hz, *df3* (165 - 240) Hz, *df4* (250 - 360) Hz.

To obtain more detailed data about whirl formation physical parametres changes in dependance to «rotor» rotation frequency and used liquid viscosity the study of these dependancies must be done while scanning in increments of $d\Omega = 1$ 1/s in the rotation rate area studied $\Omega = (5 - 30)$ 1/s.

In table 2 there are data presented concerning azimuthal wave numbers for whirl formations ($g = f_{harmonics}/f_{rotation}$) at rotation rate Ω = (15.5 and 28.1) 1/s. The data obtained show that spiral and whirl azimuthal size is usually smaller than rotor gap periphery of a circle and slightly depends on rotation rate.

Table 2. Azimuthal wave numbers.

	frequency harmonic (Hz) at Ω =28.1, 1/s	**frequency harmonic (Hz) at Ω =15.5, 1/s**	**g for Ω = 28.1, 1/s**	**g for Ω = 15.5, 1/s**
f1	1.5	0.3	0.38	0.15
f2	2.7	1.5	0.68	0.75
f3	4	4	1	2
f4	7.6	6.4	1.9	3.2
f5	11.3	7.6	2.83	3.8
f6	12.5		3.13	
f7	13.7		3.43	
f8	15		3.75	

Conclusion

- low-frequency spectra (0 – 75 Hz) slightly changes in width with rotation rate increase and attributes a lot to resistance moment;
- In the area 4 of the spectra (165 Hz – 355 Hz) there are some maximum drifts with rotation rate change, resistance moment inputs in dependance on rotation rate differ insignificantly.

The results of the experiments conducted may serve as a base for the next theoretical and experimental research while creating effective systems of alternative supplies of thermal energy generation.

References

1. A.F. Serov, V.N. Mamonov, V.I. Terekhov, A.D. Nazarov, Patent for invention "Opposite wind-head generator", № 2612237, 03.03.2017 (2017)
2. V.N. Mamonov, A.D. Nazarov, A.F. Serov, Thermophys. Aeromech. **23,** 139 (2016)
3. C.D. Andereck, S.S. Liu, H.L. Swinney, J. Fluid Mech. **164,** 155 (1986)

The study of the influence of stabilizing devices on the pressure pulsations at the free discharge of water through the turbine

Dmitry Platonov[1,2*], *Andrey* Minakov[1,2], and *Dmitry* Dekterev[1,2]

[1]Siberian Federal University, 79 Svobodny Ave., 660041 Krasnoyarsk, Russia
[2]Institute of Thermophysics, SB RAS, 1 Lavrentyev Ave., 630090, Novosibirsk, Russia

Abstract. The free discharge through the turbine is applied in the course of construction of hydro power plant or in case of excessive water inflow during floods or emergency situations. The experimental and numerical investigation of flow-induced pressure pulsations in hydraulic turbine draft tube at free discharge was performed using a special setup for suppression of draft tube surge. Such special equipment in draft tube cone as fins and "spider"-like units was considered for reducing these pressure pulsations. The paper considers influence of these units on vortex structures and pressure pulsations in the draft tube in several regimes.

1 Introduction

Low-frequency flow-induced pulsation occurs in Francis and Kaplan hydraulic turbines at off-design operation. Since early 20th century, oscillations of power of hydraulic turbines have been observed. In 1940, Rheingans [1] showed that the oscillations are associated with draft tube surges. In 1960, Deriaz [2] showed that the draft tube surges are caused by vortex rope precession. Generally, the pulsation is associated with the unsteady swirling flow in the draft tube of the hydraulic turbine. There are several special countermeasures for reducing these pulsations. Air admission is the most widely used method [3]. However, this method is too expensive in case of static pressure at the aeration point exceeding atmospheric pressure. Another way is special equipment for stable operation. The special equipment influences the flow to reduce or destroy the vortex rope. There is such equipment as fins on the draft tube walls, a spider-like unit under the runner, a hollow cylinder under the runner hub, etc. The fins, placed on the draft tube wall along the axis of the runner, are often used to reduce swirl of the flow and, consequently, to suppress the pulsations [4]. The width of the fins is usually of about 0.1 – 0.2 diameter of the draft tube cross-section. Between 2 and 8 fins are usually used [5]. The draft tube fins are almost useless in case of high-load operating points. In addition, disadvantages of the fins are loss in the efficiency of the turbine and necessity to guard the fins from cavitation erosion and structural damage [5].

* Corresponding author: platonov-08@yandex.ru

2 Experimental and numerical techniques

An experimental study was performed on an open aerodynamic stand. The configuration considered is a 61:1 scaled-down laboratory model of a Francis turbine (RO-230) of an LMZ (Leningrad Metal Industry). Air supply in the draft tube model was carried out using a high performance air blower with the maximum flow rate Q_{max} =170 m^3/h and the excess pressure of Δp=0.4 bar. An ultrasonic flow meter «IRVIS-Ultra» and a frequency converter of the blower with feedback controlled air flow rate were used. Figure 1 provides a photo of the experimental turbine model. The draft tube model, runner, spiral case and guide vanes geometry were provided by the company LMZ and were manufactured using the technology of 3D printing.

Servo drive SPSh10-3410 ensured precise rotation frequency of the runner in the range from 0 to 3000 rpm. A computer ensured the control over the experiment. Using the original software, it was possible to maintain the flow regime with an accuracy of 1.5% and 0.5% of the flow rate and the runner rotation frequency, respectively.

The experiments included measurements of pressure pulsations with special acoustic sensors. The acoustic sensors were based on high precision microphones, attached with tiny tips for capturing pressure pulsations at the local points [6]. Two of these sensors were used to conduct point-to-point measurements. The perturbation mode that corresponds to the precessing vortex core was identified by analysing the signal from the sensors located in diametrically opposite angular positions. Note that the possibility of using well-proven acoustic technology is another advantage of the air model stand.

Pressure fluctuations were recorded using a microphone Bruel & Kjaer Type 2250. The signals from the microphones were digitized by an analog-to-digital converter L-card E-440. The experiments were conducted at the runner rotation frequency from f = 0 to 2000 rpm with a step of 500 rpm and a flow rate, varying in the range from Q =100 to 150 m^3/h with a step of 50 m^3/h. At each point, the signal detected by the acoustic sensors was digitized for 5 seconds with a sampling rate of 2 kHz.

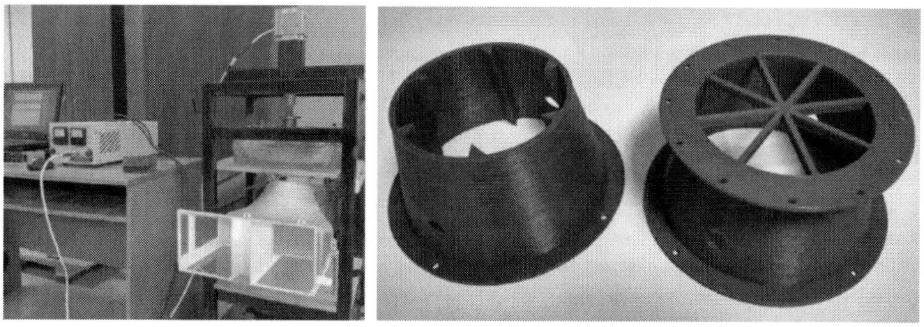

Fig. 1. The photo of the experimental stand and flow stabilization devices.

Numerically describing the flows in the flow path of water turbines one tackles several problems. The first challenge is related to modeling of turbulence in channels with complex geometry and strong swirl flow. In this paper, the used numerical technique for calculating the low-frequency pressure pulsations in a hydraulic turbine is based on the DES turbulence model (based on Menter k–ω SST model). Also in the simulation of hydraulic turbines it is necessary to consider the rotation of the runner. There are many approaches for modeling flows with rotating bodies, they include: dynamic grid, sliding mesh, moving mesh and the method based on a moving reference frame. In this work the modeling of the runner rotation was performed with rotated reference frame for the runner zone. Numerous test calculations show the correctness of this approach, both in the description of integral characteristics of flow and pressure pulsation [7-9].

3 The results

In this paper, the results of a study of nonstationary processes at free discharge of water through a turbine are presented. The free discharge through the turbine is applied in the course of construction of hydro power plant or in case of excessive water inflow during floods or emergency. The variant of free discharge with the dismounted runner of the turbine is considered. In this case, the runner is taken out of the turbine. The hydrodynamic resistance drops, and the flow of water increases substantially. However, in this case, pressure pulsations in the turbine chamber increase significantly. The reason for this are precessing vortices, formed in the spiral chamber. These vortices are clearly visible in Figure 2, which shows the results of numerical simulation. These results are given for the flow regime in which the maximum level of pressure pulsations is observed. The opening guide vanes is A_0=18mm and the flowrate Q=130 m^3/h.

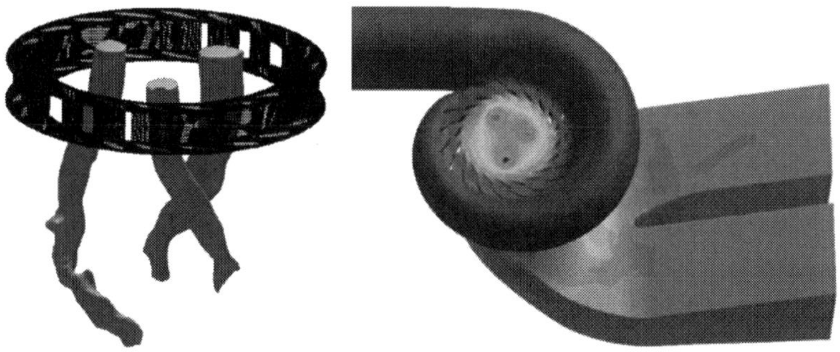

Fig. 2. Vortex structure in the turbine chamber and static pressure at the turbine wall. (For regime: opening guide vanes A_0=18mm and flowrate Q=130 m^3/h).

Pressure pulsations for this mode are shown in Fig. 3. Comparison of calculation and experiment is shown. A good agreement between calculation and experiment on the amplitude and frequencies of pressure pulsations has been obtained.

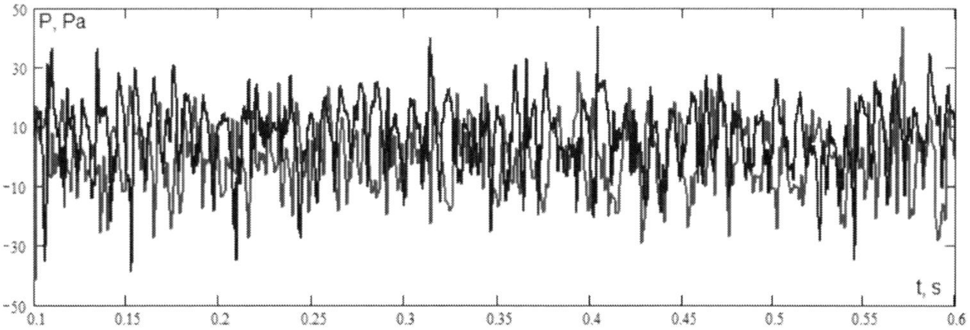

Fig. 3. Pressure pulsations in the diffuser draft tube (red color - experiment data and blue - calculation result).

To reduce pressure pulsations the stabilizing devices are examined (the fins and the cross are shown in Figure 1). These devises were installed in the turbine diffuser, and pressure pulsations were measured. The result of these measurements is shown in Fig. 4. Analysis of the conducted experiments has shown that the use of stabilizing fins may reduce pressure pulsations by 30-40%. The use of a stabilizing cross allows reducing the pressure pulsations by 50-70% for regimes with maximum pulsations.

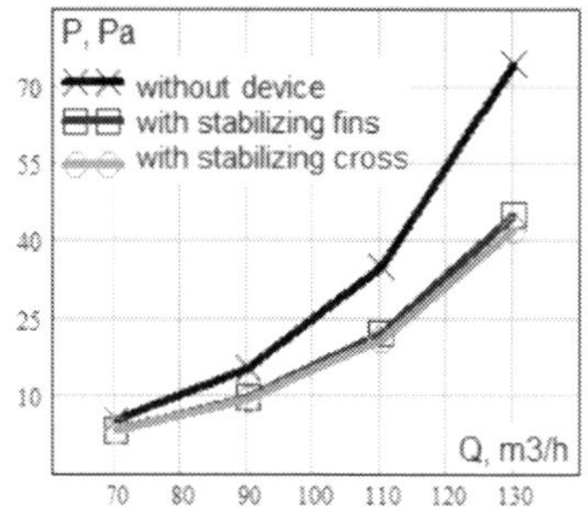

Fig. 4. Maximum amplitude of pressure pulsations in the turbine chamber.

4 Conclusions

The effect of stabilizing structures on pressure pulsations at free water discharge through the turbine is investigated. It is shown, that at free discharge the level of pressure pulsation is significantly higher than in the nominal regime. It has been shown experimentally, that the use of stabilizing devices allows reducing pressure pulsations by almost two times in the regime of maximum pulsations.

The current work is performed with partial support of the projects funded by the Russian Foundation for Basic Research and Krasnoyarsk Regional Fund for Support of Scientific and Scientific-Technical Activities (16-48-240707).

References

1. W.J. Rheingans, Power swings in hydroelectric power plants, Trans. ASME **62**, (1940)
2. P. A. Deriaz, Hydraulic Machinery and Equipment Symposium. (1960)
3. P. Dorfler, M. Sick, A. Coutu, *Flow-induced pulsation and vibration in hydroelectric machinery* (London, Springer, 2013)
4. V. Biela. IAHR 19th Symposium Hydraulic Machinery and Cavitation (1998)
5. H.T. Falvey, U.S. Bureau of Reclamation (1971)
6. I.V. Litvinov, S.I. Shtork, P.A. Kuibin, S.V. Alekseenko, K. Hanjalic, Int. J. Heat Fluid Flow **42**, 251 (2013)
7. A.V. Minakov, A.V. Sentyabov, D.V. Platonov, A.A. Dekterev, A.V. Zakharov, Comp. & Fluids. **111**, 197 (2015)
8. A.V. Minakov, A.V. Sentyabov, D.V. Platonov, A.A. Dekterev, A.V. Zakharov, Int. J. Heat Fluid Flow **53**, 183 (2015)
9. D. Platonov, A. Minakov, D. Dekterev, A. Sentyabov, A. Dekterev, J. Phys. Conf. Series **754**, 062007(2016)

Mechanochemical modification of the composition and structure of plant raw materials to control the combustion of alternative fuel

Aleksey Bychkov[1*], *Ekaterina* Podgorbunskikh[1], *Tatiana* Skripkina[1], *Anatoly* Burdukov[2], and *Oleg* Lomovsky[1]

[1]Institute of Solid State Chemistry and Mechanochemistry, SB RAS, 630128 Novosibirsk, 18 Kutateladze Str., Russia
[2]Kutateladze Institute of Thermophysics, SB RAS, 630090 Novosibirsk, 1 Lavrentyev ave., Russia

Abstract. The possibilities of mechanochemistry in processing of renewable lignocellulose raw material into solid kinds of biofuel are demonstrated in this work. A review of lignocellulose raw materials promising for our country is presented. These raw materials include wastes from agriculture and forestry, and the biomass of rapidly growing plants. The physicochemical properties of lignocellulose materials with different delignification degrees were modeled with the help of the artificial mixtures of plant raw material with purified cellulose and lignin. The data illustrating the effect of disperse state and lignin content on the reactivity of the material in subsequent combustion are presented. The tests at the combustion bench with the thermal power up to 5 MW allowed determining the optimal combustion parameters for the obtained biofuel in the autothermal mode.

1 Introduction

Much attention in the world is paid to the shift from mineral to renewable energy sources. For example, Russia has set the ambitious goal of achieving 4.5 % of renewables in 2020, and by 2030 it is planned to provide the renewable energy at least at 80-100 billion kilowatt hours per year [1]. Significant progress has been made in the industrial production of wood-based solid biofuels and in the consumption of new kinds of biomass as raw materials [2-3].

The sources of biomasses suitable for biorefinary and biofuel production are under investigation in many developing regions of the world [4]. These sources include the biomass of plants growing in a specific region, and agricultural wastes [5]. Studies are aimed at optimization of the combustion of artificial mixtures, including lignin from black liquor and various kinds of coal [6].

The urgency of the development of technologies for the treatment of lignified wastes is defined by the fact that each kilogram of produced cereals brings up to 5 kg of improperly

*Corresponding author: bychkov.a.l.@gmail.com

used or field-burnt lignocellulose wastes. One of the major biomass components is lignin; it is a polyphenol compound, and it is ballast from the viewpoint of food chemistry. At the same time, lignin has a heating value higher than that of carbohydrates, so it could be efficiently used in power engineering. However, the existing technology of lignin separation from plant raw materials (mainly by means of sulfite pulping) causes contamination with sulfur in the amount multiply exceeding the standards admissible for fuel.

The goal of the present work was mechanochemical modification of the composition and structure of plant raw material to control the combustion of fine solid biofuel.

2 Materials and Methods

Plant raw material annually renewable in the Russian Federation in large amounts and concentrated in the sites of growing or processing was used in the work. Wheat, oats, corn, and barley straw, biomass of reed, sedge, corncobs, originated from different districts of Novosibirsk Region. Husk of rice, pumpkin and sunflower seed samples were collected at the processing plants of Krasnodar Region. Pine bark and branches, birch branches were collected at the wood processing plant in Kurtamysh, Kurgan Region. Buckthorn branches were collected at the plantations in the Altay Region. Husk of pine nuts and hulls of pine strobiles were collected in the Tomsk Region.

The artificial model samples of biofuel with lignin content of 10-80 % were prepared by mixing the weighted portions of wheat straw with the calculated amounts of pure cellulose (Sigma Aldrich, CAS # 9004-34-6) and lignin (Sigma Aldrich, CAS # 8068-05-1). To average the composition, thorough mixing and grinding in agate mortar was carried out for 5 minutes. The composition of wheat straw determined according to the procedure described in [7] was: lignin – 18.3 %, cellulose – 32.9 %, hemicellulose – 33.6 %, extractives – 9.4 %, and moisture – 5.9 %. The samples with lignin content from 0 to 100 % did not contain wheat straw and were α-cellulose and lignin, respectively.

Moisture and ash content of the samples were determined using the standard procedures [8] and [9], respectively.

Carbon, hydrogen, nitrogen and sulfur content was determined with the help of an automatic analyzer EURO EA(Italy) according to procedures described in [10]. Higher heating values (HHV) were determined according to the procedure described in [11].

3 Results and Discussion

The wastes from agriculture and forestry were chosen as the promising lignocellulose raw material for obtaining the solid biofuel. The main criterion for the choice was the formation and/or accumulation of large amounts of the chosen kinds of biomass. We also considered wild plants that form localized resources in nature. Fuel with different content of hydrolytic lignin was modeled with the help of artificial mixtures of wheat straw (the basic component) and the calculated amounts of pure cellulose or lignin.

The results of the elemental analysis and determination of the ash content [12] of samples show that an increase in lignin content leads to a substantial increase in the content of the major element – carbon. At the same time, an increase in the ash content of the lignocellulose material is accompanied by the depletion of its organic composition, which must have a negative effect on the heating value. The highest ash content is that exhibited by rice husk; it may contain up to 20 % amorphous silicon dioxide in its supramolecular structure. The minimal ash content is that of the husk of pine nuts (less than 1 %), which allows one to obtain low-ash solid fuel from this kind of raw material, in particular fuel pellets of the Premium grade.

The maximal combustion heat is produced by sunflower seed husk; this value is close to those of highly lignified model samples. This fact for the natural raw material can be explained by the features of its composition. For example, sunflower seed husk contains a definite amount of lipids remaining after the removal of seeds; these lipids are characterized by high combustion heat. High ash content of the material (rice husk and oat straw) or low lignin content (eg. corncobs) leads to a substantial decrease in the heat of combustion.

From the modern point of view, the combustion efficiency of fine fuel is affected to a higher extent by two major factors: combustion rate, connected with particle size, and activation energy, connected with the chemical composition of fuel. In the present work, the kinetics of oxidation of fine fuel samples characterized by different dispersity and different lignin content was studied using the methods of thermal analysis. Wheat straw was used as a raw material to prepare the samples.

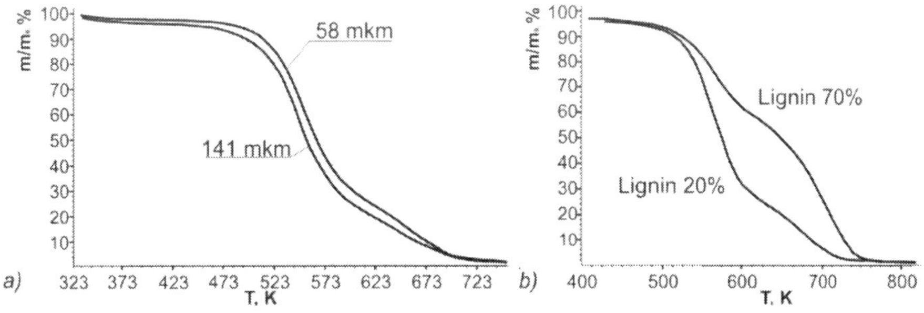

Fig. 1. Curves of oxidation of lignocellulose fuel samples: a - with different particle sizes, b – with different lignin content. Heating rate: 10 K/min.

One can see from the shapes of the curves of biofuel combustion that the nature of oxidation process does not change principally depending on dispersity (Fig. 1, *a*). In both cases (average particle sizes in the samples were 58 and 141 mkm, respectively) the major mass loss is observed at 500-575 K with a small shoulder at 590-675 K. According to the modern data [13], this situation corresponds to the reactions of hemicellulose and cellulose combustion (the reaction of lignin combustion is not revealed separately by means of thermogravimetry; instead, it forms a broad halo on the mass loss curve within temperature range 520-770 K). The rate (the slope of the tangent to the abscissa axis) for the sample with particle size 58 mkm is approximately equal to that for the sample with particle size 141 mkm. This allows us to suppose that particle size within the studied range is not a limiting factor for combustion rate. This conclusion is extremely important for the economics of the production of dispersed lignocellulose fuel because mechanochemical grinding of plant raw material to obtain particles smaller than 150 mkm requires much energy.

The curves of mass loss (Fig. 1, *b*) illustrate oxidation of the samples of fine fuel with lignin content of 20 and 70 %, respectively. One can see that an increase in lignin content in fuel changes the character of oxidation process. For instance, for the sample with lignin content of 20 %, the kinetics presented in Fig. 1 is characteristic. At the same time, for the sample with lignin content of 70 %, the major mass loss occurs at the high-temperature stage, which is likely the evidence of a shift of the contribution from lignin halo to the high-temperature region. It is logical to assume that lignin combustion at high temperature advantageously distinguishes highly lignified fuel samples from low-lignin samples. In addition, the rate of oxidation at the high-temperature stage of the process increases by 7 %.

These changes in combustion rates lead us to the conclusion that the combustion of hemicellulose and cellulose may become the limiting stage in a highly lignified sample, while lignin gets burnt rather rapidly in the high-temperature region. In this connection, it is necessary to choose the conditions of fuel combustion so that the particles could be heated as soon as possible to the high-temperature range, so that the heating value of lignin could be used in full for the afterburning of carbohydrate polymers.

Investigation of the parameters of lightning, inflammation, and stable autothermal combustion of biofuel with different dispersity and composition was carried out at the combustion bench with the thermal powder up to 5 MW. These tests allowed us to reveal acceptable thermodynamic parameters of combustion (temperature, the ratio of fuel under investigation to the oxidizer at the first and second stages of the burner) meeting the standardized ecological characteristics of combustion [14]. The optimal technological parameters are: the excess of primary air with respect to the theoretical value for complete combustion 0.25-0.3, the ratio of primary to secondary air 1.2-1.4.

Thus, the possibilities of mechanochemical methods in obtaining fine solid biofuel from renewable raw material are demonstrated in the work. We carried out the screening studies of raw materials promising for Russian Federation, namely the wastes of agriculture and forestry, and the biomass of rapidly growing plants. It is demonstrated that the nature of oxidation process does not change principally with the decrease in average particle size from 141 mkm to 58 mkm. In all cases, the major mass loss is observed at 500-575 K, with a small contribution form the second stage at 590-675 K. It is demonstrated with the help of artificial samples with different composition that an increase in the fraction of lignin leads to a substantial increase in the heat of combustion and to a shift of oxidation process into the high-temperature region.

This work was supported by the Russian Science Foundation (project no. 16-13-10200)

References

1. Roadmap EU-Russia Energy Cooperation until 2050. March 2013. https://ec.europa.eu/energy/sites/ener/files/documents/2013_03_eu_russia_roadmap_2050_signed.pdf Accessed 29 Mar 2017.
2. S. Proskurina, J. Heinimö, M. Mikkilä, E. Vakkilainen, Renew. Sust. Energ. Rev. **51**, 730 (2015).
3. P. Maryandyshev, A. Chernov, V. Lyubov, G. Trouve, A. Brillard, J-F. Brilhac, J. Therm. Anal. Calorim. **122**, 2 (2015)
4. Y. Su, P. Zhang, Y. Su, Renew. Sust. Energ. Rev. **50**, 991 (2015)
5. N. Gaurav, S. Sivasankari, G.S. Kiran, A. Ninawe, J. Selvin, Renew. Sust. Energ. Rev. **73,** 205 (2017)
6. S. Mesfun, J. Lundgren, C.E. Grip, A. Toffolo, P.L.K. Nilsson, U. Rova, Bioresource Technol. **166**, 508 (2014)
7. A.L. Bychkov, E.I. Ryabchikova, K.G. Korolev, O.I. Lomovsky, Biomass Bioenerg; **47**, 260 (2012)
8. Standard # EN 14774-3 "Solid biofuels – Determination of moisture content" (2009)
9. Standard # EN 14775 "Solid biofuels – Determination of ash content" (2009)
10. V.P. Fadeeva, T.D. Tikhova, O.N. Nikulicheva, J. Anal. Chem. **63**, 11 (2008)
11. ISO 1928:2009-06 "Solid mineral fuels – Determination of gross calorific value by the bomb calorimetric method and calculation of net calorific value" (2009)
12. A.L. Bychkov, A.I. Denkin, V.D. Tikhova, O.I. Lomovsky, J. Therm. Anal. Calorimetry (to be published in 2017)
13. H. Haykiri-Acma J. Anal. Pyrolysis. **75**, 211-216. (2006)
14. A.P. Burdukov, M.Y. Chernetskiy, A.A. Dekterev, I.S. Anufriev, P.A. Strizhak, P.Yu. Greben'kov, Therm. Eng. **63**, 1 (2016)

DOI: 10.1051/matecconf/201711506004

Vortex rope patterns at different load of hydro turbine model

Sergey Skripkin[1,*], *Mikhail* Tsoy[1], *Pavel* Kuibin[1] and *Sergey* Shtork[1]

[1]Kutateladze Institute of Thermophysics, Department of Heat Power Engineering, 630090, Novosibirsk, Russia.

Abstract. Operation of hydraulic turbines beyond optimal conditions leads to formation of precessing vortex core in a draft tube that generates powerful pressure pulsations in a hydraulic system. In case of resonance it leads to stability decreasing of hydraulic unit and electrical grid on the whole. In present work, such regimes are explored in a conical part of simplified turbine model. Studies are performed at constant flowrate Q = 70 m^3/h and varying the runner rotational speed to explore different loads of the hydroturbine unit. The experiments involve pressure measurements, high speed-visualization and velocity measurements by means of laser Doppler anemometer technique. Interesting finding is related with abrupt increasing precession frequency at low swirl parameter of flow near optimal regime.

1 Introduction

The integration of new energy sources into the existing electrical grid requires the expansion of hydraulic turbine operating range. However most of the installed hydropower units are optimized for their Best Efficiency Point (BEP). When the output flowrate of turbine is changed to execute the regulation function a residual swirl at the draft tube inlet remains, that results in emergence of phenomenon known as precessing vortex core (PVC). PVC leads to uneven pressure distribution in the draft tube that decreases the hydro turbine efficiency and safety.

Depending on the turbine discharge relative to the discharge at BEP a number of operating zones are distinguished at which the flow pattern is differed. With discharge decreasing relative to BEP there are upper part load, part load and deep part load. At increasing discharge, there is overload regime which rarely occurs in real turbines. Computational fluid dynamics (CFD) techniques and scaled down experimental modeling significantly enhances the understanding of PVC phenomena in a hydraulic turbine [1-3]. Nevertheless, CFD requires experimental data for verification but experimental data obtained on scaled model requires a valid theory for interpreting and transferring data to full-scale models that makes experimental modelling of flow structures still highly demanded.

* Corresponding author: skripkin.s.g@gmail.com

2 Experimental setup and measurement equipment

The test case is a reduced-scale simplified hydraulic turbine model, designed for an analogous test case, developed at Politehnical University of Timisoara [4]. It is designed to produce a flow similar to the one downstream a Francis turbine runner, operating at part load conditions. Detailed description of experimental set-up can be found in work [5]. Centrifugal pump with a flow rate up to 150 m3/h is used as a feed pump, and the flow rate is controlled by a feedback system that includes an electromagnetic flow meter and a frequency inverter. Runner rotational speed is controlled by electric motor through magnetic couple. Different flow regimes can be obtained changing parameters such as flow rate, runner rotational speed and geometry of stationary swirler. Commonly the flow rate is varied in a scaled hydro turbine model with a fixed rotation speed, thus outside the turbine's best efficiency point the flow in a draft tube inlet has a significant degree of swirl. In contrast, in the current study, the fixed flow rate is used with the aim to keep Reynolds number constant, and excess residual swirl is created by forcibly rotating runner.

Flow visualization techniques, laser Doppler anemometer (LDA) measurements and draft tube wall pressure measurements are employed to study the swirling flow characteristics. The flow exits the runner into the Plexiglas draft tube cone with full optical access for LDA measurements and visualizations. The axial and tangential velocity components used for the calculation of the flow swirl in the inlet cross-sections of the draft tube cone are obtained by performing LDA measurements in a 2-D mode traversing the measurement point between the cone wall and the centerline with regular steps. The velocities are measured every 5 mm, completing each measurement with a last point located on the cone centerline. Flat outer walls of the draft tube cone significantly decrease optical distortions appearing because of inner cone wall curvature. Nevertheless, tangential velocity is recalculated according methodologic proposed by Z. Zhang [6]. Maximum correction of 8% of measured velocity was applied near the central axis of the draft tube cone. Based on mean velocity distribution, the integral swirl parameter characterizing a flow regime is calculated as follows:

$$S = \frac{\int_0^R V_{tg} V_{ax} r^2 dr}{R \int_0^R U_{ax}^2 r dr} \tag{1}$$

where V_{tg} is the mean tangential velocity, V_{ax} is the mean axial velocity, and R is the radius of the cross-sectional area.

Under 70 m^3/h and runner 400 rpm the integral swirl parameter S is close to zero, i.e. the regime is defined as optimal in the turbine model. Experiments show that increasing of swirl parameter, which is directly proportional to runner rotational speed, is analogous to shifting the flow regime towards part load conditions.

3 Results

The modeling of several flow regimes is conducted with fixed flowrate and changing the runner rotational speed. Under flowrate Q = 70 m3/h and runner rotational speed N = 450 rpm (S = 0.45) the swirling flow with well-developed PVC is observed (fig 1.a). One can see the drastically different flow pattern at the same flowrate but at larger N = 700 rpm which corresponds to high swirl parameter S = 1.2 fig 1(b).

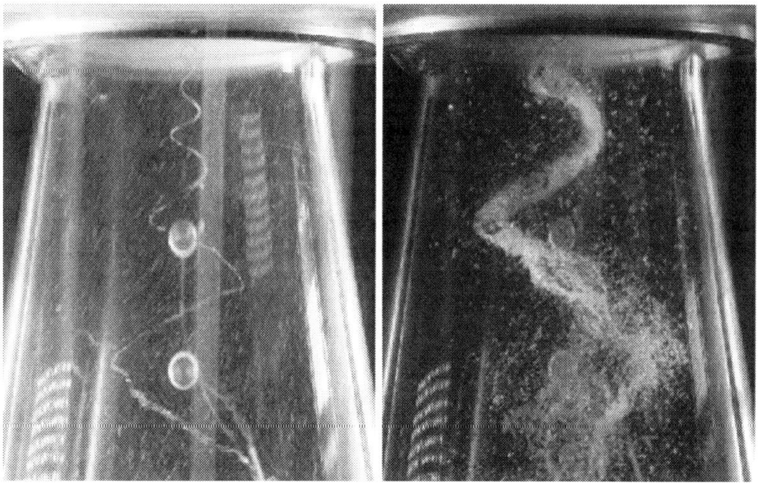

Fig. 1. Instant photo visualization, Q = 70 m^3/h, left – N = 450 rpm, S = 0.45, right N = 700 rpm, S = 1.2

At S = 0.45 the precession radius is decreased in two times; also spatial wavelength is decreased in more than two times compared with regime at S = 1.2. As a result, precession frequency registered by high-speed visualization technique at S = 0.45 is higher than at S = 1.2. The same behavior of PVC at low swirl number was observed by Fernandes et al. [7] for axial swirl generator with variable blade angles. Fernandes et al. showed that there was critical swirl number below which the precession frequency increased with swirl number decreasing while the stable PVC existed. This has a good correlation with figure 2 (left) where dependence of PVC frequency on runner rotational speed is presented. Note that the swirl number increased linearly with rotational speed from N = 400 rpm. At low swirl the precession frequency can be determined only based on high-speed visualization, performed at 500 fps because of weak periodical mode of the pressure signal, recorded at N below 500 rpm (figure 2 b). Note that the visualization technique shows a good agreement with Fast Fourier transform of pressure signal at higher N.

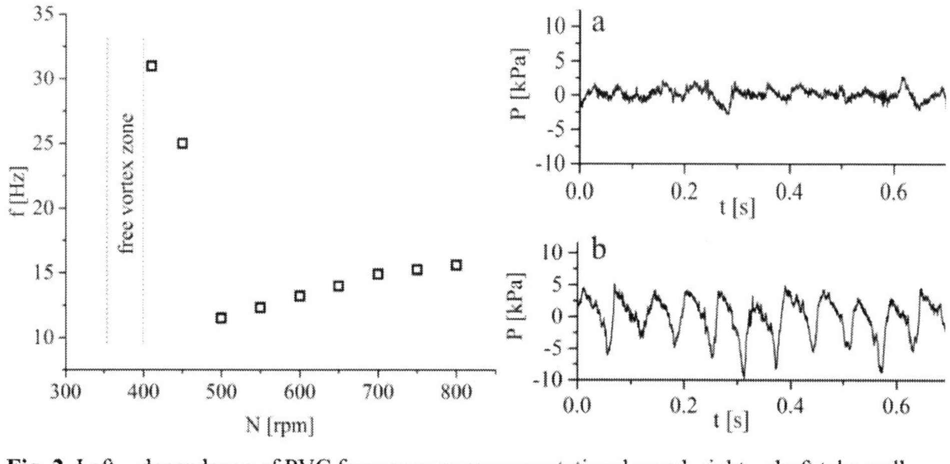

Fig. 2. Left – dependence of PVC frequency on runner rotational speed, right – draft tube wall pressure pulsations, Q = 70 m^3/h, a – 450 rpm, b – 700 rpm

Figures 3 shows the quantitative information obtained by the LDA measurements. Velocities were nondimensionilesd using bulk velocity U = 2.47 m/s. One can see in both cases there is a narrow region of the central reverse flow. At higher rotational speed the back flow zone is more developed that correlates with flow visualization on figure 1.

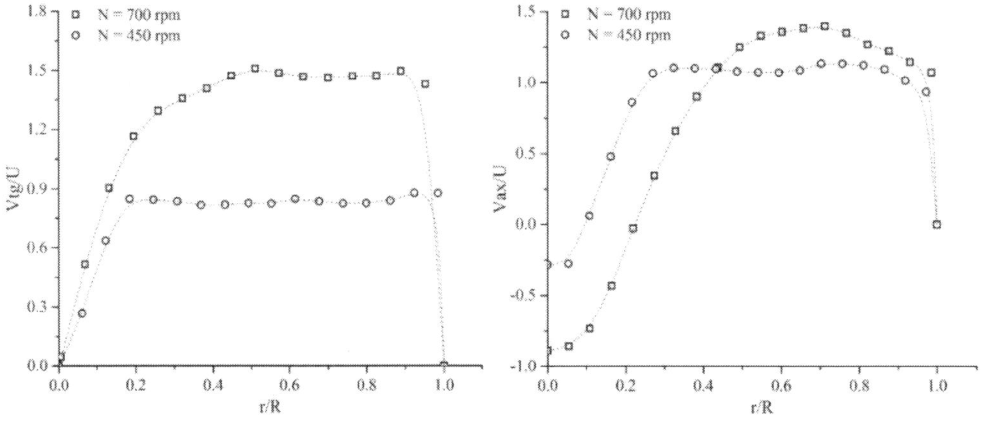

Fig. 2. Dimensionless tangential (left) and axial (right) mean velocity, Q = 70 m3/h.

4 Conclusion

Observation of the PVC in the conical draft tube model provides some new information regarding the vortex structure at low swirl parameter. As it turned out there is a narrow region below critical swirl parameter where precession frequency of the PVC grows up with decreasing swirl parameter. It is not related to an increase in tangential velocity component but can be explained by a drastical change of the PVC pattern. At low swirl the spiral vortex is formed with a smaller radius and shorter spatial wavelength.

Although the observed effect of high precession frequency with low amplitude does not pose any practical direct problem, it can become an additional source of pressure disturbances, which are sensitive to cavitation number.

This work was supported by the Russian Science Foundation, grant No 16-19-00138.

References

1. H. Foroutan, S. Yavuzkurt, ASME J. Fluids Eng., **137**, 061101 (2015)
2. M. Nishi, S. Liu, Int. J. Fluid Mach. Syst., **6**, 33 (2013)
3. C. Nicolet, A. Zobeiri, P. Maruzewski, F. Avellan, Int. J. Fluid Mach. Syst., **4**, 179 (2011)
4. A. Javadi, A. Bosioc, H. Nilsson, S. Muntean, R. Susan-Resiga, ASME J. Fluids Eng., **138**, 081106 (2016)
5. S. Skripkin, M. Tsoy, S. Shtork, K. Hanjalic´, J. Hydraul. Res., **54**, 450 (2016)
6. Z. Zhang, Exp. Fluids, **37**, 29 (2004)
7. E. Fernandes, M. Heitor, S. Shtork, Exp. Fluids, **40**, 177 (2006)

Investigation of flow parameters in a cylindrical pebble bed and in a model of a fuel assembly with microfuels

Yuliya Smorchkova*, *Alexander* Varava, *Alexey* Dedov, *Alexander* Zakharenkov, *Alexander* Komov, *and Alexander* Borozdin

National Research University "MPEI", Krasnokazarmennaya 14, 111250 Moscow, Russia

Abstract. The results of an experimental study of the hydrodynamics and heat transfer of a pebble bed in a cylindrical channel and in a model of a fuel assembly with microfuels (FA MT) with a cross flow of a coolant are presented. Data on the temperature distribution along the height and the pebble radius for volumetric heat release, as well as data on pressure losses over a wide range of mass flow rates of the coolant from 0.05 to 1.75 kg/s are obtained.

1 Introduction

Further increase of nuclear energy efficiency with simultaneous enhancement of security requires a conceptually new approach. One of the promising solutions of this problem is the use of alternative fuel rods in the form of spherical fuel elements. On the one hand FA MT have an inherent internal security, and on the other hand they will allow to improve energy efficiency.

The construction of the fuel assembly with microfuels was proposed in [1]. In such an assembly, the microfuels are placed between the perforated covers. To minimize the pressure losses, a cross flow of the heat coolant is realized. Fuel assemblies based on microfuels are likely to find application, primarily in low-power nuclear station.

2 Experimental facility

This paper is a continuation of the research presented at [2, 3]. A low-pressure model installation with two test sections was designed and manufactured, with the possibility of simulating both single-phase and two-phase flow of the coolant in a pebble bed with volumetric energy release. The description of the hydraulic circuit is presented in [3].

2.1 Test sections

The central part of the test section No. 1 (Fig. 1a) modelling a cylindrical pebble bed consists of two coaxially arranged polycarbonate tubes (1, 2) with a diameter of 32 and

* Corresponding author: SmorchkovaYV@mpei.ru

50 mm. The inner tube holds the pebble bed. The outer tube ensures the strength and tightness of the structure. The pebble bed is fixed by gratings (3) made of welded balls. The supply of coolant, the alignment of pipes and grids is carried out through the elements of the body frame (4), made of fiberglass. They also provide pressure sampling (6) and sealed feedthroughs (5) for thermocouples that measure the temperature profiles in the backfill and the temperature of the coolant at the inlet and outlet of the test section.

A detailed description of the test section 2 (Fig. 1b) is given in [3]. In this test section, pulse tubes for measuring the pressure drop were modernized. Earlier in [3], the measured pressure drop took into account the pressure losses at the inlet and outlet of the section. These losses made a significant contribution to the overall pressure loss. To exclude this effect, pressure taps have been transferred directly to the dispenser (1) and collecting receiver (5).

15 cable thermocouples are placed in the medium of the pebble bed. All of them are installed in 4 sections along the height of the pebble bed (T1-T15).

Pebble bed consists of balls with a diameter of 2.0 mm are made from steel grade AISI 420. The internal heat in the pebble bed is modelled by high-frequency induction heating by an inductor (7).

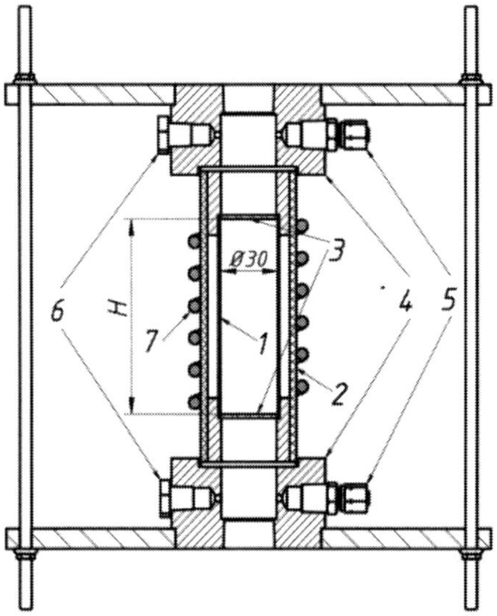

a)

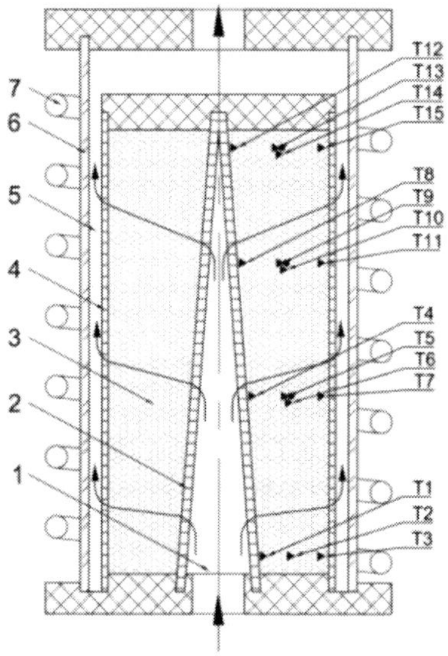

b)

Fig. 1. Test sections: a) pebble bed in the cylindrical conduit, b) model of fuel assembly with microfuels

3 Experimental data

3.1 Test section 1

Experiments to determine the pressure losses for the pebble bed with a height of 100 mm were carried out in the range of mass flow rate of water from 0.50 to 1.75 kg/s at a temperature of 25 °C. Fig. 2a shows the dependence of the coefficient of hydraulic resistance as a function of the Reynolds number. To generalize the obtained experimental data, the calculation formulas of Bogoyavlensky R.S. (1) and Avdeev A.A. (2) for the coefficient of hydraulic resistance have been applied, which had been published in [4] and [5], respectively.

$$\xi = \frac{0,54}{\varepsilon^4}\left(1 + \frac{10}{\mathrm{Re}_{str}^{0,7}} + \frac{100}{\mathrm{Re}_{str}}\right) \tag{1}$$

$$\xi = \frac{1,53}{\varepsilon^{4,2}}\left(0,335 + \frac{3}{\mathrm{Re}_{str}^{0,7}} + \frac{30}{\mathrm{Re}_{str}}\right), \quad \mathrm{Re}_{str} = \frac{0,45\,\mathrm{Re}}{\sqrt{\varepsilon}(1-\varepsilon)} \tag{2}$$

In formulas (1), (2), ε is the porosity of the pebble bed, the Reynolds number Re is determined from the filtration rate and the diameter of the ball.

The temperature field along the height of the pebble bed, with a thermal capacity of 13 kW is shown in Fig. 2b. The points are obtained by averaging the values measured by the thermocouples located in one cross section. This temperature distribution over the height of the pebble bed does not contradict the physical concepts of the upward flows in pebble beds: the enthalpy of the coolant flow increases in height, while retaining the character of the temperature distribution.

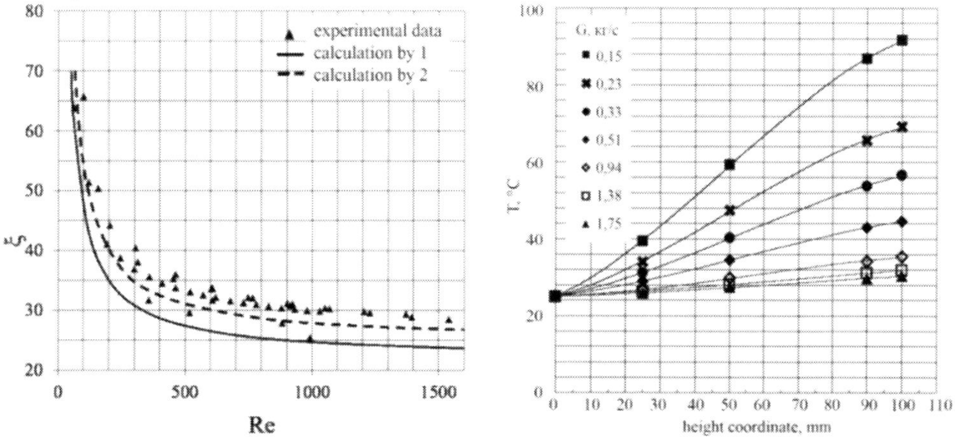

a) b)

Fig. 2. a) Dependence of the coefficient of hydraulic resistance on the Reynolds number, b) Temperature distribution over the height of the pebble bed

3.2 Test section 2

The primary experimental data in the form of the dependence of the pressure loss on the mass flow rate are shown in Fig. 3a. Mass flow rate varied from 0.05 to 0.60 kg/s, the liquid temperature was maintained at 20°C. To take into account the contribution of the inner and outer perforated covers to the total pressure loss in the test section, experiments were performed without the pebble bed. The results of these experiments are shown in the figure. Then, several series of experiments were performed with pebble bed at the same regime parameters. The results of the experiments are shown in Fig. 3a. There is good reproducibility of the results, the discrepancy between the data of different series does not exceed 10%.

When comparing the data on pressure losses with pebble bed and without pebble bed, it can be seen that the largest contribution to the total pressure loss is made by perforated covers (up to 65%).

To determine the degree of uniformity of heat release in the pebble bed, experiments were carried out with heating of the pebble bed without flow of the coolant. The results of the experiments in the form of the temperature dependence on the height coordinate of the pebble bed are shown in Fig. 3b. When analysing the data (Fig. 3b), there is a significant heterogeneity in the heating of the filling with respect to height. In the central part of the pebble bed, the temperature at different points varies within 5 °C. At the same time, the difference in the temperature of the pebble bed in the central part and at the butt ends reaches 17 °C. These differences are related to the features of induction heating. A similar effect is observed in the works of other authors, for example [6].

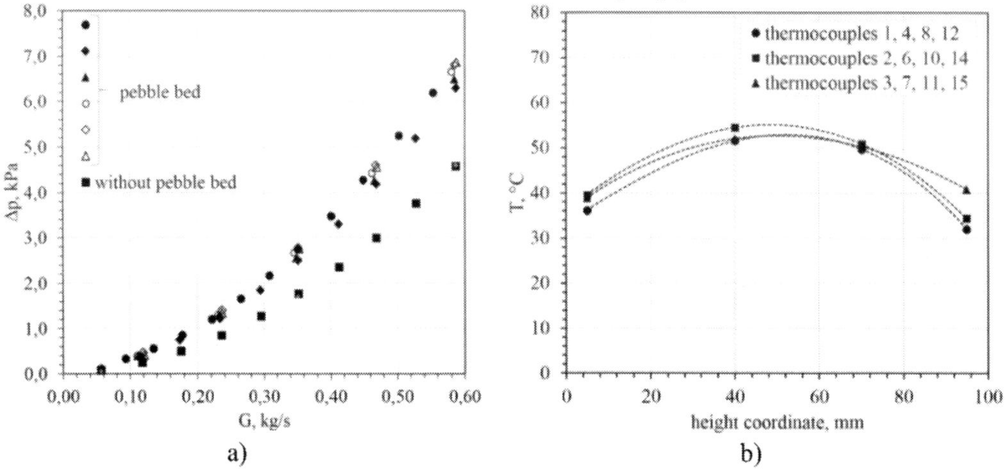

Fig. 3. a) Dependence of pressure losses on mass flow rate
b) Temperature distribution over the height of the pebble bed

4 Conclusion

- For the first time experimental data on the thermohydraulic parameters on the FA MT model have been obtained;
- Experimental data on the flow in a cylindrical backfill supplemented and expanded the available data of other authors and confirmed the correctness of calculations using the empirical formulas [4, 5].

The work is supported RSF (grant № 16-19-10457)

References

1. N.N. Ponomarev-Stepnoy, N.E. Kukharkin, A.A. Khrulev, Y.G. Degaltsev., Nuclear Energy **86** 443-449 (1999)
2. A.V. Borozdin, A.N. Varava, A.V. Dedov, A.T. Komov, S.A. Malahovskiy, Y.V. Smorchkova, Thermal processes in the engineering, **7**, 295 (2015)
3. Y.V. Smorchkova, A.N. Varava, A.V. Dedov, A.T. Komov, J. Phys.: Conf. Ser., **754**, 112008 (2016)
4. R.S. Bogoyavlensky, Atomizdat, Moscow, 112 (1978) [in Russian]
5. A.A. Avdeev, R.I. Soziev, G.A. Filippov, B.F. Balunov, R.A. Rybin, High Temp., **41**, 377 (2003)
6. X. Meng, Z.G. Sun, G.Z. Xu, Nuclear Eng. Design, **252**, 121 (2012)

Unburnt carbon and iron content in the ash and slag thermal power plants

Elena Yu. Temnikova[1,*], *Alexander* R. Bogomolov[1,2], and *Alexey* A. Lapin[1]

[1]Gorbachev Kuzbass State Technical University, 650000 Kemerovo, Russia
[2]Kutateladze Institute of Thermophysics SB RAS, 630090 Novosibirsk, Russia

Abstract. The work is devoted to the determination of the content of carbon underburn and iron in the fly ash of thermal power plants. The results of determining the mechanical underburning by fractions for three stations are presented to show that the stations have a slight underburning besides the Tom-Usinsk SDPP. In addition, there is a general regularity showing a 4-10 times higher content of remnant carbon in fly ash than in slag. Data are presented on the composition of elements of sulfur, calcium and iron in the original fly ash and in the separated magnetic fraction in a one-step separation on a magnetic analyzer in water.

1 Introduction

Ash-slag materials (ASM) of thermal power plants (TPP) include a significant amount of the basic and acid oxides, and small concentrations of oxides of zinc, titanium, nickel, lead, barium, vanadium, and others.

The mineral components of coals after flaring at TPP can be considered technogenic wastes, when they are transported by different transport modes and stored in designated areas, or complex and diverse raw materials, suitable for processing by the well known methods in order to obtain demanded products in various industries.

The extraction products are the known ash-slag materials containing silicon oxides (57-62%), aluminum oxides (20-25%) and iron oxides (3.12%), and hollow aluminosilicate microspheres (1-2%) formed at high-temperature coal flaring.

According to the data of different authors, the annual volume of ASM production by the coal plants of Russia is from 34 to 50 million tons. Only 13-14% of these materials are recycled and used in industry.

2 Problem statement

Among the ASM components, the hollow microspheres, no-microsphere fly ash and slag achieved the most common application. The important characteristic is the presence or absence of unburnt carbon and iron oxide in these separated fractions for potential industrial consumers.

The work purpose is to study the content of remnant carbon and iron of fly ash and slag.

* Corresponding author: teu.pmahp@kuzstu.ru

3 Results and discussion

The dispersion analysis of fly ash from the Kemerovo State District Power Plant (KemSDPP) (Russia, Kuzbass), the Tom-Usinsk state district power plant (TU SDPP) (Russia, Kuzbass), and the Novo-Kemerovo heat and power plant (NK HPP) (Russia, Kuzbass) allowed it to be divided into fractions, studied to determine the content of remnant carbon. Mechanical underburning (the presence of remnant carbon) was determined by burning out the organic mass from ASM according to GOST 11022-95 "Solid mineral fuel. Methods for determination of ash content". The results of ashing are presented in Table 1.

Table 1. The results of determination of the remnant carbon on fractions.

Faction, μm	The content of remnant carbon in fly ash, %, mass.		
	KemSDPP	TU SDPP	NK HPP
0-50	0.7387	9.08	1.96
50-63	1.6987	7.78	2.72
63-80	1.5196	9.25	3.7
80-100	2.0951	9.77	3.71
100-160	5.2857	14.8	5.71
160-200	4.8478	14.48	5.81
200-250	2.2152	14.16	8.51
250-300	0	0	4.98
> 300	-	-	0

Data analysis of the content of remnant carbon in fractionated fly ash shows that the ash particles formed in boiler of Kemerovo SDPP during combustion of pulverized coal have little carbon residue. The maximum occurs at a particle size of about 150 microns. Ash particles formed during combustion of pulverized coal at the Novo-Kemerovo HPP have maximal underburning degree at sizes of about 220 microns. The greatest amount of unburnt carbon in fly ash belongs to Tom-Usinsk SDPP (about 14%) with a particle size of 120-220 μm. A significant underburning is present at 25-50 microns (about 9%) in comparison with carbon underburning in the same classes of fly ash at Kemerovo SDPP and Novo-Kemerovo HPP (about 1-2%).

It should be noted that fly ash mainly consists of particles of 0-100 μm in size. The number of particles with the weight of less than 100 microns is ≈ 92% for KemSDPP, ≈ 89% for TU SDPP, and ≈ 88% for NK HPP [1]. That is, the share of unburned carbon is concentrated in this fraction.

The results of Table 1 show that the gas-dynamic and temperature operating mode in the boiler furnaces of Tom-Usinsk SDPP require scrutiny. It can be assumed that coal in the boiler is burned at high temperatures, at which the mineral component, having a low softening point, passes into liquid state. Under these conditions, part of the unburned carbon is enveloped by the liquid phase of the mineral component, which prevents oxidation and combustion of carbon located in the protective envelope.

Further, when the temperature of flue gases decreases while moving along the screen tubes and flue duct of the convective part together with the liquid particles, the mineral parts of particles transform into the solid state and non-oxidized carbon remains in the "shell". This process is similar to formation of slags and druses in the layered furnaces, where much unoxidized carbon remains under the shell of the mineral component. To reduce the formation of large slag granulate and achieve more complete combustion of carbon, it is recommended to carry out frequent shuffling and use a thin height of the coal layer on the grate in the layer furnaces.

Table 2 presents the results of instrumental measurements of the remnant carbon and iron content in ash and slag samples from four stations (here data on the Belovo state district power plant (BelSDPP) is included). Element content was determined using the scanning electron microscopy method using the scanning electron microscope JEOL JSM-6390 LA with the energy-dispersive X-ray detector JED 2300 of the Collective Use Center of the Federal Research Center for Coal and Coal Chemistry, RAS.

Table 2. Results of determining remnant carbon and iron.

Element	The content of remnant carbon and iron in fly ash/slag, %, mass.			
	KemSDPP	BelSDPP	TU SDPP	NK HPP
C	2.63/0.23	2.57/0.66	4.86/4.77	2.98/0.27
Fe	4.04/1.35	2.86/–	3.71/0.04	2.19/1.36

It can be seen from Tables 1 and 2 that there is a general dependence showing 4-10 times higher content of remnant carbon in the fly ash as compared with slag at the KemSDPP, BelSDPP and NK HPP. A slight difference, with observance of the tendency noted by the authors, was revealed at Tom-Usinsk SDPP.

Probably, this is due to the combustion regimes of pulverized coal fuel, which involve not only identical content of remnant carbon in the ash and slag of the TU SDPP, but about 2 times higher carbon content as compared to ash and slag at other plants.

Table 2 shows that there is essentially less amount of iron compounds in the slag than in the ash. Iron can be in the fly ash and in the slag as a separate element of Fe, as iron oxide (II, III) (Fe_2O_3 and/or Fe_3O_4 ($FeO{\cdot}Fe_2O_3$ in the form of magnetite or iron cinder), as iron carbonate (II) $FeCO_3$, as ironic sulphate FeSO4, which in the presence of sodium carbonate Na_2CO_3 or calcium carbonate $CaCO_3$ in an inert medium passes into iron carbonate $FeCO_3$ to form sodium sulfate Na_2SO_4 or calcium sulfate $CaSO_4$. At that, sulfuric or sulfurous oxides form solid mineral sulfates either with sodium or with calcium, or with iron, i.e., there is binding of gaseous sulfur oxides formed during combustion of coal with a small amount of sulfur. It should be noted that iron carbonates $FeCO_3$, formed during combustion at temperatures above 500°C, are converted to iron (II) oxides with emission of carbon dioxide gas CO_2. This mechanism explains the higher content of iron oxides in fly ash than in slag.

Thus, at magnetic fraction separation from the total composition of fly ash, an increase in concentration of iron-containing compounds should occur; we assume that there should be more iron sulphates in the form of $FeSO_4$, calcium sulfates $CaSO_4$ and Fe_2O_3 oxides (in the iron equivalent), as well as an increase in concentration of sulfur compounds (in the sulfur equivalent).

Data on the composition (in terms of the element) of elements of sulfur, calcium and iron in the original fly ash and in the separated magnetic fraction are presented in Table 3 in a one-step separation on a magnetic analyzer in water.

It can be seen from Table 3 that in the magnetic fraction (in the element equivalent), there are 2-3 times more elemental iron, separated by one-stage wet magnetic separation on a magnetic analyzer, than in the original fly ash. An increase in concentration of sulfur with simultaneous increase in calcium shows that sulfur can be in the form of iron and calcium sulphates with a high probability.

Table 3. Data on the composition of elements of sulfur, calcium and iron in the ash and the separated magnetic fraction.

Element	Content of elements,%, mass.					
	BelSDPP		TU SDPP		NK HPP	
	ash	magnetic fraction	ash	magnetic fraction	ash	magnetic fraction
S	0.07	0.11	0.18	0.36	0.22	0.36
Ca	2.33	8.48	2.18	5.48	1.58	4.19
Fe	2.86	4.79	3.71	10.51	2.19	7.90

4 Conclusion

Maximal carbon underburning in fly ash of the studied plants is observed in the fractions of 100-200 microns. A significant amount of remnant carbon is available in fractional composition of fly ash with the particle size of above 200 microns at Tom-Usinsk SDPP and Novo-Kemerovo HPP plants. However, the share of these fractions is insignificant in the total amount of waste (materials) stored in the ash dump that pollute the environment. It should be noted that the method of fly ash capturing at Tom-Usinsk SDPP and Novo-Kemerovo HPP differs from the Kemerovo SDPP. These differences require special attention when analyzing the share of coal underburning at the plants.

To increase the boiler efficiency, the model thermal and gas dynamic calculations with comparison of experimental results are required for the detailed analysis of the reasons for underburning of large particles of pulverized coal at the plants.

The tendency to a higher content of remnant carbon and iron in the fly ash in comparison with the slag removed by the liquid method is shown.

It has been determined that the sulphurous compounds of coal (organic, pyrite and sulphate- sulphide) migrate mainly to iron and calcium sulfates in the process of combustion.

The studies were financially supported by the Russian Foundation for Basic Research and Department of Education and Science of Kemerovo region in the framework of scientific project No. 16-48-420871, "r_a" and Agreement No. 6 of December 15, 2016.

References

1. E.Yu. Temnikova, A.R. Bogomolov, N.V. Tiunova, A.A. Lapin, Vestnik KuzSTU (2017) [in Russian]

Some approaches to numerical modelling of a phenomenon observed during steam generator tube rupture in the reactor with liquid metal coolant

Eduard V. Usov[1], *Pavel* Lobanov[1,3*], *Oleg N.* Kashinskiy[3], *Nikolai A.* Pribaturin[1,3], *Nastasya A.* Mosunova[2], *Vladimir I.* Chuhno[1,4], *Alexander E.* Kutlimetov[1,4], and *Alexander I.* Svetonosov[4]

[1]Nuclear Safety Institute of the Russian Academy of Sciences, Novosibirsk Branch, Novosibirsk, 630090, 1 Lavrentiev Ave., 1, Russia
[2]Nuclear Safety Institute of the Russian Academy of Sciences, Moscow, 115191, 52 B. Tulskaya Str., Russia
[3]Institute of Thermophysics of the Russian Academy of Sciences, Siberian Branch, Novosibirsk, 630090, 1 Lavrentiev Ave., Russia
[4]Novosibirsk State University, Novosibirsk, 630090, 2 Pirogova Str., Russia

Abstract. The presented paper contains a description of approaches to simulate processes, observed during leakage in the steam generator of the reactor with liquid metal coolant. These approaches have been implemented in thermal hydraulic code HYDRA-IBRAE/LM. To calculate motion of gas bubbles in liquid metal flow and heat transfer of gas bubbles with metal, different relations are used in HYDRA-IBRAE/LM code. The code contains models of chemical interaction between water and sodium for modelling of tube rupture in sodium cooled fast reactors, Modelling of the experiments has been made using HYDRA-IBRAE/LM code. The results of the modelling with determined main factors are presented in the article.

1 Introduction

Development of the new generation of fast reactors with liquid coolant leads to requirement of safety research of this type of reactor. One of the most important accidents is a damage of steam generator tubes. Investigation of processes with steam and gas injection into liquid metals, motion of gas bubbles in liquid metal flow and heat transfer between liquid metal and gas bubbles is necessary for understanding the specificity of reactor operation during this type of accidents. To study these processes theoretical and experimental investigations on different experimental setups were performed. For example, investigations with heavy liquid metal coolant have been provided at LIFUS test facility [2] and test facilities of the Institute of Thermophysics of Siberian Branch of Russian Academy of Science [3]. Some

* Corresponding author: lobanov@itp.nsc.ru

experiments with sodium coolant are presented in [4]. Theoretical investigation of steam generator tube rupture could be done using thermal hydraulic code HYDRA-IBRAE/LM [1]. This code allows modelling of thermal hydraulic processes in sodium, lead, lead-bismuth, Rose's alloy and water coolants during gas injection. Two-fluid model with constitutive relations for coefficients of wall friction, interphase friction and heat transfer is used for modelling of processes in heavy liquid metal and sodium coolant during gas injection. Constitutive relations depend on the flow regime type.

2 HYDRA-IBRAE/LM relations for numerical modelling of processes in a two-phase liquid metal coolant

HYDRA-IBRAE/LM code have been developed by the Nuclear Safety Institute Russian Academy of Science as a part of the Federal Target Program "Nuclear Power Technologies of the New Generation for 2010-2015 and until 2020". There is a system of closing relations in HYDRA-IBRAE/LM code for modelling of motion and heat transfer between gas bubbles and liquid metal. The system consists of correlations for calculation of interphase and wall friction coefficients, interphase and wall heat transfer coefficients and interfacial area.

Closing relations depend on flow regime type. For bubbly flow the interfacial area a_i is calculated as:

$$a_i = 6\, \varphi_g / D_b, \tag{1}$$

where D_b is a diameter of bubble, φ_g is a gas volume fraction. For slug flow the interfacial area is defined as:

$$a_i = 4.5\, (\varphi_g - \varphi_{gs}) / (1 - \varphi_{gs})\, / D_h + 6\, \varphi_{gs}\, (1 - \varphi_g) / (1 - \varphi_{gs})\, / D_h, \tag{2}$$

where D_h is a diameter of channels. The interphase friction coefficient ξ_i for the bubbly flow regime can be calculated as [5]:

$$\xi_i = 24\, (1 + 0.15\, Re_b^{\,0.687}) / Re_b + 0.42 / (1 + 4.25\, 10^4\, Re_b^{\,-1.16}). \tag{3}$$

This relation is used for spherical bubbles. For spherical caps correlation from [6] can be used:

$$\xi_i = 8/3\, Eo / (Eo + 4), \tag{4}$$

where Eo is an Eotvos number.

The interphase friction for slugs can be defined as [7]:

$$\xi_i = 1/k,\ k = 0.35\, (1 - e^{-0.01\, N/0.345})\, (1 - e^{(3.37 - Eo)/m}), \tag{5}$$

$$N = (g\, D_h\, (\rho_f - \rho_g)\, \rho_f)^{0.5} / \mu_f. \tag{6}$$

The coefficient m is equal to 10, if N more than 10; m is equal to 25, if N less than 18. And m is equal to $69 \cdot N^{-0.35}$ in the other cases.

The interphase heat transfer coefficient (α_{fi}) also depends on the flow regime type. For the bubbly flow the liquid-side heat transfer coefficient is given by:

$$\alpha_{fi} = Nu \cdot k_f / D_b,\ Nu = 2 + 0.386\, Pe^{0.5},\ [8] \tag{7}$$

The gas-side heat transfer coefficient is given by:

$$\alpha_{gi} \cdot a_i = C_{pg} \cdot \rho_g \, \varphi_g / \tau, \ \tau = D_b^2 \, C_{pg} \cdot \rho_g / (4\pi \, k_g). \tag{8}$$

Liquid-side heat transfer coefficient for slug flow can be defined by the following semianalytical relation:

$$\alpha_{fi} = 2 \cdot k_f / \delta, \ \delta = D_h \, (1 - \varphi_g^{0.5}) / 2 \tag{9}$$

The gas-side heat transfer coefficient can be defined as [9]:

$$\alpha_{gi} \cdot = Nu \cdot k_g / D_h / \varphi_g^{0.5}, \ Nu = 0.023 \, Re_g^{0.8} \, Pr_g^{0.4}. \tag{10}$$

Model of interaction between water and sodium has been implemented for modelling of processes in sodium coolant during tube rupture. Chemical reaction between water and sodium is considered as:

$$Na \cdot + H_2O \longrightarrow \tfrac{1}{2} H_2 + NaOH + \Delta h, \tag{11}$$

where Δh is an enthalpy change due to chemical reaction. The value of the enthalpy change is added to the right-hand side of the energy equation of the two-fluid model.

3 HYDRA-IBRAE/LM code validation

The experiments with gas injections have been modelled for validation of models implemented in HYDRA-IBRAE/LM code.The experiments have been provided on the HLM-FIT facility at the Institutes of Thermophysics SB RAS. A vertical circular tube has been used in the experiments. The tube diameter was equal to 48 mm (fig.1). The tube has been filled with liquid lead. The lead temperature was equal to 400^0C. Temperature measurements have been made by thermocouple.

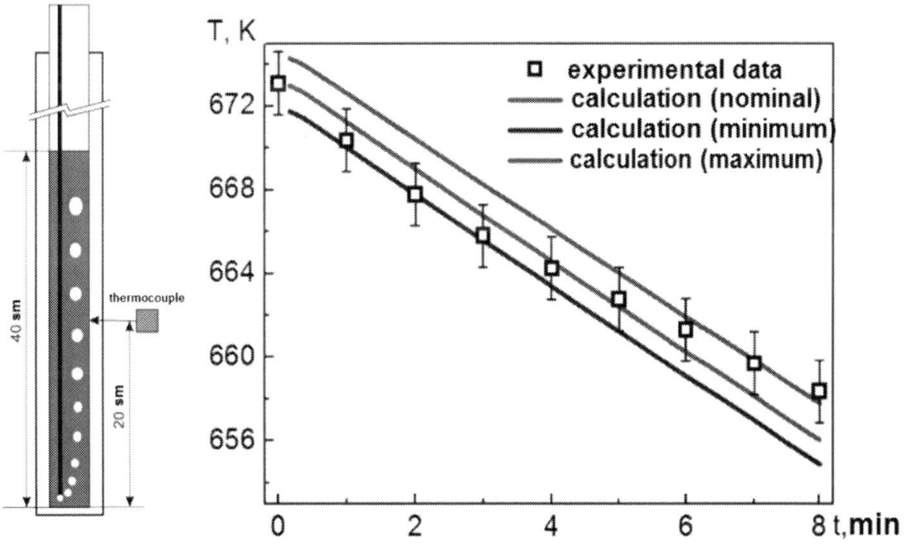

Fig. 1. Scheme of the experiments at the IT SB RAS (left) and the calculations results (right).

Comparison between experimental data and calculation results for volume flow rate equal to 180 ml/sec is shown in fig.1. Fig. 1 contains the results of calculations performed with nominal parameters and the results of uncertainty analysis.

Experiments with water–sodium interactions [4] have been modelled to validate the model of water-sodium reaction. The experimental setup is consisted of the sodium container, water storage and pressure sensor line (fig.2). Pressure evolution in the tank has been measured during the sodium-water reaction. The results of calculation in comparison with experimental data are shown in fig.2.

4 Conclusions

The models implemented in the HYDRA-IBRAE/LM code are presented in the current paper. The paper contains the relations that are used for calculation of the heat transfer and interphase friction coefficients in the bubbly and slug flow regimes. Modelling of the experiments with gas injection into lead and sodium-water interaction experiment has been performed using thermal hydraulic code HYDRA-IBRAE/LM. A good agreement between the experimental data and the calculation results has been obtained.

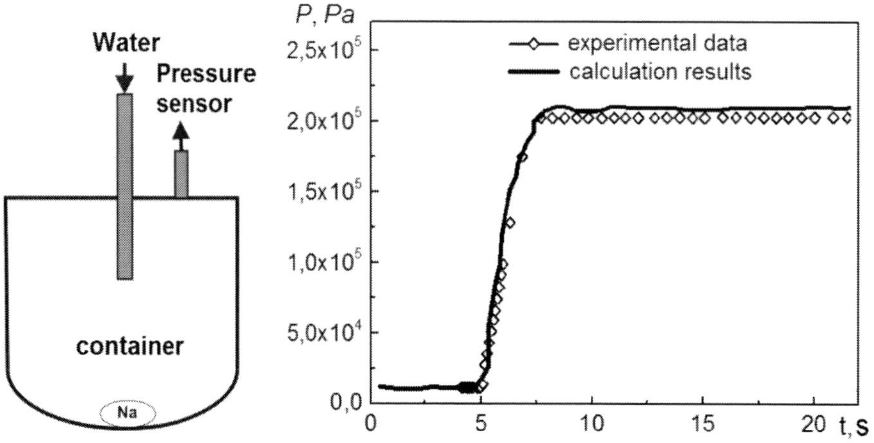

Fig. 2. Scheme of the experiments (left) and the calculation results (right) [4].

References

1. V.M. Alipchenkov, A.M. Anfimov, D.A. Afremov, V.S. Gorbunov, Yu.A. Zeigarnik, A. V. Kudryavtsev, S.L. Osipov, N.A. Mosunova, V.F. Strizhov, E.V. Usov, Thermal Eng. **63**, 130 (2016)
2. A. Ciampichetti, D. Pellini, P. Agostiny, G. Benamati, N. Forgione, F. Oriolo, Nuclear Eng. Design. **239**, 2468 (2009)
3. E.V. Usov, P.D. Lobanov, N.A. Pribaturin, N.A. Mosunova, V.I. Chuhno, A.E. Kutlimetov, J. Phys.: Conference Series. **754** 112009 (2016)
4. G. Park, S. J. Kim, H.S. Park, M.H. Kim, J. Baek, J. Nucl. Sci. Tech. **53**, 198 (2016)
5. R. Clift, W. H. Gauvin, Canadian J. Chem. Eng. **49**, 439 (1971)
6. A. Tomiyama, I. Kataoka, I. Zun, T.Sakaguchi, JSME Int. J., Series B Fluids Thermal Eng. **41**, 472 (1998)
7. G. B. Wallis, *One-dimensional two-phase flow* (New York, McGraw Hill, 1969)
8. B. Lubarsky, S.J. Kaufman. (NASA Tech. Note 3336, 1955)
9. F.W. Dittus, L.M. Boelter, University of California Publication in Engng. **2**. 13. (1930)

Vapor-dynamic cooling systems for nuclear power plants

Sergey Shcheklein*

Ural Federal University, Department of Nuclear Power Plants and Renewable Energy Sources, 620002, Yekaterinburg, st. Mira 19, Russia

Abstract. Are the results of the pilot study intensive cooling pools extracts of spent nuclear fuel at nuclear power plants using vapor-dynamic technology. The possibility of passive heat dissipation energy axially heat flow density above 10^6 W/m^2.

1 Introduction

One of the most dangerous situations leading to bdba NPP is long due. In such emergency conditions the main task of the security systems is the preservation of the integrity of the reactor and the spent nuclear fuel (SNF), located in prereactor spent fuel pool (SFP).

The need to comply with this requirement became apparent after accident and the Japanese Fukushima-1. The creators of nuclear power plant got a new challenge of removing heat from the SFP passive systems.

The loss of heat removal led to heating and evaporation of the coolant in the active zones and cooling water in the SFP, that caused the overheating nuclear fuel from its damage and the formation of large quantities of hydrogen. hydrogen Accumulation under the protective shell of the reactor, and then and in the building of the reactor compartment, leading to explosions with a significant destruction of buildings.

Existing systems that ensure the removal of residual heat from SFP, systems are "active" principle of applying cooling water by circulation pumps. Therefore, in the conditions of a long-term loss of power supply for own needs and the impossibility of connecting emergency sources of electrical energy, the provision of a normal discharge of residual heat of spent nuclear fuel is determined by the time of evaporation of water from SFP and amounts to several days to several hours, depending on the SFP loading. An increase in temperature with subsequent evaporation of water will lead to the beginning of the steam-zirconium reaction with the formation of hydrogen. When the outer surface of the shells of fuel elements of SNF is bare, the conditions for normal heat removal are violated, which leads to their damage and becomes the cause of a radiation accident.

At the same time, the need to create passive systems, usually based on the principles of natural convection, conflicted with the need to reduce the volume of buildings and structures of NPP, because by virtue of the low intensity of heat transfer processes requires

* Corresponding author: s.e.shcheklein@urfu.ru

considerable heat exchange surfaces, low hydraulic resistance tracts and significant walk-through sections cooling medium.

Interest of creators of new technologies cause more intense than natural convection heat transfer methods environment medium, based on natural principles under relatively small thermal conditions. One such method is the use of the difference of the densities of liquids and vapors in the gravitational field, implemented in designs of two-phase thermosyphons [1]. A detailed theoretical analysis of the use of this heat removal technology was carried out in [2], where the principle possibility and effectiveness of the use of the thermosyphon method. However, despite the apparent simplicity and good knowledge of the thermophysical processes taking place in these devices, the available theoretical and experimental data are insufficient to reliably determine the transmitted power, to select and create a design of heat exchange equipment capable of reliably thermally stabilizing the soak pool SNF [3-6].

The purpose of this work is the experimental substantiation of the possibility of using one of the thermosyphon method variants-the steam dynamic deviation of heat from the protected object (SFP SNF).

2 The methodology of experimental research

The experiments were carried out on two organic low-boiling heat transfer fluids-ethanol and acetone with different temperatures start steaming. The heat exchanger-evaporator of the circuit was heated by two methods:

- hot air-using the heat gun;
- water heated up to 90 ^{0} with by placing it in the tank thermostat.

The installation path was a long stretch of pipelines, the efferent formed in evaporator couples located above the heat exchanger-air cooled condenser; the resulting condensate is returned by gravity into the evaporator. Cooling heat exchanger-condenser proceeded convection air.

Installation path had significant dimensions, so the total length of the steam and condensate pipelines exceeding 15 meters, which is required for heat dissipation over considerable distances in projects of modern nuclear power plant. Processes of evaporation, condensation, steam and condensate flows develop in time and have a nonstationary nature. View the basic units of the pilot stand is shown in Figure 1.

a b c

Fig. 1. The main components of the experimental stand a-evaporation section; b-transport pipeline; c- condensation section

For information about changes of temperatures and pressures in the long path was developed by fast automated measuring complex consisting of analog-to-digital converter is compatible with the personal computer and multiplexer temperature sensors. The system allows in automatic mode to produce synchronous multi-channel measurement temperature with simultaneous formation of the array data in the computer's memory. The symbols of the temperature sensors and their location on the plant circuit are given in Table. 2.

Table 2. The designation and location of thermocouples

Designation	Location
1	The output of the evaporator
2	Upstream, at a height of 2 m from the evaporator
3	Upstream, at an altitude of 5 m from the evaporator
4	Entrance into the condenser
5	Output capacitor
6	The downward flow, at an altitude of 5 m from the evaporator
7	The downward flow, at an altitude of 5 m from the evaporator
8	Evaporator inlet

3 The results of research

Results of experimental researches for acetone are shown in Fig. 2 . Studies were carried out in non-stationary mode, when heating of the evaporator water heated to 90° C water, by putting it into the tank thermostat.

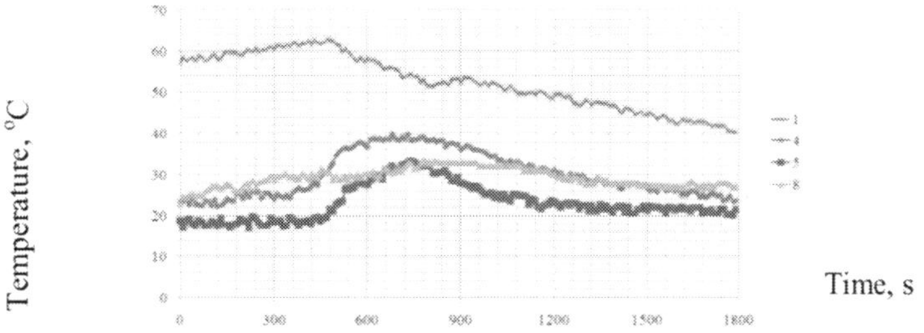

Fig. 2. Temperature dependence of characteristic points from time to time for acetone

The second series of experiments was carried out using as coolant ethanol. The studies were conducted in the stationary mode when heated by hot air, filed by using the eat gun. The results of these studies are shown in the figure. 3.

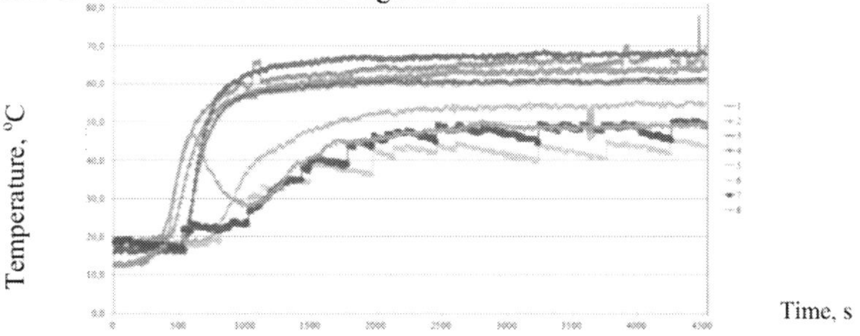

Fig. 3 Temperature dependence of characteristic points from time to time for ethanol.

4 Discussion on and analysis of the results of the study

Data experiments as for acetone and ethyl alcohol obtained both stationary and nonstationary methods show the following:

-when all studied thermal loads in parodinamicheskom circuit appeared steady circulation phases, there were no limitations of heat pipes and thermosyphons, including associated with condensate steam rising movement obstacle thread;

-after a transition process associated with warming and vaporization of stabilization occurs beginning temperature and pressure in the circuit.

Score transmitted heat output path in unsteady mode according to the number of allotted heat from the tank with heated at 90 °C with water and heat exchanger-condenser announcement:

-full power assigned to heat exchanger-evaporator by heating medium:

$$Q = V \cdot \rho \cdot c_p \cdot \Delta t / \tau, \quad W, \tag{1}$$

-heat flux density in heat exchanger-evporator:

$$q = Q / L \cdot S, \quad W/m^2, \tag{2}$$

- heat flux density in axial direction:

$$q_{oc} = 4 \cdot Q / \pi \cdot D^2, \quad W/m^2, \tag{3}$$

where: V-the volume of hot water in the tank, m^3, ρ,C_p -density and specific heat capacity of water, kg/m^3 and j/kg, Δt -water temperature change in process of heat dissipation, τ-measurement time, sec., L, S -length and perimeter pipe in heat exchanger evaporator, m, D - diameter of pipelines of heat transfer circuit, m.

The processing of the measurement results with the aid of relations (1-3) made it possible to evaluate the heat transfer capacity of the vapor-dynamic circuit. Results of processing are given in table 3.

Table 3. Thermal transfer power of the vapor-dynamic circuit

The centrality of power, w	The centrality of power without the heat loss, W	The density of heat flow in the evaporator, kW/m^2	Heat density in the axial flow the direction of a path, kW/m^2
465	445	3.44	$2.21 \cdot 10^3$

5 Conclusions

It has been experimentally shown that it is possible to divert considerable thermal power over considerable distances (over 10 m) using a developed vapor-dynamic circuit with low-boiling heat carriers. The results of axial transfers of heat are very high (above 10^6 W / m^2) and can be further increased by intensification of heat exchange (finning) both from the evaporator side (cooled zone) and from the condenser side (heat removal zone).

The work was carried out with the financial support of the Government of the Russian Federation, Decree No. 211, Contract No. 02.A03.21.0006

References

1. I.L. Pioro, L.O. Antonenko, L.C. Pioro, *Effective heat exchangers with two-phase thermosyphon* (Kiev, 1991) [in Russian]
2. M.K. Bezrodniy, I.L. Pioro, T. Kostiuk, *Transport processes in two-phase thermosiphon systems. Theory and practice*. (Kiev, 2005) [in Russian]
3. L.L. Vasilyev, V.G. Kiselev, Yu.N. Matveev, F.F. Molodkin, *Heat exchanger, heat pipes* (Minsk, Nauka i tekhnika, 1987) [in Russian]
4. D. Japikse, R.F. Winter, Int. J. Heat Mass Transfer **14**, 427 (1971).
5. A.V. Podoprigora, I.I. Sviridenko, D.V. Shevelev, Vestnik Sevntu 119 (2011)
6. S.E. Shcheklein, E.V. Starikov, Yu.E. Nemikhin, A.D. Nikitin, A.V. Zhukov, S.A. Korzhavin, Int. Sci. J. Alternative Energy Ecology 86, (2015)

Thermodynamic analysis of environmental problems of energy

Boris M. Kaganovich[1,*], *Maxim S.* Zarodnyuk[1], and *Sergey V.* Yakshin[1]

[1] Melentiev Energy Systems Institute of SB RAS, 664033 Irkutsk, Russia

Abstract. The paper discusses the problems of the ecological analysis of physicochemical processes in power units and the impact of energy systems on the nature in large territorial regions. The model of extreme intermediate states developed at the Energy Systems Institute based on the principles of classical equilibrium thermodynamics was chosen to devise specific computational methods. The results of the conducted studies are presented and directions for further work are outlined.

1 Introduction

Development of methods for the ecological analysis of the energy and other sectors of the national economy is stimulated both by the urgency of the problem of man-induced nature pollution and by the rapidly expanding possibilities for its fulfillment, which is determined by the improvement of computer engineering and information technologies. The model of extreme intermediate states (MEIS) was applied at the Energy Systems Institute for realization of the emerging possibilities [1]. This model is constructed based on simple and universal principles of classical mechanics and equilibrium thermodynamics: conservation, equilibrium and extremality. The model of extreme intermediate states is presented in the language of the mathematical programming which is a modern theory of equilibrium extreme states. Unlike the traditional thermodynamic models designed exclusively for searching for the final equilibrium point, it allows us to consider the entire region of thermodynamic attainability from the given initial state of the modeled system and to find the points corresponding to the extreme values of the quantities of interest to the researcher in this region. This gives an idea of the possible results of processes occurring in the system without determining the trajectories of their achievement. The accuracy of the created MEIS modifications significantly improved by the inclusion of kinetic constraints on individual stages of the studied processes. Circuit variants of the model of extreme intermediate states were designed to represent the objects of analysis in the one-dimensional, always potential, space of variables [1–4]. Graphs in the form of circuits can represent either real schemes of transport networks (electrical, pipeline), or conventional schemes of a wide variety of displacements (for example, rail and road transport). Application of the circuit MEISs allowed us to take advantages of the thermodynamic space of variables with monotonic characteristic functions, and the potential, one-dimensional space of variables. The next step in improving equilibrium thermodynamic modeling

* Corresponding author: thermo@isem.irk.ru

consisted in development of a method for constructing trajectories which is applicable to the analysis of both reversible and irreversible processes [3, 4]. The initial idea was to stop using the general equation of the process and to switch to the step-by-step increase in the optimal results of computations. It was decided to take the steps in space and time so small that the assumptions on the motion stationarity and the reducibility of the solved intrastep problem to the problem of the equilibrium search for an extreme state proved to be permissible.

The equilibrium thermodynamic analysis of the efficiency of energy technologies and systems which was conducted on the basis of the created MEIS modifications revealed to a great extent both the fruitfulness and the problems of further development of its ecological component. The issues of the optimal use of the gained experience are considered below.

2 The results and directions for the further development of thermodynamic analysis

The applied MEIS modifications which allow the analysis of fuel combustion and processing and formation of harmful substances in the atmospheric air found the most extensive application. In the case of fixed temperature T, pressure P and initial composition of the reagents, this modification has the form:

find $$\operatorname{extr}\left[F(\mathbf{x}) = \sum_{j \in J^{\mathrm{ext}}} c_j x_j\right] = F(\mathbf{x}^{\mathrm{ext}}) \qquad (1)$$

subject to $$\mathbf{Ax} = \mathbf{b}, \qquad (2)$$

$$D_{\mathrm{t}}(\mathbf{y}) = \left\{\mathbf{x}: \left| \begin{array}{l} \mathbf{x} \le \mathbf{y}, \qquad (3) \\ \varphi_r(x_r) \le \psi_r, r \in R^{\mathrm{lim}}, \qquad (4) \end{array}\right.\right\}$$

$$G(\mathbf{x}) = \sum_j G_j(\mathbf{x}) x_j, \qquad (5)$$

$$\mathbf{x} \ge 0, \qquad (6)$$

where $\mathbf{x} = (x_1,...,x_n)^T$ – vector of system composition (mole quantities of components); $\mathbf{y}$ – vector of initial composition, $\mathbf{y} \subset \mathbf{x}$; $\mathbf{x}^{\mathrm{ext}}$ – composition of mixture at the point of partial or full equilibrium that corresponds to the extreme value of the objective function $F(\mathbf{x})$; J^{ext} – set of indices of the components whose extreme concentration of the mixture should be determined; c_j – coefficient ranking usefulness, harmfulness or any other property of the j-th component of $\mathbf{x}$; $\mathbf{A} - (m \times n)$ – matrix of element content in the system components; $\mathbf{b}$ – vector of element mole quantities; $D_{\mathrm{t}}(\mathbf{y})$ – region (set) of thermodynamic attainability from the initial state; φ_r and ψ_r – limiting kinetic function of the r-th component of $\mathbf{x}$ and its limiting value, respectively; R^{lim} – set of indices of constraints on the macroscopic kinetics; G – Gibbs free energy of the system. In ecological problems the objective function $F(\mathbf{x})$ implies a sought concentration of harmful products of the studied process. Equation (2) is the material balance. Expression (3) is the condition that the Gibbs energy does not increase. Sign $\le$ inside the brackets is understood in a physical (thermodynamic) sense: $\mathbf{x}_2 \le \mathbf{x}_1$, if it is possible to switch from $\mathbf{x}_1$ to $\mathbf{x}_2$ by the path, along which $G(\mathbf{x})$ does not increase monotonically. The thermodynamic description of the kinetic constraint (4) applied to solve specific problems was the most complicated problem in constructing model (1)–(6). An example of such a problem is the analysis of harmful substance formation (for example, nitrogen oxides) in the flame combustion of brown coal. The traditional kinetic model can be constructed based on the information about the whole mechanism of the total process of every stage (pyrolysis, combustion of volatiles and combustion of coke), and the values of the kinetic coefficients. Application of MEIS includes only description of the limiting stages allowing such difficulties to be overcome.

The possibilities for improving the analysis of fuel combustion and processing were revealed in three main directions: 1) increase of the theoretical component share in estimating the limiting concentrations of harmful emissions, and, correspondingly, reduction of the required volume of full-scale experiments, 2) increase of the versatility of performed studies and transition from searching for the content of individual pollutants in the mixture of final reagents to searching for their complete composition; 3) determination of both the concentrations and the attainment trajectories of their limiting values. The last direction concerns improvement of the methods for control of technological processes and creation of opportunities to influence the environmental compatibility of these processes [3, 4].

In the thermodynamic analysis of atmospheric pollution, one of the successes was the substantiation of the possibility for such an analysis. After all, the opinion that thermodynamics cannot be applied to the atmospheric chemistry in principle is widely spread. It follows from the notion of the extremely low rate of chemical reactions at a relatively low air temperature and the unattainability of the point of final equilibrium. Simplicity of finding the points of both final and partial equilibria with the help of MEIS removed all difficulties associated with the slowly running atmospheric processes. This agrees well with the available experimental data of numerous results of calculations of the limiting concentrations of various atmospheric pollutants and change of these concentrations as a function of air temperature, pressure, and humidity.

The revealed influence of high-capacity wind power plants (500 and 1000 MW) on the physicochemical atmospheric processes turned out to be an interesting result [4]. The analysis was carried out with the help of the conventional circuit model for air flow distribution in the area of the wind power plant location. The performed calculations showed that the sudden fluctuations in atmospheric pressure caused by the wind turbine operation can lead to a sharp change in the ratio between the amounts of water vapor and condensate. In turn, the increase in the latter leads to an increase in the surface of the droplets, on which solutions of harmful substances are formed. As a result, the "harmful" precipitation becomes more probable.

The results achieved in the analysis of fuel combustion and processing and air pollution suggested to the authors the statement of a fundamentally new task of creating a conventional thermodynamic-hydrodynamic circuit model for propagation and transformation of energy and industrial emissions in the airspace of a large territorial region. The pollutant flows do not affect the distribution of air flows in the circuit, since they have an immeasurably smaller mass than the flows of the carrier "fluid" (air). At the same time, the material balances for these flows should be observed at the nodes of interface of the branches of the air "network", and the nodes of inflows and outflows of "dirt". In addition, it is necessary to determine changes in the composition of pollutants as a result of chemical reactions on the circuit branches. Under the accepted assumptions, the model of flow distribution takes the form:

find
$$\min\left(\sum_{i=1}^{n} z_i x_{ai}^{\beta+1} + \sum_{i=1}^{n} P_i^{W} x_{ai}\right) \tag{8}$$

subject to
$$\mathbf{A}\mathbf{x}_{a} = \mathbf{Q}, \tag{9}$$

$$\sum_{jk} x_{dki} = 0, \quad \sum_{\ell k} x_{dki} = 0, \tag{10}$$

$$\mathbf{x}_{\mathrm{d}i}^{\mathrm{end}} = f(\mathbf{x}_{\mathrm{d}i}^{\mathrm{beg}}), \quad (11)$$

where $\mathbf{x}_a$ – air flow vector on the branches; $\mathbf{P}^W$ – pressure "loss" vector at wind turbines; z_i – resistance of the i-th branch; $\mathbf{A} = [a_{ij}] - (m-1) \times n$ – matrix of connections of independent nodes and branches; $\mathbf{Q}$ – vector of external inflows and outflows; j and ℓ – indices of nodes of air and dirt inflows and outflows, respectively; $\mathbf{x}_{dk}$ – vector of flows of the k-th pollutant on the branches interfacing at the j-th node; $\mathbf{x}_{\mathrm{d}i}^{\mathrm{end}}$ and $\mathbf{x}_{\mathrm{d}i}^{\mathrm{beg}}$ – vectors of pollutant flows at the end and beginning points of the i-th branch, respectively. Transition from $\mathbf{x}_{\mathrm{d}i}^{\mathrm{beg}}$ to $\mathbf{x}_{\mathrm{d}i}^{\mathrm{end}}$ is the result of chemical reactions occurring on the branch. Tendency of the objective function to the minimum point is explained by the circuit passivity. The function $f(\mathbf{x}_{\mathrm{d}i}^{\mathrm{beg}})$ is determined using MEIS similar to model (1)–(6). In the analysis of the features of this function the pollutant inputs from the environment and the outflows into it are taken into account along with chemical reactions, In addition to pollution sources, this conventional scheme may include indirect pollutants: wind and hydraulic power stations which affect the air pressure and humidity and, consequently, the processes of harmful substance formation in it.

Development of the modifications of model (8)–(11) which could find wide practical application is surely associated with extremely great difficulties. However, the accumulated experience of equilibrium thermodynamic modeling allows us to hope for a successful solution of the emerging problems.

3 Conclusion

1. In the ecological analysis of energy development the role of equilibrium thermodynamic modeling should substantially increase, since the use of simple and universal natural regularities in it is perfectly combined with the use of modern computer engineering and information technologies.
2. In the application of thermodynamics the model of extreme intermediate states proved to be very effective. However, in further environmental studies its modifications will be equally important, representing the modeled systems and processes in both multi- and one-dimensional circuit spaces of variables.
3. In addition to estimating the efficiency of individual facilities and power plants, more attention should be paid to estimating the environmental efficiency of energy systems and energy sector as a whole.

References

1. A.N. Gorban, B.M. Kaganovich, S.P., Filippov, A.V. Keiko, V.A. Shamansky, I.A. Shirkalin, *Thermodynamic equilibria and extrema. Analisis of attainability regions and partial equilibria* (Springer, 2006)
2. B.M. Kaganovich, S.P. Filippov, A.V. Keiko, V.A. Shamansky, Adv. Chem. Eng., **39**, 1, Elsevier (2010)
3. B.M. Kaganovich, S.P., A.V. Keiko, V.A. Shamansky, M.S. Zarodnyuk, *Chemical Kinetics* (Intech, Rijeka, 2012)
4. B.M. Kaganovich, *Equilibrium thermodynamics. Problems and perspectives.* 2nd ed., (LAP LAMBERT Academic publishing, Saarbrucken, 2015)

Experience of operation of the gas turbine units in Russky Island

Dmitry Bibikov[1,*], *Anton* Polei[1], *Konstantin* Shtym[1], and *Konstantin* Tsoi[1]

[1] Far-Eastern Federal University, 8, Sukhanova Street, Vladivostok 690950

Abstract. In the present article, the heat and power supply system based on the distributed cogeneration and created as part of the APEC Summit and development of the southern Far East is considered. A brief review of the experience of operating the gas turbine units of the Co-generation plant in the Russky Island is given. The output of the heat energy over 2015 reached 142.228 thousands Gcal which is larger than that in 2014 by 18.8 %. The specific reference fuel consumption for electric supply was 272.4 g/(kW*h) while for heat supply – 195.8 kg/Gcal. In spite of partial load of the stations with regard to electric and heat energy (volumes of consumptions by the connected consumers is less than designed loads), the indices of reference fuel consumption suggest the efficiency of the generation equipment of co-generation plant.

Introduction

The development of the Russky Island infrastructure for the 2012 APEC Summit in Vladivostok has begun from the construction of electric-power and heat-power facilities. From the inland Vladivostok, the cable of 220kV and gas pipeline for the fuel supply of three co-generation plants ("North" (Ne = 5.6 MW, Nh = 12.4 MW), ("Central" (Ne = 33 MW, Nh = 143.4 MW) and "Oceanarium" (Ne = 13.2 MW, Nh = 34.3 MW) were laid through the Bosfor-Vostochny strait. The basic equipment of the co-generation plant includes the water-heating fire-tube boilers and gas turbine units (GTU) with waste heat recovery units.

Basic part of report

The used GTU was GPB-70D with synchronous, containerized module generator GFV8 made by «Fuji Electric» firm with nominal power of 6.6 MW when operated at gas and 5.8 MW when operated at diesel fuel (at an ambient temperature of 10°C). The gas turbine installation (Fig.1) was made in the single-shaft arrangement with axial once-through medium circulation operating according to Brayton thermodynamic cycle.

The compressor of GTU has 11 steps, and first three compressor blades are equipped with swing function by means of the leverage mechanism and electrodrive. At GTU starting, these blades provide a minimum angle in order to lower starting power and to

* Corresponding author: dibibikov@yandex.ru

avoid the compressor swinging. At the outlet section of the compressor, near the combustion can, the air outlet port is mounted to open at a startup for reducing pressure in the compressor. The coefficient of pressure in compressor is 1:16 at air flow rate of 26.7 kg/s. A part of compressed air at the output is used after passing through the rotor for cooling the turbine first stage. After the compressor fifth stage, air is discharged to the bearing housings for compaction of the lubricating oil. At the moment of start and runout of turbine, the sealing air under the pressure of 0.3 MPa is drawn in from the compressor plants located in the engine room.

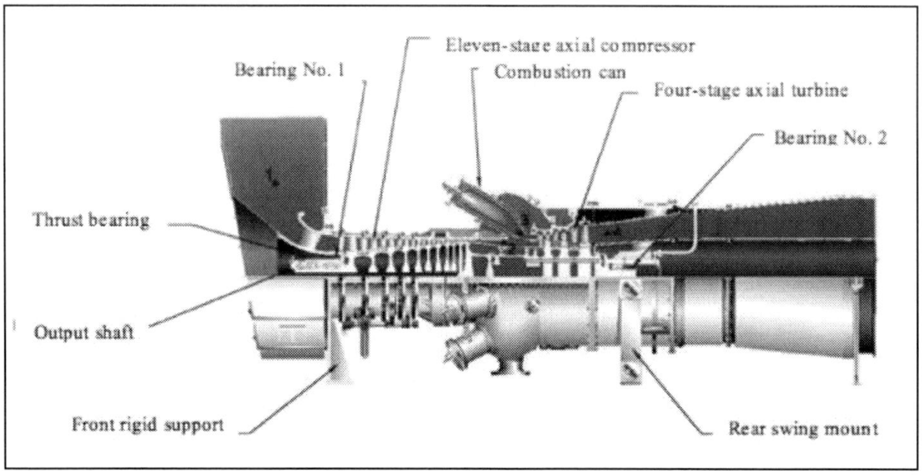

Fig.1. Longitudinal section of GTU Kawasaki GPB 70D

The natural gas pressure for operation of the installation should be supported at a level of 2.1Mpa (+- 0.1 MPa), therefore to increase of the main gas pressure (1.2 MPa), the piston, gas booster compressor VWFD-5,1/8-23 (LTB) made by Bengbu Hongshen firm (China) is used. The maximum discharge of natural gas per one GTU is 3000 NCMH.

The turbine driving a shaft consists of four stages with input temperature of 1200°C; the first stage is cooled by the compressor-bleed air и and made of high-temperature resistant metal with thermal barrier coating. The generator shaft is connected with the turbine through couplings and planetary transmission of gear-reduction box, reducing the rotating speed from 13790 rpm to 1500 rpm which is also serves as a driving actuator for main oil pump and diesel fuel pump.

The combustion cans (CC) (Fig.2) have 4 injectors for gaseous fuel: control (P), primary (M1), secondary (M2), auxilliary (S) as well as 2 injectors for the fluid fuel: primary fluid (Pr), main fluid (M), at that, the control, primary, primary fluid and main fluid injectors form one fuel tube.

The presence of the secondary and auxilliary injectors allows operating in DLE (Dry Low Emission) mode, to burn the diluted premixed air mixture, used while working on natural gas in parallel with network and at load over 50%.

The startup of GTU and its operation at starting modes are realized using the normal diffusion flame from the control injector (Fig. 2) in the range of «pilot zone», and burning is performed with a large excess of air (α=3.5). Beginning with 50% of nominal power, the fuel starts passing through the swirler and, in the range of 60-70%, the injectors switch over, the control injector is shut down and, instead of it, the primary, secondary and auxilliary injectors provide burning. Hence, burning is divided into the «pilot» and «basic» zones. When the installation operates in the DLE mode, a level of NOx does not exceed 50 mg/m^3 which corresponds to GOST 23290-78.

 DOI: 10.1051/matecconf/201711506010

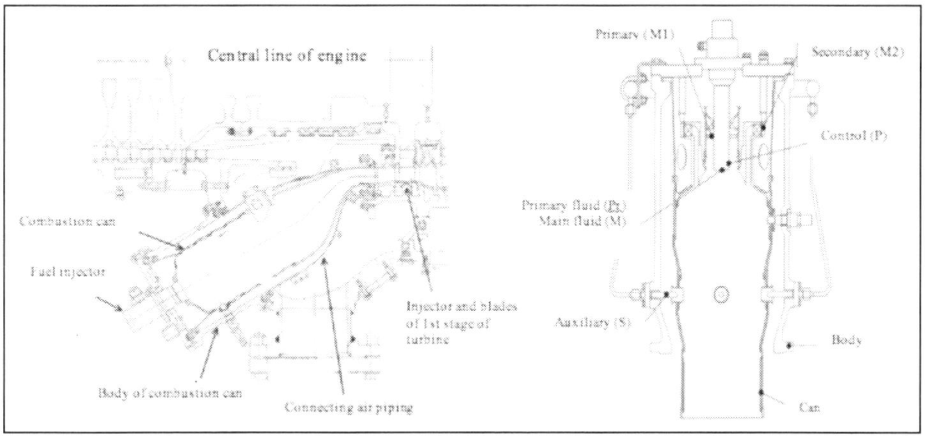

Fig. 2. Combustion can of GTU Kawasaki GPB 70D (in section)

After mounting and checking the assembly diagrams, the commissioning works with adjustment of GTU (natural gas) operating modes were accomplished (Fig. 3). In Fig. 3 shows such parameters as efficiency factor, power, mass flow of exhaust gases, flue-gas temperature, consumption of fuel depending on the ambient air temperature.

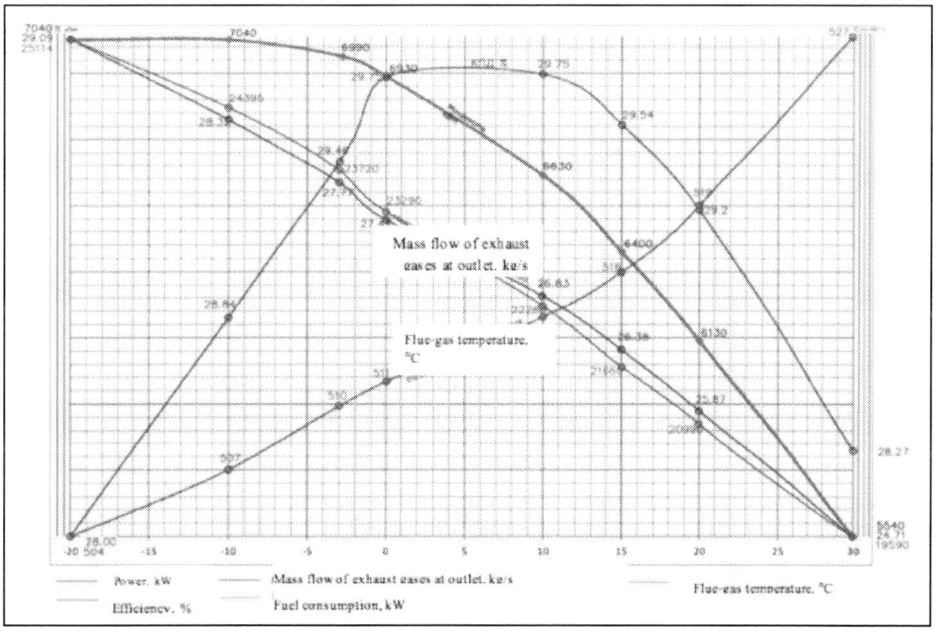

Fig. 3. Dependence of maximum power on initial temperature of GTU No/ 1 (gas) of co-generation plant «Oceanarium»

From the diagram, it follows that when increasing the ambient air temperature, the power of GTU decreases from 7040 to 5540 kW due to the growth of flue-gas temperature and a decrease the absolute thermal efficiency of GTU is also observed. The GTUs Kawasaki GPB 70D proved themselves as the highly automated and reliable under all operation conditions. This is really very high level of power production technology to which it is necessary to reach after.

In February, 2016, the boroscopic inspection of the combustion cans (CC) of the GTU No. 4 of the co-generation plant « Central » was carried out which showed the burning through of the deflector body (about 70% of surface area) and burn-off of edges of materials in the first section at two points with dimensions of 100x15mm and 40x10mm in combustion can No. 2.

The inspection of the dismounted combustion cans No. 2 and 4 (Fig. 4) was performed. The transition ducts, nozzle diaphragm blades of the first and second stages and impeller blades of the first stage of turbine were examined through the flanges of dismounted combustion cans. In two sections of the combustion can No. 4, there are segments with detachment of fireproof enamel which can result in burnings-out in case of further operation. During the following boroscopic inspections, it is needed to monitor the possible development of defect.

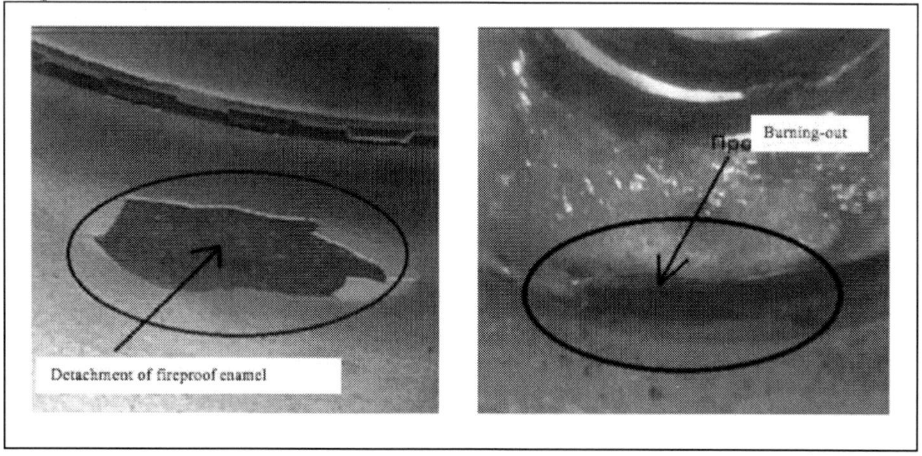

Fig. 4. Combustion can No. 2 and 4

Conclusion

The installed equipment of the distributed energetics based on co-generation is the first experience in the Far-Eastern region and holds great prospects for development. Their reliability and efficiency will serve to provide consumers of the Russky Island with heat and electric energy. In 2016, mean life time of machines reached 3500 after which the boroscopic inspection of all machines was carried out and the CCs and nozzle blocks were subject to the non-destructive testing. As of April, 2017, the mean time of the GTU of co-generation plant is 7000 equivalent hours, it means that the following stage of maintenance (inspection of high-temperature part which will show the state of transition duct and blades of turbine) is needed in order to timely determine and eliminate defects which can arise in operation.

References

1. Kawasaki GPB70D. Generator plant of gas turbine. Operation manual. Kawasaki Heavy Industries, LTD. (to be published)
2. S.V. Tsanev, V.D. Burov, A.N. Relizov, Gas-turbine and combined cycle installations of power plants **584** (2002)
3. Y.B. Goncharenko, A.K. Polei, SB of RAS **225** (2014)
4. K.A. Tsoi, A.K. Polei, M.A. Makarov, Collection of Scientific Papers **646** (2016)

Numerical study of aerodynamics and brown coal combustion in the vortex furnace with air excess variation

Denis Krasinsky [1,*]

[1]Kutateladze Institute of Thermophysics SB RAS, 630090 Novosibirsk, Russia

Abstract. The results of numerical modelling of 3D turbulent two-phase reacting flow with account for all the principal heat and mass transfer processes during the pulverized brown coal combustion in the vortex furnace of a power plant boiler unit have been presented. For two computational cases where air excess coefficient α was varied (set to 1.15 and 1.25), the detailed aerothermochemical 3D structure of reacting flow in the furnace volume has been revealed. The comparison of integral heat engineering parameters and NOx emissions obtained in the two cases has shown a slightly improved vortex furnace performance in the case α=1.25.

1 Introduction

Development of modern furnace devices matching the demands for clean and efficient coal combustion at thermal power plants is vital for heat power engineering [1]. With this aim, it is necessary to apply prospective scientific approaches to elaboration of projectible coal-fired furnaces. In this context, taking into account the need to utilize the brown and low-rank coals, one of the promising approaches is the technology of pulverized coal combustion in vortex flow, implemented in a so-called vortex furnace [2]. The vortex furnace basic design, developed in earlier years at the Central Boiler and Turbine Institute (CBTI-design) [2], consists of the following main elements (see Figure 1): the vortex combustion chamber located in the lower part of the furnace, the diffuser, and the cooling chamber in the upper part of the furnace device. The input stream of coal-air mixture is supplied through the rectangular nozzle located over combustion chamber (before the diffuser). Thus the primary swirl (vortex) flow in the combustion chamber is formed by the tangential injection of fuel-air input jet. In this vortex furnace design a clear functional distinction is assumed between the combustion chamber (with refractory-lined walls) where the high-temperature pulverized coal combustion regime with liquid slag removal takes place, and the upper-part cooling chamber packed with heat-receiving water-walls.

In the previous research works the complex experimental and numerical studies on aerodynamics and turbulent flow features have been carried out at laboratory-scale vortex furnace model in isothermal formulation [2, 3]. It has been concluded there that aerodynamic features of the swirl flow inside the vortex combustion chamber provide the

* Corresponding author: dkr@itp.nsc.ru

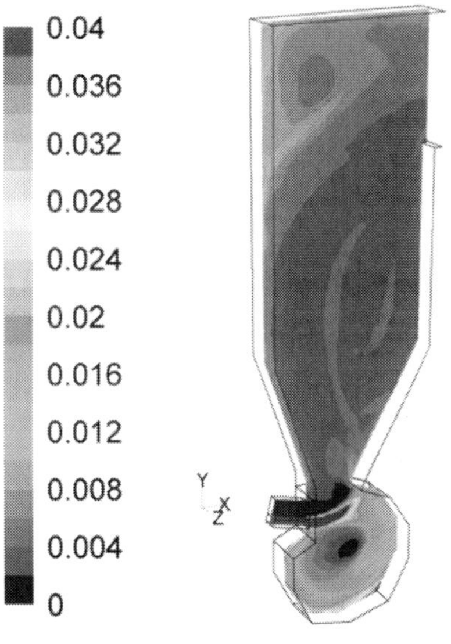

Fig. 1. 3D-view of the vortex furnace and the field of disperse phase concentration in XY-section across nozzle center, kg/m3.

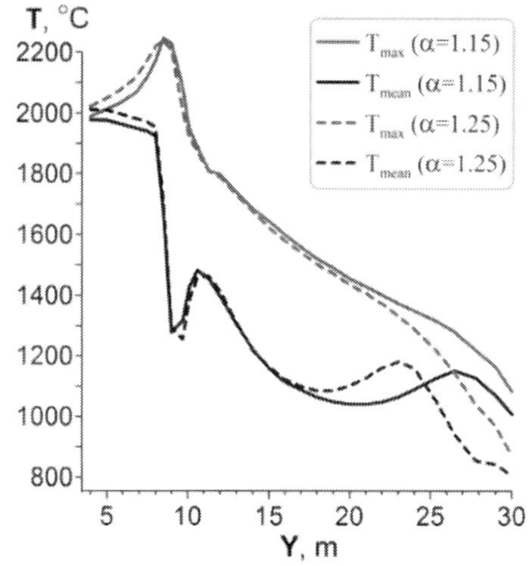

Fig. 2. Distributions of maximum temperature (red lines) and the temperature averaged over *XZ*-sections (black lines), along the vortex furnace height, °C. Cases of air excess α=1.15 (solid lines) and α=1.25 (dashed lines).

following advantages during the coal-firing process: - stable structure of two-phase reacting flow; - prolonged residence time of coal particles, thus facilitating the completeness of coal burnout; - intensified heat- and mass-transfer due to enhanced mixing and high temperature level (this also allows to reduce the size of the furnace and entire boiler unit at the furnace design stage); - mitigation of NO_x emissions due to the effect of recirculation of combustion products towards reaction zones inside the vortex chamber. Numerical simulation of the flow and pulverized-coal combustion processes in the full-size vortex furnace has been performed in [4] on the basis of Euler-Euler description for the both carrier gas and disperse phases, however, only monodispersed coal particles were considered in [4]. The numerical modelling of coal-firing processes in the full-size vortex furnace has also been carried out in [3] with the use of Euler-Lagrange approach which enables to represent in details (e.g. in 100 fractions by size, or more) the polydispersity of coal particles fed into the furnace.

Two new modifications of the basic vortex furnace design were proposed and studied [5–7] at the Institute of Thermophysics SB RAS. These enhancements (called a dual-nozzle design) are characterized by additional tangential-injection nozzles mounted at the bottom or at the upper part of the vortex combustion chamber, allowing for more flexible control of the aerodynamic structure and the furnace regime parameters. Nevertheless, a principal improvement of the working performance hasn't yet been achieved for the new dual-nozzle vortex furnace modifications, compared to the single-nozzle CBTI-design of the vortex furnace, and further research is still necessary.

2 Mathematical model and computational results

The presented work is a continuation of the previous research [3] and is devoted to numerical investigation of two operational regimes of combustion of Mongolian brown coal from Shiwei-Ovoo deposit [8] differing by air excess coefficient α in the vortex furnace of CBTI-design. Mathematical formulation of 3D turbulent two-phase reacting flow is based on Euler-Lagrange approach. The Favre-averaged governing equations of momentum, enthalpy and species concentration transport for carrier gas phase are closed with "realizable k-ε" model of turbulence [9]. The gas phase mixture is represented by the following species: O_2, CH_nO_m, CO, CO_2, H_2O, N_2. The radiation energy transport equation is based on the use of P_1 approximation and WSGGM model for absorption coefficients. The coal composition is presumed to consist of volatiles, ash and free carbon. During Lagrangian tracking of polydispersed coal particles the heterogeneous reactions of the volatiles pyrolysis and the coke combustion are taken into account. The numerical solution was obtained on unstructured computational grid consisting of 254848 hexahedral cells. The interphase exchange of mass, momentum and heat transfer is accounted following the Particle-Source-In-Cell method [10]. Detailed description of the used models, assumptions, boundary conditions and numerical algorithm is given in [3, 6]. Computations were performed with the use of CFD package FLUENT.

For the studied full-size vortex furnace, the diameter of its combustion chamber is DF = 6.2 m and the bounding dimensions of computational domain are: Xmax= 8.2 m, Ymax= 30.6 m, Zmax=2 m. The diffuser neck width ratio is Hx/DF =0.4. The burner nozzle inclination angle to horizon (XZ plane) is 15° and the nozzle inlet cross-section size is 1.2×0.7 m (where the latter dimension is the width in z-direction). The prescribed load of dry Shiwei-Ovoo coal [8] was 3 kg/s (per computed half-section of the vortex furnace), the corresponding heat release rate is 68.6 MW. To assign the fractional composition of pulverized coal, the known Rosin-Rammler formula was applied, where the grinding polydispersity degree was presumed as 1.2 and the coal grinding fineness parameter R90 =12%. The two computational variants were performed for the furnace air excess coefficient set to α=1.15 and α=1.25.

Detailed aerothermochemical 3D structure of a reacting flow in the vortex furnace volume was obtained from variant numerical predictions, including the fields of velocity, temperature, species and dispersed phase concentrations, and heat fluxes. Some of these results are presented in Figures 1–3. The field of disperse phase concentration in XY-section (z=1 m) across the nozzle center, shown in Figure 1, represents the evolution of pulverized-coal polydisperse stream released from the nozzle. It is observed that intensive heating of the lower "border" of this two-phase stream occurs due to the hot swirl flow below it (in the vortex combustion chamber). Then the volatiles burnout takes place nearby or downstream this lower "border", leading to the ignition of free carbon content in the coke particles. From Figure 2, where the profiles of maximum temperature Tmax and the temperature Tmean averaged over horizontal XZ-sections are depicted along the vortex furnace height for the two cases of air excess, the high-temperature level inside the vortex combustion chamber can be seen (with Tmax reaching 2200°C due to adiabatic conditions at the combustion

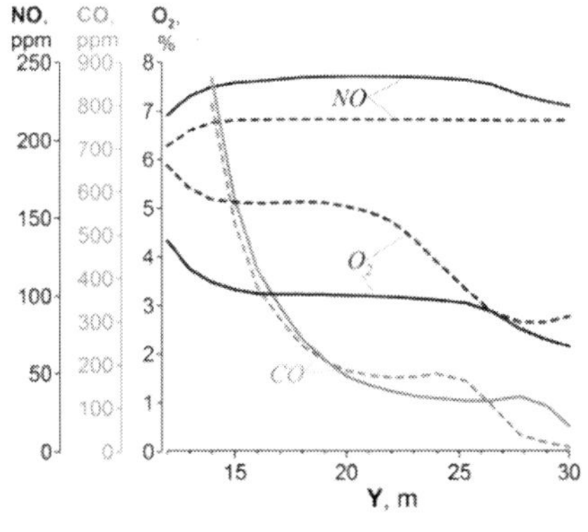

Fig. 3. Distributions of O_2, CO, NO concentrations (by volume, averaged over *XZ*-sections) along the height of the vortex furnace cooling chamber. Cases of air excess α=1.15 (solid lines) and α=1.25 (dashed lines).

chamber walls) – this provides the stable regime of liquid slag removal. For the case α=1.25, a more rapid decrease of maximum temperature *T*max can be observed inside the cooling chamber. The profiles of O_2, *CO*, *NO* concentrations averaged over *XZ*-sections are plotted in Figure 3 along the height of the vortex furnace cooling chamber, thus indicating the effect of air excess variation. The integral heat engineering and ecological parameters obtained in the outflow section of the studied vortex furnace are given in Table 1 for the two computational cases.

Table 1. Parameters at the vortex furnace outlet (averaged over the outflow section x=8.2 m).

Parameter	Case α=1.15	Case α=1.25
Mean temperature T_{mean}, °C	1045	988
O_2 concentration, vol. %	2.63	3.34
CO concentration, ppm	56.0	62.6
Coke burnout incompleteness coefficient q_4, %	1.20	1.17
NO_2 emission (at 6% O_2), mg/N m^3	390	372

3 Conclusions

The comparison of integral parameters of pulverized brown coal combustion in the vortex furnace, based on the numerical predictions for two cases of air excess, has shown (see Table 1) that the higher air excess coefficient leads to lower temperature at the outflow, slightly better carbon burnout, and also to somewhat lower NO_x emission. This allows to conclude on a slightly improved vortex furnace performance in the regime with α=1.25. Nevertheless in both variants the obtained heat engineering and ecological parameters of the studied vortex furnace meet the requirements used in thermal engineering practice.

This work was financially supported by the Russian Science Foundation (project no. 14-29-00093).

References

1. V.E. Nakoryakov, J. Eng. Thermophys. **23**, 171 (2014)
2. V.V. Salomatov, D.V. Krasinskii, Yu.A. Anikin, I.S. Anufriev, O.V. Sharypov, Kh. Enhzhargal, J. Eng. Phys. Thermophys. **85**, 282 (2012)
3. S.V. Alekseenko, I.S. Anufriev, V.G. Glavniy, D.V. Krasinsky, V.V. Rakhmanov, V.V. Salomatov, E.Yu. Shadrin, Heat Transfer Research **47**, 653 (2016)
4. D.V. Krasinsky, Int. Rev. Mech. Eng. (I.RE.M.E.) **9**, 507 (2015)
5. D.V. Krasinsky, V.V. Salomatov, I.S. Anufriev, O.V. Sharypov, E.Yu. Shadrin, Yu.A. Anikin, Thermal Eng. **62**, 117 (2015)
6. D.V. Krasinsky, V.V. Salomatov, I.S. Anufriev, O.V. Sharypov, E.Yu. Shadrin, Yu.A. Anikin, Thermal Eng. **62**, 208 (2015)
7. D.V. Krasinsky, O.V. Sharypov, J. Eng. Thermophys. **24**, 348 (2015)
8. A.S. Zavorin, A.Yu. Dolgikh, V.V. Salomatov, S. Batmunkh, Kh. Enkhzhargal, Bulletin of Tomsk Polytechn. Univ. **324**, 47 (2014)
9. T.-H. Shih, W.W. Liou, A. Shabbir, Z. Yang, J. Zhu, Computers & Fluids **24**, 227 (1995)
10. C.T. Crowe, M.P. Sharma, D.E. Stock, ASME J. Fluids Eng. **99**, 325 (1977)

Investigating the forced convection of magnetic nanofluids

Dmitriy Guzei[1,2,*], *Andrey* Minakov[1,2], *Maxim* Pryazhnikov[1,2], and *Konstantin* Meshkov[1]

[1]Siberian Federal University, 660041, Krasnoyarsk, Russia
[2]Institute of Thermophysics SB RAS, 630090, Novosibirsk, Russia

Abstract. Experimental study of the magnetic nanofluids forced convection in a constant magnetic field was carried out. Dependence study of the magnetic nanofluid heat transfer coefficient on the effect of a constant magnetic field was made. The effect of a constant magnetic field on the pressure drop during the magnetic nanofluid flow was shown.

1 Introduction

One of the solutions to the problem of heat transfer performance enhancement could be the use of so called nanofluids, which are fluids containing nanoparticles of various composition. The use of electromagnetic fields is a promising way to control the properties of nanofluids. The magnetic field can be influenced by nanofluids prepared from ferromagnetic materials (ferromagnetic nanofluids), for example, from iron nanoparticles, its oxide (Fe_3O_4), nickel etc. Such fluids can be extremely useful in many applications.

There are lots of works study of nanoparticles transport in the magnetic field in the channels [1-3]. Sundar et al. [1] studied turbulent forced convection heat transfer and friction factor of Fe_3O_4 magnetic nanofluid in a tube in the absence of magnetic field and developed correlations for the estimation of Nusselt number and friction factor. Their results show that the heat transfer coefficient is enhanced by 30.96 % and friction factor by 10.01 % at 0.6 % volume fraction compared to the base fluid. Recently, Ghofrani et al. [2] investigated the effect of constant and alternating magnetic fields on the forced convection heat transfer in a short tube. They showed that increasing the alternating magnetic field frequency increases the heat transfer up to of 27.6 % in low Reynolds numbers. Lajvardi, et al.[3] experimentally studied the convective heat transfer of water – Fe_3O_4 ferrofluid flowing through a tube in laminar regime in the presence of magnetic field perpendicular to the flow direction. They performed their experiments on different ferrofluids concentration and at different magnetic positions in magnetic fields. They concluded that by improving the thermo physical properties of the ferrofluids, heat transfer can be attributed in the presence of magnetic field.

2 Description of the experimental setup

Experimental study of the forced convection of magnetic nanofluids in a constant magnetic field was carried out. Experiments with nanofluid forced convection were performed at the

* Corresponding author: gudimas@yandex.ru

experimental stand of the Department of Thermophysics of the Siberian Federal University. The experimental unit is a loop with a circulation coolant. The experimental unit diagram is shown in Fig 1a.

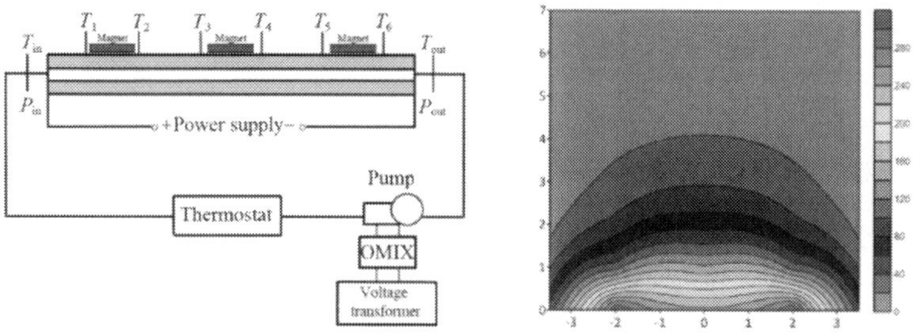

a) b)

Fig. 1. a) The experimental unit diagram; b) The magnetic field induction over the magnet (mT).

The working fluid is pumped to the heated test section from the accumulator tank. The gear pump WT30001 was used for pumping a nanofluid. Pump drives provide the ability to pump different fluids at a flow rate from 85.7 to 2571.4 ml/min.

After passing the heated section, the liquid enters the heat exchanger and gives heat to the thermostat. The working fluid flow rate in the circuit is regulated by control valves. Monitoring of the flow rate in the circuit is carried out by flowmeter. The heated section is a stainless steel tube with a diameter 10 mm and a length of 1m. The wall thickness is 1 mm. Electric current is supplied to the wall for heating. This ensures the conditions of constant heat flow on the wall. Heating power is regulated by LATR. 6 copper–constantan thermocouples are fixed on the wall at an equal distance from each other to measure the local temperature of the tube. The temperature is read by the TPM-200 meters. The temperature at the inlet and outlet of the heated section is measured by thermocouples. In this case, the thermocouple for measuring the ambient temperature at the circuit outlet is located at a considerable distance from the heated section end to ensure ambient temperature uniformity at the measurement site. The circuit section from the heater to the ambient temperature measurement site was heat insulated. The drop pressure study at the inlet and outlet of the heated section was also measured. The pressure drop over the length of the working section is measured by a differential pressure gauge OWEN PD200. The experimental unit for measuring heat transfer coefficient was tested using the known empirical data for the heat transfer of pure water.

Three permanent neodymium magnets with dimensions 50x30x10mm were used to create a magnetic field. The magnetic field intensity around the magnet was measured with a milliteslameter TPU-02. The distribution of the magnetic field induction over the magnet is shown in Fig. 1b. The maximum value of the magnetic induction near the magnet is of the order of 0.3T. In experiments, the magnets were located in the immediate vicinity of the heated section at the same distance from each other.

Iron oxide nanoparticles Fe_3O_4 were used to prepare the nanofluid. The average particle size was 100 nm. Distilled water was used as a base fluid. The volume density of nanoparticles was 0.25%. A standard two-step method was used to prepare the nanofluid. Suspensions were processed in an ultrasonic bath "Sapphire TC-10338". The nanofluids were free of surfactants.

The nanofluid viscosity was measured at a Brookfield DV2T rotary viscometer with a low viscosity ULA adapter (0). The measurement accuracy for the viscosity coefficient was not less than 2%. The viscometer was calibrated with distilled water and ethylene glycol before the measurements. The results were in good agreement with the reference data. The viscosity coefficients of water and nanofluid is shown in Table 1.

Table 1. Viscosity of water and nanofluid.

	Water	Nanofluids Fe_3O_4 0.25%
Viscosity, cP	0.902	0.929

The coefficient of thermal conductivity of nanofluids was measured using a non-stationary hot wire method. The detailed description of the installation and its testing is given in [4].

3 Experimental results

The average and local values of the heat transfer coefficient at the channel walls were measured as a result of experiments. The dependences of the heat transfer coefficient on the Reynolds number for a nanofluid with a volume density of 0.25% were obtained with and without the magnetic field action. A strong influence of a constant magnetic field on the forced convection of the magnetic nanofluid was shown during the experiments. The graph of the average heat-transfer coefficient versus time (Fig. 2) shows that the heat transfer coefficient increases with the constant magnetic field action. Measurements were made without the magnetic field action in the interval from 0 to 100 s. After that, three permanent magnets were installed near the working section wall, and the recording of the measured parameters was not interrupted. The average heat transfer coefficient was shown to increase.

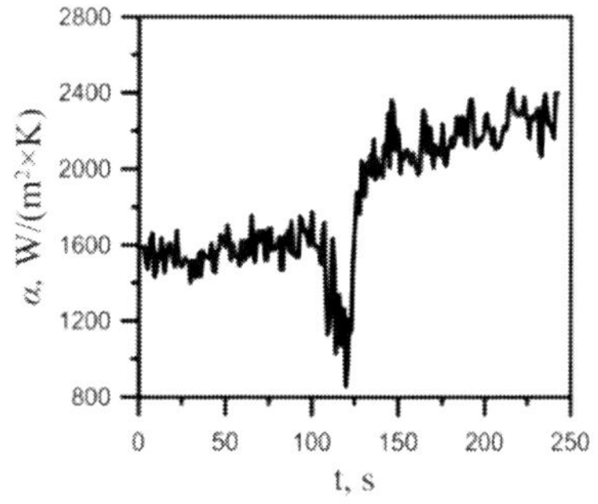

Fig. 2. Dependence of the average heat transfer coefficient on time.

The magnetic field action on the nanofluid heat transfer coefficient with a nanoparticle volume density of 0.25% was investigated. The influence of the magnetic field intensifies the average heat transfer coefficient by 13% relative to the base fluid at a fixed flow rate, while the nanofluid heat transfer coefficient without the magnetic field influence is 10% below pure water. The dependence of the heat transfer coefficient on the flow rate is shown in Fig. 3a. At a fixed Reynolds number the average heat transfer coefficient of a magnetic nanofluid with a particle concentration of 0.25% increases by 35% relative to the base fluid with the static magnetic field action produced by three magnets (Fig. 3b).

The increasing effect of the heat transfer coefficient takes place for a fixed value of the Reynolds number and the coolant flow. While the heat transfer coefficient with increasing particle concentration can decrease with a fixed value of the coolant flow rate in turbulent heat exchange of nonmagnetic nanofluids or nanofluids without the magnetic field influence, investigated earlier [5].

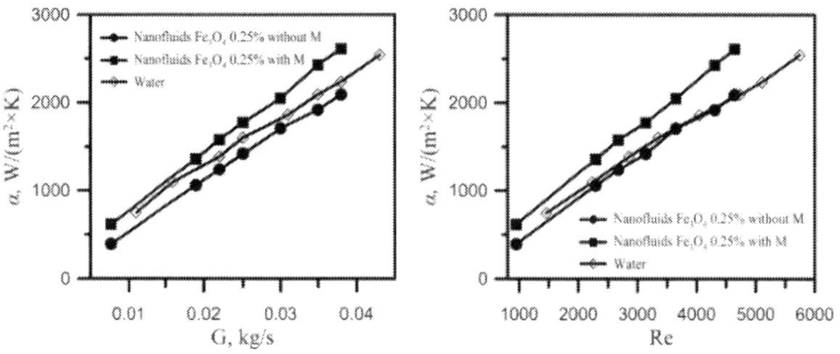

a) b)

Fig. 3. Dependence of the average heat transfer coefficient on flow rate a) and Reynolds number b) with and without the magnetic field action (Fe_3O_4 100nm, 0.25%).

Experiments have established that the magnetic field has a significant effect on the pressure drop in the working section. The graph of the pressure drop in the working section depending on the Reynolds number is shown in Fig. 4. In the experiments it was shown that three magnets located along the tube increase the pressure drop in the channel by 21% compared to the pressure drop without the field for a nanofluid with a volume density of Fe_3O_4 particles 0,25%. The pressure drop magnitude for a nanofluid with a particle volume density of 0.25% without a magnetic field at a constant Reynolds number is 30% higher the pressure drop for the base fluid.

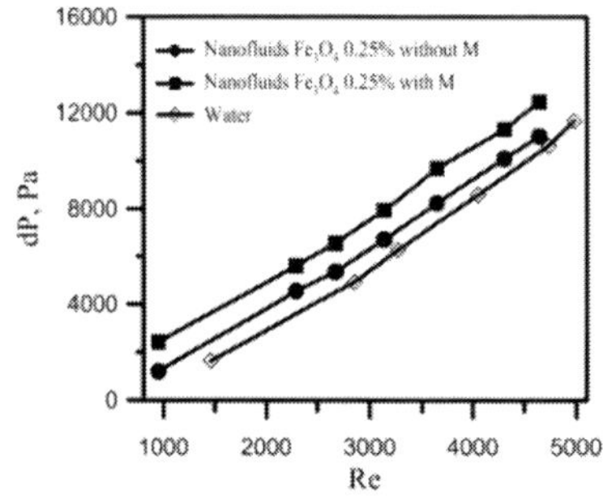

Fig. 4. Dependence of the pressure drop on Reynolds number with and without the magnetic field action.

The reported study was funded by Russian Foundation for Basic Research, Government of Krasnoyarsk Territory, Krasnoyarsk Region Science and Technology Support Fund, the research project №16-48-243061.

References

1. L.S. Sundar, M.T. Naik, K.V. Sharma, M.K. Singh, T.Ch. Reddy, Exp. Therm. Fluid Sci. **37**, 65 (2012)
2. A. Ghofrani, M.H. Dibaei, A. Hakim Sima, M.B. Shafii, Exp. Therm. Fluid Sci. **49**, 193 (2013)
3. M. Lajvardi, J. Moghimi-Rad, I. Hadi, A. Gavili, T.D. Isfahani, F. Zabihi, J. Sabbaghzadeh, J. Magn. Magn. Mat. **322**, 35083 (2010)
4. A.V. Minakov, V.Ya. Rudyak, D.V. Guzei, M.I. Pryazhnikov, A.S. Lobasov, J. Eng. Phys. Thermophys **88**, 149 (2015)
5. A.V. Minakov, D.V. Guzei, M.I. Pryazhnikov, V.A. Zhigarev, V.Ya. Rudyak, Int. J. Heat Mass Transfer **102**, 745 (2016)

Density effect on the mixing efficiency and flow modes in T-shaped micromixers

Alexander Lobasov[1,2,3*], and *Andrey* Minakov[1,2,3]

[1]Siberian Federal University, 660074, Krasnoyarsk, Russia
[2]Institute of Thermophysics SB RAS, 630090, Novosibirsk, Russia
[3]Novosibirsk State University of Architecture and Civil Engineering (Sibstrin), 630008, Russia

Abstract. Flow patterns and mixing of liquids with different densities in T-shaped micromixers are numerically investigated at Reynolds number range from 1 to 250. The density ratio of the mixing media varies from 1 to 2; its effect on the flow structure and the mixing is studied. The dependences of the mixing efficiency and the pressure difference in this mixer on the density ratio and the Reynolds number are obtained. It is shown that the density ratio has a considerable effect on the flow structure, especially before the transition from the symmetric to the asymmetric flow pattern.

1 Introduction

The miniaturization of technological processes has been actively promoted in recent years in the chemical industry, and thus micro-mechanics has become a rapidly developing and promising research area. Microchannel devices are widely used in various fields of science and technology as microreactors, micro-scale heat exchangers, micro-mixers, etc. Many studies have noted that the use of microdevices allows significantly enhancing the physicochemical processes in comparison to classical space consuming reactors [1-3]. Control over pressure, temperature, reaction time and flow velocities in reactors with small volumes is now realized much easier and more efficiently. This implies the main undeniable advantages of microreactor microsystems: safety of highly exothermic reactions and work with toxic or explosive reagents, possible reactions in supercritical conditions, significant reduction of research costs, as well as implementation and scaling of chemical processes. In this paper a T-shaped micromixer, which is one of the simplest in the manufacture, but at the same time, quite effective form of a microfluidic mixer was considered. Thus, flow regimes depending on the Reynolds number were investigated in many works [4,5]. At that, quite an interesting hydrodynamic phenomenon, namely the flow reversal or engulfment flow regime [5-8] upon reaching the critical Reynolds number of 130-160 was observed in the T-shaped mixer. The aim of the present work is a systematic study of the effects of miscible fluids densities ratio on the flow and mixing regimes in the T-shaped micromixer.

* Corresponding author: perpetuityrs@mail.ru

2 Mathematical model and numerical algorithm

We consider incompressible flows of multi-component Newtonian fluids, which are described using a hydrodynamic approach based on the solution of the Navier-Stokes equations. Currently, numerous experiments show that such description for fluids works well up to the channel size of 1 micron.

In general, the Navier-Stokes equations system has the following form:

$$\frac{\partial \rho}{\partial t} + \nabla(\rho \mathbf{v}) = 0$$

$$\frac{\partial \rho \mathbf{v}}{\partial t} + \nabla(\rho \mathbf{v}\mathbf{v}) = -\nabla p + \nabla \mathbf{T} \quad (1)$$

The density of the mixture is expressed in terms of the mass concentrations f_i of flow components and partial densities ρ_i in the following way:

$$\rho = \left[\sum_i (f_i / \rho_i) \right]^{-1}$$

Here the evolution of mass concentrations is determined by the equation:

$$\frac{\partial \rho f_i}{\partial t} + \nabla(\rho f_i \mathbf{v}) = \nabla(\rho D_i \nabla f_i) \quad (2)$$

where D_i is the diffusion coefficient of the i-th component.

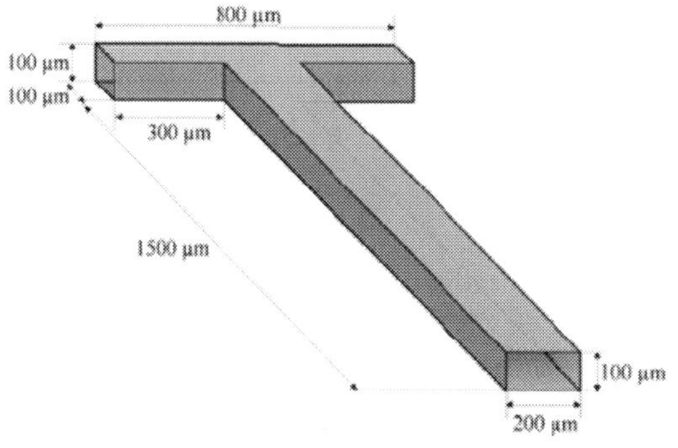

Fig. 1. Geometry of the problem.

In our calculations we used a three-block grid consisting of 9.5 million grid nodes. Preliminary calculations have shown that this level of grid particularization is acceptable in terms of calculation accuracy. The fluid flow rate determined at the inlet of mixer channels was constant with steady-state velocity profile. At the mixing channel outlet, Neumann conditions were set, meaning that the derivative normal to the outlet surface, taken of all scalar quantities, is equal to zero. The walls of the mixer were considered as insulated. For velocity vector components, the no-slip condition was taken as boundary condition on the channels walls. The applicability of this type of boundary conditions for channels with a size of 50 μm had been demonstrated in [1,4].

The channel dimensions are shown in Fig. 1. The channel thickness is 200 μm, while the width of its narrow and wide parts is 200 μm and 400 μm, respectively.

In our calculations we determined the pressure drop between one of the mixer inlets and the outlet, as well as mixing efficiency. In the literature, mixing efficiency is usually quantified using the parameter $M = 1 - \sigma/\sigma_0$, where $\sigma = V^{-1}\int_V \left(f - \langle f\rangle\right)^2 dV$ is the root-mean-square deviation of the mass fraction of mixture component f from its average value $\langle f\rangle$, $\sigma_0 = \langle f\rangle \cdot \left(1 - \langle f\rangle\right)$ is the maximum root-mean-square deviation, and V is the volume of the computational domain.

3 Results and discussion

The effect of density on flow regimes in the microchannel was studied. Water, tinted with rhodamine was supplied through one of the channel inlets. Fluid with viscosity and diffusion coefficient similar to those for pure water but with a changed density was supplied at the same flow rate through another inlet to the mixer. We have performed three series of calculations with densities 1.25, 1.5 and 2 times greater than the density of water. Because of this fact the viscosities of both fluids were set constant and equal to $\mu = 0.001$ Pa·s in such methodical research. The change in the flow regimes in the microchannel is characterized by Reynolds number, determined as $\mathrm{Re} = \rho U d_h/\mu$, where $U = Q/(2\rho H^2) = Q_{in}/(\rho H^2)$ is the superficial velocity in the mixing channel, $H = 200$ μm is the channel height, and $d_h = 267$ μm is the hydraulic diameter.

Increasing the density of one of the miscible fluids leads to significant changes in the flow pattern. For very small (less than 10) Reynolds numbers the flow structure and mixing of fluids with different densities is in general similar to the flow of two identical fluids. Therefore, the mixing efficiency in this regime slightly depends on the density of fluids (see Fig. 2a). The increase of Reynolds number in the channel causes formation of vortices, and the difference in densities starts its effect. It is clear that in contrast to the case of equal densities of miscible fluids, the interface between media is not flat. Fluid with a higher density pushes a less dense fluid from channel walls and somewhat envelops it. When fluids density ratio increases, this effect is amplified; that results in an increase of the interface area of miscible fluids and consequently the enhancement of the mixing efficiency in this regime (see Fig. 2a).

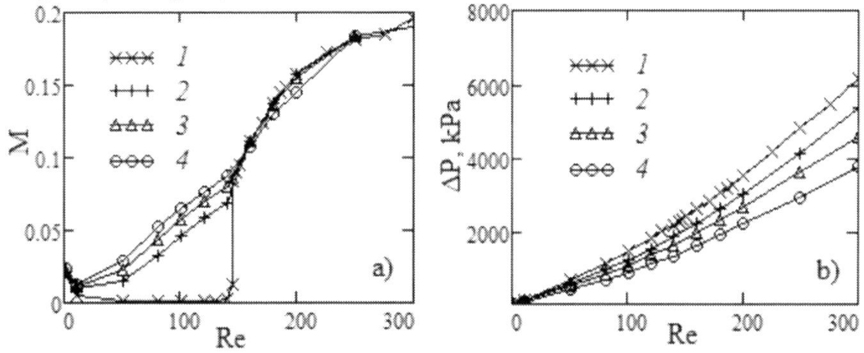

Fig. 2. Mixing efficiency (a) and pressure drop between the mixer inlet and the outlet (b) depending on Reynolds number: $\rho/\rho_0 = 1$ (1), 1.25 (2), 1.5 (3), 2 (4).

In addition, an interesting behavior is observed in horseshoe vortices (Dean vortices), which are generally preserved when increasing the density of one of the fluids. Though the increase in density ratio leads to appearance of asymmetry with respect to the central axis of the channel. The horseshoe vortex in the mixing channel is growing faster in a fluid with a lower density than in the fluid with a greater density. And the larger the density ratio the more pronounced this process. A similar asymmetry is observed in the region of secondary vortices formation at the junction of inlet channels and the mixing channel. At that, the

Dean horseshoe vortex, which is located in the more dense fluid, changes significantly with increasing the density of this fluid, in contrast to the vortex located in the water. With further increase in the Reynolds number we observed flow reversal and the beginning of engulfment regime. In the course of numerical simulations analysis, it was revealed that this transition occurs within the region of Reynolds numbers from 144 to 144.5, regardless of the density ratio at the inlet of the T-shaped channel. Flow reversal and transition from symmetric to engulfment flow regime are accompanied by a sharp increase in the mixing efficiency. In addition, the mixing efficiency jump in the transition region becomes almost unpronounced up to its complete nonoccurrence, as it is observed for mixing of fluids with density ratio equal to two (Fig. 2a).

Also, we analyzed a pressure drop between the mixing channel inlet and outlet. Fig. 2b presents the pressure drop between the inlet to the mixing channel with a more dense fluid and the outlet from the T-shaped microchannel. As is obvious, similarly to the case of $\rho/\rho_0=1$, the pressure drop in fluids with higher density is not characterized by any peculiarity in the transition region from the symmetric flow regime to the asymmetric pattern. It is interesting that with increasing density the pressure drop decreases.

4 Conclusions

The conducted numerical simulation allowed identifying the following regimes for incompressible fluid flow in T-shaped micromixers: stationary vortex-free flow; stationary symmetric vortex flow; intermittent transition from symmetric to asymmetric flow (engulfment flow); and stationary asymmetric vortex flow. It is shown that the change in the density of one of the miscible fluids does not result in a shift of flow regimes. However we observed a significant flow asymmetry in the region before its transition to the engulfment flow mode. It is established that in the symmetric flow region, the mixing efficiency increases with the increase in density ratio because of the same reason. This is caused by an increase in the interface between miscible media due to the flow around a more dense fluid over the less dense fluid.

The work is performed at partial financial support of the projects funded by the Russian Foundation for Basic Research and Krasnoyarsk Regional Fund for Support of Scientific and Scientific-Technical Activities (Contract No. 16-48-243061) and by the Ministry of Education and Science of the Russian Federation Government contract with Siberian Federal University in 2017 (No. 16.8368.2017).

References

1. P. Tabeling, *Introduction to Microfluidics* (Oxford University Press, 2005)
2. G. Karnidakis, A. Beskok, N. Aluru, *Interdisciplinary Appl. Math.* (Springer Science+Business Media Inc., 2005)
3. R. Karnik, *Encyclopedia of microfluidics and nanofluidics* (Springer, 2008)
4. A. Minakov, A. Yagodnitsina, A. Lobasov, V. Rudyak, A. Bilsky, La Houille Blanche, **5** 12 (2013)
5. A.S. Lobasov, A.V. Minakov, V.Ya. Rudyak, Fluid Dynamics. **51**, 381 (2016)
6. T. Andreussi, C. Galletti, R. Mauri, S. Camarri, M.V. Salvetti, Chem. Eng. J. **76** 150 (2015)
7. A. Fani, S. Camarri, M.V. Salvetti, Phys. Fluids. **25**, 064102 (2013)
8. A. Fani, S. Camarri, M.V. Salvetti, Phys. Fluids. **26**, 074101 (2014)

 DOI: 10.1051/matecconf/201711507003

Modeling the flow of activated H_2 + CH_4 mixture by deposition of diamond nanostructures

Mikhail Plotnikov, *Alexey* Rebrov, *Ivan* Yudin*

Kutateladze Institute of Thermophysics SB RAS, 630090 Novosibirsk, 1 Lavrentiev Ave.

Abstract. Algorithm of the direct simulation Monte Carlo method for the flow of hydrogen and methane mixture in a cylindrical channel is developed. Heterogeneous reactions on tungsten channel surfaces are included into the model. Their effects on flows are analyzed. A one-dimensional approach based on the solution of equilibrium chemical kinetics equations is used to analyze gas-phase methane decomposition. The obtained results may be useful for optimization of gas-dynamic sources of activated gas diamond synthesis.

The method of Hot-Wire Chemical Vapor Deposition (HWCVD) is widely used for diamond deposition from an activated precursor gas. Its main feature is activation of a gas precursor on metallic surfaces: wires, placed in a virtually resting gas [1] or in high-speed flows [2], as well as cylindrical channels [3]. The presented work is devoted to the numerical study of gas-dynamic aspects of the rarefied gas flow in a hot tube with dissociation and recombination on the surface. The algorithm of direct simulation Monte Carlo (DSMC) method [4] with majorant frequency scheme [5] is used for the numerical analysis and optimization of approaches to deposition of diamond-like films using the hot tubes. In [6-8] the importance of taking into account dissociation and recombination in the analysis of flows in the tubes was shown. The use of a cylindrical channel as an activator provides better gas dissociation due to multiple collisions with the surface [6-8] in comparison with the case of interaction of the supersonic flow with a wire obstacle [9,10]. A one-dimensional approach based on the solution of equilibrium chemical kinetics equations was used to analyze gas-phase methane decomposition.

The main goal of the presented work is to analyze the influence of the determining parameters on the composition and energy state of the gases flowing to the substrate. Particular attention is paid to the behavior of atomic hydrogen, one of the main components that determine the synthesis of diamond structures.

1 Computation methods

The mixture of molecular hydrogen with methane [1, 11] is used as the initial gas. The VSS molecular model [4] was adopted for simulation of molecular hydrogen, atomic hydrogen, and methane. Data for molecular hydrogen and methane are taken from [4]. To describe

* Corresponding author: yudinib@gmail.com

 DOI: 10.1051/matecconf/201711507003

interaction of the gas particles with the surface, the model of specular-diffuse reflection was used [4]. To describe the processes of dissociation and recombination, we use coefficients α_d and α_r: the hydrogen molecule splits with probability α_d into two atoms, which are reflected from the surface with complete accommodation of momentum and energy, and with probability $(1 - \alpha_d)$, it remains a molecule. Similarly, the hydrogen atom recombines with probability α_r to the molecule, which is reflected from the surface with complete accommodation of momentum and energy, and with probability $(1 - \alpha_r)$, it does not recombine. The particles that have not entered into chemical reactions are reflected from the surface according to the accommodation coefficient.

Different accommodation coefficients were specified for the cold and hot parts of the channel. The choice of dissociation and recombination coefficients for hydrogen was carried out based on a numerical technique [7] using experimental data [12, 13]. A more detailed description of the DSMC method parameters is given in [14]. Only recombination was simulated in hydrogen interaction with cold channel surface, while dissociation simulating was included in the interaction with the hot one. The analysis showed that under the involved conditions, gas-phase hydrogen dissociation and recombination make a negligible contribution to the result in comparison with the heterogeneous reactions. Consequently, gas-phase reactions of hydrogen can be neglected.

Obtaining more detailed information about the concentrations of particles approaching the substrate requires data on the degree of methane decomposition into fragments. The flow analysis shows that the flow macroparameters are distributed fairly uniformly over the radius. In this case, it is possible to use a one-dimensional approach for calculating the changes in the mixture composition along the flow axis by means of solving a system of chemical kinetics equations. Following [15], a mixture of 11 components is considered in solving the equations of equilibrium chemical kinetics: H_2, H, CH_4, CH_3, CH_2, $CH_2(S)$, CH, C, C_2H_2, C_3, and C_2H. The computations are performed for 13 forward and backward chemical reactions. The rate constants of reactions are calculated on the basis of the data obtained in [16].

2 Numerical experiments

Gas mixture of $H_2 + CH_4$ with temperature of 1200 K flows through the cylindrical channel into a chamber with a substrate (Fig. 1). The channel consists of two parts: inlet part L_{cold} = 10 mm with the cold wall (T = 1500 K) and outlet part L_{hot} = 15 mm with the wall temperature of 2400 K. The cylinder diameter is 3 mm, and the substrate diameter is 14 mm. The substrate temperature is 1300 K, and the distance to the substrate is L_{sub} = 10 mm. The hydrogen flow rate is 1500 sccm. Methane concentration is 10%. The chamber is initially filled with molecular hydrogen with background pressure P = 0, 2 and 20 Torr. Such conditions correspond to typical experiments on gas-jet deposition of diamond-like films [17-19].

The number density and streamlines of methane for P = 0 are shown in Fig. 2. The spatial distributions in Fig. 2 give a general notion about the features of the flow under consideration. Here and below, planes X = 0.06 and X = 0.07 correspond to the cutoff of the hot tube and to the substrate position, respectively. Four areas can be distinguished: the cold and the hot parts of the channel, the outflow into the vacuum, and the compressed layer near the substrate. An increase in the background pressure up to 2 Torr has a negligible effect on the flow in the tube. The presence of a background gas leads to the formation of the denser compressed layer. Axial distribution of the degree of hydrogen dissociation $K = n_H/(2n_{H2}+n_H)$ is shown in Fig. 3. Here n_{H2} and n_H are the number densities of molecular and atomic hydrogen. The distribution of relative concentration of mixture components along the flow axis for the case of outflow to vacuum is shown in Fig 4.

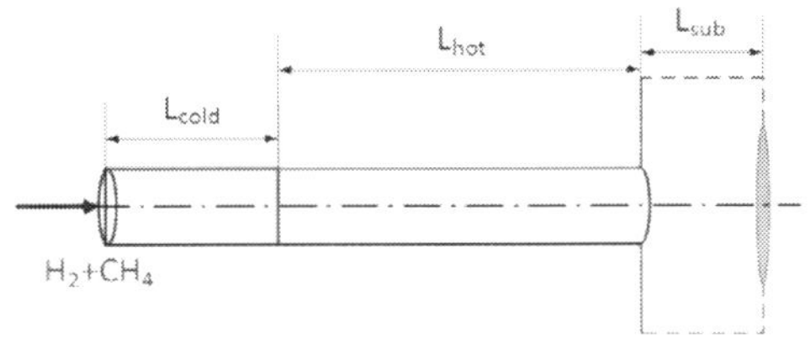

Fig. 1. The computational domain.

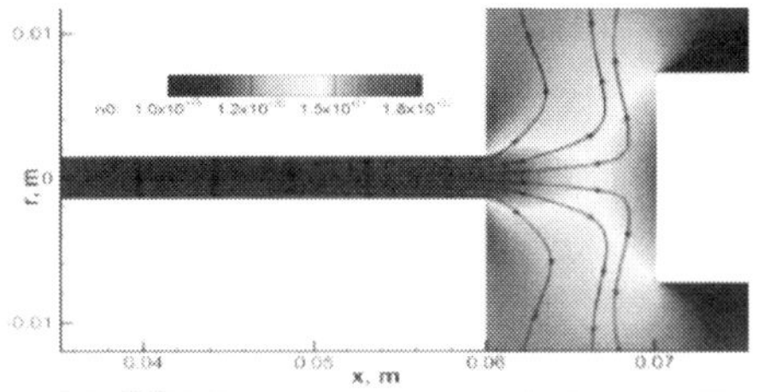

Fig. 2. The number density and streamlines of methane for the background pressure $P = 0$.

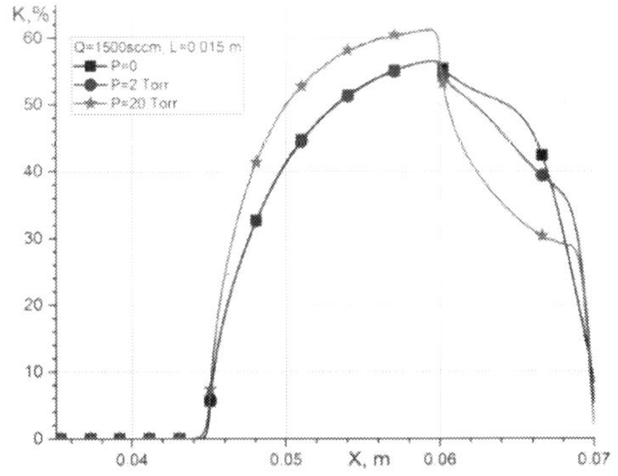

Fig. 3. Axial distribution of the degree of hydrogen dissociation for different background pressure.

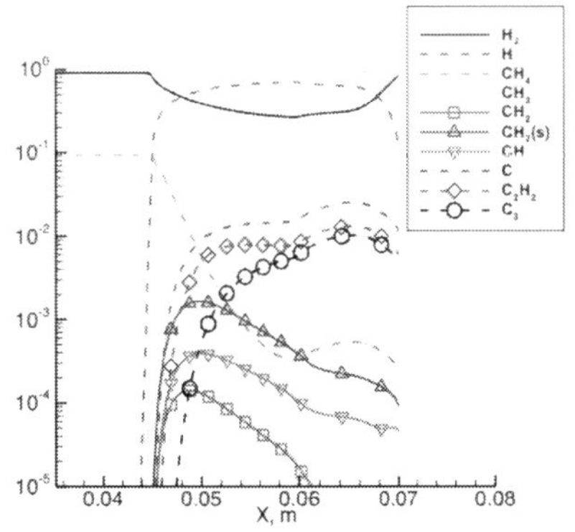

Fig. 4. Axial distribution of relative concentration of mixture components.

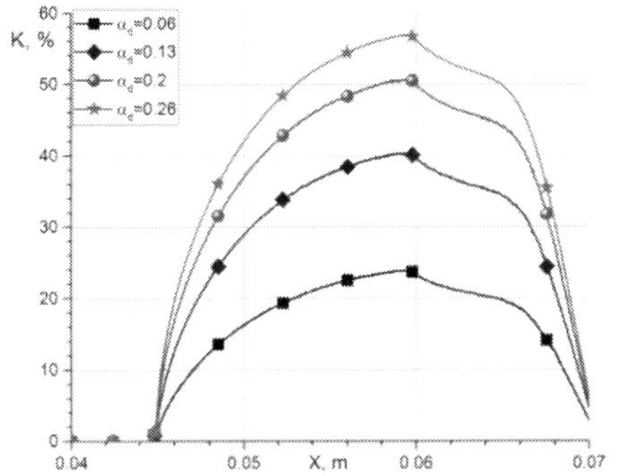

Fig. 5. Axial distribution of the degree of hydrogen dissociation for the vacuum outlet and different dissociation coefficients on tube surface.

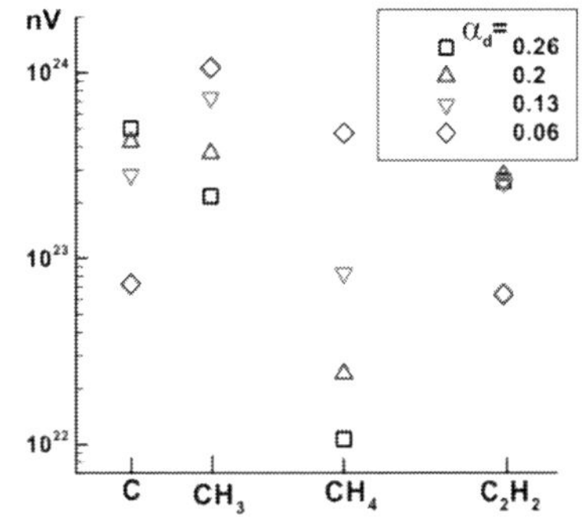

Fig. 6. nV for the four components of the mixture for the vacuum outlet and different dissociation coefficients on tube surface.

Data are obtained from the one-dimensional approach. Methane rapidly decomposes into separate fragments in the hot part of the tube. For the case under consideration, the predominant fragments are carbon, methyl, acetylene, and C_3. There is some stabilization of the mixture composition in the region of gas expansion in the chamber. One of the problems of the deposition technique under consideration is associated with the possibility of carburizing of the tungsten channel [20, 21]. In this case, catalytic activity of the tungsten surface decreases, while the rate of methane decomposition significantly depends on the amount of atomic hydrogen. Therefore, numerical experiments were conducted to assess the influence of the "catalyticness" of tungsten surface on the degree of hydrogen dissociation and methane decomposition. Axial distribution of the degree of hydrogen dissociation is shown in Fig. 5. The probability of dissociation $\alpha_d = 0.26$ corresponds to a "clean" surface. The fluxes of various mixture components near the substrate are presented in Fig. 6. The decrease in dissociation probability leads to the expected decrease in the hydrogen dissociation degree at the channel outlet. It results in a decrease in the methane decomposition, and its amount in the mixture grows up. Calculations show that in the cases under consideration atomic hydrogen is sufficient to form a large amount of methyl. However, the decrease in the amount of atomic hydrogen leads to a decrease in further fragmentation of methyl, which results in a decrease in the amount of carbon and acetylene.

3 Conclusions

In this study, the algorithm of DSMC method was used to study the hydrogen and methane mixture in a cylindrical channel. The effect of heterogeneous reactions on flows has been analyzed. A one-dimensional approach based on the solution of equilibrium chemical kinetics equations was used to analyze gas-phase methane decomposition. The obtained results may be useful for optimizing gas-dynamic sources of activated gas by diamond synthesis. They are related to problems of heat transfer by non-equilibrium flows in channels.

This work was supported by the Russian Scientific Foundation 15-19-00061.

References

1. H. Matsumura, A. Masuda, H. Umemoto, Thin Solid Films **501**, 58 (2006)
2. A.K. Rebrov, R.V. Maltsev, A.I. Safonov, N.I. Timoshenko, Thin Solid Films **519**, 4542 (2011)
3. C.B. Lukas, *Atomic and Molecular Beams. Production and Collimation* (CRC Press, 2014)
4. G.A. Bird, *Molecular Gas Dynamics and the Direct Simulation of Gas Flows* (Oxford: Clarendon Press, 1994)
5. M.S. Ivanov, S.V. Rogasinsky, Russian J. Numer. Anal. Math. Modelling **3**(6), 453 (1987)
6. A. Rebrov, I. Yudin, Phys. Status Solidi C **12**, 886 (2015)
7. M. Yu. Plotnikov, E. V. Shkarupa, Vacuum **129**, 31 (2016)
8. A.K. Rebrov, I.B. Yudin, Dokl. Phys. **61,** 223 (2016)
9. M.Yu. Plotnikov, Appl. Mech. Tech. Phys. **53, 482** (2012)
10. M.Yu. Plotnikov, A. K. Rebrov, Appl. Mech. Tech. Phys. **54, 517** (2013)
11. A.K. Rebrov, Phys. Usp. **60** (2), 179 (2017)
12. H. Koschmieder, V. Raible, Rev. Sci. Instrum. **46**, 536(1975).
13. J.N. Smith, Jr., W. L. Fite, J. Chem. Phys. **37**(4), 898 (1962)
14. A.A. Morozov, M.Yu. Plotnikov, A.K. Rebrov, I.B. Yudin, AIP Conf. Proc. of 30th Intern. Symp. on Rarefied Gas Dynamics, University of Victoria, Canada, **1786**, 050015 (2016)
15. P.W. May, M.N.R., Ashfold, Yu.A., Mankelevich, J. Appl. Phys. **101**, 053115 (2007)
16. G.P. Smith, D. M. Golden, M. Frenklach, N.W. Moriarty et. al., GRI-Mech, http://www.me.berkeley.edu/gri–mech/
17. A. Rebrov, Diamond & Related Materials **72**, 20 (2017)
18. A.A. Emelyanov, A.K. Rebrov, I.B. Yudin, Tech. Phys. **61**, 1821 (2016)
19. A.K. Rebrov, M.N. Andreev, T.T. B'yadovskiy, K.V. Kubrak, I.B. Yudin, Rev. Sci. Instrum. **87,** 103902 (2016)
20. M. Sommer, F. W. Smith, J. Mater. Res. **5**, 2433 (1990).
21. H.F. Winters, H. Seki, R.R. Rye, M.E. Coltrin, J. Appl. Phys. **76**, 1228 (1994).

Synthesis of diamond structures from the coaxial flows of H_2 and H_2+CH_4 mixture

Alexey Emel'yanov, *Alexey* Rebrov and *Ivan* Yudin*

Kutateladze Institute of Thermophysics SB RAS, 630090 Novosibirsk, Lavrentiev prosp. 1.

Abstract. The work presents the results of experiments on the diamond gas-jet synthesis from methane and hydrogen mixture, flowing from coaxial channels under the following conditions: temperature of the hydrogen activating surface of 2400 K, pressure in the deposition chamber of 20 Torr, different gas fluxes, and different CH_4 concentrations. The Direct Simulation Monte Carlo method for numerical analysis of experiments was applied.

1 Experiment

Spitsyn [1] was among the first researchers who studied chemical crystallization of diamond during the separate feed of interacting flows at the beginning of 1970s. A growth rate of 1 μ/h was attained on a diamond nucleus crystal using the flows with activated H_2 and CH_4. The low rate can apparently be explained by insufficiently high temperature of the tungsten activating surfaces and by inevitable effect of carbonization, viz., the deterioration of catalytic properties of the tungsten surfaces. The information on the effect of carbonization of the surface on the synthesis of diamond coating can be found in [2]. In the case of hot-wire chemical vapor deposition (HWCVD), the activated mixture of hydrogen and hydrocarbons is formed in single collisions of initial gas molecules with hot surfaces of heated wires. When the initial gas flow is activated on an extended high-temperature surface, multiple collisions lead to deep dissociation of hydrogen and decomposition of methane. This concept was realized in [3–7].

For the gas-jet deposition of diamond structures, the reactor with a high degree of dissociation of the hydrogen flow was elaborated and manufactured. The detailed description of this reactor (Figure 1) is presented in [7]. The experiments were performed under the following conditions: the pressure in chamber was 20 Torr, and temperatures of the tungsten tube, substrate, and tungsten capillary were 2400, 1300, and about 2400 K, respectively. The hydrogen flow rate in the annular channel of the tube 1 was 1500 sccm; the flow of mixture $H_2 + CH_4$ with CH_4 concentration of 0.3-1% was delivered through capillary 5 with hydrogen flow rate of 500–1500 sccm.

* Corresponding author: yudinib@gmail.com

 DOI: 10.1051/matecconf/201711507004

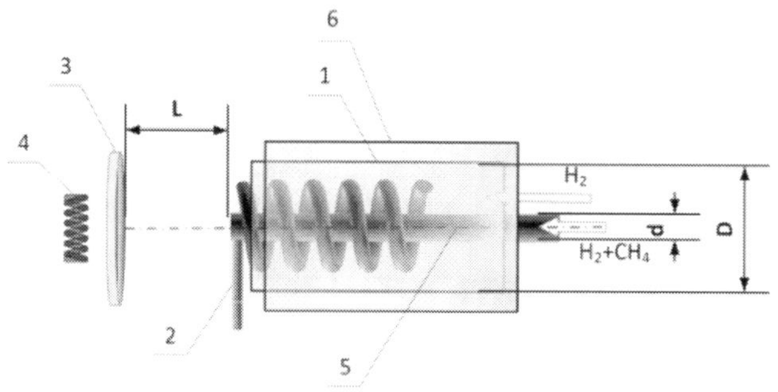

Fig. 1. Two-channel activator with a heating spiral in the hydrogen channel: (1) activator tube; (2) tungsten spiral; (3) substrate; (4) heater; (5) tungsten capillary, and (6) heat screen.

The example of crystals when the flow of methane is 8 sccm is shown in Fig. 2a. The increase of the methane concentration leads to the growth of diamond crystals (Fig. 2b). These results of experiments were obtained for hydrogen flow through a capillary 1500 sccm.

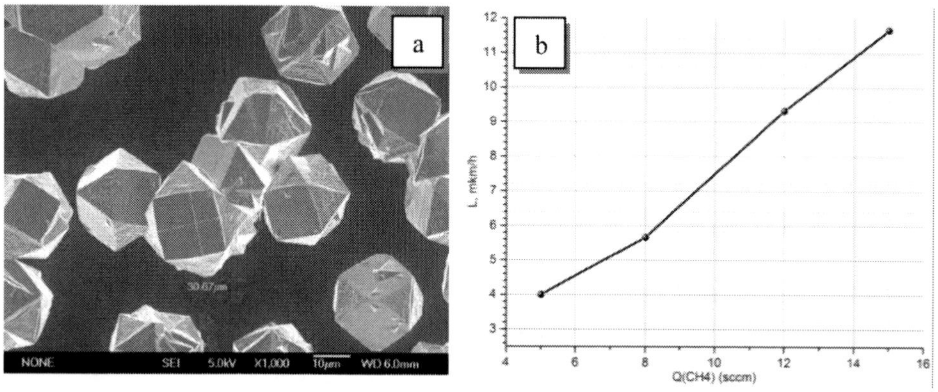

Fig. 2. a) SEM photo of crystals; b) the growth rate of crystals.

2 Gas-dynamic modelling

To understand peculiarities of deposition from such coaxial source the simulation of the flow was performed by Direct Simulation Monte Carlo method [8, 9] for conditions of experiments. One of the main results was the data on atomic hydrogen flows in the space from source exit to the substrate. The calculations were performed for the mixture of H_2, H and CH_4. Algorithm parameters were chosen according to recommendations of [10].

The calculation domain includes channels formed by cylinders with diameters of 3 and 13 mm, and space surrounding the substrate, which is located at the distance of 10 mm (Figure 3a). The used temperatures were 2400 K for spiral and internal cylinder, 1800 K for external cylinder, 1300 K for the substrate. Hydrogen flow rate was 1500 sccm in both channels, and methane flow rate was 5 sccm in a capillary. Data for heterogeneous reactions were used from [11- 13]. The hydrogen dissociation in the capillary was neglected because of carbidization of internal surface [2, 3].

Three stream lines in Figure 3a give contours of the atomic hydrogen flow in the computational domain. The radial distribution of the dissociation rate ($K = n_H/(2n_{H2}+n_H)$), where n_{H2} and n_H are the number densities) (Figure 3b) shows very fast smoothing K inside the jet with approaching a substrate.

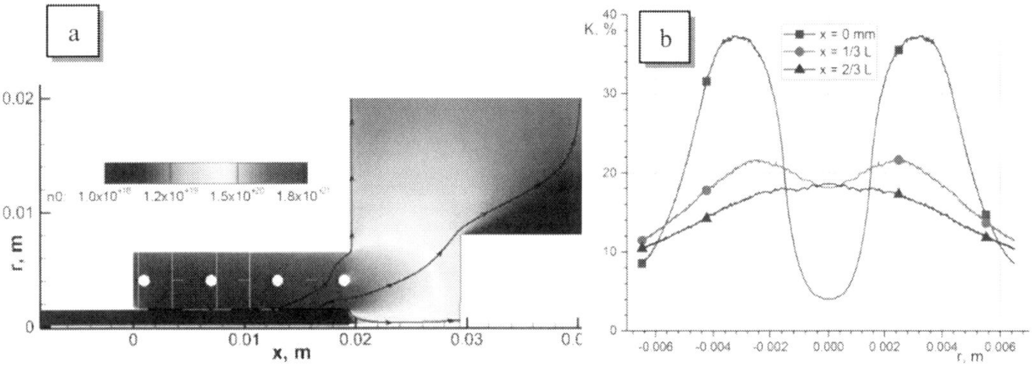

Fig. 3. a) Distribution of density(m^{-3}) and streamline of H; b) radial distribution of hydrogen dissociation at different distances from cylinders exit.

It is explained by high diffusion velocity of atomic hydrogen. The illustration of the radial distribution of specific flow nV and density for H_2 and H is given in Figure 4 a, b.

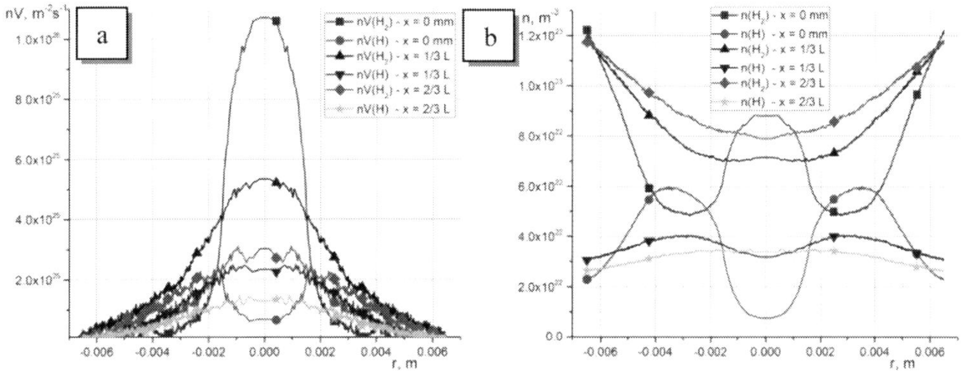

Fig. 4. Radial distribution of nV (a) and densities (b) of H_2 and H.

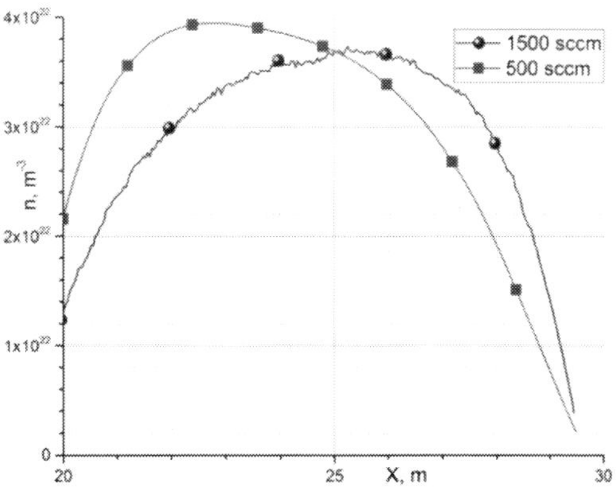

Fig.5. Atomic hydrogen distribution along the flow axis by different H_2 flux through the capillary.

Peculiarities of diamond deposition become clear from the comparison of axial distribution of atomic hydrogen density at different flow rates (and velocities) through the central channel. In Figure 5 one can see high enrichment of the zone upstream the substrate by atomic hydrogen at higher hydrogen flux through the capillary. It explains more intensive diamond deposition at high mass flux of H_2 through the central channel in performed experiments.

This work was supported by RFBR 15-08-01949 and by the grant of Presidium of the RAS II.2П/III.22-1.

References

1. B.V. Spitsyn, *Handbook of Crystal Growth* (Elsevier, Amsterdam, 1994)
2. M. Sommer, F. W. Smith, J. Mater. Res. **5** (11) (1990)
3. A. Emelyanov, A. Rebrov, I. Yudin, Phys. Status Solidi (A) **211**, 2279 (2014)
4. A. Rebrov, A. Emelyanov, S. Kosolobov, I. Yudin, Phys. Status Solidi (C) **12**, 931 (2015)
5. A.A. Emel'yanov, A.K. Rebrov, I.B. Yudin, J. Appl. Mech. Tech. Phys. **55**, 270 (2014)
6. A.K. Rebrov, A.A. Emelyanov, I.B. Yudin, Thin Solid Films **575**, 113 (2015)
7. A.A. Emel'yanov, A. K. Rebrov, I.B. Yudin, Technical Physics **61**, 12 (2016)
8. G.A. Bird, *Molecular Gas Dynamics and the Direct Simulation of Gas Flows* (Oxford: Clarendon Press, 1994)
9. M.S. Ivanov, S.V. Rogasinsky, Russian J. Numer. Anal. Math. Modelling **3**, 6 (1987).
10. M.Yu. Plotnikov, E.V. Shkarupa, Comput. Fluids **58**, 102 (2012)
11. A.A. Morozov, M.Yu. Plotnikov, A.K. Rebrov, I.B. Yudin, AIP Conference Proceedings of 30th Intern. Symp. on Rarefied Gas Dynamics **1786**, 050015
12. A.K. Rebrov, I.B. Yudin, AIP Conference Proceedings of 30th Intern. Symp. on Rarefied Gas Dynamics **1786**, 150012 (2016)
13. M.Yu. Plotnikov, E.V. Shkarupa, Vacuum **129** 31 (2016)

Heat and mass transfer at sublimation of a single chromium (III) and zirconium (IV) β-diketonate particle in the inert gas mixture

Igor Igumenov[1], *Maksim* Makarov[2], and *Sofya* Makarova[1,2,*]

[1]Nikolaev Institute of Inorganic Chemistry, SB RAS, 3 Lavrentiev ave., Novosibirsk 630090, Russia
[2]Kutateladze Institute of Thermophysics, SB RAS, 1 Lavrentiev ave., Novosibirsk 630090, Russia

Abstract. The paper proposes a physical and mathematical model of nonstationary sublimation of single spherical particles of volatile chromium (III) and zirconium (IV) β-diketonates, floating in the flow of a binary helium-argon mixture. The influence of the carrier gas composition on the kinetics of sublimation has been analyzed. The addition of helium to the carrier gas is shown to increase the intensity of sublimation.

1 Introduction

Heat and mass transfer at sublimation of solid particles is important for optimizing the operation modes of cryogenic systems, for intensifying the low-temperature drying of pharmaceuticals and food products, as well as in the development of new technologies of chemical vapor deposition (CVD). The technology of applying ceramic thermal barrier coatings is one of the key challenges when creating gas turbine engines. A method of chemical deposition from the gas phase using volatile compounds of metals with organic ligands (MO CVD) is promising for obtaining such coatings [1, 2]. At that, the technology has to ensure a sufficiently high growth rate of the coating while maintaining the columnar structure. The necessary high concentration of precursor in the vapor phase can be achieved through sublimation under the conditions of intensive heat and mass transfer. High intensity of heat supply to the surface of particles suspended in an inert carrier gas may be obtained in a centrifugal fluidized bed of large inert particles in the vortex chamber [3]. For the optimum design of the vortex chamber the essential is the kinetics of substance sublimation, depending on the thermodynamic and transport properties of the precursor and carrier gas.

This paper presents the physical and mathematical model of sublimation of a single particle of the metalorganic compound, floating in a binary mixture of inert gases. The influence of the carrier gas composition on the dynamics of the sublimation process has been analyzed.

* Corresponding author: snmakarova@itp.nsc.ru

2 Problem statement

The considered phenomenon is nonstationary sublimation of a single spherical particle of the precursor (chromium (III) acetylacetonate $Cr(aa)_3$ or zirconium (IV) dipivaloyl-methane $Zr(dpm)_4$). The sublimation is believed to occur uniformly over the entire surface of the solid particle, and the bulk sublimation inside the pores of the material is not taken into account. The precursor particle does not lose its spherical symmetry over time.

At low velocities of streamlining (diffusion mode of sublimation) the processes of heat and mass transfer in the gas phase are described by a system of parabolic partial differential equations of heat conduction and diffusion of the second order:

$$c_p^g \frac{\partial(\rho^g T)}{\partial \tau} = \frac{1}{r^2}\frac{\partial}{\partial r}\left(r^2 \lambda^g \frac{\partial T}{\partial r}\right) + \sum_{i=1}^{n}\left(-J_{ir} c_{pi}^g\right)\frac{\partial T}{\partial r} \text{ and } \rho^g \frac{\partial K_i^g}{\partial \tau} = \frac{1}{r^2}\frac{\partial\left(-r^2 J_{ir}\right)}{\partial r}.$$

For multicomponent gas mixtures (three or more components) the diffusion flux of the i-th gas in the mixture depends not only on its own concentration gradient, but also on concentration gradients of remaining components of the mixture. A number of approaches were developed to describe this effect [4, 5]. The most accurate results are ensured by the approaches based on Maxwell-Stefan relations, namely, models of Krishna, Lapin-Strelets, etc. However, such approaches require significant computational resources. Simplified models based on the concept of effective diffusion coefficient are more economically efficient, but often have a limited scope of application. One of the most effective models of multicomponent diffusion is Ramshaw model [6], which is used in this work. The diffusion flux of each component of the mixture is determined by correlations below:

$$J_{ir} = -\rho^g D_i \frac{\partial K_i^g}{\partial r} + \delta_{Jir};\quad \delta_{Jir} = \rho^g \frac{D_i K_i^g}{M}\frac{\partial M}{\partial r} - \rho^g K_i^g \sum_{k=1}^{n}\left(-D_k \frac{\partial K_k^g}{\partial r} + \frac{D_k K_k^g}{M}\frac{\partial M}{\partial r}\right).$$

The effective diffusion coefficient is calculated according to the formula:

$$D_i = \sum_{\substack{k=1\\k\neq i}}^{n} \frac{K_k^g}{M_k} \Bigg/ \sum_{\substack{k=1\\k\neq i}}^{n} \frac{K_k^g}{M_k D_{ik}}.$$

At high velocities of steramlining (convective mode of sublimation) the heat and mass transfer intensity on the particle surface is determined by the Ranz-Marshall similarity equations [7]:

$$\mathrm{Nu} = 2 + 0.6\mathrm{Re}_d^{0.5}\,\mathrm{Pr}^{1/3},\quad \mathrm{Sh} = 2 + 0.6\mathrm{Re}_d^{0.5}\,\mathrm{Sc}^{1/3}. \tag{1}$$

The temperature of the solid particle regardless of the sublimation mode is described by the heat conduction equation in the form:

$$c_p^s \rho^s \frac{\partial T}{\partial \tau} = \frac{1}{r^2}\frac{\partial}{\partial r}\left(r^2 \lambda^s \frac{\partial T}{\partial r}\right).$$

The main difficulty in the transition from diffusive to convective approximation is that the gas mixture composition on the particle surface, necessary for calculating the similarity numbers in (1) cannot be determined from the boundary conditions of the problem. For a ternary mixture, a solution to this problem is reduced to defining a separation factor of inert gases. In this work, the separation factor is determined by the results of the precursor sublimation modeling in the diffusion approximation. The obtained data are approximated by a rational polynomial of the second degree, depending on the mass fraction of the precursor vapor on the particle surface, and then are used to calculate the equations (1).

Initial and boundary conditions, determining the size of calculation domain, and testing the solution method for the binary system $Cr(aa)_3/Ar$ are described in detail in [8]. The mass fraction of vapor of the precursor $K_{1R_w^+}^g$ at the phase boundary is determined by the temperature of the particle surface and the curve of vapor saturation (sublimation curve).

According to [9] for chromium (III) acetylacetonate, the sublimation curve is described by the formula: $\ln p = 39.197 - 15308.5/T$; for zirconium (IV) dipivaloylmethane the data of work [10], supplemented by results of specially conducted research in the IIC SB RAS for a wider range of temperatures, are approximated by the formula: $\ln p = 26.652 - 11006.4/T$. The pressures and temperatures in these correlations are determined in Pa and K, correspondently.

3 Method of solution

The main feature of the solution method is integration of the written differential equations system in moving coordinates (τ, η), where τ is the time, η is the relative radial coordinate for the area of gas-vapor mixture $\eta^g = (r - R_w)/(R_e - R_w)$ and the area of a solid particle $\eta^s = r/R_w$. The position of the phase change boundary R_w is determined by the sublimation intensity and is time-dependent. Explicit boundary detection allows minimizing the time of computation and improving the accuracy of the results.

4 Results of modeling

Fig. 1 presents data for the separation factor of argon-helium mixture, depending on the precursor vapors fraction at the phase change boundary, obtained in the diffusion approximation. It is shown that there are sublimation modes with a constant ratio of inert components in the gas phase. In the case of sublimation of $Cr(aa)_3$ the consistency of ratios of mass fractions of Ar and He is observed at $K_e^{Ar} = 0.9$, for $Zr(dpm)_4$ – at $K_e^{Ar} = 0.8$. At a high concentration of argon in the environment ($K_e^{Ar} = 0.95$) the vapor-gas mixture at the phase change boundary is more enriched in helium, and at low concentration it is the other way around. The factor of the mixture separation does not depend on the environment gas temperature.

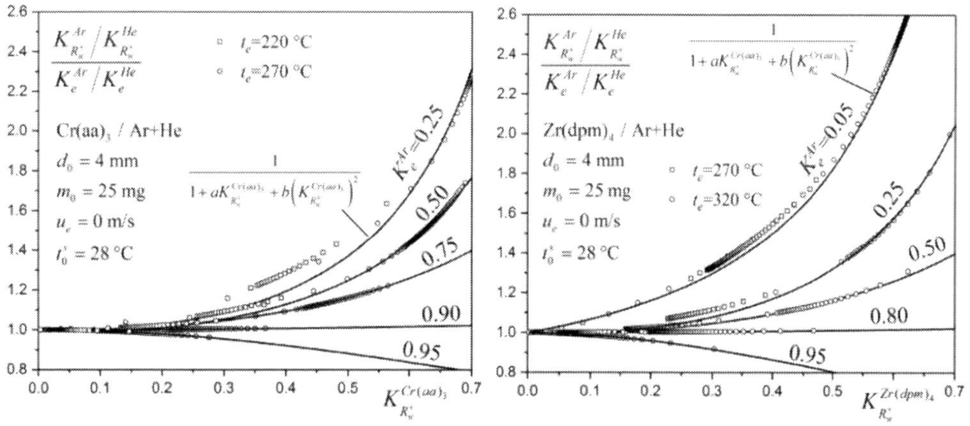

Fig. 1. The ratio of inert components Ar and He at sublimation of the particles of $Cr(aa)_3$ and $Zr(dpm)_4$: points – simulation data, and lines – approximation for subsequent calculations in the forced convection mode.

Fig. 2 presents the results of calculations of mass flow rates of precursor vapors depending on time. It is seen that a small weight addition of helium to argon leads to a significant increase in the intensity of the mass transfer and a decrease of the full time of sublimation. For example, at $K_e^{He} = 0.05$ the full time of sublimation is twofold less than at sublimation in pure argon for both precursors.

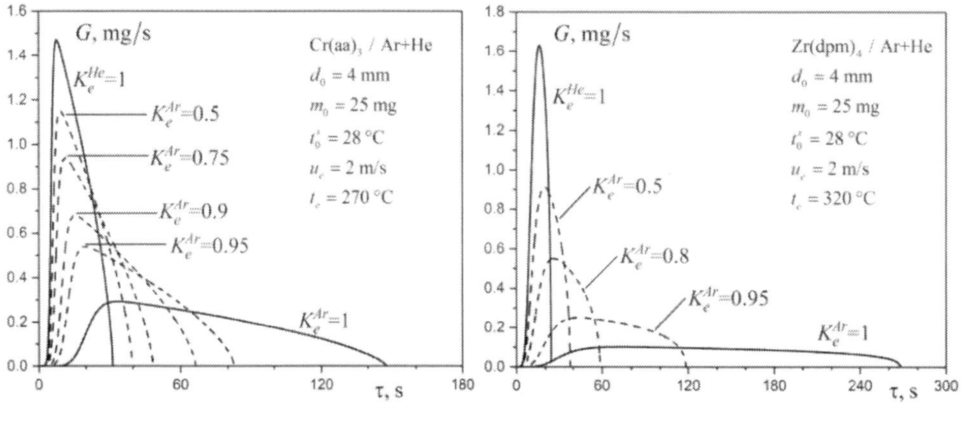

Fig. 2. The mass flow rate of $Cr(aa)_3$ and $Zr(dpm)_4$ vapors at the time.

5 Conclusion

Physical and mathematical model of nonstationary sublimation of a single particle of the MO CVD precursor floating in the gas-carrier flow with average pressure is presented. Numerical simulation of heat and mass transfer with $Cr(aa)_3$ and $Zr(dpm)_4$ precursors sublimation in the mixture of inert gases Ar and He has been carried out. The addition of helium to argon leads to a significant increase in the mass flow intensity and a decrease of the full time of sublimation.

The work was supported by the Russian Science Foundation (project No. 16-19-10325).

References

1. V.V. Krisyuk, A.E. Turgambaeva, I.K. Igumenov, Chem. vap. deposition **4**, 43 (1998)
2. R.G. Parkhomenko, N.B. Morozova, G.I. Zharkova, Y.V. Shubin, S.V. Trubin, V.V. Kriventsov, B.M. Kuchumov, T.P. Koretskaya, I.K. Igumenov, Chem. vap. deposition **18**, 336 (2012)
3. E.P. Volchkov, N.A. Dvornikov, V.V. Lukashov, R.K. Abdrakhmanov, Thermophysics and Aeromechanics **20**, 663 (2013)
4. R.C. Reid, J.M. Prausnitz, T.K. Sherwood, *The Properties of gases and liquids* (McGraw-Hill, New York, 1977)
5. Yu.V. Lapin, M.Kh. Strelets, *Internal flows of gas mixtures* (Nauka, Moscow, 1989)
6. J.D. Ramshaw, Journal of non-equilibrium thermodynamics **15** (1990)
7. M.E. Ranz, W.R. Marshall, Chem Eng Prog **48**, 141 (1952)
8. M.S. Makarov, S.N. Makarova, A.A. Shibaev, Proceedings of the Jubilee conference of National committee for heat and mass transfer of RAS and XXI School-seminar of young scientists and specialists under leadership of RAS academician A.I. Leontiev **2** (2017)
9. P.P. Semyannikov, I.K. Igumenov, S.V. Trubin, T.P. Chusova, Z.I. Semenova, Thermochim. Acta **432** (2005)
10. M. Fulem, K. Růžička, V. Růžička, T. Šimeček, E. Hulicius, J. Pangrác, Journal of Crystal Growth **264** (2004)

Study on dynamics and structure of evaporation front in ethanol depending on pressure and subcooling

Mikhail Moiseev*, and *Vladimir* Zhukov

Kutateladze Institute of Thermophysics, 630090, Novosibirsk, 1 Acad.Lavrentiev ave., Russia

Abstract. This paper is dedicated to study of self-sustained evaporation front in subcooled ethanol at stepwise heat generation. The data on evaporation front velocity and microstructure at different subcoolings and pressure are presented. Experiments show that development of the evaporation front drastically depends on studied parameters. At low pressures initial vapour bubbles grow without losing interface stability and initiation of evaporation fronts doesn't occur. At higher pressures the fronts propagate along the heater, and its structure depends on temperature.

1 Introduction

Under some conditions (nonstationary heating, low reduced pressure, etc.) onset of boiling crisis occurs via self-sustaining evaporation fronts, bypassing nucleate boiling stage. Critical heat flux in this case can be much lower than the first critical heat flux. Evaporation front is a complex nonequilibrium phenomenon which includes interrelated hydrodynamic and heat transfer processes at the interface, and also needs consideration of the arising hydrodynamic front instability, structure formation, evaporation, and thermocapillary effects. There are a few papers dedicated to model description of a self-sustained evaporation front [1-3]. But none of these models can describe front behaviour within a wide range of experimental parameters. Main flaw of most of the models is that they don't take into account the microstructure of the front, namely small-scaled hydrodynamic instability, which greatly influences heat transfer through the interface [3-5]. The lack of experimental data on evaporation front structure at various conditions also hinders its strict and complete mathematical description.

This paper describes results of experimental study on evaporation front in subcooled ethanol at stepwise heat generation and presents data on front velocity and microstructure at different subcooling and pressure.

2 Experimental setup and procedure

Experiments were carried out in ethanol at pressures P = 0.0075…0.101 MPa and subcooling ΔT_{sub} = 14-43 K. For experimental studies of heat transfer and transient

* Corresponding author: moiseevmikhail@gmail.com

dynamics at non-stationary heat release the setup, shown schematically in Fig. 1, was developed.

The frame 2 with refill tank 3 and working chamber 4, housing the working section 10, is hinged on bed 1. The working chamber is a sealed cylindrical stainless steel vessel. The inner diameter of the chamber is 250 mm, and the height of the workspace is 250 mm. The working chamber is equipped with four windows 5. The bellows 6 with an adjusting screw 7 allows creating the required pressure in the working chamber at the closed valve 8. Heat exchanger 9 located in the bottom of the working chamber allowed setting the desired temperature of liquid.

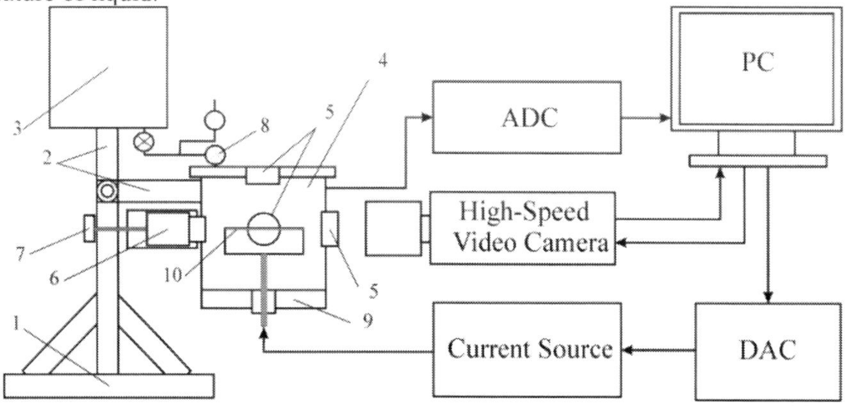

Fig. 1. Scheme of experimental setup.

The working section was made of the stainless steel tube with outer diameter of 8 mm, wall thickness of 0.2 mm and length of 50 mm. The rectangular current impulse of 420 A was supplied to the working section. Temperature of the heater was determined by the current impulse duration. After current was turned off, the wall temperature remained almost constant at the times of the evaporation front propagation. Numerical solution of heat equation has shown that for the time of front passage (much lesser than the time of convection development) the wall temperature before the front dropped no more than by 0.2 K. Thus in each experiment, the propagation of self-sustained evaporation front was investigated at the pre-set constant wall superheat.

The working section temperature until the moment of vapor phase emergence was measured using temperature dependence of the heater resistance. For this purpose, two conductors of 0.05-mm diameter were welded in the middle of the tube to measure the voltage drop in the region of 30-mm length. In each measurement, calibration was performed by the temperature of undisturbed fluid. This technique allowed measuring the average temperature of the working section with an error less than ±1.5 K. After the vapor phase formation, the wall temperature was determined numerically based on the non-stationary heat equation. This approach is valid under conditions of our experiments since the times of the evaporation front propagation (about 50 ms) were much less than the time of convection development (about 200 ms).

Visual observations of vapor phase formation and propagation on the heating surface were performed using a high-speed digital video camera Phantom v7.0. The shooting rate was up to 45 000 frames per second with the exposure of 19 microseconds. When analyzing the results of high-speed digital video-shooting, the line of the interfacial boundary was determined on each frame using specially developed software. The obtained data was used to study dynamic properties of evaporation front boundary.

3 Experimental results

We have obtained the dependance of the front velocity on wall superheat over saturation temperature T_{sat} (Fig. 2). In the studied range of pressure and subcooling the front velocity is almost independent of the mentioned parameters, except the series with lowest pressures. At the lowest pressure 0.0075 MPa initiation of evaporation fronts didn't occur. At pressure 0.026 MPa the front velocity dependence on wall superheat becomes stronger after reaching certain temperature. This effect is due to development of the front interface instability [3-5].

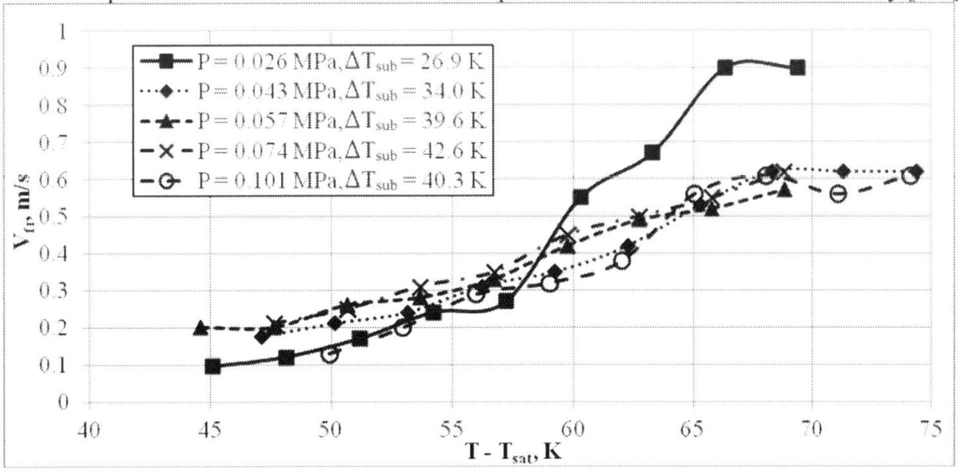

Fig. 2. Dependence of the front velocity on wall superheat at different values of pressure and subcooiling (ΔTsub).

3.1 Mechanisms of front initiation

The data obtained using macro videoshooting allows describing the front initiation mechanism as follows. Vapor bubble, beginning to grow at the heater surface, has smooth interface while it is within the metastable termal layer near the heater wall (Fig. 3a). At the lowest pressure 0.0075 MPa vapor bubbles grow up to big size without losing interface stability, until they merge, and vapor layer covers the whole heater. At higher pressures, when a bubble grows out of the thermal layer into the subcooled area, condesation processes begin to compete with evaporation, and the bubble interface loses stability (Fig. 3b). After that evaporation front starts to propagate with its headmost point within the thermal layer (Fig. 3c). Size of the interface perturbations is comparable with the thermal layer thickness.

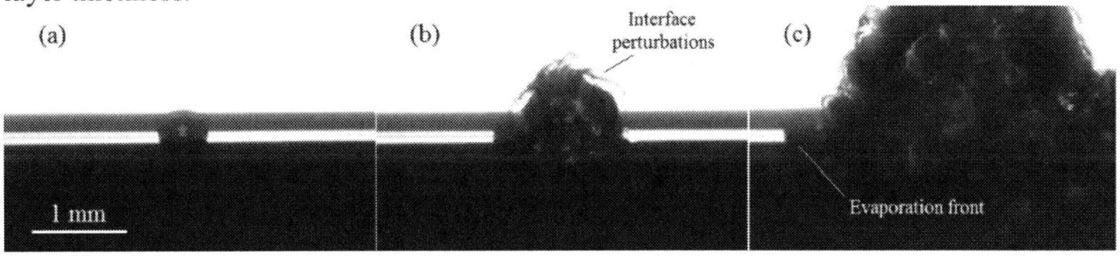

Fig 3. Evolution of a vapor bubble and emergence of the evaporation front. P = 0.101 MPa. ΔTsub = 40.3 K. Heating time was 40 ms (T-Tsat = 56 K). (a) – 60.15 ms after beginning of heat release; (b) – 62.62 ms; (c) – 66.88 ms.

Experiments in reflected light show that development and structure of the front significantly depend on pressure and wall superheat. As mentioned above, initiation doesn't occur at lowest pressure front. At higher pressure and low wall superheat the evaporation front propagating along the heater covers small caverns on the heater surface. In these caverns micro-bubbles nucleate and rapidly grow, expanding the front interface (fig. 4a). When a bubble grows out of the thermal layer, its interface loses stability, and the front moves inside the thermal layer in form of small vapor ejections until it meets new cavern, and the process repeats. Thus, the front propagates as a chain of unstable bubbles in a fractal pattern. At high wall superheat separate vapor bubbles become indistinguishable (fig. 4b). The front interface becomes quasi-flat, with only small-scaled (with wavelength 100-200 mcm) perturbations.

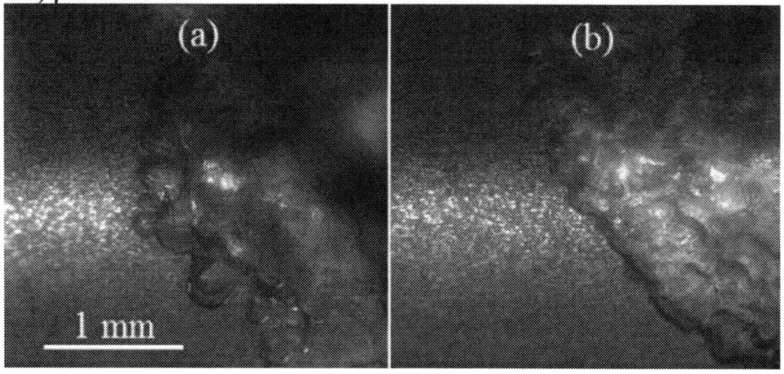

Fig 4. Evaporation front interface. P = 0.101 MPa. ΔTsub = 40.3 K. (a) – T-Tsat = 53 K. (b) – T-Tsat = 68 K.

4 Conclusions

Detailed study of evaporation front dynamics and microstructure in subcooled ethanol at stepwise heat generation shows the following results. Front velocity is almost independent on pressure and subcooling within the pressure range 0.043-0.101 MPa, and has linear dependence on wall superheat. At very low pressure (0.0075 MPa) evaporation fronts don't emerge. At moderately low pressure (0.026 MPa) the dependence of front velocity on wall superheat becomes stronger after reaching certain critical superheat, which is corresponding to development of evaporation front instability.

Structure of the evaporation front significantly depends on wall superheat. At low superheats the front propagates as a chain of unstable bubbles in a fractal pattern. At high superheats the front interface becomes quasi-flat with small-scaled perturbations.

This research has been performed at IT SB RAS with the support of the RFBR (project No. 15-08-01359)

References

1. S.P. Aktershev, V.V. Ovchinnikov, J. Eng. Thermophys. **20**, 77 (2011)
2. P.A. Pavlov, V.E. Vinogradov, High Temp. **48**, 683 (2010)
3. A.N. Pavlenko, V.V. Lel, Thermophys. Aeromech. **6,** 251 (1999)
4. A.N. Pavlenko, E.A. Tairov, V.E. Zhukov, A.A. Levin, M.I. Moiseev, J. Eng. Thermophys. **23**, 173 (2014)
5. V.E. Zhukov, A.N. Pavlenko, M.I. Moiseev, D.V. Kuznetsov, High Temp. **55**, 79 (2017)

Effect of the structured packing height on efficiency of freons mixture separation in a large-scale model of distillation column

Aleksandr Pavlenko*, *Vladimir* Zhukov, *Nikolay* Pecherkin, *Aleksandr* Nazarov, *Xingang* Li, *Mingyan* Liu, *Hong* Sui and *Hong* Li

Kutateladze Institute of Thermophysics, Siberian Branch, Russian Academy of Sciences
1 Lavrentyev Ave., Novosibirsk, 630090, Russia

Abstract. Results of experimental studies of heat-and-mass transfer and hydrodynamic processes at distillation on a regular packing are presented. The mixture of freons R114–R21 at the pressure of 0.3 MPa was used as a working mixture. The mixture was separated on the *Mellapak 350Y* structured packing with the diameter of 0.9 m under the conditions of complete reflux (L/V = 1) at different packing heights. A specially designed liquid distributor with a possibility to change the density and pattern of drip points was used to irrigate the packing. The experimental data on the efficiency of mixture separation (height of transfer unit *HTU*) and distribution of the local flow rate density over the column cross-section were compared. It is shown that an increase in the height of the structured packing from 2.1 m to 4.0 m leads to a significant decrease in the efficiency of mixture separation in the distillation column.

1 Introduction

Results of research and operation of industrial distillation columns with the regular packing have shown that often separation efficiency can decrease with an increase in their height and diameter, and this affects productivity and purity of the finished product. The fundamental phenomena responsible for a decrease in separation efficiency are the effects caused by transverse mal-distribution and longitudinal mixing. These effects relate to the vapor and liquid flow within the column, and they are presented by deviation of productivity and product purity from the calculated values at uniform distribution. The results published for laboratory samples of a small packing demonstrate an efficiency of about 20 or more theoretical stages per 1 meter of the packing height, while the efficiency of the best industrial samples is 5-6 theoretical stages per 1 meter of the height [1, 2]. Setting the optimal structure of initial irrigation allows using the maximal packing height without significant deterioration of the efficiency of mixture separation caused by the negative factors in formation of large-scale mal-distribution of local mixture flows within the packing. The use of periodic packing irrigation, for example, allows destructing the stationary large-scale structures of uneven distribution of the local flows inside the packing

* Corresponding author: pavl@itp.nsc.ru

and, accordingly, increasing separation efficiency of mixtures [3]. The purpose of this work is to investigate the effect of the regular packing height on the efficiency of Freon mixture separation, formation dynamics of large-scale mal-distribution of the local flow parameters over the packing cross-section along the column height for the given drip point pattern.

2 Experimental setup and methods

The experiments were carried out at the setup «Large Freon Column», whose detailed description is given in [2]. The mixture of freons R114–R21 at the pressure of 0.3 MPa was used as the working mixture. This mixture was chosen to simulate the processes of cryogenic mixture separation. The mixture was separated on the regular *Mellapak 350Y* packing with the diameter of 0.9 m under the conditions of complete reflux ($L/V = 1$). The setups operated in the closed cycle. Two series of experiments were carried out with the packing height of 2.1 and 4.0 m, respectively. A specially designed liquid distributor with the possibility to change the number and structure of drip points was used to irrigate the packing [3]. In these experiments, the rectangular drip point pattern with the density of 131 m^{-2} was applied. During the experiments, the efficiency of mixture separation (height of transfer unit *HTU*) was determined in the range of superficial vapor velocities of $0.017 < K_v < 0.035$ m/s and the temperature field distributions in the packing cross-sections were registered at three levels along the column height. Distributions of the local liquid flow density, formed during the liquid film flow over the packing of various heights, were also obtained.

3 Discussion

Dependence of the mixture separation efficiency (*HTU*) on superficial vapor velocity (K_v) is shown in *Fig.* 1 for the column with the packing height of 4.0 and 2.1 m. As it can be seen from the figure, the efficiency of mixture separation in the studied range of variations of superficial vapor velocity $0.017 < Kv < 0.035$ m/s increases significantly with increasing K_v. An increase in the value of K_v leads to a monotonic decrease in the height of transfer unit by the factor of 2.4-2.6, approximately. A significant decrease in the efficiency of mixture separation of the mixture at low values of superficial vapor velocities (and superficial liquid velocities, respectively) is caused by formation of local zones of deficient irrigation of the mass transfer surface inside the packing, and, as a result, its only partial wetting [4, 5]. For equal values of K_v, the efficiency of mixture separation is considerably lower on the packing with the height of 4.0 m. This result is explained by formation of large-scale maldistribution of the local mixture flow along the packing height, which causes a nonlinear effect (the so-called pinch effect [2]), which significantly reduces the efficiency of mixture separation. The main role in this negative influence is associated with formation of a large-scale maldistribution of the local liquid flow over the cross-section of the packing along its height, and, consequently, maldistribution of the local vapor flow.
The topograms of distribution of the local flow density of liquid flowing down from the lower edge of the packing are shown in *Fig.* 2, at its different heights and close values of K_v. As it follows from the analysis of experimental data presented in *Fig.* 2, distribution of the local liquid flow density at the outlet of the packing with the height of 4.0 m is slightly more uniform in comparison with distribution for the packing with the height of 2.1 m. The values of standard deviation for the presented distributions are 11.5 and 14.5%, respectively. At the outlet of the packing with the height of 2.1 m, a higher density of the local liquid flow is observed at the packing periphery and in its central part. According to

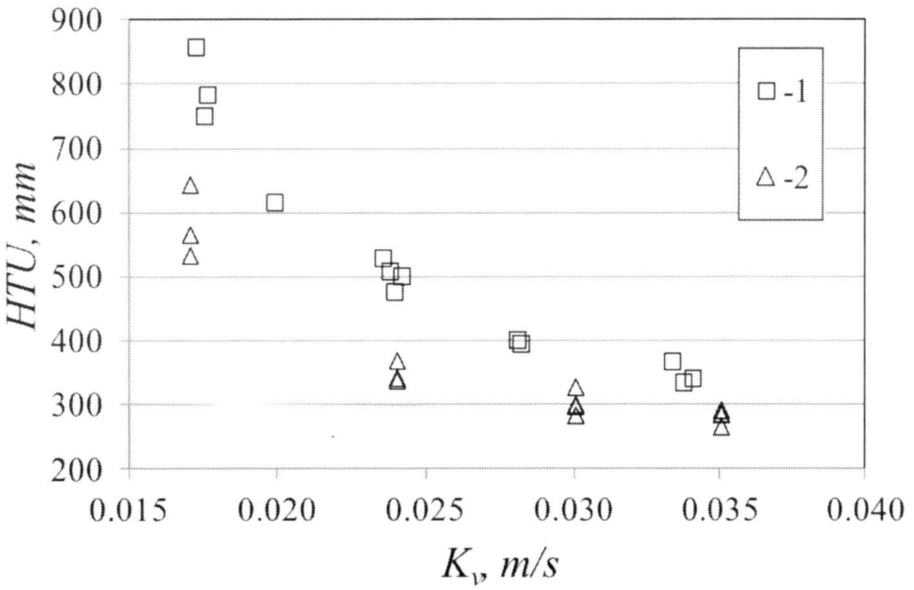

Fig. 1. Dependence $HTU\ (K_v)$; *1* – H = 4.0 m; *2* – H = 2.1.

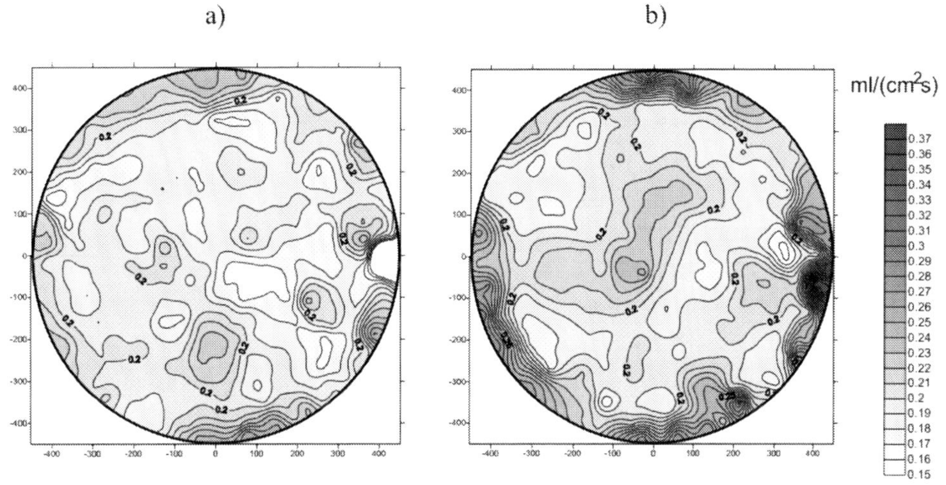

Fig. 2. Distribution of the local flow rate densty of liquid flowing from the lower edge of the packing. K_v ~ (0.027- 0.030) m/s; *a* – H = 4.0 m; *b* – H = 2.1.

the estimates obtained by the authors of [2] on the basis of the model of so-called divided column [6], the same degree of the local liquid flow mal-distribution over the packing cross-section has considerably greater negative effect on the height of transfer unit *HTU* for the higher packing. This can explain the fact that for the close mal-distribution degrees determined in experiments for the local flow rates of liquid on the packing with different heights, the efficiency of mixture separation on the packing of 4.0 m height is much lower than that on the packing with the height of 2.1 m.

4 Conclusions

It is shown that an increase in the height of the structured packing *Mellapak 350Y* from 2.1 to 4.0 m leads to a significant decrease in the efficiency of mixture separation in the distillation column with the diameter of 0.9 m. Experimental data show close values of a degree of the local flow rate mal-distribution over the column cross-section for the packing of different heights. As it follows from the analysis, a significant increase in the height of transfer unit is caused by the fact that the same mal-distribution degrees of the local flow parameters for the packing of different heights have a more significant negative effect on the efficiency of mixture separation on the higher packing.

The obtained basic experimental data will be used for comparative analysis in further studies with varying density and structure of the drip points under the conditions of stationary and dynamically controlled laws of regular packing irrigation.

The study was performed at the IT SB RAS due to the grant of the Russian Science Foundation (project No. 14- 49- 00010).

References

1. V.S. Leontiev, S.I. Sidorov, Khim Prom. **82** 347 (2005)
2. A.N. Pavlenko, V.E. Zhukov, N.I. Pecherkin, V.Yu. Chekhovich, O.A. Volodin, A. Shilkin, C. Grossmann, AIChE J. **60**, 690 (2014)
3. A.N. Pavlenko, V.E. Zhukov, N.I. Pecherkin, X. Li, H. Sui, J. Physics: Conference Series **754,** 042012 (2016)
4. A.N. Pavlenko, X. Li, H. Li, X. Gao, O.A. Volodin, A.S. Surtaev, V.S Serdyukov, Techn. Phys. Lett. **41**, 774 (2015)
5. A.N. Pavlenko, O.A. Volodin, A.S. Surtaev, Appl. Thermal Eng. **114,** 1265 (2016)
6. M.J. Lockett, J.F. Billingham, *Distillation and Absorption 2002 Proceedings* (Baden-Baden, Germany), (2002)

Thermophysical problems of laser cutting of metals

Anatoliy Orishich*, *Victor* Shulyatyev, *Alexander* Golyshev *and Alexander* Malikov

Khristianovich Institute of Theoretical and Applied Mechanics SB RAS, 630090, 4/1 Institutskaya Str., Novosibirsk, Russia

Abstract. Variety and complex interaction of physical processes during laser cutting is a critical characteristic of the laser cutting of metals. Small spatial and temporal scales complicate significantly the experimental investigations of the multi-phase fluid flow in the conditions of laser cutting of metals. In these conditions, the surface formed during the cutting is an indicator determining the melt flow character. The quantitative parameter reflecting the peculiarities of the multi-phase fluid flow, is normally the roughness of the forming surface, and the minimal roughness is the criterion of the qualitative flow [1 – 2]. The purpose of this work is to perform the experimental comparative investigation of the thermophysical pattern of the multi-phase melt flow in the conditions of the laser cutting of metals with the laser wavelength of 10.6 μm and 1.07 μm.

During the laser cutting, the local material melting and melt removal by an assistant gas jet take place. Cutting of low-carbon steel sheets is usually carried out in an oxygen jet (oxygen-assisted laser cutting). In this case, the exothermic reaction of iron oxidation is an additive energy source. The power balance for the oxygen-assisted laser cutting and the cutting with a chemically inert gas is written respectively as:

$$AW + W_{ox} = W_m + W_{cond} \tag{1}$$

$$AW = W_m + W_{cond} \tag{2}$$

where A is the absorption coefficient; W is the laser power; W_{ox} is the power released at the exothermic reaction; W_m is the power consumed for melting; and W_{cond} is the power lost from the cut region due to thermal conductivity.

In [2, 3] the authors calculated the energy losses from the thermal conductivity in the laser cutting conditions $W_{cond} = \lambda_m\, t\, \Delta T\, f(\mathrm{Pe})$, where $f(\mathrm{Pe})$ is the dimensionless function of the Peclet number $\mathrm{Pe} = (V\, b\, \rho_m\, C_m) / \lambda_m$, $\Delta T = T^* - T_0 = 2000\ ^0\mathrm{C}$ [4], λ_m is the metal thermal conductivity, C_m is the metal heat capacity, ρ_m is the metal density, t is the steel sheet thickness, and b is the cut width. Passing to dimensionless variables, we have the general expression for the energy balance: $Q + Q_{ox} \cong \mathrm{Pe}\,[1 + L_m / (C_m\, \Delta T)] + f(\mathrm{Pe})$, where the following designations are used: dimensionless laser power $Q = A\, W / (\lambda_m\, t\, \Delta T)$,

* Corresponding author: laser@itam.nsc.ru

dimensionless consumptions for metal melting $Q_m = W_m / (\lambda_m t \Delta T) = \text{Pe} [1 + L_m / (C_m \Delta T)]$, and dimensionless energy of iron oxidation $Q_{ox} = W_{ox} / (\lambda_m t \Delta T)$. The energy balance expression, similarly to the calculations of [2, 3], does not include the dependence of thermophysical parameters (C_m and λ_m) on the temperature; instead, their value at the average temperature $T = (T^* - T_0) / 2$ was used.

Two types of lasers were used, and their wavelengths were 10.6 and 1.07 μm. The ytterbium fiber laser IPG/IRE-Polus was used; its power reached 2 kW, and the BPP parameter was 3.8 mm·mrad (BPP – Beam Parameter Product). The lens focal distance was 200 mm. Also, the CO_2 laser [5] with a self-filtering resonator was applied; its power reached 8 kW, and the BPP = 4.7 mm·mrad at the focal distance of the ZnSe lens of 127 and 190 mm.

The sheets of low-carbon steel (St. 20) were cut; their thickness was 3, 5, 10, and 16 mm, the laser power varied from 0.5 to 4 kW, and oxygen was used as a technological gas (pressure 0.25 MPa). Under study was also cutting of stainless steel (12X18H10T) sheets, their thickness 3, 5, and 8 mm, nitrogen pressure 1.3 - 1.6 MPa.

Note that, to have experimental data for each energy flux, the optimal laser power value W, cutting speed V and cut width b, were found for each thickness t, absorption coefficients A and power losses due to the thermal conductivity were measured for these values (measurement techniques are presented in [1, 5]. Obtained data enabled to determine the values W_m, AW, W_{cond}. The value W_{ox} is the power of iron oxidation exothermic reaction and was found from the expression $W_{ox} = W_m + W_{cond} - AW$. All optimizations involved the criterion of minimal cut roughness [1, 5]. Experimental values of the energy fluxes determining the optimal laser cut are shown in Fig. 1. Evident that these values do not depend on the cut material thickness and are close to each other. It can be assumed that such an energy balance of the oxygen-assisted laser cutting of steel is a condition of the minimal surface roughness. Breach of this symmetry resulting from the non-optimal choice of cutting parameters, domination of any process, for example, excessive power of the exothermic reaction of the oxygen oxidation of iron, or jump of the power lost from the cut region due to the thermal conductivity into the surrounding metal may cause the degradation of the surface quality because of uncontrolled burning of cut region overheating.

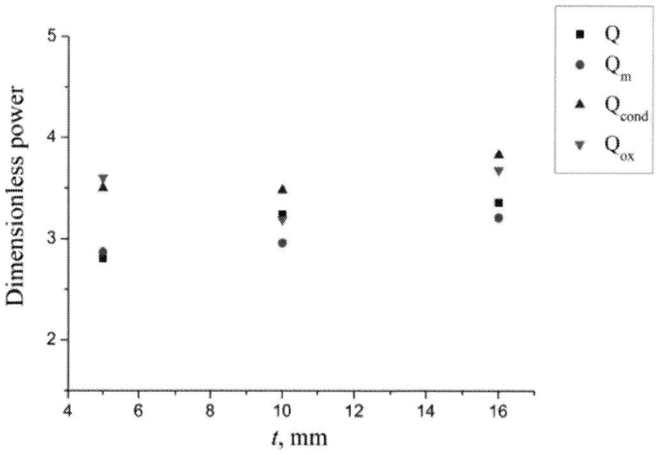

Fig. 1. Experimental values of the fluxes of dimensionless power for thicknesses 5, 10, and 16 mm.

During the cutting with a chemically inert gas, only the utmost cutting speed guarantting the complete sheet cutting, depends on the laser power. Thus, the maximal for the fiber laser power of 2 kW was applied in each experiment. Opposite to the oxygen-assisted laser cutting, wherein there are the power, cut speed and width optimal for each sheet thickness, the laser cutting with a chemically inert gas gave the qualitative cut within a wide range of speed parameters. In this case, it was necessary only to preset the optimal position of the focal spot in respect to the sheet surface at the high gas pressure (about 15 atm), for which the laser beam absorption coefficient was measured.

Figure 2 presents the generalized energy data of the cutting with an inert gas (nitrogen) as the dependence of the dimensionless parameters $Q_{abs\_opt} = A_{opt}\ W\ /\ (\lambda_m\ t\ T_m)$ on $\mathrm{Pe}_{opt}=V_{opt}\ b_{opt}/\ \gamma$ (γ – material thermal diffusivity). The same Figure shows the results of analysis of the energy balance of low-carbon steel cutting by the lasers with wavelengths 1.07 and 10.6 μm. The sum of the absorbed laser energy and energy of iron oxidation exothermic reaction was used as the absorbed energy.

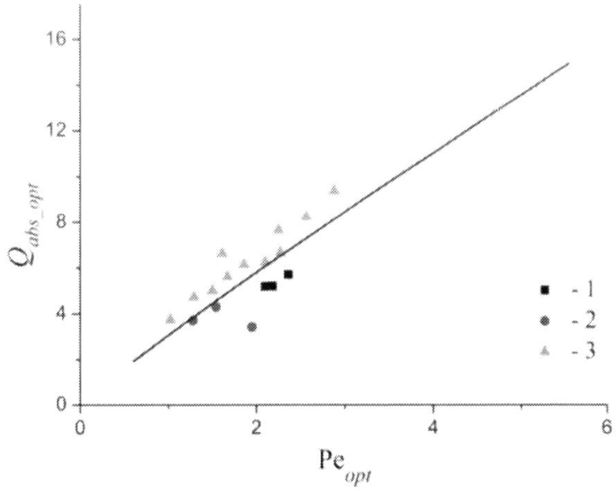

Fig. 2. Optimal value of the dimensionless power versus the optimal Peclet number during cutting of low-carbon steel by the CO_2 (dots 1) and fiber (dots 2) lasers, and stainless steel by the fiber laser (dots 3). The line is plotted by the formula (3).

Evident that utilization of the dimensionless parameters permit generalizing all variety of the experimental data regardless the metal thickness, laser power and type, into one common dependence approaching to $Q_{abs\_opt} \sim \mathrm{Pe}_{opt}$. The inclination of the experimental dependence in Fig. 2 enables to evaluate the average energy contribution in the volume unit of the metal removed from the cut region: $E_{abs\_opt} \approx 26$ J/mm^3. Note that, during the cutting with the inert gas, all this energy should be provided by the laser radiation, whereas in the presence of oxygen, 50 % of this contribution is provided by the laser beam (13 J/mm^3) and 50 % - by the exothermic reaction of iron oxidation.

Let us consider the applicability of the theoretical energy balance to describe the experimental data. According to [3], the expression $W^*_{cond}=f(\mathrm{Pe})=W_{cond}/(\lambda_m t\Delta T)=3.2(\mathrm{Pe}/2)^{0.868}$ can be used for the evaluation of thermal losses. In [2], another dependence of the thermal losses on the Peclet number was found: $W^*_{cond}=f(\mathrm{Pe})=W_{cond}/(\lambda_m\ t\Delta T)=3.2\ (\mathrm{Pe}/2)^{0.3}$. Evident that only the model proposed in [3] can satisfy the experimental dependence $Q_{abs\_opt} \sim \mathrm{Pe}_{opt}$. Substituting theoretical thermal losses in the energy balance, we have:

$$Q_{abs\_ox} + Q_{abs\_inert} = 1.3\ \mathrm{Pe} + 3.2\ \ (\mathrm{Pe}/2)^{0.868} \tag{3}$$

where $Q_{abs\_ox} = Q + Q_{ox} \cong 2\ Q$.

Opposite to [3], the Peclet number in the resulting expressions was calculated with the channel diameter. Material properties are regarded by the parameter $L_m\ /\ (C_m\ \Delta T)$, which equaled to 0.284 and 0.316 for the low-carbon and stainless steels, respectively. During the calculation of the power balance (5), the average value 0.3 was used. Figure 2 shows with the solid line the results of calculation of the power balance. Thus, comparison with the experiment demonstrates that the model of the thermal losses proposed in [3] describes the experimental data more adequately.

References

1. A.A. Golyshev, A.G. Malikov, A.M. Orishich, V.B. Shulyatyev, Quantum Electronics **45**, 873 (2015)
2. W. Schulz, D. Becker, J. Franke, R. Kemmerling, G. Herziger, J. Phys. D: Appl. Phys. **26,** 1357 (1993)
3. J.M. Prusa, G. Venkitachalam, P.A. Molian, Int. J. Machine Tools and Manufacture **39,** 537 (1999)
4. J. Powell, D. Petring, R.V. Kumar, S.O. Al-Mashikhi, A.F.H. Kaplan, K.T. Voisey J. Phys. D: Appl. Phys. **42,** 015504 (2009)
5. V.M. Fomin, A.A. Golyshev, A.G. Malikov, A.M. Orishich, V.B. Shulyatyev, J. Appl. Mech. Techn. Phys. **56**, 726 (2015)

Experimental study of water evaporation of sessile droplets on a solid substrate with different thermal conductivities

Vladimir Borodulin, *Vladimir* Letushko, *Michail* Nizovtsev, and *Alexey* Sterlyagov*

Kutateladze Institute of Thermophysics, 1 Ac. Lavrentiev ave., Novosibirsk, Russia

Abstract. The results of experimental studies of water evaporation of sessile droplet on solid substrates with different thermal conductivities are presented. In experiments during droplet evaporation the temperature of its surface was determined using the infrared thermography method. The obtained results showed that interfacial temperature was higher than the adiabatic evaporation temperature for all substrates. As thermal conductivity of the substrate decreased, the droplet temperature decreased and the evaporation lifetime increased significantly. As a result it was established that the thermal conductivity of the material has a significant effect on the evaporation of droplets.

1 Introduction

Evaporation of sessile droplets on solid substrates has been widely discussed in the literature due to the wide occurrence of this physical phenomenon in a variety of practical applications including formation of nanostructures [1], inkjet printing [2], fuel burning [3] and bio analyses [4]. Thus the understanding of the evaporation process of sessile droplets is important for such applications. Recently there have been many active studies of evaporation of sessile droplets on the hydrophobic textured surfaces [5, 6].

The process of evaporation of sessile droplets is significantly affected by the substrate material. This task is a subject of many publications in the world. Sefiane et al. [7, 8] showed the strong influence of the substrate conductivity on evaporation of sessile drops. In recent work Lopes et al. [9] used glass and silicon substrates with different conductivity and thermal diffusivity. They found that the thermal properties of the substrate have a strong influence on evaporation lifetime of a sessile droplet. In most of these experimental works the changing of the droplet shape was investigated, and the droplet temperature was measured using contact methods. However, the use of these methods disturbs the integrity of the drop and induces errors in results.

The aim of this paper is an experimental study of evaporating sessile drops on the substrates with different thermal conductivities. In experiments the temperature of the droplets was obtained by a non-contact method of infrared thermography. Similar experiments were previously carried out by D. Brutin et al. [10], however, in these

* Corresponding author: sterlyagov@itp.nsc.ru

experiments infrared camera was used for visualization of thermal motion inside the evaporation sessile drop.

2 Experimental setup

Experimental setup was constructed in the Institute of Thermophysics to investigate sessile droplet evaporation on horozontal surfaces of substrates (see Fig.1).

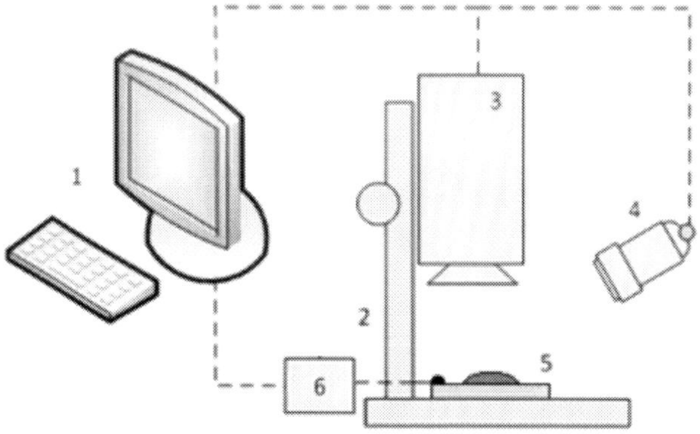

Fig. 1. Experimental setup: 1 – PC; 2 – support; 3 – infrared imaging camera; 4 – digital microscope; 5 – substrate; 6 – digital thermo-hygrometer

The infrared camera (3) was vertically located above the working area. Digital microscope (4) was located on the side of the working area. Temperature and relative humidity of ambient air were measured by thermohygrometer (6). Experimental data were collected using a computer (1).

Experiments on the evaporation of droplets were carried out as follows: The drop was first formed on the surface of substrate by mechanical pipette «Thermo Scientific» with accuracy of ~ 0.1 μL. Then the local temperature distribution along the interface of an evaporating droplet was determined at different times with fixed step by infrared thermography. Infrared camera NEC TH 7102WV was used in experiments. It provided temperature measurements with accuracy of about 0.1 ^{0}C. In order to calibrate and to determine the surface emissivity, contact measurements of the substrate temperature were additionally performed.

During the experiment, the shape of a drop of liquid was continuously recorded by a digital microscope. It allowed us to determine the change in geometric dimensions of the drop during evaporation. Initial droplet diameters were measured by digital microphotography method with an accuracy of ± 10 μm.

In experiments, the copper plates with thermal conductivity coefficient λ=401 W/m·K, teflon plates with λ=0.25 W/m·K and extruded polystyrene plates (EPS) with λ=0.03 W/m·K were used as substrates. The thickness of the plates was 3 mm. A distilled water droplet with initial volume of 5 μL was studied. The ambient temperature and relative humidity were set at 24 °C and φ =24 %, respectively. At these conditions, the temperature of adiabatic evaporation or wet bulb temperature was t_{wb} = 12.7 ^{0}C.

3 Experimental results

Interfacial temperature as function of time of water droplets sessile on the surfaces of different materials was obtained using the method of infrared thermography (Fig. 2).

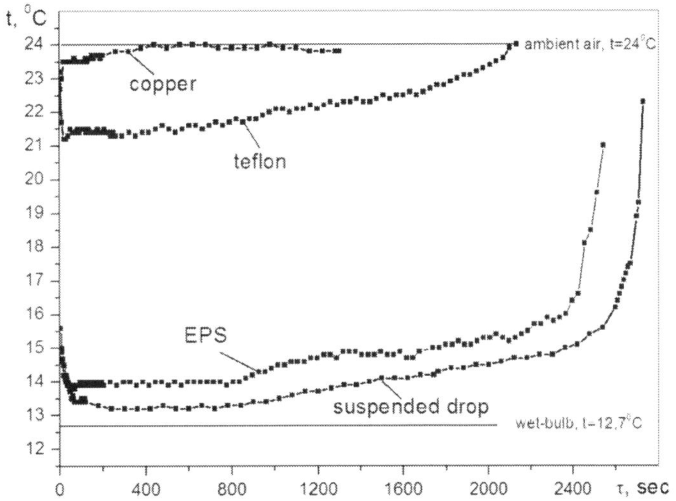

Fig. 2. Average droplet temperature on the substrate as a function of time

The behavior of temperature is seen to depend on the thermal conductivity of substrate. After initial cooling of the water drop on the copper plate, there was a fast increase of temperature up to a plate one, equal to the temperature of ambient medium. Therefore, temperature of the droplet practically did not change until the end of evaporation lifetime. After fast cooling by 2.5 ^{0}C, the temperature of water droplet on the teflon substrate gradually increased to ambient temperature. Firstly droplets on the surface of EPS sharply cooled by 10.3 ^{0}C. Then there was a long stage with gradually increase of temperature. At the final stage of evaporation, there was a sharp increase in temperature to temperature of ambient.

A more significant decrease of the temperature to the wet bulb temperature t_{wb} = 12.7 ^{0}C was observed under similar conditions for a free droplet. The presented data agree with experimental and computational data of other investigators [7, 8]. In contrast to these studies, the used method of infrared thermography allowed determining the droplet temperature without contact and obtaining data on temperature dynamics with good accuracy. As a result, the minimum droplet temperature is presented on Fig. 3 as a function of thermal conductivity of the substrate. The obtained data show that the greater the thermal conductivity of the substrate material, the smaller the minimum droplet temperature during evaporation and, respectively, the longer the evaporation time.

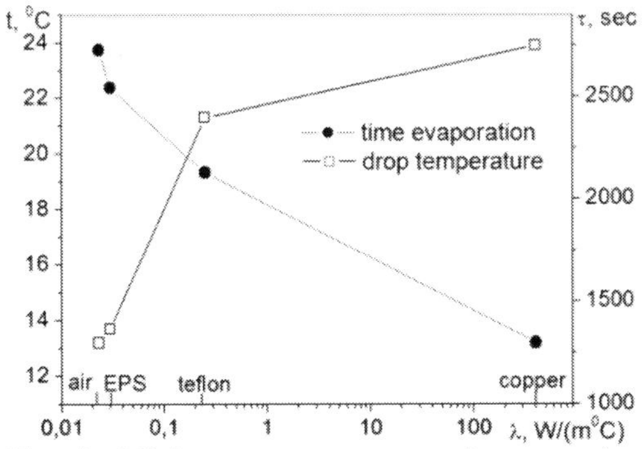

Fig. 3. Minimum temperature and evaporation lifetime of a water drop versus thermal conductivity of the substrate

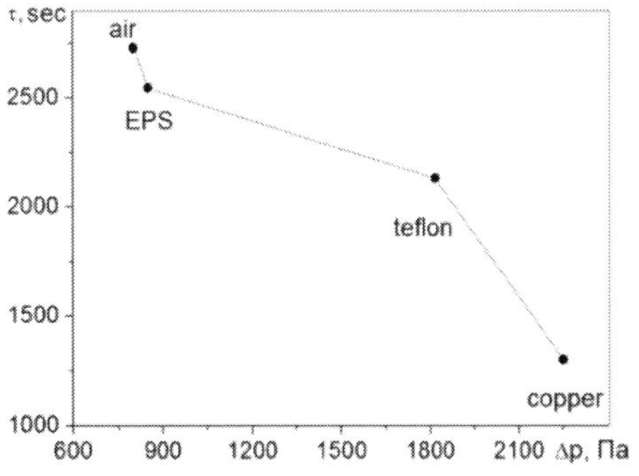

Fig. 4. Evaporation lifetime of water droplet on a substrate versus the difference in vapor pressure near the drop and in the ambient medium

The vapor flow during droplet evaporation was determined by its diffusion into the surrounding air. The rate of evaporation depends on the difference in vapor pressures near and far from the drop surface [11]. Near the droplet surface, water vapor may be considered saturated at the surface temperature. According to the measured minimum temperatures of the droplet surfaces, the saturation pressure of the vapor near the drop was determined. The dependence of the evaporation lifetime of droplets on the difference of vapor pressure near and far from the drop is shown in Fig. 4. The obtained data show that the higher thermal conductivity of the substrate material, the greater pressure drop of vapor and, hence, the greater the intensity of evaporation.

4 Conclusions

As a result of experimental studies, the droplet surface temperature during evaporation was obtained for substrates with different thermal conductivity. The use of the contactless method of infrared thermography allowed obtaining data on the dynamics of temperature changes on the surfaces of evaporating droplets with high accuracy. The obtained results show that for all the considered materials, the interfacial temperature was higher than the adiabatic evaporation temperature. Moreover, the lower the thermal conductivity of the material, the lower the temperature of the drop surface. The closer this temperature to the wet bulb temperature, the longer the evaporation lifetime. The coefficient of thermal conductivity of the substrate material was found to have a significant effect on the process of evaporation of sessile droplets.

The reported study was funded by RFBR under the research project № 17-58-53168.

References

1. P.V. Lebedev-Stepanov, S.P. Molchanov, M.V. Alfimov, R.M. Kadushnikov, N.I. Rubin, N.A. Shturkin, Nanotech. **6**, 79 (2011) [in Russian]
2. S.T. Beyer, K. Walus, Langmuir **28**, 8753 (2012)
3. Y. Gan, L. Qiao, Int. J. Heat Mass Transfer **54**, 4913 (2011)
4. D. Brutin, B. Sobac, C. Nicloux, Trans. ASME J. Heat Transfer **134**, 061101 (2012)
5. Y.P. Li, M.K. Lei, J. Mater. Sci. Technol. **30**, 965 (2014)
6. Z. Pan, S. Dash, J.A. Weibel, S.V. Garimella, Langmuir **29**, 15831 (2013)
7. S. David, K.Sefiane, L. Tadrist, Colloids Surfaces A: Physiochem. Eng. Aspects **298** 108 (2007)
8. G.J. Dunn, S.K..Wilson, B.R. Duffy, S. David and K. Sefiane, J. Fluid Mech. **623**, 329 (2009)
9. M.C. Lopes, E. Bonaccurso, T. Gambaryan-Roismana, P. Stephan, Colloid Surface A: Physiochem. Eng. Aspects **432**, 64 (2013)
10. D. Brutin. B. Sobac, F. Rigollet, C. Le Niliot, Exp. Thermal Fluid Sci. **35**, 521 (2011)
11. R. G. Picknett and R. Bexon, J. of Colloid and Interface Science **61**, 336 (1977)

Heat and mass transfer process modeling in the packing columns and tubular absorbers

Mikhail Shilyaev[1,*], *Elena* Khromova[1], and *Aleksandr* Bogomolov[2,3]

[1]Tomsk State University of Architecture and Construction, 634003, Tomsk, Russia
[2]Kutateladze Institute of Thermophysics SB RAS, 630090, Novosibirsk, Russia
[3]Gorbachev Kuzbass State Technical University, 650000, Kemerovo, Russia

Abstract. The mathematical models of heat and mass transfer in the packing columns (PC) and tubular absorbers (TA) with the square and triangular tube arrangement along their axes are derived for fully open flow cross-section of the device and closed tube space. The paper shows the energy advantage of TA in comparison with PC with the same extraction efficiencies of target gas component from the stream.

1 Introduction

The known empirical generalized volumetric mass transfer coefficients for the packing columns (PC) and tubular absorbers (TA) cannot adequately reflect the actual process for the following reasons. All of the above suggests that optimizing the regime and geometrical parameters of the PC and the TA based on the experimental factors β_{iv}, is impossible, like it is impossible to compare different vehicles (in terms of their performance, power, and size) with each other. The development of mathematical models for PC and TA will make it possible to include these devices in the general system of calculation and selection of optimal operating conditions.

2 Problem statement

A granular layer of thickness l and porosity ε, consisting of the balls with diameter d_a, irrigated by liquid with irrigation coefficient q_l, m^3/m^3, and initial temperature T_{l0} is considered **for the packing columns (PC).** A vapor-gas flow is fed into the layer together with liquid with velocity u_0 and temperature T_{00}, which interacts simultaneously with the liquid film on the surfaces of layer granules in the pore channels during the heat and mass transfer processes; so, the gas components are completely absorbed from the vapor-gas flow. Under the certain thermodynamic conditions occurring locally in the pore channels, the film can evaporate or it can condense liquid vapor. The stationary, one-dimensional equations of conservation are written for the packing, taking into account the liquid film thickness on its elements, and for the vapor-gas flow along the pore channels.

The equation of heat transfer between the layer and liquid is:

* Corresponding author: shmi@mail.tomsknet.ru

$$\frac{d}{dx}\left[c_l\rho_l\varepsilon_l\left(T_l-T_0\right)+c_m\rho_m\left(1-\varepsilon\right)\left(T_m-T_0\right)\right]=q+\Sigma r_i g_i\,, \tag{1}$$

$$T_l=T_{l0},\ T_m=T_{m0}\ \text{при}\ \ \tau=0,\quad 0\le x\le l\,,$$

where c_l and c_m, ρ_l and ρ_m, T_l and T_m are thermal capacities, densities, mean-mass temperatures of liquid and ball packing elements; T_0=273 K; r_i are specific heats of evaporation-condensation of liquid and absorption-desorption of the target gas components, J/kg; q, W/(m^3 s), and g_j, kg/(m^3 s) are heat and mass fluxes between the liquid film, balls and filtration vapor-gas flow in the packing pores:

$$q=-\alpha a\left[T_w-T\right], \tag{2}$$

$$g_i=-\beta_i a\left[\rho_{wi}-\rho_i\right], \tag{3}$$

where α is the heat transfer coefficient, W/(m^2 K); a is the specific packing surface, 1/m:

$$a=\frac{6}{d_e}\left(1-\varepsilon'\right),\quad \varepsilon'=\varepsilon-\varepsilon_l\,, \tag{4}$$

d_e is the equivalent diameter of the packing elements, for balls $d_e=d_a$; T_w is the temperature on the film surface, K ; it takes into the account of the smallness of the liquid film thickness T_l, Moreover, the temperature of the ball surface T_w it taken equal to temperature T_m; T is the temperature of the vapor-gas flow in pores, K; β_i is the mass transfer coefficient, m/s; ρ_{wi} and ρ_i are partial densities of the vapor-gas components near the film surface at the saturation temperature and in the filtration flow in pores, kg/m^3, calculated from the state equations; ε_l is the volumetric concentration of liquid in the pore channels.

The equation of heat transfer between the filtration flow and the packing is:

$$\rho\frac{d\left(c\left(T-T_0\right)\right)}{d\tau}=-q\,,\quad T=T_{00}\quad at\ \tau=0,\quad 0\le x\le l \tag{5}$$

where c is heat capacity of the vapor-gas flow:

$$c=\frac{\sum\rho_i c_i}{\rho},\quad \rho=\sum\rho_i\,. \tag{6}$$

The equation for liquid concentration is:

$$\frac{d\varepsilon_l}{d\tau}=\frac{1}{\rho_l}\sum\beta_i a\left(\rho_{wi}-\rho_i\right),\quad \varepsilon_l=\varepsilon_{l0}\quad at\ \tau=0,\quad 0\le x\le l \tag{7}$$

At that, the liquid film thickness is

$$\delta=\frac{1}{6}\frac{\varepsilon_l}{\left(1-\varepsilon\right)}d_a\,.$$

The film velocity and its thickness were calculated using the solution of the hydrodynamic problem of gravitational film flow, when it is subjected to an external filtration flow at direct and counter flow, and continuity equation for liquid, considering evaporation-condensation processes. The local heat transfer coefficients were determined by the limit criteria theoretical dependence of M.A. Goldshtik, the local mass transfer coefficients were determined using the experimental criteria dependencies of M.E. Aerov, O. M. Todes, D. A. Narinskiy ($\beta_{i э}$), and the known theoretical dependence for film mass transfer (β_{iT}). The conditions of phase equilibrium were set at the interface (the film surface – the vapor-gas flow) according to the Henry's law, and for water vapor, they were set using the well-known approximation of experimental data on the partial pressures at the state of saturation for the film temperature. The temperature of ball surface and film temperature were taken equal; at that, the first was calculated from the approximate balance equation by the heat fluxes on ball surfaces without solving the conjugate problem. The heat effect of water evaporation-condensation was taken equal to r=2500 kJ/kg, for the process of gas absorption, it was taken according to V.M. Ramm. The pressure drop on the packing layer ΔP was calculated by the Ergan formula considering film thickness δ on the ball surface and energy spent for initiation of the liquid film motion. The recovery ratio of the ith gas component from the vapor-gas flow to liquid film η_i was calculated from the ratio of mass flow rates of this component at the inlet and outlet of the pore channel:

$$\eta_i = 1 - \frac{\left[(\rho_i u_f)(\varepsilon - \varepsilon_l)\right]_l}{\left[(\rho_i u_f)(\varepsilon - \varepsilon_l)\right]_0}. \quad (8)$$

Here u_f is the velocity of the vapor-gas flow in the pore channels:

$$u_f = u_{f0}\frac{\rho_{g0}}{\rho_g}\frac{\varepsilon - \varepsilon_{l0}}{\varepsilon - \varepsilon_l}, \quad u_{f0} = \frac{u_0}{\varepsilon - \varepsilon_{l0}}, \quad (9)$$

ρ_g and ρ_i were calculated from continuity equation

$$\frac{\partial \rho_i}{\partial \tau} + \operatorname{div}\left(\rho_i \vec{u}_f\right) = -g_i, \ \rho_i = \rho_{i0} \quad at\ \tau = 0, \quad 0 \le x \le l \quad (10)$$

For ρ_g, the right part of equation (10) is zero.

Similarly, as for the PC, for a **tubular absorber (TA)** the equations of conservation are written for the tubular absorber (TA) in the same manner as for the PC. At that, the mathematical model of the process of gas absorption in a TA is considered for two patterns of tube arrangement: square cross-section and maximally compacted, triangular, according to the centers of inner diameters of tubes d_{aT} (Fig. 1).

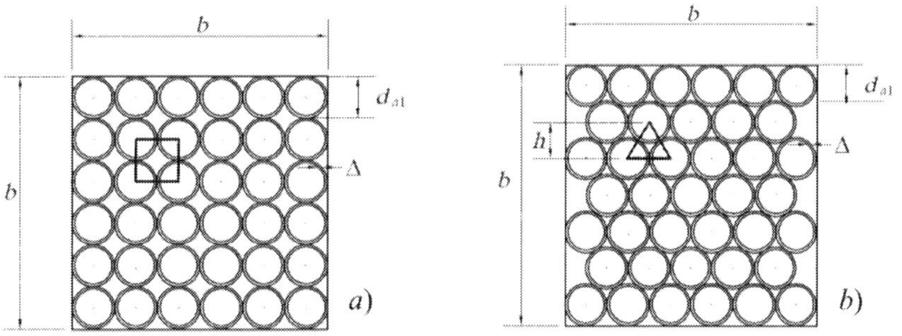

Fig. 1. Two patterns of tube arrangement in the cross-section of the working volume of TA: *a*) square location of tube axes; *b*) triangular location of tube axes.

3 Results and discussion

Results of calculations of mass concentration of SO_2 in the water film on the grains along the pore channel length are shown in Fig. 2, *a*.

According to this figure, in this case, the process reaches the state of thermodynamic equilibrium at a distance from the inlet of 0.2 m, and it is unreasonable to obtain a thicker layer. It is also clear that in the state of thermodynamic equilibrium, calculations of the film and locally volumetric mass transfer mechanisms give the same result at the same distance from the inlet.

 DOI: 10.1051/matecconf/201711508006

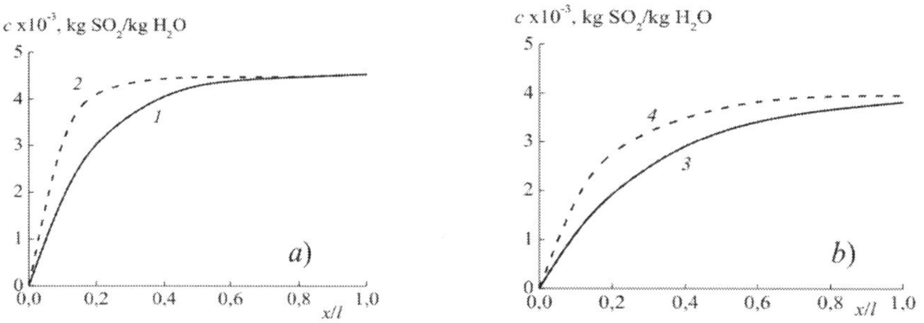

Fig. 2. Calculation results on SO_2 absorption on water in a PC at the direct flow (*a*) and comparison of parameters of SO_2 absorption in TA with a closed intertube space and triangular tube arrangement and in a PC for corresponding conditions (*b*): l=0.2 m, T_{00}=293 K, T_{l0}=288 K, u_0=2 m/s, $d_{wv,0}$=0.01 kg/kg of dry air (for water vapor), $d_{SO_2,0} = 0.1$ kg/kg of dry air, q_l=0.00953 m^3/m^3; *1* – $\beta_{iэ} = Nu_{ei}(D_i/d_{ep})$, *2* – $\beta_{iT}=3.41(D_i/\delta)$, d_a=0.02 m, ΔP=978 Pa, $\eta_{SO_2} = 0.13986$, δ=158 μm; *3* – Pc (used ($\beta_{iэ}$)): d_{ae}=0.0097836 m, ε=0.734 m^3/m^3, $\eta_{SO_2} = 0.2257$, ΔP_1=138 Pa; *4* – TA (used (β_{iT}): d_{aT}=0.02 m, $\eta_{SO_2} = 0.22466$, ΔP_2=1.94 Pa.

Calculations show that from the energy consumption viewpoint for gas cleaning, the direct and counter flow schemes with the same efficiencies of gas component extraction are equal. However, the preference should be given to the direct-flow scheme, which excludes flooding regime and allows absorption in the PC at higher velocities of the cleaned gas, i.e. at significantly higher flow rates through a single apparatus.

Calculation results of the absorption process in TA are shown in Fig. 2, *b* in comparison with ND under the appropriate conditions. At that, the transition is performed by the formula for the spherical elements of packing with the equivalent diameter:

$$d_{ae} = \frac{6}{a}(1-\varepsilon).$$

As seen from the figure, calculation results for PC and TA are in satisfactory agreement with each other in terms of the efficiency of SO_2 extraction on water. However, in terms of the pressure drop, the advantage of TA over PC is obvious. It is shown that it is extremely difficult, almost impossible to distribute evenly the liquid in the form of a film over the inner and outer tube surfaces, when the efficiency of pattern with the outer intertube space prevails over the pattern with the closed intertube space. The second pattern is more preferable. The stable and uniform distribution of the film over the inner surfaces of tubes can be easily arranged by rotating the liquid, sprayed at the inlet, with a screw device inside the tubes, which creates rotational movement of the vapor-gas flow. Due to the action of the arising centrifugal forces, the dispersed liquid droplets are thrown to the walls of tubes, forming there a rotating film.

In addition, the intertube space could be used for tubes cooling to maintain their temperatures at the lowest level contributing to more efficient extraction of the target gas component on the absorbent.

4 Conclusion

The proposed models can be used in engineering calculations of heat and mass transfer of PC, TA and gas-cleaning devices.

Dynamics of propagation of the self-sustaining evaporation front in a mixture of Freons under the natural convection conditions

Vladimir Zhukov [*], *Mikhail* Moiseev

Kutateladze Institute of Thermophysics SB RAS, Lavrentiev ave. 1, 630090, Novosibirsk, Russia

Abstract. The results of experimental studies on dynamics of propagation of the self-sustaining evaporation front at nonstationary heat release are presented in the paper. The experiments were carried out in a large volume at liquid saturation temperature and subcooling on a horizontal cylindrical heat-releasing surface with the diameter of 3 mm under nonstationary heat release. Freons R21, R114 and their mixtures were used as the working liquid. The experiments carried out in the range of concentrations of high-volatile component $0 < C < 1$ showed that in the range of temperature heads corresponding to the loss of hydrodynamic stability of evaporation front interface, a higher evaporation front velocity corresponds to the higher content of volatile component R114. The temperature head corresponding to the initiation of the evaporation front has a minimal value at mixture concentration corresponding to the point of azeotrope ($C = 0.62$). At a fixed value of the temperature head in the range of high values of ΔT, liquid subcooling relative to saturation line leads to a decrease in the velocity of evaporation front propagation.

1 Introduction

Nonstationary heat release at the local section of the heat exchanger can cause significant overheating of liquid heat carrier contacting with a heated wall, and formation and propagation of a continuous vapor film. This effect can be observed in the systems of film and droplet irrigation and in a large volume. There are a number of models describing propagation of undisturbed self-sustaining evaporation front, as well as individual attempts to simulate the front dynamics taking into account small-scale perturbations of the interface at intense evaporation [1]. In recent years, there has been a tendency to use the mixtures instead of a one-component fluid as a working medium in the refrigeration machines, heat pumps and thermal transformers. The ozone-safe non-azeotropic mixture refrigerants R32/R134a and R32/R152a are considered in [2] as well as heat transfer at boiling in horizontal tubes.

The purpose of this work is the experimental study of evaporation front dynamics and development of hydrodynamic instability on the interface in freons R21, R114 and their mixtures.

* Corresponding author: zhukov@itp.nsc.ru

2 Experimental setup and methods

The experimental setup is a cylindrical vessel with the diameter and height of 250 mm with the built-in windows for visualization of transient processes. The mixtures of Freons R114 and R21 on the saturation line at the pressure of 0.27 - 0.28 MPa and under the subcooling conditions were used as the working liquid. A horizontally oriented cylindrical test section made of a stainless steel tube 3-mm in diameter and wall thickness of 0.5 mm was used in experiments. The roughness of the section with the diameter of 3 mm was presented by multidirectional grooves with the width of not more than 10 μm and separate smoothed cavities with the diameter of 30-50 μm. The temperature of the working section wall under the conditions of nonstationary heat release was determined from the current-voltage characteristic and calibration curve, and by numerical calculations using the equation of nonstationary heat conductivity. The accuracy of temperature determination was ± 1.5 K. To observe the evaporation front dynamics, a high-speed video camera was used. The shooting speed was 25000 frames per second with an exposure of 25 microseconds. Heat-releasing surface was heated by a rectangular current pulse of given duration and amplitude. To illuminate the object in transmitted and reflected light, the LED assemblies with a light flux of 9000 lm were used, their switching-on time was synchronized with the point in time corresponding to begining of heat release. The detailed description of setup and experimental procedure is given in [3].

3 Discussion

The experiments were carried out with Freons R21 and R114 and their mixtures, including the azeotropic mixture of 62% R114. Dependences of evaporation front velocity V_{fr} on temperature head relative to saturation temperature ΔT are shown in *Fig.* 1 for different molar concentrations C of high-volatile component R114. As seen from the diagram, $V_{fr}(\Delta T)$ dependence has two characteristic regions – the region of a weaker dependence (*region* 1) and region of a stronger dependence (*region* 2). In [1, 3], the presence of *region* 2 is associated with development of Landau's hydrodynamic instability of. For all series of experiments carried out in *region* 1, there is no significant difference in the front propagation velocity. In *region* 2, the experimental data are stratified depending on concentration of R114 component in the mixture. At the same values of temperature head ΔT, the propagation velocity of evaporation front is larger in the mixture with higher concentration of component R114, and for pure Freon R114, the front velocity is maximal.

The temperature head with respect to saturation temperature (ΔT_0), corresponding to initiation of evaporation front for the mixtures with different concentrations, varies from 34 to 41 K (*Fig.* 2). The dependence has a nonmonotonic character with minimal values of ΔT_0 at concentrations C = 0.35 - 0.62.

DOI: 10.1051/matecconf/201711508007

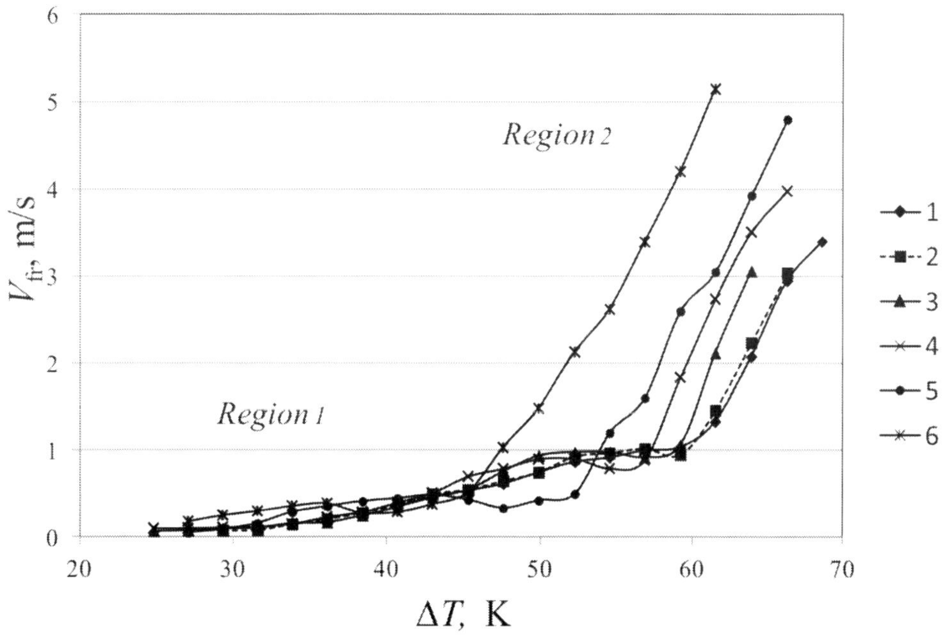

Fig. 1. Dependence of evaporation front velocity V_{fr} on temperature head ΔT for Freons R21, R114 and their mixtures. Conditions of saturation: *1* – R21, $P = 0.267$ MPa; *2* – $C = 0.1$, $P = 0.27$ MPa; *3* – $C = 0.2$, $P = 0.27$ MPa; *4* – $C = 0.35$, $P = 0.27$ MPa; *5* – $C = 0.62$, $P = 0.28$ MPa; *6* – R114, $P = 0.28$ MPa.

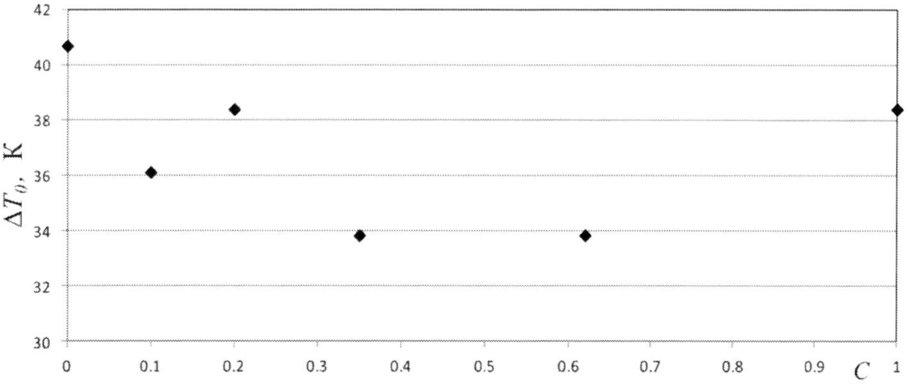

Fig. 2. Dependence of ΔT_0 on molar concentration of high-volatile component R114 in the mixture. $P = 0.27$ MPa

Data obtained in experiments with saturated liquid and liquid subcooled with respect to saturation temperature by η are shown in *Fig.* 3. Dependences of evaporation front velocity V_{fr} on the temperature head with respect to saturation temperature ΔT are given for molar concentration C = 0.22 for different subcoolings. As seen from the diagram, the pressure and subcooling of liquid affect the velocity of the self-sustaining evaporation front. Under the conditions of liquid being at the saturation line, an increase in pressure from 0.266 to 0.355 MPa led to a shift in $V_{fr}(\Delta T)$ dependence to the region of lower temperature heads by a value of about 3 K. Under the conditions of liquid subcooling relative to the saturation line, V_{fr} (ΔT) dependence shifts to the region of high values of the temperature head approximately by the same value. At a fixed temperature head in *region* 2, liquid subcooling leads to a decrease in the propagation velocity of evaporation front.

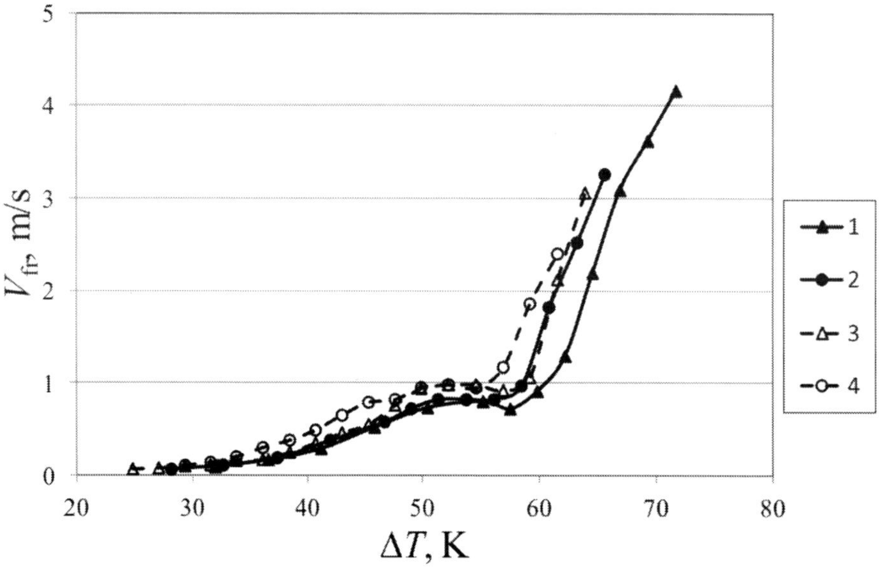

Fig. 3. Dependence of evaporation front velocity V_{fr} on temperature head ΔT for mixture with C = 0.22 : *1* – η = 6.4 K, P = 0.27 MPa; *2* – η = 14.9 K, P = 0.351 MPa; *3* – η = 0 K, P = 0.266 MPa; *4* – η = 0 K, P = 0.355 MPa.

4 Conclusions

The experiments carried out on the mixture of Freons R114 and R21 within the range of concentrations of high-volatile component $0 < C < 1$ showed that in the region of temperature heads corresponding to the loss of hydrodynamic stability of evaporation front interface, a higher content of volatile component R114 corresponds to a higher velocity of evaporation front. The temperature head corresponding to initiation of evaporation front at mixture concentration corresponding to the point of azeotrope has a minimal value. In the region of high ΔT at a fixed value of the temperature head, liquid subcooling below the saturation line leads to a decrease in propagation velocity of evaporation front.

The work was financially supported by the Program of Fundamental Research of SB RAS (project No. F3 III.18.2.3)

References

1. A.N. Pavlenko, V.V. Lel, Thermophys. Aeromech. **6**, 105 (1999)
2. N.N. Mezentseva, I.V. Mezentsev, A.V. Meleshkin, MATEC Web of Conf. **23**, 01027 (2015)
3. A.N. Pavlenko, E.A. Tairov, V.E. Zhukov, A.A. Levin, A.N. Tsoi, J. Eng. Thermophys. **20**, 380 (2011)

Surface tension determination using data of the evolution of thermocapillary deformations in a locally heated liquid layer

Ella Barakhovskaia[1,2,*] and *Igor* Marchuk[1,2]

[1]Novosibirsk State University, Pirogova Str. 2, Novosiborsk, 630090, Russia
[2]Kutateladze Institute of Thermophysics, Russian Academy of Sciences, Prosp. Lavrentyev 1, Novosibirsk, 630090, Russia

Abstract. Identifying liquids in chemical technologies and in the oil industry is an actual problem. Such identifications can be made using individual characteristics of thermocapillary flows and deformations in the horizontal layers of the investigated liquids. Numerical algorithms and calculating code have been developed for solving inverse problems for identifying properties of liquid. As additional information for inverse problem a function of thermocapillary response is used. The algorithm for solving inverse problem is based on solving problem of minimizing the residual function. The minimum is obtained numerically by the Newton's method. Numerical calculations have been made for silicone oil. Thermocapillary responses have been calculated for different values of the coefficients. The developed algorithms have been tested on synthetic data, which have been obtained from the solution of direct problems. The developed algorithms have shown a good convergence and efficiency.

1 Introduction

Identifying liquids in chemical technologies and in the oil industry is a topical issue in the present days. Such identification can be made according to individual characteristics of thermocapillary flows and deformations in the horizontal layers of the investigated liquids. To a large extent the thermocapillary effect occurs while thin liquid layers are locally heated [1]. Thermocapillary effect can be successfully used to determine the properties and the thickness of the thin locally heated liquid layer [2].

The inverse problem of thermocapillary convection consists in determining the surface tension coefficients σ and the temperature coefficient of the surface tension σ_T according to thermocapillary response that has been measured in the experiment. Thermocapillary response is the time dependence of the laser beam diameter reflected from the surface of the liquid and projected onto the screen, Fig. 1a.

A method for measuring the thickness of a liquid layer using a laser-induced thermocapillary effect was proposed in paper [2], and a method for identifying liquids was proposed in paper [3]. This paper presents a method for calculating σ and σ_T coefficients

* Corresponding author: ella94@bk.ru

based on the data about the dynamics of deformations in a thin locally heated horizontal liquid layer.

2 Problem statement

Solving the inverse problem of thermocapillary convection is based on results of studying the direct problem of thermocapillary deformation of thin horizontal film [5].

The model considers a thin horizontal liquid layer on the substrate with local heater. Liquid is viscous and incompressible. Heater is thin and uniform. Geometry and system conditions are axisymmetric. At the time t=0 the heater is activated and starts producing the pulse heating power from the substrate side during 1 second. Shear stress occurs on the liquid surface caused by the heterogeneity of its temperature. This causes thermocapillary flow and deformations of the liquid surface.

Two-dimensional unsteady problem is considered. The direct and inverse problems have been solved numerically in axisymmetric statement. The thin layer approximation model is used [4]. Capillary pressure, viscosity and gravity are taken into account. Evaporating rate is neglected because of chosen type of liquid (silicone oil – dimethylpolysiloxane, PMS-5). Initially the liquid layer has flat surface and uniform constant temperature. The initial thickness of the liquid layer varies from 100μm to a few millimeters. Schematic of the system is shown in Fig. 1a.

3 Measuring the thermocapillary response

The inverse problem of thermocapillary convection consists on determining the surface tension coefficients σ and σ_T. A function of thermocapillary response is used as additional information for solving inverse problem:

$$dSp(t) = \max_{r\in[0,d_l/2]} \left| 2 \cdot \left| r - (d - h(t,r)) \cdot tg\left(2arctg\left(h'_r(t,r)\right)\right) \right| \right| \quad (1)$$

where dl – is the diameter of a laser beam, $h(t,r)$ – is the film thickness, $h'_r(t,r)$ – is the first derivative of thickness, d – is the distance to the screen.

Thermocapillary response function is a special experimentally measured time dependent function of laser beam diameter, reflected from the liquid surface and projected on a screen, Fig. 1a.

Liquid flow and deformations of its surface and, as a result, thermocapillary response are determined by the liquid properties, as well as the properties of the cuvette material, heater and heating characteristics. For different liquids at other factors being equal, the kind of thermocapillary response is significantly different.

4 Mathematical formulation

Dynamics of thin films is well described by the evolution equation, which is obtained in the lubrication approximation theory [5, 6]:

$$h_t + divq = 0 \quad (2)$$

where, h_t – is the velocity of the surface, $\boldsymbol{q} = \frac{h^3}{3\mu}\boldsymbol{f} + \frac{h^2}{2\mu}\boldsymbol{\tau}$ - is the liquid flow rate along the r direction, $\boldsymbol{f} = grad\left(\rho gh + \sigma H\right)$ - is the gradient of the modified pressure, $\boldsymbol{\tau} = \sigma_T gradT$ - is the thermocapillary tangent stress, μ - is the dynamics viscosity, ρ - is the density, **g** - is the gravity vector, h - is the film thickness, σ - is the surface tension, σT - is the linear coefficient, which determines the dependence of the surface tension on temperature, T - is the temperature, $H = \frac{h_{rr}}{\left(1+h_r^2\right)^{3/2}} + \frac{h_r}{r\left(1+h_r^2\right)^{1/2}}$ - is the double mean curvature.

The temperature distribution in the liquid layer is determined by the energy equation in cylindrical coordinates:

$$\rho C_p\left(\frac{\partial T}{\partial t}+u\frac{1}{r}\frac{\partial (rT_r)}{\partial r}+v\frac{\partial T_z}{\partial z}\right)=\frac{1}{r}\frac{\partial}{\partial r}\left(\lambda\left(r\frac{\partial T_r}{\partial r}\right)\right)+\frac{\partial^2 T_z}{\partial z^2}+Q \tag{3}$$

where, $Q = Q(t,r,z) = \begin{cases} Q = 0, \text{ outside the heater} \\ Q = const > 0, \text{ in the heater} \end{cases}$ - is the bulk density of the heat sources, λ - thermal conductivity coefficient; Cp - is the specific heat; ρ - is the density of the liquid.

System of equations (2)-(3) is the direct problem of thermocapillary convection. Equation (3) is a nonlinear differential equation of the first order with respect to time, and a fourth order in spatial variables with respect to the unknown function $h(t,r)$. Numerical simulation of inverse problem is based on results obtained from solving the direct problem of thermocapillary convection [4, 5, 6].

5 Solving algorithm

To find the numerical solution of the direct problem (2) - (3) we used the algorithm for joint solving the energy equation (3) and the evolution equation (2). Stationary solutions are obtained by the establishment method [5, 6].

The algorithm for solving inverse problem is based on solving problem of minimizing the residual function, which expresses the integral deviation of the measured response, and subsequent calculations. The minimum is found numerically by the Newton's method.

Residual function is given as:

$$J(\sigma^k) = \frac{1}{t_1}\int_0^{t_1}\left[dSp(\sigma^*) - dSp(\sigma^k, t)\right]^2 dt \geq 0 \tag{4}$$

where dSp – is the function of thermocapillary response, σ^* - is the exact value of surface tension coefficient, measurement time t_1=5s.

The problem of function minimization is formulated as:

$$J(\sigma^k) \rightarrow \min_{\sigma^k \in [\sigma_{min}, \sigma_{max}]} \tag{5}$$

6 Numerical results

Numerical calculations have been made for silicone oil (dimethylpolysiloxane) type PMS-5. Silicone oil PMS-5 is a non-volatile liquid, so evaporation was not taken into account. Figures 1(b, c) show calculated thermocapillary responses for different values of the coefficients σ and σT. It can be seen that for different σ or σT thermocapillary responses (the diameters of the reflected laser beam) are significantly different. It means that the system is quite sensitive to the liquid properties σ and σT.

The results of the calculated functions $J(\sigma)$ and $J(\sigma_T)$ for layer PMS-5 are shown in Fig.2. It can be seen that the functions are "good" in terms of solving the inverse problem: the shapes of functions are convex and have unique minimum. This allows finding the minimum functions by Newton method for several iterations with a sufficiently high accuracy.

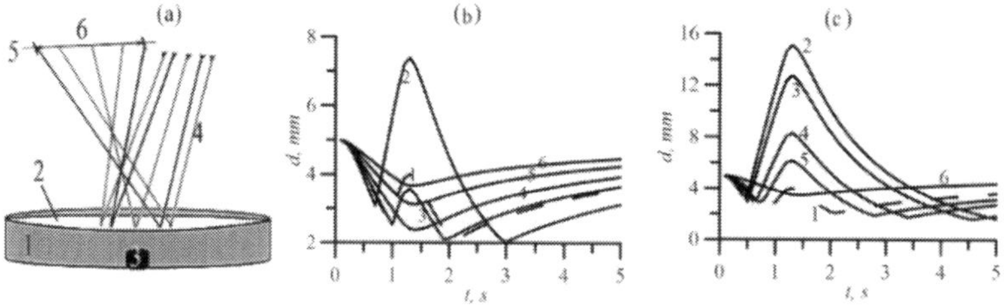

Fig. 1. (a) System scheme: 1 – Teflon cuvette, d = 32 mm, 2 - PMS-5, 3 – heater, r = 1.6 mm, Q=0.00165 W, 4 – laser beam, 5 – screen, 6 – diameter of the light spot. (b) Thermocapillary response depending on different σ for PMS-5, h_0 = 238 μm: 1 – exact solution σ=σ*=0.01827 N/m, 2 – σ=0.01 N/m, 3 – σ=0.02 N/m, 4 – σ=0.03 N/m, 5 – σ=0.05 N/m, 6 – σ=0.08 N/m. (c) Thermocapillary response depending on different σ_T with the same conditions: 1 – exact solution σ = σ_T*=-0.066x10^{-3} N/(mxK), 2 – σ_T = -0.16x10^{-3} N/(mxK), 3 – σ_T = -0.141x10^{-3} N/(mxK), 4 – σ_T = -0.104x10^{-3} N/(mxK), 5 – σT=-0.08x10-3 H/(mxK), 6 – σ_T=-0.03x10^{-3} N/(mxK).

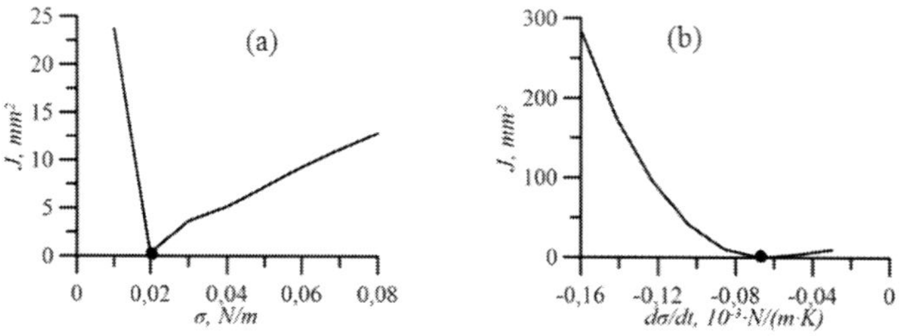

Fig. 2. Residual function for PMS-5, h_0=238μm, (a) $J(\sigma)$, •- exact σ*, (b) $J(\sigma_T)$, •- exact σ_T*

7 Conclusion

We have developed algorithms and created calculating code for solving inverse problem to identify properties of liquid: σ and σ_T. Thermocapillary response function is used as additional information for solving inverse problem. Calculation results show that the thermocapillary response is quite sensitive to the liquid properties σ and σ_T. It means that using of the thermocapillary response as additional data of this inverse problem is justified and reliable. The developed algorithms have been tested on synthetic data, which has been obtained from the solution of direct problems. The developed algorithm has shown a good convergence and efficiency.

The study was financially supported by the Russian Science Foundation (Project No. 15-19-20049).

References

1. I.V. Marchuk, O.A Kabov, Doklady Physics **61**, 1 (2016)
2. B.A. Bezuglyi, A.A. Fedorets, O.A. Tarasov, N.A. Ivanova, Surface, X-Ray, Synchrtronic and Neytronic Studies **6**, 80 (2004) [in Russian]
3. B.A. Bezuglyi, A.A. Fedorets, Technical Physics Letters **27**, 359 (2001)
4. E.V. Barakhovskaia, I.V. Marchuk, A.A Fedorets, J. Physics: Conference Series, **754** (2016)
5. I.V. Marchuk, J. Eng. Thermophys. **24**, 381 (2015)
6. A. Oron, S.H. Davis, S.G. Bankoff, Rev. Modern Physics **69**, 3 (1997)

Suspended nanofluid droplet evaporation

Aleksandr Borisov[1,*], *Aleksandr* Nazarov[1,2], *Anatoly* Serov[1,3], and *Valery* Mamonov[1]

[1]Kutateladze Institute of Thermophysics SB RAS, Novosibirsk, Russia
[2]Novosibirsk State University, Novosibirsk, Russia
[3]Novosibirsk State Technical University, Novosibirsk, Russia

Abstract. The experimental research of suspended droplet evaporation process in air flow was conducted for distilled water (DW) with different silicon oxide (SiO_2) nanoparticle concentration. The data obtained allowed comparing droplet volume and temperature alteration dynamics with that in pure DW at different temperatures and flow rates.

Introduction

The evaporation process is used in technology for liquid mixtures separation and substance purification. It serves a basic process in steam power engineering, refrigerators, evaporation, and other systems, as well as material drying processes. Study of these processes is of great importance to highlight process enhancement methods and increase electric power installation efficiency.

New fundamental knowledge concerning gas-droplet flows with phase transformation is required for development of high-performance heat exchange apparatus, increase spray cooling effectiveness, liquid fuel combustion, and other practical applications. It is necessary to understand the heat and mass exchange mechanisms during nanofluid evaporation of flows with phase changes. There are a significant number of articles concerning complex droplet evaporation problem [1], but there is no universal theory concerning complex mixture droplet evaporation which could combine all principles experimentally obtained.

The aim of experiments was to obtain the data regarding solid admixture concentration impact on liquid carrier evaporation rate from the droplet surface in heated air flow.

Experimental setup

The experimental equipment consists of vertical cylindrical air channel with a stainless steel confuser (l = 0,5 m, d = 50 mm, outlet nozzle d = 10 mm), and heat exchanger with electrical heater distributed along the entire air channel with thyristor power regulator [2]. To decrease thermal loss the heat exchanger and the air channel are covered with thermal insulation. The heat exchanger combined with power regulator is capable of air flow heating within the range from room temperature (20 ^{0}C) up to 150 ^{0}C. The air flow is regulated with reducing gearbox at the start of the heating system within the range from 0,1 up to 5 m/s. After heating section, the air flow passes through initial lining by honeycomb

* Corresponding author: aborisov.librilian@gmail.com

(cell step 0,5 mm) placed next to an adapter between the heat exchanger and the vertical air channel. Secondary linearization is performed with a grid (cell step 0,2 mm) installed in front of the outlet nozzle. A liquid droplet is suspended on a carrier at a distance of 20 mm from the center of confuser outlet. A polytetrafluoroethylene (PTFE) cylinder (d = 0,1 mm) serves as the carrier. A liquid droplet is formed with the BP96131 pipette (0,5 – 5 μl, 3 % accuracy). A mixture was prepared using DW and SiO_2 nanoparticles ~ 200 nm size with mass fraction m_{SiO2}/m_{H2O} equal to 2, 3, 4, 5, 6, and 7 %. Base liquid and nanoparticle mass was measured with weighing machine (with accuracy of 1%). Prior to an experiment nanomixture was placed into the ultrasonic bath for 30 min. After the ultrasonic mixing nanofluid was left to cool down to room temperature. A nanofluid droplet with diameter 2 ± 0,3 mm was the research object.

Air flow rate was measured with anemometer at the nozzle outlet and with rotameter placed in air line. When the flow temperature exceeded the maximum anemometer temperature (more than 50 ^{0}C), the flow rate was controlled with the rotameter, with air parameters dependent on temperature that was taken into consideration. Anemometer ("KIMOinstruments" VT110) and rotameter precision was 3% for the flow rates ranged from 0,15 to 3 m/s.

Air flow temperature was recorded by means of three detectors. Two platinum resistant thermometers were stationary placed in the upper and lower parts of vertical channel in front of the air flow linearizing grids. The third thermometer (thermocouple) was temporary inserted into the air flow before the experiment to control the temperature near the outlet nozzle. Temperature measurement accuracy for each of the sensors was 1,5 %. Relative air flow humidity was measured with the hygrometer Model 872 (with accuracy of 4 %) before the experiment.

Droplet diameter was captured during the evaporation process using DigiScopeIIv3 microscope. Droplet diameter was determined after the experiment processing digital images. Droplet diameter determination accuracy was 7,5 %. Droplet surface temperature on the thermal images, acquired by Thermo Tracer thermoimager, was determined with a use of special software. Average droplet surface temperature accuracy for this method was found to be 10 %.

Experiments were carried out at ambient pressure. The ambient temperature changed from 20 up to 30 ^{0}C. Air flow rate around the droplet ranged from 0 to 3 m/s depending on selected regime, temperature ranged from 20 up to 100 ^{0}C with relative air flow humidity not exceeding 1 % during each experiments.

Results

Experiments were conducted to estimate air flow rate and temperature influence on suspended nanofluid droplet evaporation rate (DW + SiO_2 mixture, nanoparticle mass fraction from 2 to 7 %, three air flow rates). Nanofluid droplet evaporation dynamics is shown on Fig. 1 for DW + SiO_2 in comparison with DW droplet evaporation (initial conditions: air flow rate u_0 = 0,2 m/s (a) and 1,5 m/s (b), ambient and air flow temperature $t_{0g} = t_{0s}$ = 23,4 ^{0}C, air flow relative humidity φ = 1%, ambient pressure P = 1 atm).

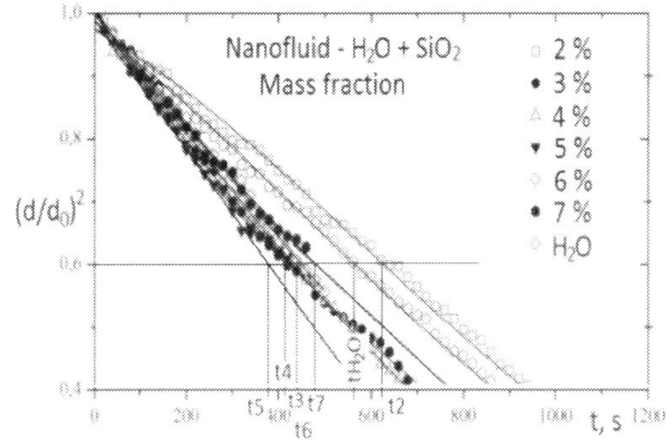

a.

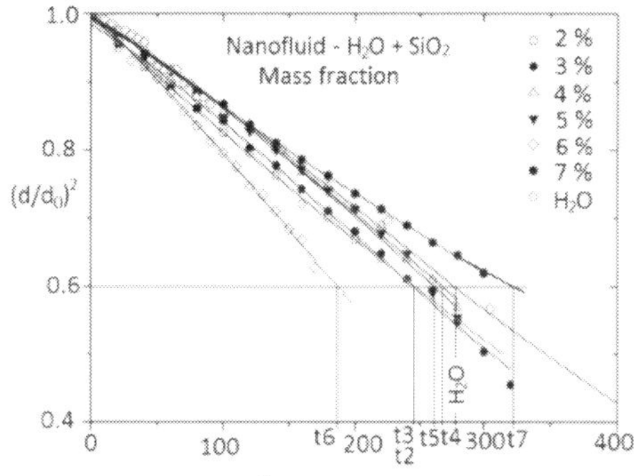

b.

Fig. 1. Nanofluid droplet evaporation dynamics. Experimental conditions: u_0 = 0.2 m/s (a), 1.5 m/s (b), temperature $t_{0g} = t_{0s}$ = 23 ^{0}C, φ = 0 %, P = 1 atm. t2, t3, t4, t5, t6, t7, tH_2O – evaporation time when droplet diameter was 60 % of initial droplet diameter.

As shown on Fig. 1, water evaporation time for the mixtures (t2, t3, t4, t5, t6, t7) is significantly different from that for the DW(t_{H2O}) prior to checkpoint (d/d_0 = 0,6). It is necessary to emphasize, that droplet volume measurement accuracy did not exceed Δd ~ 3 %, resulting in evaporation time detection Δt_{ev} ~ 5 % at the time point when 40 % of liquid evaporated. Ratios acquired let us analyze nanomaterial admixture influence on liquid evaporation rate with sufficient accuracy. Applying current analysis, it can be noted, that increase in air flow rate (1,5 m/s) significantly reduced nanofluid droplet evaporation time, while evaporation rate differences were present (Δt ~ 40 %).

Nanofluid droplet evaporation dynamics is shown on a Fig. 2 (air flow temperature raised up to t_{0g}= 100^{0}C, u_0= 1 m/s, t_{0s}= 23^{0}C, φ = 0%, P = 1atm).

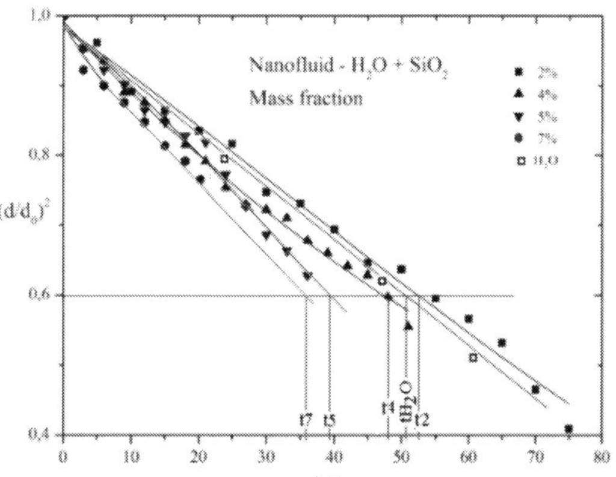

Fig. 2. Nanofluid droplet evaporation dynamics. Experimental conditions: u_0 = 1 m/s, T_{0g} = 100 ^{0}C, T_{0s} = 23 ^{0}C, φ = 0 %, P = 1 atm. t2, t4, t5, t7, tH_2O – evaporation time when droplet diameter was 60 % of initial droplet diameter.

Experimental results of droplet evaporation of water with suspended carbon nanotubes (mass fraction ~ 0,1 %) are presented in an article [3]. Based on these data it can be concluded that nanoparticle addition to base liquid (DW) slightly affects heat-and-mass transfer mechanism. Submitted experimental data concerning nanomaterials (different SiO_2 concentrations in base liquid) demonstrate that nanofluid droplet evaporation dynamics curve has an interval of accelerated evaporation when flow rate increases. The dependence of the ratio between relative nanofluid evaporation time and base liquid evaporation time (60 % of initial diameter) on nanofluid concentrations is shown in Figure 3, where f = (tn –

tH_2O) / tH_2O, tn – is nanofluid evaporation time. It can be noted, that air flow with temperature lower than 100 ^{0}C affects evaporation intensity given that admixture concentration ranges from 3 up to 7 %.

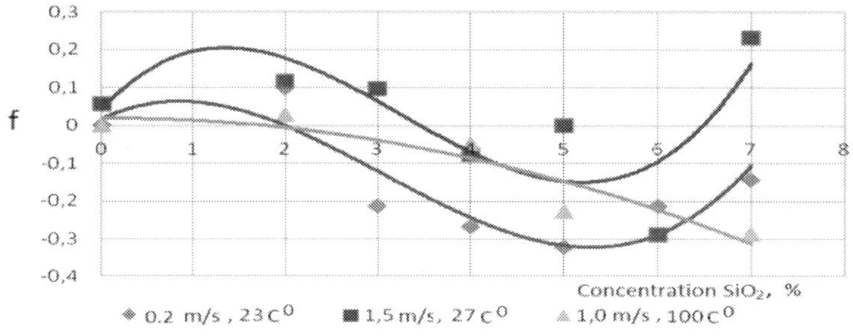

Fig. 3. Nanofluid droplet evaporation dynamics.

With the air flow temperature being ~ 100 ^{0}C, the opposite effects can be detected, so that droplet evaporation rate increase is observed within the entire concentrations range (Figure 3). These data correlate with results published in [4].

Conclusion

SiO_2 nanoparticle supplementation up to 3% concentration does not lead to remarkable reducing of nanofluid droplet evaporation time in comparison to the base liquid when air flow rate and temperature increase. Nanoparticle concentration rise over 3 % decreases evaporation time when air flow rate (temperature range from 20 up to 30 ^{0}C) increases in comparison with base liquid droplets. Moreover rise of the evaporation rate is not a monotone value in the current concentration range. Monotone decrease of nanofluide droplet evaporation time appears when air flow temperature is 100 ^{0}C given that nanoparticle concentration is over 3 %. The results obtained show, that in distillation processes there is an opportunity to use more effective mixture separation (up to 30 %).

The work was conducted under the financial support of the grant of the President of the Russian Federation No. SS-8780.2016.8.

References

1. P. Talbot, B. Sobac, A. Rednikov, P. Colinet, B. Haut, Int. J Heat Mass Transfer **97**, 803 (2016)
2. E.M. Bochkareva, V.V. Terekhov, A.D. Nazarov, A.A. Borisov, *V Int. Seminar with elements of scientific school for young scientists (ISHM-V) "Topical issues of heat and mass transfer at phase transitions and multiphase flows in modern chemical technology and energy equipment"* 96 (2016)
3. V.I. Terekhov, N.E. Shishkin, Modern Science: Researches, Ideas, Results, Technologies **2,** 197 (2011)
4. R.-H Chen, T.X. Phuoc, D. Martello, Int. J. Heat Mass Transfer **54**, 2459 (2011)

An experimental and theoretical study of the evaporation of non-ideal solutions droplets

Elena M. Bochkareva*, *Vladimir V.* Terekhov, *Aleksandr A.* Borisov and *Nikolay B.* Miskiv

Institute of Thermophysics SB RAS, 630090, Novosibirsk, 1 Lavrentiev Ave., Russia

Abstract. This work presents the results of an experimental study of the dynamics of evaporation of suspended droplets in the air stream. Droplets of pure liquids (water, ethanol) and their mixtures of various compositions are considered. Dependences of the size of droplets and their surface temperature on time in a wide range of velocities of the flowing stream and its temperature, as well as the variation of the composition of the droplets, are obtained. The paper presents a mathematical model used to compare the experimental data.

1 Introduction

The evaporation of liquid droplets is applied in thermal power engineering, mechanical engineering, constructing, food processing and chemical industries. It is the foundation to steam power engineering, working refrigeration, evaporators, heat exchangers and other installations, and to all processes of drying materials. Investigations of the process and attempts to intensify and improve the efficiency of power units are extremely important [1-4]. At the same time, the solution of the problems that arise when designing devices using drop evaporation is a difficult task because it requires solving the problems of gas dynamics and heat and mass transfer in a multiphase flow with phase transformations. Evaporation of a single droplet has been studied by many authors [1-5], but establishment of the laws that allow precise calculation is an urgent task. There is no unified theory of evaporation of droplets of multicomponent mixtures that would reflect all the patterns and effects, observed in experiments.

In the present work, the process of evaporation of non ideal solutions droplets, streamlined by the gas flow, was studied theoretically and experimentally.

2 Experimental

The experimental part of this work was carried out on the setup (Fig. 1). The gas flow was formed by laminar flow due to the structural elements of the experimental stand. The flow velocity and temperature can be set in a wide range of values (0.1…5 m/s, 22…200 ^{0}C) and are constant during the experiment.

* Corresponding author: prefous-lm@yandex.ru

 DOI: 10.1051/matecconf/201711508010

At the height of about 2 cm from the nozzle, a droplet was suspended on a holder, which was a thread of glass fiber impregnated with wax of about 100 μm. Data on dynamics of the droplet size were obtained by the digital microscope, and the surface temperature of a droplet was measured by the thermal imager. The size of the droplets was ~ 2 mm, and the measurements were performed up to 0.3d.

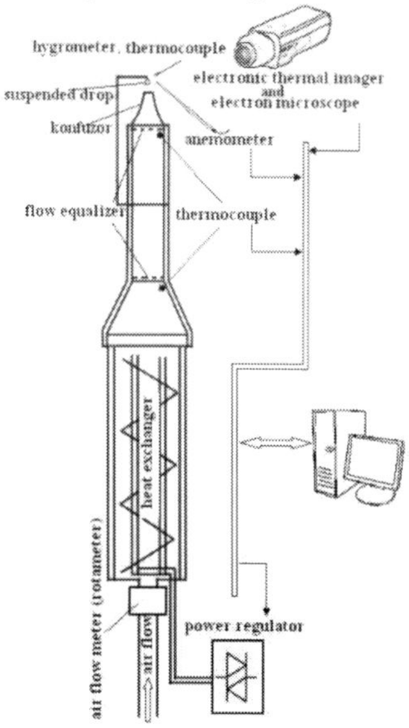

Fig. 1. The scheme of the experimental setup.

The experiments were carried out at atmospheric pressure, and the relative humidity of the air in the stream did not exceed 3%. Mixtures of working fluids were prepared by means of electronic weights using the mass ratio of the components.

3 Theoretical

As described above, in the present experiments, the drop was located on the holder, which was the main source of heat loss from the drop. The purpose of the simulation was to estimate the magnitude of these losses. The mathematical model was based on the balance of heat and mass at the drop surface. It was assumed that the temperature inside the droplet is uniform. In adiabatic conditions, the change in temperature during evaporation is the supplied heat from the gas minus the heat of the phase transformation.

$$\rho_l c_{pl} \frac{dT_l}{dt} = j \frac{c_{pg}\left(T_g - T_l\right)}{B_T} - jL \ , \tag{1}$$

where B_T is the thermal Spalding number. In the case of non-adiabatic evaporation, an additional term is added to the right-hand side of (1) to account for heat losses. In this paper, it was made in the form of $\alpha\left(T_g - T_l\right)$. Note that for all the calculations performed, the heat transfer coefficient α was assumed to be constant.

Mass flux of vapor is evaluated as:

$$j = 2\pi\rho_g D_g ShR \ln\left(1 + B_M\right), \tag{2}$$

 DOI: 10.1051/matecconf/201711508010

where B_M is the mass Spalding number. In order to determine the heat, Spalding used hypothesis about the similarity of heat and mass transfer:

$$B_T = (1 + B_M)^{\phi} - 1 \ , \ \phi = \frac{c_{pf}}{c_p} \frac{Sh}{Nu} \frac{1}{Le} \tag{3}$$

Note that the Sherwood and Nusselt numbers are functions of the Spalding numbers, so they are evaluated by iteration. The Nusselt and Schmidt numbers were determined from the known dependences $Nu=Nu(Re, B_T)$ for the sphere [5]:

$$Nu = 1 + \frac{(1 + \mathrm{Re\,Pr})^{1/3} \max\left[\mathrm{Re}^{0.077}, 1\right] - 1}{2F(B_T)} \tag{4}$$

Identical depending $Sh=Sh(Re, Sc, B_M)$ used to determine the rate of mass transfer. The system of equations was closed using the law of liquid-vapor equilibrium. The partial vapor pressure at the surface could be evaluated using an ideal model or non ideal one, taking into account the activity coefficients.

Thus temporal changes of droplet size:

$$\frac{dR}{dt} = -\frac{j}{4\pi\rho_l R^2} \tag{5}$$

4 Results and discussion

The obtained results of evaporation of water droplets are compared with the experimental work [3] in Fig. 2. The experimental data correlate well with the data of [3].

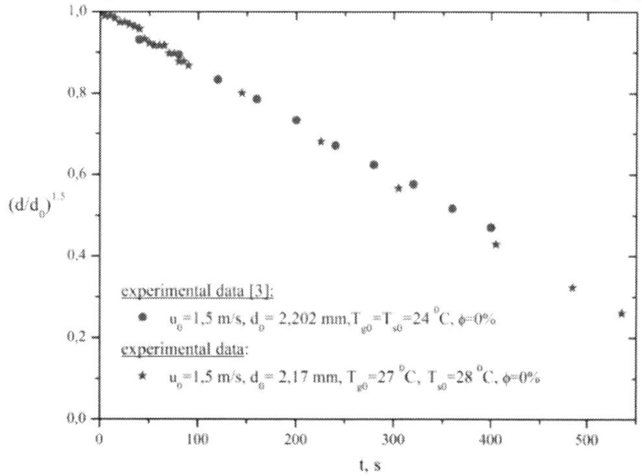

Fig. 2. The dependence of water droplet diameter on time.

Experiments and numerical simulation for water, ethanol and their mixtures in various proportions have been carried out. As it may be expected, evaporation of water occurs much slower than ethanol, evaporation of mixtures water and ethanol correspondingly are between them (Fig. 3). In this case, the surface temperature of the drop of ethanol, at the initial stage of evaporation (less than 50 s), respectively, is lower. Later, the temperature increases, due to the supply of heat through the holder.

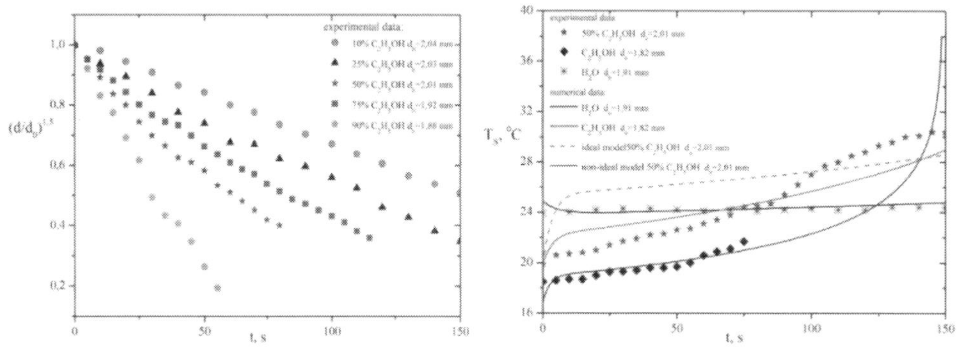

Fig. 3. Evaporation characteristics of droplets in experimental conditions: u_0=0.1 m/s, t_{0g}=47.6 ^{0}C, t_{0s}=22 ^{0}C, φ =3%.

From the temperature distribution it can be seen that the heat losses in the holder have a significant effect. This effect is most noticeable when the droplet size is reduced (see the Fig. 3a). Note that due to heat losses, the effect is observed: the temperature of the drop surface of the ethanol-water mixture and ethanol exceeds the temperature for the water drop. This is due to the fact that at the same time a drop of the mixture (or ethanol) is much smaller than a drop of water (see the Fig. 3a). In this case, the effect of heat supply through the holder is more significant, since the magnitude of the heat flux does not depend on the characteristics of the droplet.

This work was supported by the grant Russian Foundation for Basic Research No. 17-58-53168 and the grant of Russian President No. NSh-8780.2016.8.

References

1. S.S. Sazhin, Progress Energy Combust. Sci., **32** 162 (2016)
2. C.K. Law, T.Y. Xiong, C.H. Wang, Int. J. Heat Mass Transfer **7**, 30 (1987)
3. V.I. Terekhov, V.V. Terekhov, N.E. Shishkin, K.Ch. Bi, J. Eng. Phys. Thermophys. **3**, 83 (2010)
4. H. Zhu, Y. Zhang, M. Xu, R.J. Kee, Proceedings Combust. Institute **34** 1545 (2013)
5. B. Abramzon, W.A. Sirignano, Int. J. Heat Mass Transfer **32**, 9 (1989)

Movement of water droplets in a layer of thermally decomposable forest fuel

Roman S. Volkov, *Alyona* O. Zhdanova*, *Geniy* V. Kuznetsov, and *Pavel* A. Strizhak

National Research Tomsk Polytechnic University, 634050 Tomsk, Russia

Abstract. This article contains results of an experimental study of the process of droplets gravitating in a layer of a typical forest fuel during its thermal decomposition. Also similar experiments were carried out in conditions of moderate temperature conditions (about 300 K). Pine needles were considered as a material for study. The initial droplets' volume was varied from 90 to 900 μL. After analysis of experimental results we established the depths, on which water droplets can penetrate into the layer of forest combustible material (FCM). The characteristic velocities of water droplets' motion at different distances from the FCM surface were calculated. Also we determined the fraction of the mass of water consumed in the FCM layer for vaporization and cooling of the material to temperatures below the values corresponding to thermal decomposition.

1 Introduction

Extinguishing forest fires using aviation help is carried out mainly [1-3] due to the suppling of large amounts of water in the combustion area. In all weather conditions, suppression of pyrolysis of forest combustible material (FCM), which is named as one of the main mechanisms, is provided. At the present time the investigation of the processes occurring during the interaction of FCM with water droplets heightens the greatest interest. Currently there are no published experimental data, which describe the regularities of the heat and mass transfer processes in the porous structure of thermally decomposing FCM in conditions of water droplets' motion through a layer of such material (material is heated up to temperatures, which substantially exceed the temperature of the onset of thermal decomposition of FCM).

The aim of this work is revealing the mechanism of moving water droplets in a porous structure passing through the thermal decomposition stage of FCM based on the results of experimental studies.

2 Experimental methods and procedures

The experimental setup was used in carrying out of experimental research (Fig.1). We used thin-walled quartz glass cylinder with diameter $d_f \approx 100$ mm and height $h_f \approx 120$ mm for creating the model of fire seat. Pine needles were chosen as forest fuel. The density and

* Corresponding author: zhdanovaao@tpu.ru

humidity of the FCM sample were selected as averages of possible values for fires in the boreal zone. The density was 30–35 kg/m^3, calculated using formula $\rho_f = m_{f0} / (h_f \cdot S_f)$, where $S_f = \pi \cdot (d_f/2)^2$) is the cross-sectional area of the sample FCM, m^2. Relative humidity of spruce needles was 7–10% calculated using formula $\gamma_f = (m_{fw} - m_{fd}) / m_{fw} \cdot 100$, where m_{fw} is the sample's weight before thermal drying, g; m_{fd} – is the sample's weight after thermal drying, g.

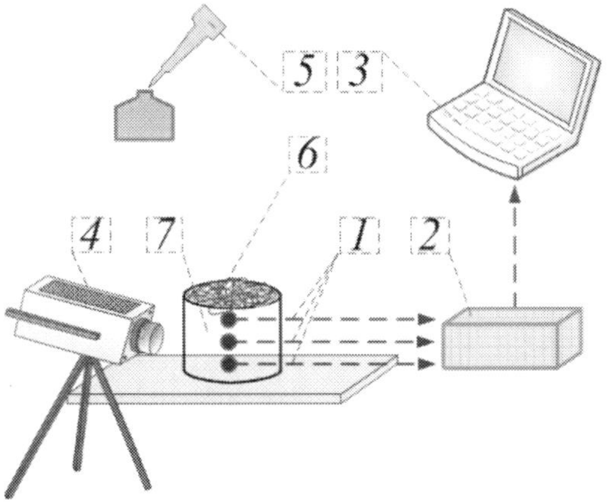

Fig. 1. A scheme of experimental setup: *1* – three low-inertia thermocouples; *2* – high-speed analog input module; *3* – workstation (PC); *4* – high-speed video camera; *5* – dispenser; *6* – FCM; *7* – hollow cylinder with a hinge of FCM.

Experimental studies were carried out in two stages. At the first stage water droplets were placed on the surface of a FCM's sample that was not subject to pyrolysis. The second stage assumed similar experiments with thermally decomposed FCM. In this case, the initiation of FCM's thermal decomposition was realized using three piezo igniters. These ignited the top of the sample. After 15-20 seconds the temperature in the FCM's layer throughout the entire thickness became higher than the temperature of the onset of thermal decomposition (400 K).

Measurements of the distances *h* passed by the drops and the corresponding times *t* were carried out using two registration schemes. According to the first one, we used the high-speed analog input module *2* "National Instruments NI 9213" and three type K thermocouples installed in the FCM's layer (with temperature measurement range of 223-1473 K, systematic error of ± 3 K, and thermal inertia not exceeding 0.1 s). When a drop of water hits the sensing element (or passes in the immediate vicinity of the element), the thermocouple recorded a significant temperature drop (this time indicated the achievement by a drop of the corresponding FCM's section). Such a scheme was used only in experiments with pyrolysis processes.

The second scheme for the *h(t)* determination involved finding distances *h* that were passed by the water drops, and the corresponding times *t* during the analysis of the obtained video recordings. We used high-speed digital *4* CMOS video camera "Phantom V411" (with shooting frequency up to 105 frames per second and maximum resolution of 1280 × 1280 pixels). Video shooting process was carried out with a frequency of 200 frames per second. The FCM layer was lighted using a specialized illumination system based on an LED matrix. We used the Tema Automotive software [5, 6] for continuous tracking of water drop's motion.

Random experiments' errors were determined in a way similar to the experiments [7, 8] and amounted to the mean values with the identical conditions *h* - 15%. We conducted 8-12 experiments for getting accurate results.

Also, we controlled the mass of water drops after passing through the FCM layer. For this purpose we used ViBRA HT 84RCE scales. The droplets were weighed before dropping onto the FCM sample (m_{d0}) and after passing through it (m_{de}) for moderate temperature conditions and during thermal decomposition. The random errors of determining m_{de} did not exceed 5%.

3 Results and discussion

In our experiments we recorded the distances passed by the drop in the FCM's layer (progress depth h). The main significant factors of the studied processes were the initial droplets' sizes (volumes) and the FCM's temperature. Table 1 shows approximate expressions for the experimental results, which were obtained using thermocouple measurements and high-speed video recording.

Table 1. Approximation correlations to calculate distances that were passed by water droplets of different volumes in the FCM layer depending on time.

Droplets' volumes	During pyrolysis process	Without pyrolysis process	
	Registration scheme		
	Thermocouple measurements	Thermocouple measurements	High-speed video recording
V_d≈90 μL – R_d≈2.8 mm	h=3.5911·ln(t)+7.1641	h=3.4618·ln(t)+6.9894	h=4,3785·ln(t)+15.424
V_d≈200 μL – R_d≈3.6 mm	h=19.77·ln(t)+20.066	h=21.163·ln(t)+17.697	h=9.3762·ln(t)+44.685
V_d≈800 μL – R_d≈5.7 mm	h=44.166·ln(t)+32.69	h=48.808·ln(t)+30.694	h=27.983·ln(t)+98.761

It should be mentioned that the process of high-temperature pyrolysis of FCM does not practically effect on penetration depth of droplets in FCM. It can be explained by high porosity of this kind of forest fuel (relative porosity of pine needles can reach 70-90%). Gaseous products, which were formed during pyrolysis of needles, move in the opposite direction to the droplets and practically do not influence on their movement. It can also be concluded that the effect of mass forces is more significant than the effect of resistance forces to the drops' motion. Drops with volume of 800 µl can pass through the entire layer of FCM (with thickness up to 60) losing up to 20-30% of their initial mass. Analyzing the approximation correlations (Table 1), we can say, that large droplets have much higher velocities than the small ones. Small droplets are inhibited by the FCM structure elements. It happens because transverse and longitudinal dimensions of the elements are comparable with the diameters of relatively small droplets (with size up to 3 mm).

Table 2 shows experimental results of droplet mass estimations after their passing through a layer of FCM during the thermal decomposition process. Parameter Δm ($\Delta m = m_{d0} - m_{de}$) characterizes the mass changing during the passage through the FCM layer, taking into account the losses due to wetting of the pine needles and evaporation.

 DOI: 10.1051/matecconf/201711508011

Table 2. Dependence of the parameter *Δm* on the initial mass of droplets.

m_{d0}, g	Δm, g	
	The passage of water droplets through FCM layer (without pyrolysis process)	The passage of water droplets through FCM layer (during pyrolysis process)
0.225	0.139	0.141
0.325	0.137	0.181
0.425	0.154	0.227
0.525	0.189	0.279
0.625	0.243	0.337
0.725	0.315	0.401
0.825	0.407	0.472

The experimental results in Table 2 show the amount of water, which is consumed on the endothermic phase conversion. It is worth mentioning that this amount increases with initial water mass growth. Such fact can be explained not so much by the intensification of the vaporization process (most likely, the water mass increase in the layer leads to the FCM's temperature decrease and a drop in evaporation rates), but because of growth of contact area between water and the FCM structure elements.

4 Conclusions

The obtained result leads to conclusion that suppression of the thermal decomposition processes of FCM are most effective at a certain (most likely, less than 4-5 mm) characteristic of droplets not very large in size. Also it was clarified that the most efficient water use for reducing the temperature in the layer of thermally decomposing FCM is achieved by using fine droplets.

The research was funded by the Russian Federation President Program (project No. MK-1684.2017.8).

References

1. X.K. Xiao, B.H. Cong, X.S. Wang, K.Q. Kuang, Richard K.K. Yuen, G.X. Liao, J. Fire Sci., **29**, 339 (2011)
2. S. McAllister, I. Grenfell, A. Hadlow, W.M. Jolly, M. Finney, J. Cohen, Fire Safety J. **51**, 133 (2012)
3. Meenakshi Gupta, R. Rajora, S. Sahai, R. Shankar, Anjan Ray, S.R. Kale, Fire Safety J. **54**, 130 (2012)
4. O.P. Korobeinichev, A.G. Shmakov, V.M., Shvartsberg, A.A. Chernov, S.A. Yakimov K.P. Koutsenogii, V.I. Makarov, Fire Saf. J. **51**, 102 (2012)
5. J. Janiszewski, Metrol. Measur. Sys., **19**, 797 (2012)
6. J. Janiszewski, Int. J. Solids Struct., **49**, 1001 (2012)
7. Vysokomornaya O.V., Kuznetsov G.V., Strizhak P.A. Fire Safety J. **70,** 61 (2014)
8. Glushkov D.O., Kuznetsov G.V., Strizhak P.A., Volkov R.S. Thermal Science, **20** 129 (2016)

Characteristics of the behavior of a liquid film in a pulsed gas-droplet flow

Pavel Karpov[1,*], *Nikolay* Miskiv[1], *Anatoly* Serov[1]

[1]Kutateladze Institute of Thermophysics SB RAS, Lavrentiev av. 1, Novosibirsk, 630090, Russia

Abstract. The results of the investigation of the influence of a multi-jet impact air flow on the process of formation of a liquid film from a drop flow on the vertical surface of a heat exchanger with a low temperature are presented. Characteristics of the film flow and the effect of this flow on heat transfer, shown in the paper, are related to the interaction of the boundary layer of the wall air flow and the perturbation induced by the droplets under the pulsed regime of introducing the cooling mass into the flow of impact air jets. The paper determines the contribution of the impact air flow to the heat transfer during the cooling of the vertical surface.

1 Introduction

The development of modern power equipment is aimed at increasing their specific power, on the one hand, and reducing the size of the unit, on the other hand. With the increase in the power of small size electronic devices, the accumulation of heat leads to an increase in the operating temperature that affects the stability of the device and leads to a decrease in its life-time. Standard methods of single-phase convection do not meet the requirements for maintaining an acceptable operating temperature, compared to methods using latent heat of vaporization: boiling, evaporation, etc. Pulse cooling of dispersed liquid is a quick way to remove large heat fluxes having many advantages, such as for example, low coolant flow and high dissipation.

It is known that in order to realize effective cooling, it is necessary to ensure an even distribution of the mass concentration of the liquid not only on the surface of the heat exchanger, but also during the movement of the cooling mass [1]. The density of the heat flux will be the greater, the higher the mass concentration of the liquid in the air flow. The overwhelming number of works on cooling by means of gas-drop flows is devoted to various boiling regimes Well-studied modes of film cooling are widely used depending on the temperature of the cooled surface, its properties and geometry. In this paper we present data on the behavior of a liquid film formed on a large underheated surface by a pulsed gas-droplet flow.

* Corresponding author: Flags712008@yandex.ru

2 Experimental setup and procedure

The experimental installation is a multifunctional stand consisting of the following units: a programmable multi-jet source of pulsed gas-droplet flow, an automated calorimeter, an automated system for recording the parameters of the flow, which includes original capacitive sensors for recording the thickness and velocity of the liquid film waves [2]. The controlled source of the gas-droplet jet is structurally designed as a two-chamber unit: for air and water. Flat part of the source is provided with 16 liquid injectors. The liquid injector is a sprayer of four nozzles with a diameter of 125 μm, connected by one solenoid valve. On this same surface, 25 gas nozzles with an outlet opening diameter of 0.5 mm are installed to create a gas-droplet air flow. The design of the multi-nozzle source was performed subject to the condition that the liquid nozzle is surrounded by 4 air nozzles, i.e. is in the "box". The distance between the source and the heat-loaded surface is 230 mm.

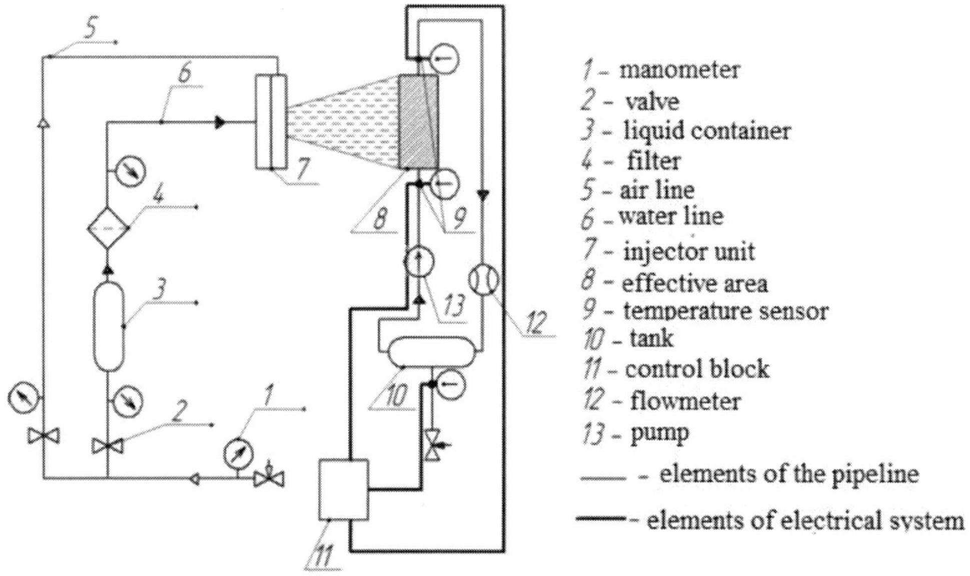

Fig. 1. Schematic of experimental setup

The film flow registration system, built on video surveillance, high-speed photography and data from a multichannel dielkometric registrar on the local film thickness, made it possible to perform measurements of the induced flow on the surface of the heat exchanger. Figure 2 shows the general velocity map obtained from video surveillance data. In the flows displaced to the periphery, several different types of three-dimensional instabilities occur at low Reynolds numbers in different zones of the heat exchanger. Optical observations showed the existence of stable two- and three-dimensional waves of small amplitude (dh/h = 0.05) with transverse modulation along the flow outside the isolated regions A, D, C, and D. Especially are highlighted the characteristic zones A, B, C, and D, where the measurements showed a weak current and the absence of large waves.

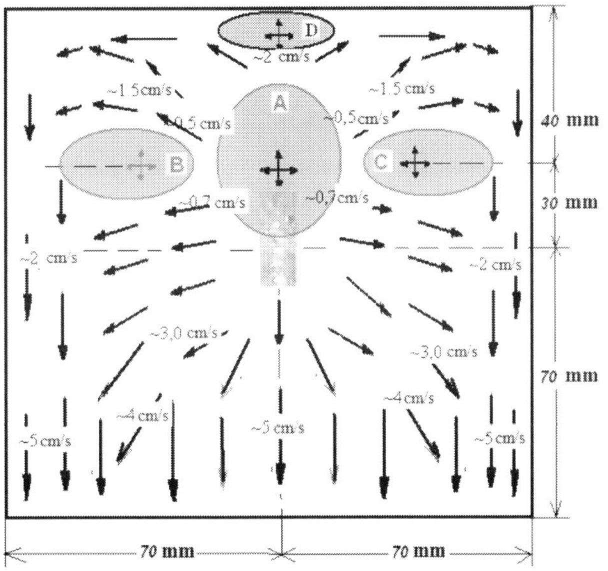

Fig. 2. Wave current projection map for cooling film

Figure 3 shows the film thickness variation (h_{max} = 0.92 mm) in zone A within dh = 0.06 mm on probe No. 1 and dh = 0.02 mm on probe No. 2. In addition to slow waves, the recorder showed an induced pulsation, which can be attributed to the motion of very weak slow waves with an amplitude dH = 0.06 mm from the center to the edges of the heat exchanger at a velocity V = (0-0.01) m/s, Re = (4...12) and the oscillation frequency F = (0.2...1) Hz.

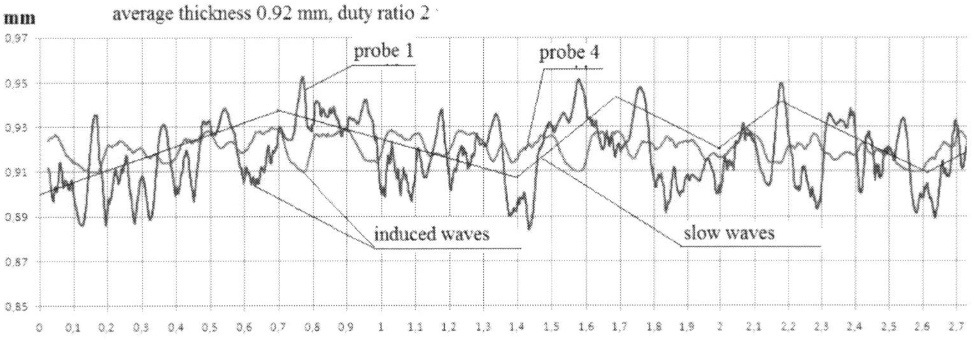

Fig. 3. Local film thickness

3 Conclusions

The performed studies have shown that when comparing the data on the structure and projections of the wall air flow velocities with the local flow of the film, we can conclude:

1. A wide multi-jet impact flow destroys the standard formation mechanism of a gravitational cooling film.
2. In the inner region under the impact flow, local stagnant zones of the cooling film are formed with induced pulsations.
3. The condition for the formation of these zones is the equality of multidirectional capillary and gravitational forces in the film as well as frictional forces from the "wall" streams of impact air jets on the film.

4. In stagnant zones, provided that there is no film flow and active "turbulence" with drops, maximum heat transfer conditions are created due to convective and evaporative cooling.
5. The most energy-efficient cooling mode is the "single" pulse mode, where the heat exchanger surface is irrigated by an ensemble of drops.

The work was financially supported by RFBR (project No. 16-38-00853) and the grant of Russian President No. NSh-8780.2016.8.

References

1. A.L.N. Moreira, J. Carvalho, M.R.O. Panao, Int. J. Heat Fluid Flow 2007 **28**, 191 (2007)
2. A. Nazarov, A.Serov, M. Bodrov, Technical Phys. **80**, 724 (2010)

Experimental investigation of explosive vaporization of liquids on the flat microheater

Igor Kozulin[1,2 *]

[1]SB RAS, Institute of Thermophysics, Prospect Akademika Lavrentjeva 1, Novosibirsk 630090, Russia
[2]Novosibirsk State University, Pirogova str.,1, Novosibirsk 630090, Russia

Abstract. The initial stage of controlled decay of metastable liquid was studied under pulsed heating on the surface of the flat microheater using the optical method of reflected laser beam. Four major stages of fluids explosive boiling and phase explosion on microheater were determined. The dynamics of filling the heater surface with the vapor phase, the lifetime of the main vapor bubble, and the satellite bubble are estimated.

1 Introduction

Researches of heat and mass transfer with phase change in microfluidic systems are rapidly developed. This is related to a growing interest in the MEMS technology (microelectromechanical systems). The MEMS technology makes it possible to obtain extremely high temperatures without the breakdown of the microheater. This is also concerned with studying the vapor phase nucleation, spinodal liquid decomposition, bubble dynamics, etc. Experimental investigations of the explosive boiling on the flat microheater were performed in [1–4]. In this case, the decay of liquid metastable state occurs near the temperature of limit overheating. The regularities of water micro volumes at explosive boiling were numerically established in [5] at the high-impact thermal pulses using multilayer microheater coated submicron silicon carbide layer. The aim of this work is an experimental study of the initial stage of explosive boiling of liquids (water, propanol-1) on a microheater using an original optic system.

2 Experimental setup

The experimental setup for studying the dynamics of explosive boiling on the microheater is shown in Fig. 1, (a). The microheater is a multilayer thin-film resistor of the Hewlett Packard ThinkJet printhead [6]. The microheater's photo is presented in Fig. 1, (b). The size of microheater is 100x110 μm. The microheater is a four-layer film deposited layer-by-layer by the PECVD method in a thin glass substrate. The first layer is a silica 1.1 μm, thick followed by a 0.5 μm heat-emitting TaAl layer, whose conduction is weakly dependent on

[*] Corresponding author: kozulin@itp.nsc.ru

the temperature, and a 0.25 μm SiC layer. This sandwich-like structure of the heater ensures high growth rates of the water layer temperature under pulse heating.

The microheater (1) is immersed in the working fluid (2), and its surface is lit by laser (3). The initial temperature of the liquid was determined before the experiment and ranged from 18 to 21^{0}C, the experiments were conducted at atmospheric pressure. The single rectangular current pulse is supplied to microheater from the 214B Hewlett Packard pulse generator. Each pulse has sufficient duration for the explosive boiling of liquid near the surface of the microheater. The initial stage of explosive boiling was studied using the optical registration, based on measuring the intensity of the laser beam (3) reflected from the mirrored surface of the microheater (1). The reflected laser beam from the heater enters to the microscope (4). The diaphragm (5) selects an area with the microheater surface. The intensity of the laser beam is measured by the photodiode (6). The signal from photodiode is registered with the high-speed ADC (7). The integral coefficient of the reflection of laser light begins to decrease during the appearance of microbubbles on the microheater. The signal from the photodetector provides the information about the history of the heater surface coverage by vapor bubbles.

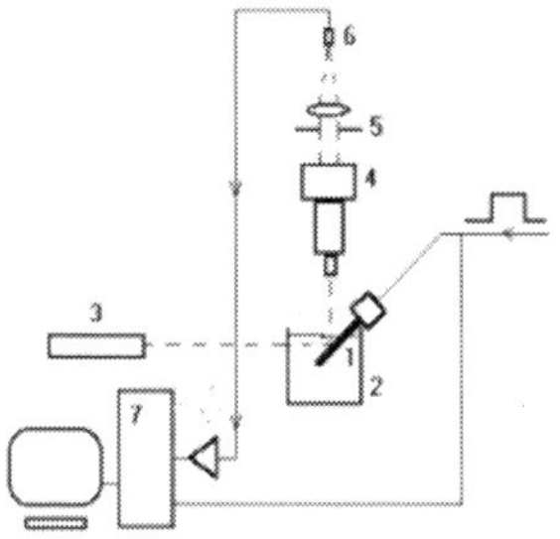

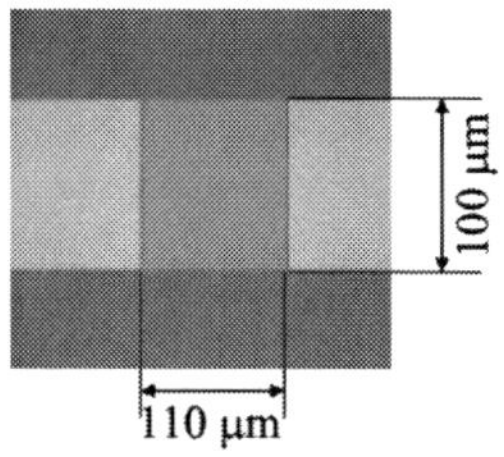

(a) (b)

Fig. 1. Experimental setup (a), photo of the multilayer thin-film microheater (b).

3 Experimental results

Figure 2 (a, b) shows the dynamics of surface coverage by the bubbles during the explosion boiling of water on the microheater. The normalized signal from the generator is also shown in Fig. 2 by a thin line. The dark line shows the dependence of reflected light intensity from the microheater according to the time during the explosion boiling of water. Figure 2 (a) shows the dynamics of surface coverage during the explosion boiling of water for the effective heat flux q_{eff}=579.18 MW/m^2, the growth rate of temperature dT/dt=51.89 MK/s and duration of the heating pulse t=4.144 μs. Figure 2 (b) shows the dynamics of surface coverage for the effective heat flux q_{eff}=2906.2 MW/m^2, the growth rate of temperature dT/dt=450.42 MK/s and duration of the heating pulse t=0.192 μs. The zero time in Fig. 2 corresponds to a supply of heat pulse from the generator.

Letters A, B, C, D and E in Fig. 2 (a, b) identify four basic stages of explosive boiling of liquid. The first stage from point A to point B corresponds to the heating of the liquid by heating pulse from the generator. The signal increased in this area because of the change of the optical properties of the liquid near the microheater surface. The second stage starts from point B which corresponds to the first appearance of bubbles on the surface of microheater and continues to point C. It corresponds to the growth of number of the bubbles on the microheater, which causes a sharp decrease of the reflected laser beam intensity scattering on the bubbles. The exponential growth of the bubbles number is determined by the frequency of homogeneous nucleation [2].

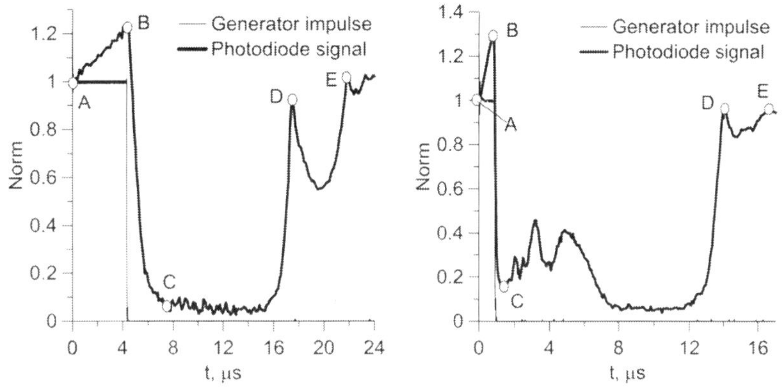

(a) (b)

Fig. 2. The intensity of the reflected laser beam depending on the time during the explosion boiling of water: (a) q_{eff}=579.18 MW/m^2, dT/dt=51.89 MK/s, t=4.144 μs, (b) q_{eff}=2906.2 MW/m^2, dT/dt=450.42 MK/s, t=0.912 μs.

Figure 3 (a) shows the intensity of the reflected laser beam depending on the time during the explosion boiling of water for an increased effective heat flux. The normalized signal from the generator is presented by the thin line. The dark line shows the dependence of intensity of reflected light according to the time during the explosion boiling of water for the effective density of heat flux q_{eff}=4319.8 MW/m2, the growth rate of temperature dT/dt=702.49 MK/s and duration of the heating pulse t=0.692 μs. The signal detailing from the photodiode is shown in Fig. 3 (b). It is seen that for a temperature growth rate higher than 500 MK/s, the nature of the signal changes. There are oscillations in a signal from the photodiode, which may indicate an earlier formation of the vapor cavity due to the unification of the bubbles. A comparison of the reflected light intensity and photographs obtained using a pulsed laser, showed that this behaviour is caused by the appearance of a powerful evaporation wave at the initial stage of explosion and transition to a continuous vapor film that reflects light well.

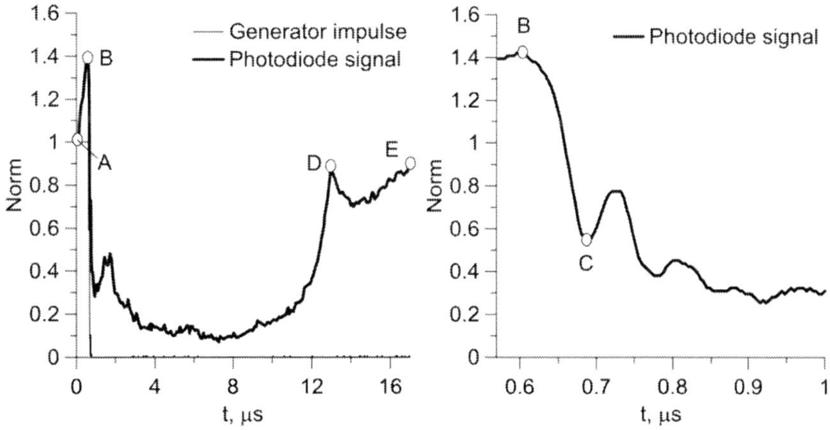

(a) (b)

Fig. 3. (a) The intensity of the reflected laser light depending on the time during the explosion boiling of water q_{eff}=4319.8 MW/m^2, dT/dt=702.49 MK/s, t=0.692 μs, (b) the signal detailing.

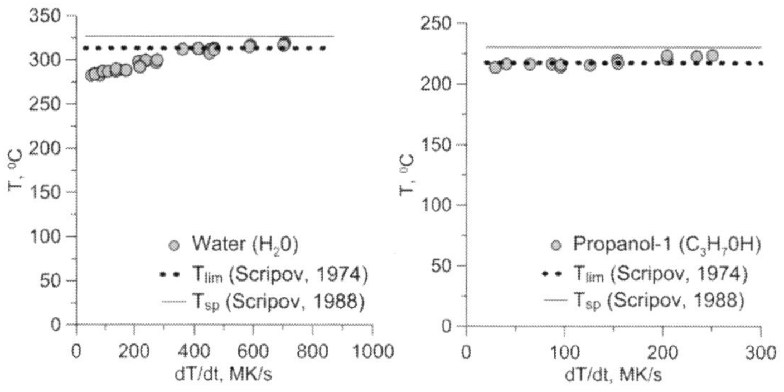

(a) (b)

Fig. 4. The dependence of start temperature of the explosion boiling from the growth rate of temperature: (a) water, (b) propanol-1.

The dependence of temperature of the explosion boiling initiation from the temperature growth rate for water and propanol-1, is presented in Fig. 4 (a, b). Here, the dotted line presents the solution of an equation $T_{lim}/T_{cr}=0.905+0.095.(T_{sat}/T_{cr})^8$ for limiting overheating of liquid based on the data of [7]. The solid line shows the spinodal line calculated according to [8].

4 Conclusion

The analysis of dynamics of the explosive boiling of liquids (water, propanol-1) on the microheater with dimensions of 100x110 μm, is presented in this paper. The investigation has shown that the used method ensures a high accuracy of measuring a part of the heater surface filled with vapor bubbles. It allows measuring the time point corresponding to beginning of explosive vaporization, the duration of the bubble liquid decomposition, and the lifetime of the main vapor bubble and the satellite bubble. The determined temperature growth rate that corresponds to the change in the mode of vapor phase generation shows a change of the mechanism of the liquid decay in the vicinity of spinodal at a high liquid temperature growth rate.

This work was performed at Kutateladze Institute of Thermophysics SB RAS supported by a grant from the Russian Science Foundation (project No. 16-19-10519).

References

1. Y. Hong, N. Ashgriz, J. Andrews, J. Heat Transfer **126**, 259 (2004)
2. V.V. Kuznetsov, I. A. Kozulin, J. Eng. Thermophys. **19**, 102 (2010).
3. V.V. Kuznetsov, V.I. Oreshkin, A.S. Zhigalin, I.A. Kozulin, S.A. Chaikovsky, A.G. Rousskikh, J. Eng. Thermophys. **20**, 240 (2011)
4. V.V. Kuznetsov, I.A. Kozulin, O.V. Vitovsky, J. of Eng. Thermophys. **21**, 136 (2012)
5. V.V. Kuznetsov. J. Physics: Conference Series. **754**, (2016)
6. E.V. Bhaskar, J.S. Aden, Hewlett-Packard J. **36**, 27 (1985)
7. V.P. Skripov, *Metastable liquids* (John Wiley & Sons, 1974)
8. V.P. Skripov, E.N. Sinitsyn, P.A. Pavlov, G.V. Ermakov, G.N. Muratov, N.V, Bulanov, V.G. Baidakov, *Thermophysical properties of the liquids in the metastable (superheated) state (*Gordon and Breach Science Publishers, New York, *1988)*

Bubble departure diameter at transient heat release

Anatoliy Levin*, and *Polina* Khan

Melentiev Energy Systems Institute SB RAS, 130 Lermontova str., Irkutsk, Russian Federation

Abstract. This paper presents results of the experimental study of the subcooled water boiling on the surface of the cylinder heater. The initial stages of formation of the developed nucleate boiling were investigated. The field temperature of heated liquid layers at the moment of bubble departure was described by numerical modelling at non-stationary condition. A high-speed digital video camera was used to capture the bubble dynamics. The experimental data on the bubble dynamics was compared with Prodanovic's and Song's models. A poor agreement between the predicted departure diameter and the experimental one was obtained for both of the models.

1 Introduction

Due to the large heat fluxes feasibility, the subcooled liquid boiling has many practical applications. There exist many research results on the subcooled liquid boiling at the constant heat flux or constant heater surface temperature conditions. However, it is still promising to investigate the non-stationary heating, when the surface temperature is gradually increased up to the Leidenfrost point and the heating process goes through the stages of nucleate, transitional, and film boiling. The models predicting bubble departure diameter, nucleation site density, nucleation frequency, and other values are required for the quantitative description of complicated transitional processes.

At present, there exist many models of the bubble evolution based on the various approaches [1–5], which were verified by using the experimental data from such publications as [1,2]. In general, these models can be divided into the following groups:

- Mechanistic models, based on the balance of forces applied to a bubble. This approach allows one to predict the bubble behaviour in a vertical channel: lift-off or sliding. However, too many tuning parameters make it difficult to generalize the model.
- Variations of the Zuber model, where the heat exchange and hydrodynamic aspects are expressed in terms of the Jacob and Bond criteria (*Ja* and *Bo* numbers). The large number of expressions for the departure diameter and the wall surface temperature clearly indicate that this kind of approach lacks distinctness. The remarkable improvement of the predictive ability of such models was achieved in [3] by introducing the characteristic temperature.

* Corresponding author: lirt@mail.ru

- Models describing the bubble evolution on the base on the heat balance, including the recent research [4]. The effect of hydrodynamic forces is not included; therefore, such models can predict the maximum diameter without considering the possibility of sliding or lift-off.

In existing publications, all those models were employed under the assumption that the heat flux or heater surface temperature is constant, which predetermined the experimental conditions. Therefore, the dynamics of the superheated liquid layer, as well as the formation and evolution of bubbles in non-stationary heating conditions, is a particular case that requires special experiments and the study of applicability of predictive models.

2 Results and discussion

Our experiments with rapidly increasing heating power were conducted on the water flowing in a channel with a cylindrical type 321 stainless steel heater in the center. The heater has the following characteristics: 12 mm outer diameter, 10 mm inner diameter, 4 μm roughness. The channel has optically transparent windows and is equipped with the temperature, voltage and pressure measuring probes. Heating power was generated by conducting three-phase rectified electric current through the heater with the pulse duration of 60–300 ms and the temperature growth rate of 1000–7000 K/s. The water flowed along the vertical channel upwards with the average velocity of 0.2 m/s, the inflow pressure of 0.11 MPa, and the inflow temperature of 30, 60 and 90°C, corresponding to the subcooling by 12–72 K with respect to the saturation temperature of 102°C.

We performed the numerical analysis and obtained the water temperature near the heater surface. We only modelled heating of the homogeneous liquid phase, and, therefore, the temperature field remained adequate until the first bubbles appeared. The axisymmetric 3D model of the steel heater and the water tank was built and verified against the thermocouple data. The laminar flow, the convective and the conductive heat transfer in water, the conductive heat transfer and the joule heat in solid were analyzed using temperature-dependent material properties. The metal temperature at the bottom and at the top was the same as the inflow temperature, taking three values of 30, 60 and 90°C, corresponding to subcooling equal to 72, 42 and 12 K. Both in the physical experiment and in the numerical model, the heater temperature was increased due to electric current. The actual voltage history was measured during the physical experiment and used as an input data during the numerical simulation.

According to [2], the maximum diameter can be predicted as follows

$$D_m = \frac{\rho_L \alpha_L^2}{\sigma}\left[A Ja^b \theta^c \left(\frac{\rho_L}{\rho_V}\right)^d Q^e \right], \tag{1}$$

where $Ja = \frac{\rho_L C_{pL}(T_w - T_s)}{\rho_V r}$, $\theta = \frac{T_w - T_0}{T_w - T_s}$, $Q = \frac{q}{\rho_L v r}$, ρ is the density, α is the thermal diffusivity, σ is the surface tension, C_p is the heat capacitance at constant pressure, T is the temperature, r is the latent heat, q is the heat flow rate, v is the liquid velocity, index L referes to the liquid phase, V – to the vapor phase, temperature index w refers to the heater wall, 0 – to the liquid bulk, s – to saturation condition. Figure 1 represents the application of expression (1) to the present data. The calculation results for the coefficients $A,b,c,d,e,$ given in [2], were compared with those resulted from our experiment

$$A = 236.75;\ b = 2.59;\ c = -0.35;\ d = -1.91;\ e = -1.68. \tag{2}$$

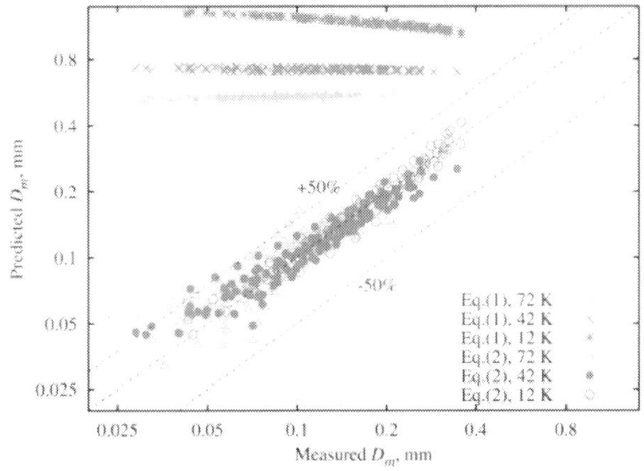

Fig. 1. Results predicted by Prodanovic's correlation against the present data on maximum diameter, obtained using initial and present coefficients.

The difference of correlations in Fig.1 can be explained by the bigger maximum diameter values (about 1 mm) in [2] for the same wall temperatures and heat flow rates, which is typical for the stationary heat release conditions.

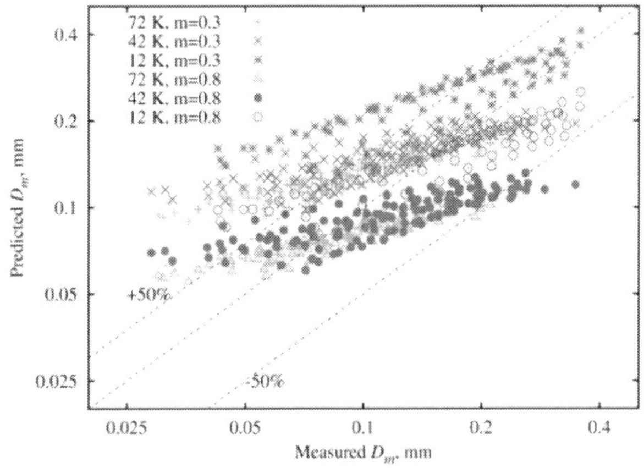

Fig. 2. Results predicted by Song's correlation against the present data on maximum diameter, obtained using initial and present *m* values.

Another expression for the maximum diameter, given in [4] is

$$D_m = 1.21ab^{-0.5} \tag{3}$$

where $a = \dfrac{(1-m)k_L(T_w - T_s)}{\rho_V r\sqrt{\pi a_L}}$, $b = \dfrac{m\varphi C(T_s - T_0)}{1-\rho_V/\rho_L}$, k is the thermal conductivity, and m is the fraction of the bubble surface area covered by the subcooled liquid at the temperature T_0, as proposed in [4]. Following [4], the coefficients φ and C for the present condition is equal to 1 and 65, correspondingly. The cooling area fraction m of 0.3 was indicated as the best one for approximating the experiment data in [4]. However, m=0.8 gives better results when applied to our data (Fig. 2).

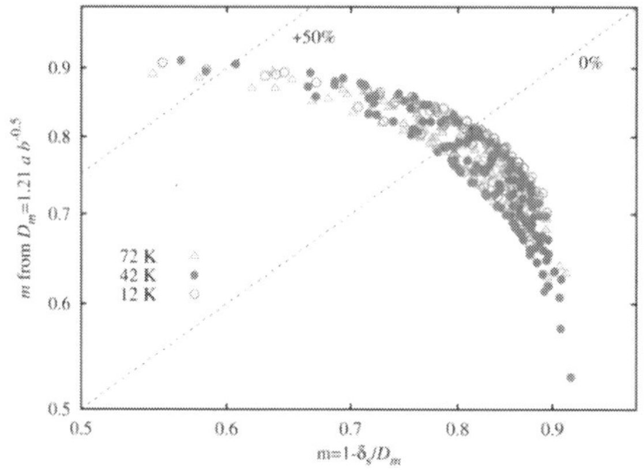

Fig. 3. Comparison of the cooled area fraction obtained from model [2] and from the numerically calculated location of the saturation temperature isotherm.

By definition, m is related to the location of T_s isotherm as follows:

$$m = 1 - \delta_s / D_m \,, \tag{4}$$

where δ_s is the thickness of the superheated liquid layer. In the present work, δ_s was found numerically, together with the temperature field. Also, m can be calculated by using the inverse function (3). The cooling area fractions found with the use of (3) and (4) are represented in Fig. 3; it can be readily seen that the factor m loses its physical meaning when applying (3) to our experimental data. It should be noted that for our experimental conditions the temperatures between T_s isotherm and the bubble top is bigger than T_0.

3 Conclusions

The present experiment data on the maximum bubble diameter was compared with the theoretical predictions by using both the initial and modified coefficient values. The approach based on the heat balance proved to be more promising for approximating both stationary and non-stationary data without major changes, because this model does not include the arbitrary coefficients. However, the physical meaning of the parameter m seems to be ambiguous. The heat balance equation can be improved by using the information about the actual temperature field around the bubble.

This research was performed at the Melentiev Energy Systems Institute of the Siberian Branch of the Russian Academy of Sciences with the use of equipment of the multi-access scientific center "High Temperature Circuit".

References

1. I.C. Chu, H.C. No, C.H. Song, J. Nucl. Sci. Technol. **48** 936 (2011)
2. V. Prodanovic, D. Fraser, M. Salcudean, Int. J. Multiphase Flow **28** 1 (2002)
3. J. Kim, M.H. Kim, Int. J. Multiphase Flow **32** 1269 (2006)
4. N.H. Hoang, I.C. Chu, D.J. Euh, C.W. Song, Int. J. Heat Mass Transfer, **94** 174 (2016)
5. M. Colombo, M. Fairweather, Int. J. Heat Mass Transfer, **85** 135 (2015)

Modeling the outflow of liquid with initial supercritical parameters using the relaxation model for condensation

Sergey Lezhnin[1,2], *Maksim* Alekseev[1,*], *Ivan* Vozhakov[1,2], and *Nikolay* Pribaturin[1]

[1]Kutateladze Institute of Thermophysics SB RAS, Novosibirsk, Russia
[2]Novosibirsk State University, Novosibirsk, Russia

Abstract. The two-temperature model of the outflow from a vessel with initial supercritical parameters of medium has been realized. The model uses thermodynamic non-equilibrium relaxation approach to describe phase transitions. Based on a new asymptotic model for computing the relaxation time, the outflow of water with supercritical initial pressure and super- and subcritical temperatures has been calculated.

1 Introduction

Depressurization of vessels and pipelines in power plants, filled with water under high pressure, results in superheated water outflow.

Such a sharp decrease in pressure leads to a generation of compression waves in the area around the separation zone and further formation of the jet of the boiling coolant [1]. If in the stationary mode both pressure and temperature are supercritical ($p > p_{CR}$, $T > T_{CR}$), it is possible for fluid parameters to approach the binodal (boundary of "single-phase and two – phase" region) from the condensation curve [2]. The process of adiabatic expansion, as a rule, is non-equilibrium, and thermodynamic non-equilibrium two-temperature model of steam-water mixture motion is frequently used for its description [3]. Simulating vaporization in real conditions requires either setting the parameters of the heterogeneous mechanism of bubbles formation (e.g. [4]), or using a different (phenomenological) approach (e.g., relaxation one [5]). In particular, in [4] it is proposed to specify the number of heterogeneous nucleation sites as an empirical parameter. In addition, asymptotic models of bubble growth are applied. Due to the use of bubble structure alone to describe the two-phase medium, this approach loses its adequacy at large ($\alpha > 0.3$-0.5) volumetric void fractions. Earlier in [6] the authors, based on the performed analysis, proposed a simple relaxation model to describe thermal and dynamic processes, accompanied by water boiling-up at depressurization of high-pressure vessel. The model implies the droplet structure of the liquid-vapor flow, which is typical for high differential pressure between the vessel and atmosphere. In this paper we generalize this model for the case of non-equilibrium condensation.

* Corresponding author: almaxcom@mail.ru

2 Governing equations

Numerical simulation is carried out in the no-slip phase condition and given the fact that the vapor and liquid phases can be in a metastable state. The system of the model equations includes the Euler equations in the axisymmetric approximation:

$$\frac{\partial}{\partial t}(\rho r)+\frac{\partial}{\partial z}(\rho r u)+\frac{\partial}{\partial r}(\rho r v)=0 \tag{1}$$

$$\frac{\partial}{\partial t}(\rho r u)+\frac{\partial}{\partial z}\left(\rho r u^2\right)+\frac{\partial}{\partial r}(\rho r u v)=-r\frac{\partial p}{\partial z} \tag{2}$$

$$\frac{\partial}{\partial t}(\rho r v)+\frac{\partial}{\partial z}(\rho r u v)+\frac{\partial}{\partial r}\left(\rho r v^2\right)=-r\frac{\partial p}{\partial r} \tag{3}$$

$$\frac{\partial}{\partial t}(Er)+\frac{\partial}{\partial z}\left(ur(E+p)\right)+\frac{\partial}{\partial r}\left(vr(E+p)\right)=0 \tag{4}$$

added with the equation for interphase mass transfer:

$$\frac{\partial(\rho r X)}{\partial t}+\frac{\partial(\rho r u X)}{\partial z}+\frac{\partial(\rho r v X)}{\partial r}=\Gamma \tag{5}$$

$$\Gamma=\frac{\rho(X_S-X)}{\theta} \tag{6}$$

Here, z, r are the axial and radial coordinates, u, v are the axial and radial velocity components, ρ is the density, p is the pressure, E is the specific internal energy, X is the steam quality, and X_S is the equilibrium steam quality (at given p and E), Γ is the vapor generation rate, and θ is the relaxation time. The system of equations (1)-(6) is closed by the equation of state (relationship between p, E, ρ, X).

3 Phase transition relaxation model

The work [6] proposed the analytical relaxation model, based on the analysis of interfacial heat fluxes "steam – superheated liquid". The model, designed only to study the non-equilibrium boiling, is based on calculation of the nonstationary temperature field in the superheated water drops. Due to low vapor density the droplet structure of the two-phase medium is characteristic even at small steam quality (but moderate and large volumetric void fractions) [7]; and it may be used in the modeling of intensive evaporation of liquid. The average droplet size is determined from the simplified analysis of instability of the "water-steam" interface; and the size of droplets corresponds to the wave length, the fastest in destroying the interfacial surface, and is determined by a critical Weber number $We_{CR} \approx 2\pi$ (water, liquid metals) [7, 8]. According to the model, at the micro-level of drops, heat required for vaporization is supplied from the superheated liquid. The temperature of the droplet surface is equal to steam temperature [9].

To calculate the relaxation time the following formula is proposed [6]:

$$\theta = \begin{cases} \dfrac{\sigma^2 \rho_L^2}{4 a_L \rho_V^2 (P-P_S)^2}, & P-P_S > \dfrac{\rho_L L^2}{2\left(\dfrac{\rho_L}{\rho_V}-1\right)^2 C_{P,L} T} \\ \dfrac{\sigma^2 \left(\dfrac{\rho_L}{\rho_V}-1\right)^4 C_{P,L}^2 T^2}{a_L \rho_V^2 L^4}, & P-P_S \le \dfrac{\rho_L L^2}{2\left(\dfrac{\rho_L}{\rho_V}-1\right)^2 C_{P,L} T} \end{cases} \quad (7)$$

Here, σ is the surface tension coefficient, a is the thermal diffusivity, and L is the enthalpy of phase transition.

For the relaxation model to be used not only at evaporation, but condensation as well, it was necessary to substantially modify the equation of two-phase medium state, which closes the system of equations (1)-(6) in case of the existence of not only saturated, but super-cooled steam.

Although the size of drops formed in the condensation may differ from the case of the jet breakup [7,8], the test calculations were based on the formula (7).

4 Modeling results and discussion

In the work, the calculations of water outflow at initial supercritical parameters conducted according to the equilibrium and non-equilibrium relaxation model of phase transition were compared. Thermodynamic equilibrium model of two-phase medium was, in fact, obtained from the full system (1)-(4) with its closure by the equation of state of the equilibrium two-phase mixture A more precise analysis showed that in the considered range of the calculated parameters, the equilibrium model was adequate with sufficient accuracy at $\theta < 10^{-5}$ s.

The value θ is proportional to the square of the droplet size, i.e. is very sensitive to the value of this parameter. The size of the drop itself, according to the model, is determined by the value of Weber number We_{CR} and, formally, has to reach this value asymptotically at large times. At initial design parameters the formula (7) gives clearly lower value $\theta < 10^{-8}$ s (corresponding droplet radius $R_{DROP} < 10^{-7}$ m). Therefore, in numerical modeling, higher fixed relaxation times were taken.

Figure 1 shows calculation results for axial profiles of steam quality for the case of depressurization at the end of pipe with water coolant in the model liquid-vapor atmosphere [6]. Profiles are calculated at time 60 μs after depressurization at different values of relaxation time θ.

As noted above, when $\theta = 10^{-5}$ s the results of calculation based on the relaxation approach do not significantly differ from those of the equilibrium thermodynamic model. As it was shown earlier [2] a small variation of initial parameters of the supercritical region can lead to a change in direction (sign) of phase transformation. Figure 1a) shows the process of coolant vaporization, while Fig. 1b) presents the process of condensation.

In numerical computation, the "local" value of relaxation time θ was determined for control, and parameters (density, temperature...) included in the expression (7) were considered as variables. It turned out that both in the case of evaporation and condensation in these calculations, the relaxation time ranges from 10^{-6} to 10^{-3} s, and in the main array, from 10^{-5} to 10^{-4} s. These values are in good agreement with experimental data on relaxation time at steam condensation in Laval nozzles. In particular, in [10] it was shown that the characteristic relaxation time is of the order of 10^{-5} s.

The proposed model needs further refinement. Its main disadvantage is "rough estimation" of the most important parameter, the radius of the droplet. Apparently, the drop radius is an empirical parameter that must be determined from the analysis of appropriate experimental data. There is also a need in a more reasoned approach to the choice of drop sizes to describe the condensation process.

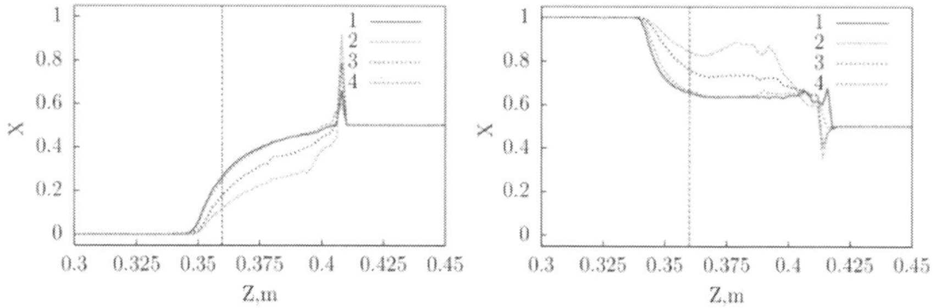

a b

Fig. 1. Profiles of steam quality at time t=60 μs for different values of relaxation time. a) P = 23 MPa, T = 643 K, b) P = 23 MPa, T = 653 K (1 – equilibrium model of phase transition, 2 – θ = 10-4 s, 3 – θ = 5·10-5 s, 4 – θ = 10-5 s), the vertical line is the coordinate of the channel end.

5 Conclusion

The non-equilibrium relaxation model describing the processes of superheated coolant boiling-up has been generalized to describe non-equilibrium condensation. Results of numerical simulation of the supercritical fluid outflow on thermodynamic equilibrium and non-equilibrium relaxation models of phase transition with different relaxation times have been compared. It is shown that the application of the proposed model at this stage leads to qualitatively correct results. This allows us specifying the ways for further development and improvement of the model.

The work was financially supported by the Russian Science Foundation (grant No. 14-29-00093).

References

1. M.V. Alekseev, I.S. Vozhakov, S.I. Lezhnin, N.A. Pribaturin, Thermophys. Aeromech. **23**, 869 (2016)
2. I.S. Vozhakov, M.V. Alekseev, S.I. Lezhnin, N.A. Pribaturin, J. Physics: Conference Series **754** 032021 (2016)
3. V.I. Artemov, K.B. Minko, G.G. Yankov, Thermal Eng. **62** 897 (2015)
4. R.H. Bolotnova, V.A. Buzina, Computational Continuum Mechanics **7**, 14 (2014) [in Russain]
5. P. Zapolski, Z. Bilicky, L. Bolle, J. Franco, Int. J. Multiphase Flow **22**, 473 (1996)
6. S.I. Lezhnin, I.S. Vozhakov, M.V. Alekseev, A.R. Bogomolov, Vestnik of Kuzbass State Technical University **5** (2016)
7. R.I. Nigmatulin, *Dynamics of multiphase media* (1987) [in Russain]
8. V.G. Levich, *Physico-chemical hydrodynamics* (1959) [in Russain]
9. L.D. Landau, E.M. Lifshits, *Hydrodynamic* (1986) [in Russain]
10. G.A. Saltanov, *Supersonic two phase flows* (1972) [in Russain]

The experimental investigations of peculiarities of metalorganic compounds sublimation

Elena Bochkareva[1], *Igor* Igumenov[2], and *Vladimir* Lukashov[1,2*]

[1]Kutateladze Institute of Thermophysics, 630090 Novosibirsk, 1 Lavrent'iev ave., Russia
[2]Nikolaev Institute of Inorganic Chemistry, 630090 Novosibirsk, 3 Lavrent'iev ave., Russia

Abstract. The paper presents the results of experimental investigations of convective heat and mass transfer by sublimation of a single particle of metallorganic compounds mixture in an argon flow. The gas temperature is 180-290 °C, the flow velocity is up to 2 m/s. The influence of the $Zr(dpm)_4$ and $Y(dpm)_3$ proportions on the characteristics of the transport processes was considered. An increase in the fraction of the more fusible component reduces the temperature of the particle during the sublimation process.

1 Introduction

At present, chemical vapor deposition of metal complexes with organic ligands (MO CVD - Metal-Organic Chemical Vapor Deposition) is one of the most promising methods for the nanostructured materials formation. The coating is formed by the decomposition of an isolated molecule of a volatile metal compound with an organic ligand (precursor) on a heated surface. This makes it possible to realize practically any coating variants starting from a system of unbound nanoparticles up to continuous micron layers by changing the experimental parameters. The low saturated vapor pressure at the operating temperature of the sublimation process and the low thermal stability of the compounds do not allow the concentration of the precursor vapor in the precipitation zone to be sufficient to produce coatings of the required thickness at a high velocity when applying protective coatings using metal beta-diketonates. Currently, one of the problems is to increase the sublimation intensity of the precursor particles in the carrier gas flow, which needs to be solve in order to develop the MO CVD technology. The thermodynamic aspects of the processes of sublimation of precursors and their relationship to the structural characteristics of the compound have been studied quite widely [1, 2] but the kinetics of vaporization (sublimation rate) of this class of compounds has not been studied. There are a limited number of studies related to heat and mass transfer under such conditions [3, 4] and less studied transport processes during the sublimation of binary mixtures. The results of such experiments are extremely necessary when developing a 'single-source delivery system' for MO CVD reactors.

The paper presents the results of experimental studies of the kinetics of sublimation of a single spherical particle of $Zr(dpm)_4/Y(dpm)_3$ mixtures in an argon flow. The choice as a research object such combination of precursors is determined by the breadth of their

* Corresponding author: luka@itp.nsc.ru

application in MO CVD deposition processes of high-temperature ceramic coatings stabilized with yttrium oxide (YSZ). A feature of this system is the significantly different melting points of the investigated compounds -179 °C and above 310 °C for $Y(dpm)_3$ and $Zr(dpm)_4$, respectively.

2 Experimental procedure

Samples of $Zr(dpm)_4$ and mixtures of $Zr(dpm)_4/Y(dpm)_3$ (zirconium tetrakis-dipivaloylmethanate and yttrium tris-dipivaloylmethane respectively) were prepared for the study of heat and mass transfer during sublimation. The samples were spheres with a diameter d_0 = 4 mm formed from small (20-50 μm) crystals with the thermocouple jammed into the center of the sphere. To measure the temperature the K- type thermocouples with a wire diameter of 100 μm were used. The raw material was wetted with heptane and pressed to obtain a spherical bead using metallic mold. The mixture "1:1" consisting of equal mass fractions of $Zr(dpm)_4$ and $Y(dpm)_3$ additionally reinforced with glass fiber.

A sample with an initial normal temperature was set in a flow of argon at a constant temperature. The gas flow preparation system was placed inside a thermo stated housing (see Figure 1), equipped with PID control systems. The temperature was set in the range t_0=210...290 ^{0}C. The gas flow was a free jet flowing out of a profiled nozzle 12 mm in diameter. Special measurements were carried out using a gradient radiant flow sensor which showed that the contribution of the radiant flux to the total heat flux on the jet axis at a distance of 12 mm from the nozzle cut did not exceed 20%. The test sample was placed at this point. Thus, the experimental setup made it possible to investigate convective heat and mass transfer during the sublimation of a single spherical particle.

The gas supply in our experiments was carried out by the digital flow controllers and was set to be equal constant value 8 standard liters per minute. When the studied particle introduced into the flow, its size was record with a Digi Scope II v3 digital microscope. The method for determining the particle size was to find the effective diameter of the sample. The effective diameter was determined from the volume of the ellipse because the particle shape is more similar to elliptical in the sublimation process.

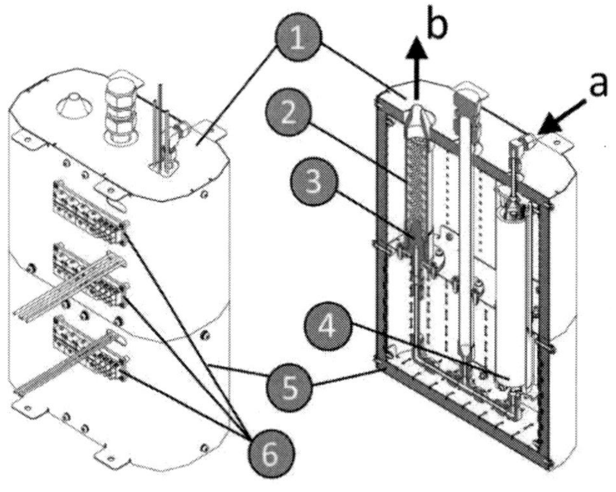

Fig. 1. The scheme of the device for the flow preparation: a – supply of inert gas, b – outlet of heated gas; 1 – upper part of the thermo-stat, 2 – nozzle apparatus filled with quartz beads, 3 – gas distributor, 4 – tubular heater, 5 – lower part of the thermostat, 6 – electrical commutation unit.

3 Results

Photos of the samples are show in Figure 2. The exposure time was 1200 s for the 1:1 mixture and 2400 s for other mixtures, and the argon temperature was 240 ^{0}C. The results of the experiments showed that during the first $\tau \sim 100$ seconds after the material was set in the argon flow, the sample was heat to a certain temperature ts which in the future remained practically constant. At the argon temperature t_0=210...240 ^{0}C the "1: 1" mixture changed to a liquid state. In this case the glass fiber reinforcement was necessary to avoid the drop falling off from the thermocouple. The visualization showed that the mixtures with a large content of $Zr(dpm)_4$ - "3:1" or "10:1" remained in a solid state in the entire investigated range of argon temperatures.

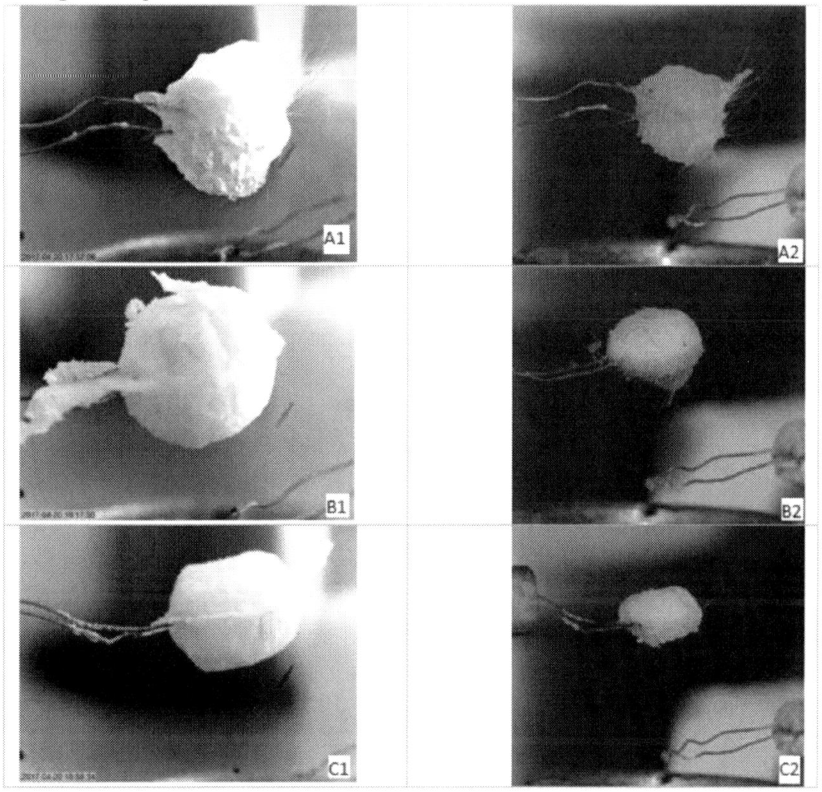

Fig. 2. Particles of different $Zr(dpm)_4/Y(dpm)_3$ mixture during sublimation in an argon flow. A1, A2 – «1:1»; B1, B2 – «3:1»; C1, C2 – «10:1».

The readings of the thermocouple built into the sample are show in Figure 3. The melting of the sample occurred during the warm-up, after the establishment of a stationary regime any features related to the preferential sublimation of the more fusible $Y(dpm)_3$ were not observed.

The intensity of mass transfer can be estimate using the time dependence of the relative particle diameter. In the temperature interval t_0 = 210-240 ^{0}C the increase of the $Y(dpm)_3$ fraction increases the intensity of the process. When the temperature of the gas flow rises to 290 ^{0}C, the kinetics of the processes changes (Figure 4). When the sublimation of $Zr(dpm)_4/Y(dpm)_3$ mixture takes place the rate of change of the relative diameter becomes markedly less than with the sublimation of pure $Zr(dpm)_4$ substance.

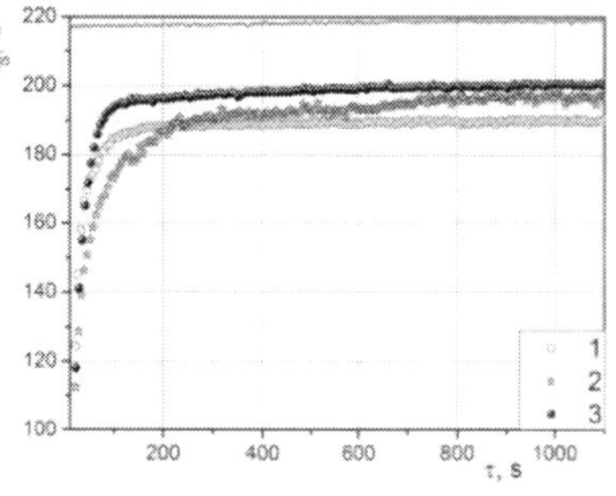

Fig. 3. The temperature of binary mixture sample during sublimation in Ar flow with t_0=219 °C. 1– «1:1»; 1– «3:1»; 1– «10:1» mixture.

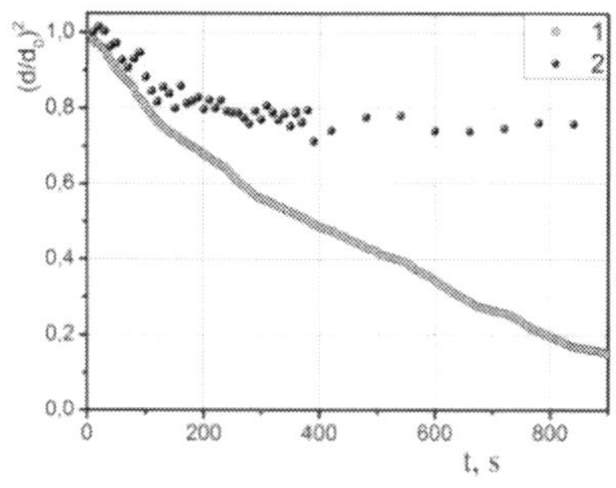

Fig. 4. Dynamics of the diameter of a precursor particle upon sublimation into an argon flow Ar, t_0=290–291°C, U_0=1.2 m/s. 1– $Zr(dpm)_4$; 2– $Zr(dpm)_4/Y(dpm)_3$ «10:1» mixture.

Perhaps this may be due to the transition from phase transition from surface of sphere to the sublimation inside the surface of the pores. In the last case the time dependence of the relative particle diameter ceases to characterize the sample mass transfer.

4 Conclusion

Experimental data on the heat and mass transfer processes during the sublimation of $Zr(dpm)_4/Y(dpm)_3$ mixtures for various compositions are obtained. At argon temperature up to 240 °C, the addition of a more fusible $Y(dpm)_3$ lowers the temperature of the sample and increases the rate of change in the relative diameter of the particle. As the temperature of the gas flow rises to 290 ^{0}C, the kinetics of the processes change, this may be related to the sublimation process inside the pores.

This work was supported by the Russian Science Foundation (grant No. 16-19-10325).

References

1. A.M. Igoshkin, I.F. Golovnev, V.V. Krisyuk, I.K. Igumenov, J. Struct. Chem. **57**, 6 (2016)
2. V.V. Krisyuk, I.A. Baidina, A.E. Turgambaeva, Kyzy S. Urkasym, I.V. Korolkov, T.P. Koretskaya, I.K. Igumenov, J. Organom. Chem. **819** (2016)
3. A.N. Cherepanov, V.P. Shapeev, L.G. Semin, V.K. Cherepanova, I.K. Igumenov, A.N. Mikheev, N.V. Gelfond, N.B. Morozova, J. Adv. Manufacturing Techn. **44**, 4 (2003)
4. N.V. Gelfond, A.N. Mikheev, N.B. Morozova, N.E. Gelfond, I.K. Igumenov, Int. J. Therm. Sci. **42**, 8 (2003)

Low-temperature purification of the synthesis gas

Anatolii V. Korobeinikov[1*], and *Sergey* A. Shevyrev[2]

[1]Gorbachev Kuzbass State Technical University, 650000 Kemerovo, Russia
[2]National Research Tomsk Polytechnic University, 30, Lenin Avenue, Tomsk, 634050, Russia

Abstract. In this paper, studies are made of the most economical method of low-temperature purification of synthesis gas, which is obtained during steam gasification of carbonizate. The analytical calculation is based on the cooling and phase transition of sulfur and carbon dioxide compounds and their separation. The paper presents the comparison of process energy costs depending on consumption characteristics and process parameters.

1 Introduction

Synthesis gas obtained in the process of steam oxygen-free gasification [1] is characterized by the content (% by vol.) of the target H_2 components equal to 62, while CO - to 26. In addition, it contains gases such as CO_2 up to 12, H_2S and SO_x up to 1% by vol., which adversely affect the catalytic processes of obtaining various hydrocarbons. For the Fischer-Tropsch process, most catalysts, including iron and cobalt, are extremely sensitive to "poisoning" with sulfur that is reflected in the shortening of their working life and deterioration of the process selectivity. As a result, it is impossible to reach the calculated technical and economic indicators and to ensure a low cost price of the products. In this connection, the most important is the purification of synthesis gas from sulfur compounds (SO_x and H_2S) and CO_2. Despite the relatively low concentration of sulfur compounds in the composition of synthesis gas (up to 1% by vol.) their content should be ensured up to 2 mg/m^3 [2].

2 Statement of the problem

The purpose of these studies is the energy analysis for the proposed method of purification of synthesis gas produced in the steam gasification carbonizate. It is proposed to purify the synthesis gas by selective desublimation/condensation of the components by discrete cooling. The implementation of such a method can compete with chemical methods of purification.

* Corresponding author: targenur@mail.ru

3 Process and schema description

The gas is cooled to the phase transition temperature of its components at atmospheric pressure: CO_2 - below 95K, H_2S - below 213K.

In order to investigate the physical methods of purification of synthesis gas, we used a principal scheme, shown in Fig. 1.

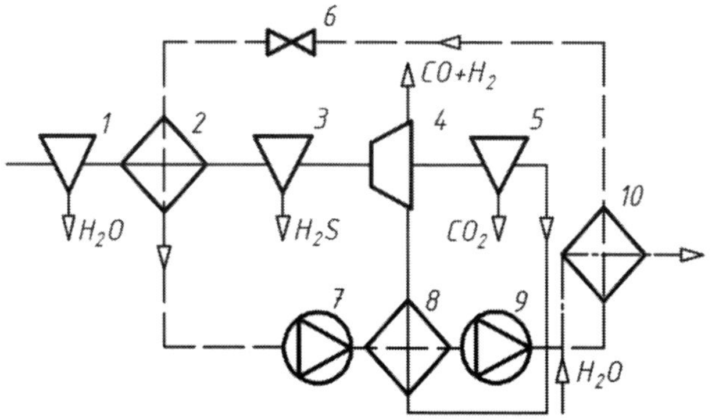

Fig. 1. The principal diagram of low-temperature purification of synthesis gas. The solid line is synthesis gas, the dashed line is the refrigerant cycle, the dash-dot line is water. *1, 3, 5* - cyclone separators; *2, 8, 10* - recuperative heat exchangers; *4* - turboexpander with a compressor on one shaft; *6* - a throttle washer; *7, 9* – 1st and 2nd compression stages.

The calculation is carried out by the thermal balance method. Synthesis gas at the inlet to the unit is wet, the temperature is 323 K, and the pressure is 101.3 kPa, without mechanical impurities. Dehumidification is carried out in the cyclone separator *1*. The gas is cooled to 203 K in the heat exchanger *2*, i.e., 10K below the condensation temperature of H_2S at atmospheric pressure. As a result, other sulfur compounds such as SO_x are condensing in case of their availability [4]. To purify sulfur compounds in separator *3*, the gas must be cooled to 185 K. At this temperature and atmospheric pressure, CO2 passes from the gas phase directly to the solid state, bypassing the liquid state, as a result of that the so-called "dry ice" is formed (Fig. 2).

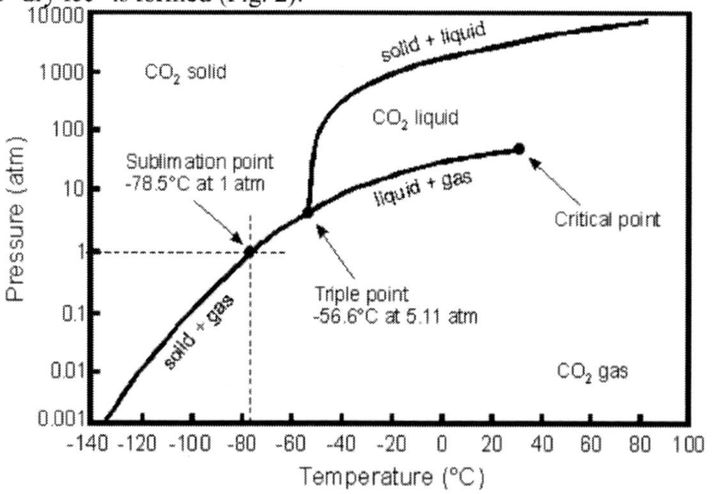

Fig. 2. Pressure-Temperature phase diagram for CO_2.

A turbo expander *4* must be used to achieve a temperature of 185 K. The synthesis gas purified in a separator *5* is compressed in a compressor and sent to a Fischer-Tropsch reactor.

A refrigeration cycle is used for the heat removal in the heat exchanger *2*. It includes two-stage compression *7* and *9*, an intercooler *8*, a condenser *10* and a throttle plate *6*. It is proposed to use R407a as a refrigerant.

4 Results and discussion

The synthesis gas flow rates are taken discretely equal to 5, 50, 500 and 5000 l/min, at a pressure of 101.3 kPa. The heat calculation of the heat exchanger *2* and the calculation of the refrigeration cycle were made according to the scheme (Fig. 1), the corresponding results are summarized in Table. 1.

Table 1. Dependence of performance indicators on increase in the synthesis gas consumption.

The synthesis gas flow rate, l/min	The amount of heat removed from the synthesis gas, kW	Heat which can be utilized usefully, kW	Power consumption, kW
5	0.0244	0.05063	0.0266
50	0.2439	0.5063	0.266
500	2.4391	5.063	2.66
5000	24.3915	50.63	26.6

It can be seen from the Table that 24.4 kW of heat should be removed for the separation of sulfur compounds from synthesis gas, at a flow rate of 5000 l/min, with a cooling coefficient of 1.09. It is necessary to use lower temperatures to remove CO_2 that can be achieved in a turbo-expander. In addition, the consumed power of the refrigerant compressors is 26.6 kW, and it is necessary to raise the pressure of the synthesis gas feeding into the reactor to 2-3 MPa. It is proposed to carry out the purification process at an elevated pressure corresponding to the Fischer-Tropsch reaction process. The parameters of the gas purification process are presented in Table 2, while the schematic of the purification process is shown in Fig. 3.

Table 2. Phase transition temperatures depending on pressure.

Pressure, MPa	Condensation temperature, K	
	Carbon dioxide – CO_2	Sulfur hydride – H_2S
0.1	195 (desublimation)	212.7
2	254	298
3	268	315.3

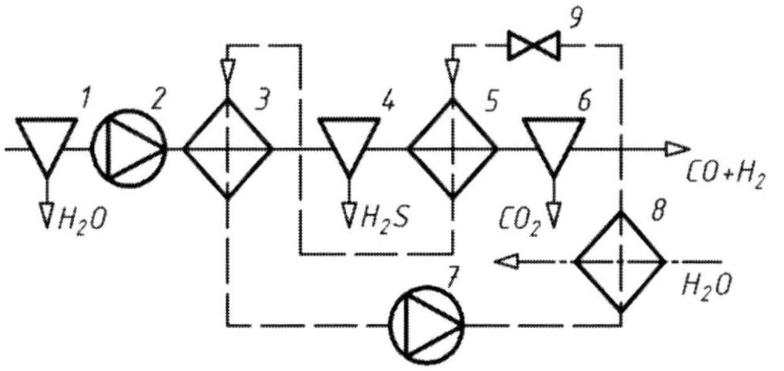

Fig. 3. The principal diagram of low-temperature purification of synthesis gas. The solid line is synthesis gas, the dashed line is the refrigerant cycle, the dash-dot line is water. *1, 4, 6* - cyclone separators; *2, 7* - compressors; *3, 5, 8* - recuperative heat exchangers; *9* - throttle washer.

There is no need to use a turboexpander, since the condensation temperature of CO_2 is above 254 K that significantly reduces the capital costs of the installation. Calculations of the energy costs for the purification of synthesis gas against CO_2 and H_2S, depending on the gas flow rate at different pressures of the cooling process, are shown in Fig. 4, and at a flow rate of 5000 l/min in Table 3.

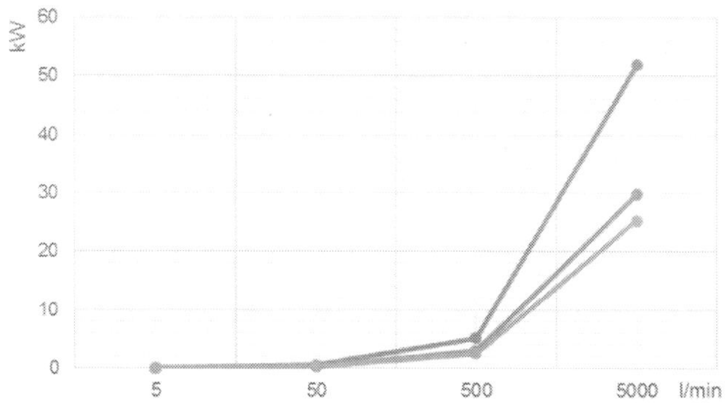

Fig. 4. Dependence of cooling capacity on consumption. Blue line (top) - cooling capacity at 0.1 MPa, orange (in the middle) - at 2 MPa, gray (bottom) - at 3 MPa.

Table 3. Dependence of performance indicators on pressure increase at a consumption of synthesis gas equal to 5000 l/min.

The synthesis gas flow rate, l/min	The amount of heat removed from the synthesis gas, kW	Heat which can be utilized usefully, kW	Power consumption, kW
0.1	51.91	50.63	26.6
2	29.79	48.7	18.9
3	25.34	35.3	9.94

4 Conclusion

Analysis of the calculations shows that the increase in the process pressure and temperature results in decrease in the necessary energy costs for purifying the synthesis gas. The refrigerating coefficient at the pressure of synthesis gas of 2 MPa is 1.58, while it is 2.55 at a pressure of 3 MPa.

References

1. A.R. Bogomolov, S.A. Shevyrev, M.V. Alekseev, Thermal Power Eng. **2**, 76 (2013)
2. G. Storch, Synthesis of hydrocarbons from carbon monoxide and hydrogen, 257 (1954)
3. A.V. Korobeinikov, Materials of the II All-Russian Youth Scientific and Practical Conference "Introduction to Energy", 23-25 November, Kemerovo (2016)
4. D. Leung, Renewable and Sustainable Energy Rev. **39**, 426 (2014)

Aerodynamic numerical calculations of air-cooled condensers and dry cooling towers at different wind velocities and multilevel configuration of sections

Oleg Milman[1,*], *Sergey* Lenev[2], *Pavel* Golov[2], *Boris* Shifrin[1], *Anna* Kartuesova[1]

[1]SPC Turbocon, 248010, Kaluga, Komsomolskaya st., 43, Russia
[2]PAO Mosenergo, 119526, Vernadskogo av., 101, corp.3, Russia

Abstract. Multilevel configuration of heat exchange sections of air-cooled condensers and dry cooling towers with a multi-row arrangement are offered. Results of a research of the combined operation of fans and V-shaped heat exchangers by methods of numerical model operation at wind and without it for three-row sections are given. It is shown that in the presence of wind multilevel configuration of heat exchange sections has advantages over single-level configuration.

1 Introduction

To make air-cooled condensers (ACC) and dry cooling towers (DCT) compact, when designing them, multi-row arrangement of heat-exchange sections is used. Internal rows of sections work in the inferior conditions that result in nonuniformity of air flow rate through ACC and DCT. Existence of wind creates hindrances for fans and increases nonuniformity of air flow rate [1]. One solution is the multilevel arrangement of sections when some ranks of sections are established at different height.

The purpose of this work is studying processes of the combined operation of the fan and its network (heat exchangers) at wind load capacity and without it for one- and a multilevel arrangement of heat exchange sections, which have V-shaped configuration [2].

2 Calculation models and boundary conditions

For realization of numerical calculations on a basis of the design consisting of three rows of sections in the SOLIDWORKS 3D CAD the corresponding 3D models were created. Two configuration options were considered: single-level, where all three sections are located at one height (fig. 1a) and two-level, where sections 1 and 3 are located above section 2 (fig. 1b).

Aerodynamic calculations of air flow were carried out using the Flow Simulation program, intended for thermal and hydraulic calculations. When performing calculations, it

* Corresponding author: turbocon@kaluga.ru

was accepted, that in each row are a large number of sections. Calculations were carried out in 2D-statement: in the XY plane, on axis Z the chosen simple size of 1.025 m was set (1/8 from the 3D model size). The numerical model (Fig. 1) represents calculation area with dimensions 35×65×1.025m with three sections and the given boundary conditions for a wind stream.

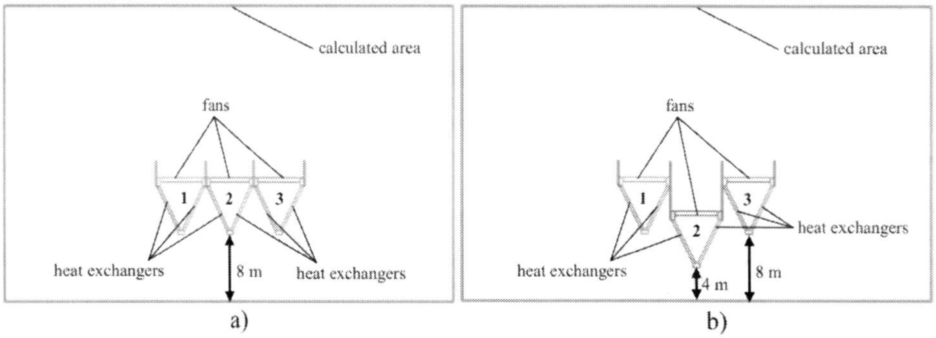

Fig. 1. Calculated model for single-level and two-level configurations of three sections.

Each section includes the fan and two heat exchange modules on which boundary conditions of "the internal fan" and "porous medium" were imposed. The "the internal fan" boundary condition is set by two surfaces imitating an inlet and outlet of the fan. For the fan it is set flow-pressure characteristic according to its model (Fig. 2a). The air stream at the exit of the fan is set without a curling.

The aerodynamic drag of heat exchange modules in the Flow Simulation program is set by a "porous medium" boundary condition, with a transmittivity in the normal direction to it and with the characteristic $\Delta p = f(G)$, where G - is the mass flow, kg/s (fig. 2b). The porous medium in calculation models is presented by parallelepipeds.

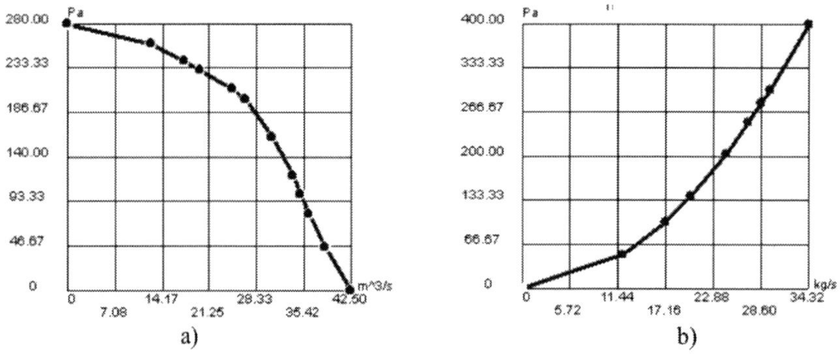

Fig. 2. Flow-pressure characteristic of fan (a) and aerodynamic drag of heat exchanger (b).

For surfaces and volumes, where "the internal fan" and "porous medium" boundary conditions are set, we constructed the initial local grid.

Also the following boundary conditions are set:

- The walls between sections are adiabatic;
- The sizes of calculated area limited to a test area;
- On a lateral face of a test area the normal velocity of a filling wind stream ranges from 0 to 10 m/s, on other sides the size of the static pressure equals to atmospheric (100 kPa;)
- Temperature of the air is 20°C;
- Turbulence model corresponds to k-ε model;
- Flow is both laminar and turbulent, fully developed.

The grid for the calculated model having single-level configuration consists of 506 520 cells; while for multilevel configuration – 509 320 cells.

3 Calculation results

3.1 Streamlines and features of a current of an airflow through sections

For single-level and multilevel configuration of three sections numerical calculations with the wind and without wind were preformed. Streamlines are shown in fig. 3-for single-level configuration and in fig. 4-for multilevel configuration.

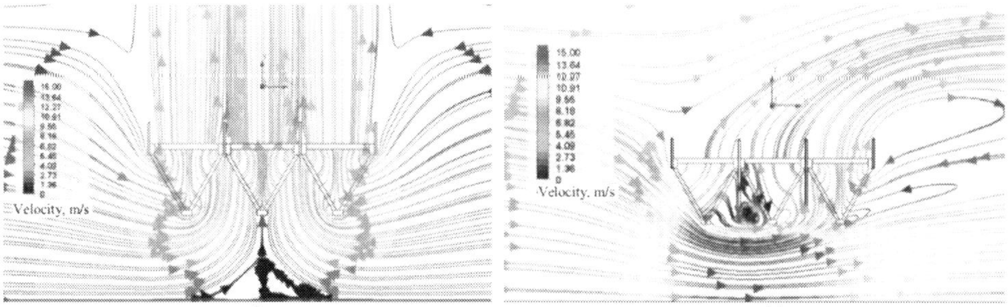

a) b)

Fig. 3. Streamlines at single-level configuration at wind of 0 m/s (a) and 6 m/s (b).

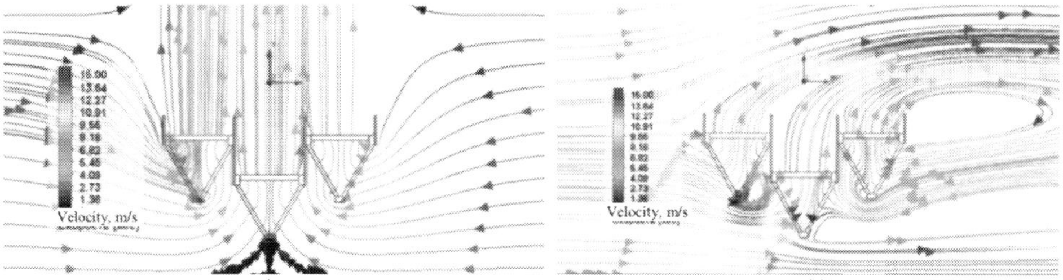

a) b)

Fig. 4. Streamlines at multilevel configuration at wind of 0 m/s (a) and 10 m/s (b).

3.1.1 Single-level configuration

In the absence of wind we observed the symmetric picture of a flow (Fig. 3a). The larger recirculation zone is formed under wind of 3 m/s on the left heat exchanger of the section 1. Because of emergence of depression field between inlets of sections 1 and 2, air flow rate through them decreases. The summary air flow rate decreases by 20% in comparison with a flow rate in the absence of wind. With increase wind velocity up to 6 m/s (fig. 3b) the recirculation on section 1 disappears, however less intensive flow appears in section 3. The depression field at the inlet between sections 1 and 2 increases. The air flow direction in the right heat exchanger of section 1 is reversed. The summary air flow rate decreases by 35%.

3.1.2 Multi-level configuration

At a velocity of wind of 3 m/s in the section 1 there is also a recirculation zone, however, less intensive, than at single-level configuration. The summary air flow rate decreases by 13%. With increase in wind velocity up to 10 m/s the recirculation zone in section 1 disappears (Fig. 4b). In the absence of wind, summary air flow rate having reached a minimum at 3-4 m/s, increases to 92.5%.

3.2 Generalization of results of research

As a result of calculations, we have obtained mass flows on fans in the sections in the presence and absence of wind, which are presented in Table 1.

Table 1. Mass flows through all fans of sections at a different velocity of wind.

Velocity of wind, m/s	Mass flows through all fans, kg/s	
	single-level	multilevel
0	113.3	113.7
3	89.71	98.86
6	74.20	—
8	—	103.11
10	—	105.11

Based on the results obtained, we constructed the dependence of mass flow through sections on wind velocity, shown in Fig. 5.

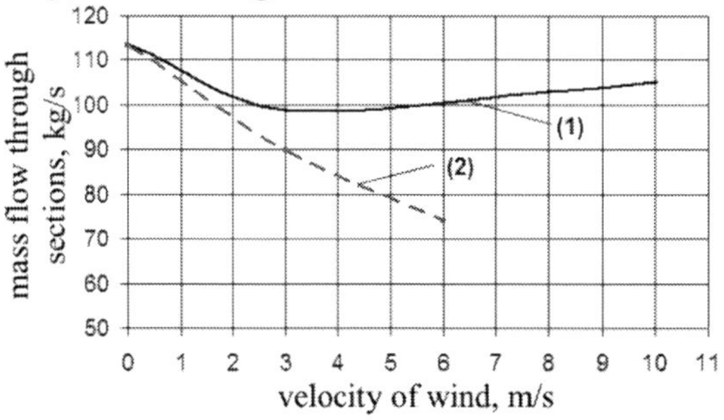

Fig. 5. Dependence of mass flow of air on wind velocity through sections for multilevel (1) and single-level (2) configurations.

The results show that a multilevel configuration is preferable as compared to a one-level configuration when developing ACC and DCT under continuous wind impact.

3 Conclusion

The executed numerical simulation showed the existence of positive effect of multilevel sections configuration:

- More uniform distribution of air flow rates in multilevel sections as compared to single-level configuration;
- Increase air flow rate through multilevel sections when increasing wind velocity.

References

1. A.A. Jinov, D.V.Shevelev, *Bulletin of MSTU of N.E.Bauman. Series: mechanical engineering*, **1**, 108-118 (2015) [in Russian]
2. O.O. Milman, V.A. Fedorov, *Air-Condensation Plants*, (MEI, Moscow, 2002) [in Russian]

Numerical modeling of energy-separation in cascaded Leontiev tubes with a central body

Maksim Makarov* and *Sofya* Makarova

Kutateladze Institute of Thermophysics, SB RAS, 630090 Novosibirsk, 1 Lavrentiev Ave., Russia

Abstract. Designs of two- and three-cascaded Leontiev tubes are proposed in the paper. The results of numerical simulation of the energy separation in such tubes are presented. The efficiency parameters are determined in direct flows of helium-xenon coolant with low Prandtl number.

1 Introduction

The prospect of increasing the degree of gas cooling due to the cascading arrangement of energy separation tubes appeared almost simultaneously with the very idea of energy separation in a supersonic flow [1, 2]. In one of frequent conversations about thermophysics academician of RAS A.I. Leontyev and future academician E.P. Volchkov started discussion about the effectiveness of the proposed method. It was obvious that air cooling in such devices (hereinafter referred to as Leontiev tubes) is less effective in comparison with Rank vortex tubes. However, Leontiev tubes have a significant advantage. A cooled (subsonic) part of the flow almost keeps the total pressure at the tube outlet, which allows using the cooled flow at further stages of energy separation in cascading arrangement of tubes. A significant increase in efficiency may be also achieved, using gas mixtures with low Prandtl numbers as coolants [2 – 4].

In this work, we propose the designs of two- and three-cascaded Leontiev tubes, present the results of numerical simulation of the effect of energy separation in such tubes, and determine the parameters of effectiveness in case of using the helium-xenon gas mixture with Prandtl number of 0.23.

2 Problem statement

The main design feature of the proposed cascaded tubes of energy separation is the formation of supersonic flows in annular channels. One of the supersonic channels is formed by a profiled central body and the cylindrical inner wall of the subsonic channel. The second one is formed by a profiled outer wall of the subsonic channel and the housing. Fig. 1 presents a flow diagram in a two-cascaded tube, the main geometric and flow rate parameters, and photos of fabricated elements of a prototype sample.

For the analysis we have chosen the optimum geometry of the tube with flow characteristics that are similar to those of a single Leontiev tube, studied experimentally in

* Corresponding author: msmakarov@itp.nsc.ru

[5, 6]. The tube length L was 450 mm, the distance to the first critical section (L_1) was 45 mm, and the distance to the second critical section (L_3) was 250 mm. The external diameter

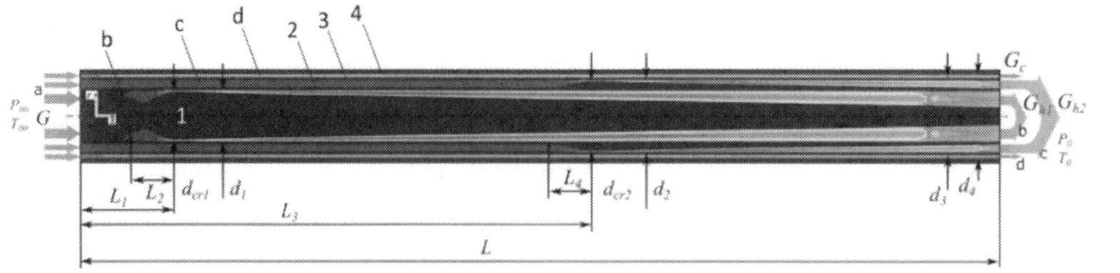

Fig. 1. The draft of two-cascaded Leontiev tube and photograph of the elements fabricated for prototyping: central body (1), inner (2) and outer (3) heat-transfer wall, outer wall (4), all elements in assembly (5); inlet flow (a), internal (b) and external (c) heating flow, cooling flow (d).

of the inner supersonic nozzle (d1) was 24 mm, the diameter of the central body at the critical section (d_{cr1}) was 22 mm, the outer diameter of the annular nozzle with the shifted critical section (d_2) was 34 mm, and the outer diameter of the separation wall 4 in critical section (d_{cr2}) was 32.5 mm. Inner and outer diameters of the subsonic channel (d_3) and (d_4) were 36 mm 39.5 mm, respectively. The calculated Mach numbers at the outflow of helium-xenon mixture were 3.2 and 3.4 for central and peripheral nozzles, respectively. The results of numerical simulation taking into account the influence of viscosity have shown that the maximum Mach numbers did not exceed 2.14 and 2.33, respectively, which, as follows from previous studies, is optimal for the process of energy separation.

A three-cascaded tube has a length of 450 mm. The critical sections are located at a distance of 45 mm, 180 mm and 315 mm from the inlet. The flow characteristics and the design modes of nozzles were selected close to the characteristics of nozzles of the two-cascaded tube.

3 Method of numerical modeling

We solved the problem by the implicit numerical method, based on the 2D axisymmetric RANS equations for the steady compressible flow with heat transfer. ANSYS Fluent solver licensed for SSCC SB RAS and self-made code were used for mesh generation and data analysis. The S-A turbulence model and the Kays-Crawford model [7] for the turbulent Prandtl number were used for modeling. As it was obtained in [8], this numerical model allows solving the convective heat transfer problems for the turbulent low-Prandtl gas flows. The turbulent intensity was 2 %, and the turbulent length scale was 1 m at the channel inlet. The adhesion conditions for motion equations, the temperature and heat flux coupling for energy equation were set as boundary conditions on the heat-conducting walls. The meshes contained from 90000 to 300000 nodes. The convergence for all variables and mass-flow rates in each channel were provided with error of 0.001%.

4 Results and discussion

Fig. 2 shows distributions of Mach number, thermodynamic temperature, total temperature and total pressure in the longitudinal section of the two-cascaded tube for the outflow of helium-xenon mixture with the maximum obtained cooling degree of -17 °C.

Fig. 3a presents data on the distribution of specific density of heat flux through the separating wall of the two-cascaded tube. As expected, in this configuration, the subsonic flow of a part of gas in the channel with shifted critical section (c in fig. 1) in the initial section (before critical section) is cooled by heat transfer to the central supersonic flow of another part of gas (b in fig. 1). After the critical section, the gas accelerates up to supersonic speeds and cools the third portion of gas in the outer subsonic channel (d in fig. 1). For the three-cascaded tube, the process of heat transfer from the outer subsonic channel to inner supersonic channels is the same (fig. 3b). Regardless of the number of cascades, it may be noted that in the area of interaction of subsonic flows, there is almost no heat transfer through the separation walls since the reduction temperature on the wall of the subsonic channel is little different from the initial total temperature.

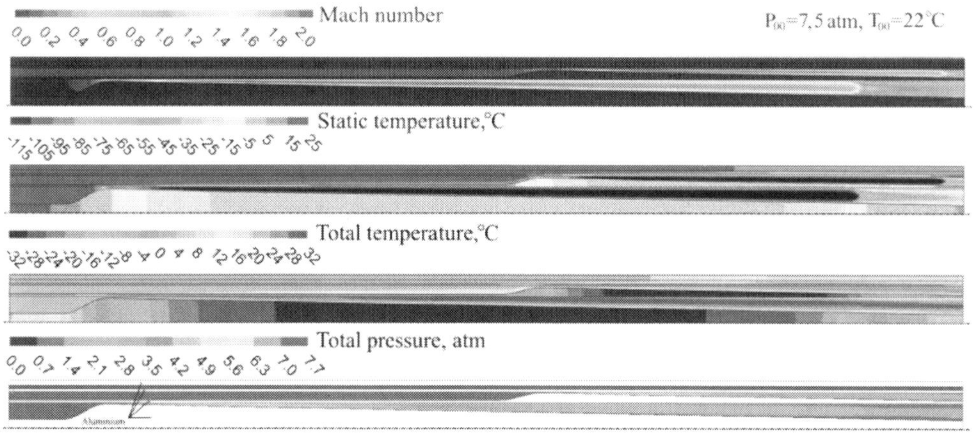

Fig. 2. Distribution of the main parameters in the longitudinal section of two-cascaded energy-separation tube with central body and separated location of critical cross-sections in annular supersonic nozzles (the data presented for the case with maximal cooling effect and G_c/G=0.09).

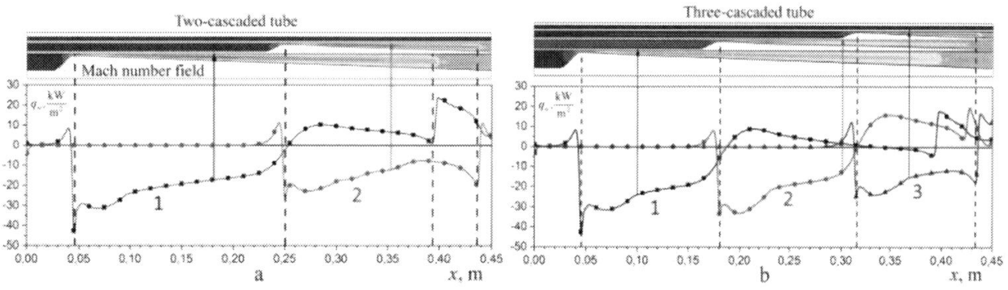

Fig. 3. Heat flux distribution at the internal heat-conducting walls for the case of two-cascaded (a) and three-cascaded (b) Leontiev tubes: data for the first cascade (1), second cascade (2), and third cascade (3) – cooling the subsonic flow; dash lines indicate critical sections of supersonic nozzles.

As can be seen in fig. 2 and fig. 3, in channels with shifted critical sections in the area of the supersonic flow, the heat flux changes its sign. Heat is transferred from the central channels to the peripheral ones, which significantly reduces the efficiency of energy separation in the cascaded tubes. This effect may be explained by the lower value of the reduction temperature in the flow, cooled in advance in the field before the critical section. The temperature of the separating wall becomes lower than the reduction temperature from the side of the central uncooled channel. Assumingly, this effect can be avoided or significantly reduced by one of the following methods. The first method is the thermal insulation of the supersonic part of the channel with shifted critical section from the side of the central annular nozzle. The second way is an increase in the calculated Mach numbers in central channels compared to the peripheral channels. As it turned out, without these or similar measures, the effectiveness of energy separation in the cascaded tubes of the proposed configurations does not exceed the efficiency of single tubes under the same stagnation parameters and at the same ratio of flow rates of cooled gas to the total gas flow through the tube (μ). Fig. 4 presents data on the degree of cooling, temperature efficiency factor and adiabatic efficiency [4] of energy separation in single and cascaded Leontiev tubes, depending on the ratio of cooled air or helium-xenon mixture.

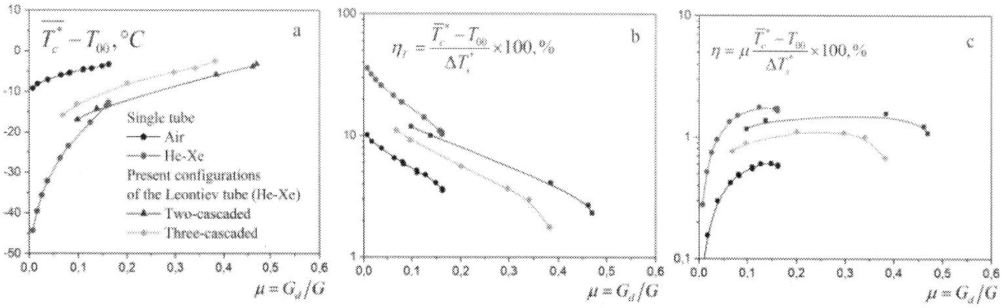

Fig. 4. Cooling effect (a), temperature efficiency factor (b) and adiabatic efficiency (c) depending on the subsonic mass flow ratio for single, two-cascaded, and three-cascaded Leontiev tubes (helium-xenon, inlet total pressure 7.5 atm, inlet total temperature 22 °C).

5 Conclusion

Energy separation of the compressible gas flow in the two-cascaded and three-cascaded Leontiev tubes has been analyzed. The parameters of energy-separation efficiency (temperature efficiency factor and adiabatic efficiency) for the cascaded Leontiev tubes are found to be less than the same factors for the single Leontiev tube, but the studied cascaded tubes work with a higher subsonic mass flow ratio. The disadvantages of this method of energy separation in the cascaded tubes with an optimum Mach number in supersonic channels, leading to a decrease in efficiency, have been identified. Ways to address them have been proposed, in particular, by increasing the estimated Mach number from the peripheral channels to the center and by thermally insulating a part of the separating wall from the central channel.

The reported study was funded by RFBR according to the research project № 15-08-04203.

References

1. A.I. Leontiev High Temperature **35** 1 (1997)
2. S.A. Burtsev, A.I. Leontiev Isv. AN. Energy **5** (2000)
3. M.S. Makarov, S.N. Makarova Thermophysics and Aeromechanics **20** 6 (2013)
4. M.S. Makarov, S.N. Makarova, A.A. Shibaev JoP:CS **754** (2016)
5. Yu.A. Vinogradov, A.G. Zditovets, M.M. Strongin Fluid Dynamics **48** 5 (2013)
6. A.G. Zditovets, Yu.A. Vinogradov, A.A. Titov IHTC15-8965 (2014)
7. B. Weigand, J.R. Ferguson, M.E. Crawford Int. J. Heat Mass Transf. **40** 17 (1997)
8. V.E. Nakorjakov, M.S. Makarov, Yu.I. Petukhov, O.V. Vitovsky, S.L. Elistratov, *Heat transfer in a low-Prandtl gas mixtures flows* (Novosibirsk, Akademizdat, 2015) [in Russian]

DOI: 10.1051/matecconf/201711509002

Studying the thermal effect in a swirled acoustic flow

Olga Mitrofanova*, and *Irina* Pozdeeva

National Research Nuclear University "MEPI", 115409, 31 Kashirskoye shosse, Moscow, Russia

Abstract. The mechanism of generation of acoustic oscillations associated with the formation of stable vortex structures in a moving medium is considered on the example of impact swirl flow. A self-regulation effect was detected for condition, when the limiting flow rate of a swirl flow has been reached. This effect is expressed in resonance amplification of the amplitude of the natural frequencies of the hydro-mechanical system due to the absorption of the component of the spectrum of acoustic oscillations generated by the vortex structure of the flow. A redistribution mechanism for the energy of an open system at resonance, associated with the influx of air from the external medium is considered. Temperature distributions on the lower surface of the obstacle were obtained using the thermal imager, and the energy losses associated with the flow cooling due to expansion were estimated.

An experimental investigation of the thermal effect in a swirled acoustic flow was carried out within the framework of studying the effect of self-regulation of acoustic oscillations in a hydromechanical system in the pre-resonance and resonance regimes [1]. A detailed description of the experimental setup is given in [2]. The vortex chamber was a rigid metal vessel with plexiglass upper surface with central hole through which a swirled air stream was flowing out. To organize the impact character of the flow above the outlet opening, made in the form of a confuser, an unfixed barrier in the form of a flat disc was placed. During the measurements, the amplitude-frequency characteristics of acoustic oscillations were recorded; they carried information about the change in the local pressure field, visualization patterns of the vortex structure and the inhomogeneities of the flow, as well as the temperature field on the lower surface of the barrier [3].

With the help of thermal imaging measurements, it was found that the formation of a stable large-scale spiral-vortex structure in the form of a torus with a double rotation in the region of the expiration of an impact swirled jet occurs with the conversion of thermal energy into mechanical energy (Fig.1).

* Corresponding author: omitr@yandex.ru

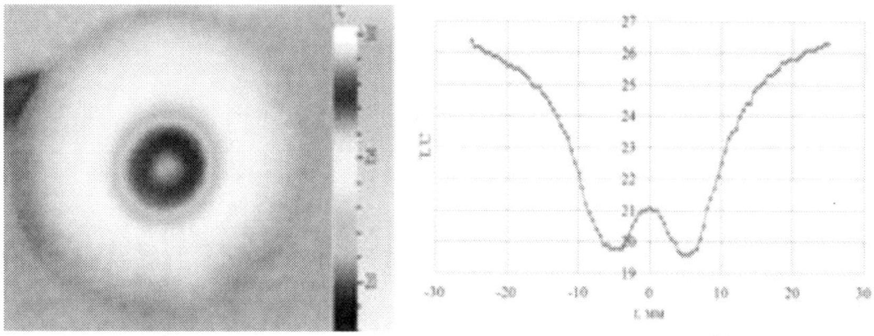

a) b)

Fig.1. Temperature field on the bottom surface of the plexiglass obstacle with a diameter D = 50 mm (a) and the temperature dependence on the radius (b) with the outlet port diameter of the vortex chamber d_0 = 5 mm and the flow rate $G = 1.3\cdot10^{-3}$ m^3/s.

The calculation shows that the temperature drop in the region of the toroidal vortex formed above the edge of the nozzle outlet from the vortex chamber (the dark annular layer in the central part of the barrier in Fig. 1a) is 6.4^0 C, which corresponds to a local decrease in temperature in the region of the vortex ring with a change of the radius in the range: 4 mm $\leq r \leq$ 10 mm (Fig.1b).

The registration of the amplitude-frequency characteristics of the flow under critical flow conditions allowed clearly distinguishing two characteristic frequencies of harmonic oscillations corresponding to the angular frequency of rotation of the swirled flow and the rotation frequency in coherent screw-helical vortices. In this case, the swirled jet, leaving the vortex chamber opening in the region of sharp rotation and radial spreading of the flow due to the presence of an obstacle, was divided into thin spiral vortices, which is confirmed by the picture of the visualization of a vortex track (Fig. 2a). A sharp decrease in pressure in this region, thanks to the vortex formation, provided the transition condition through the dew point (Fig. 2b) with the formation of drops of condensed moisture.

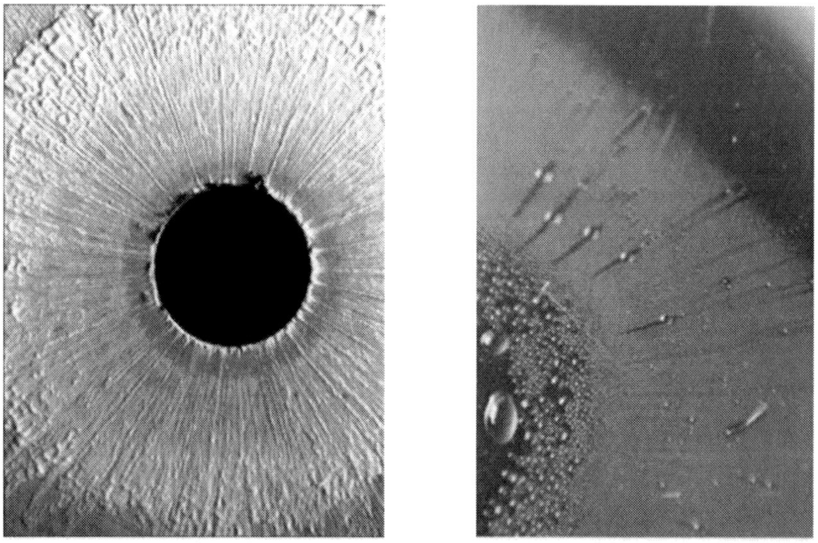

a) b)

Fig. 2. Visualization of the flow field in the resonant flow regime: a) formation of longitudinal helical vortices on the bottom surface of the gap; b) condensation of water droplets on the lower surface of the obstacle (the upper surface of the gap) in the regions of vortices localization.

The theoretical analysis carried out in [1], using of Lissajous figures corresponding to the ratio of two clearly defined harmonic oscillations, allowed proposing a physical model explaining the mechanism of energy redistribution in an open system at resonance. The first harmonic oscillation with frequency f_1 is associated with the formation of longitudinal helical vortices whose transverse scale is determined by the width of the slit. The frequency of the second fundamental harmonic oscillation is the frequency of rotation of the flow around the axis of the vortex chamber f_2. The air flow can perform several oscillation cycles until its energy becomes sufficient to release into the ambient medium.

Experimental observations have shown that, since the air before the vortex chamber entrance is preliminarily dried, the complicated topology of the vortex formation upon the expiration of the impact swirl flow is associated with the influx of air from the external medium. It can be seen from Fig. 2b that the arrangement of water droplets resulting from condensation after spreading of the flow corresponds to the arrangement of thin spiral air vortices (Fig. 2a), which ensures the air flow evacuation from the vortex chamber during the apparatus operation. Since, as already mentioned above, moist air in the work area could come only from the outside, a logical explanation for the observed condensation effect can be the assumption of a two-layer structure of spiral vortices (Fig. 3b). Under conditions of resonance regime of the impact flow, such a structure continues to exist for some time after the flow of the working fluid through the vortex chamber is cut off. The existence of the inlet airflow from the external medium into the space under the obstacle was experimentally proved by visualization with the help of a smoke jet (Fig. 3a).

a)

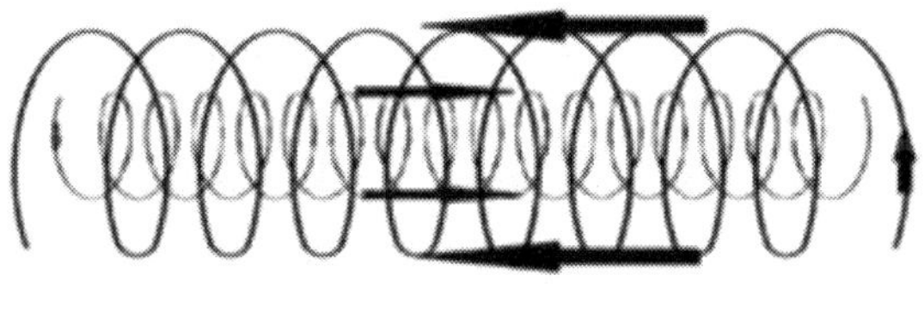

b)

Fig. 3. Visualization of air inflow into the space under the obstacle from the external medium by means of a smoke jet (a) and a flow pattern in the longitudinal helical vortex (b).

As can be seen in the photo (Fig. 3a), the smoke jet is drawn under the obstacle, thus demonstrating that an additional inflow of mass from the external medium is observed when the critical airflow from the vortex chamber outflows. To explain the mechanism by which additional flow entrance carried out under the obstacle, it is assumed that the vortex jets to which the impacted swirling flow is divided have a complex two-layer structure (Fig. 3b).

The vortex chamber is a rigid metal cylindrical vessel with a glued elastic plexiglas cover, through the hole in which a swirl air jet flows out. Having a large kinetic energy the air flow from the vortex chamber, excites vibrations of the elastic surface, which leads to fluctuations in the volume of air under the obstacle. As a result, when the speed of rotation in a spiral vortex coincides with the natural oscillation frequency of the elastic cover, the air outflow leaves the space under the obstacle in one cycle of oscillations, without returning to the center of the obstacle. The described process corresponds to the condition of acoustic resonance, accompanied by a sharp increase in acoustic oscillations at a frequency that coincides with the natural frequency of the oscillatory system.

The time sweep of the acoustic wave presented in Figure 4a also indicates the existence of a two-layer structure in a separate spiral vortex. The amplitude-frequency response corresponding to this time sweep is shown in Fig. 4b. This picture illustrates the experimental dependence, where the first peak with the low amplitude corresponds to the frequency of the external vortex f_1 in Fig. 3b, while the second peak with the higher amplitude corresponds to the frequency f_2 and rotational speed of the internal vortex with the opposite direction of rotation.

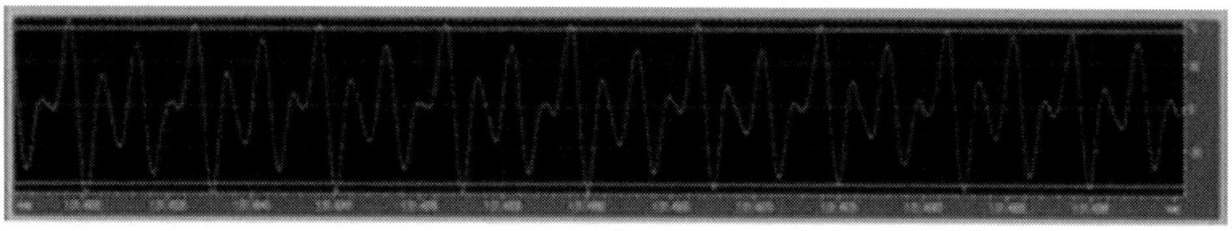

a)

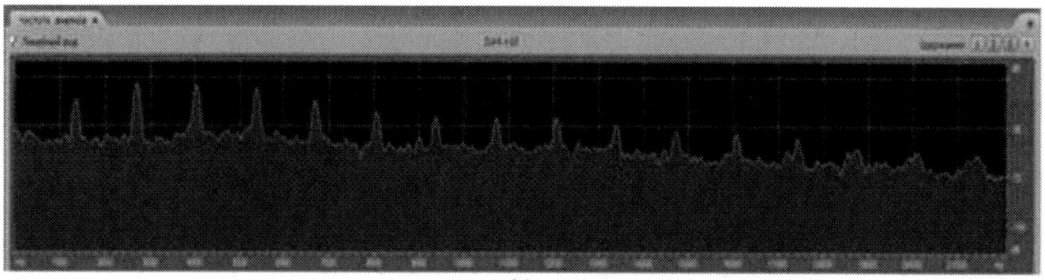

b)

Fig. 4. Time scan of the acoustic wave (a) and frequency spectrum (b) for the obstacle diameter D = 30 mm and the flow rate $G = 2.85 \cdot 10^{-3}$ m^3/s.

Thus, the recording of the shape of the sound wave, shown in Fig. 4a is the resultant graph of two antiphase oscillations.

This work was supported by RFBR grant 16-08-00687 and the program of increasing the competitive ability of National Research Nuclear University MEPhI (agreement with the Ministry of Education and Science of the Russian Federation of August 27, 2013, project no. 02.a03.21.0005).

References

1. O.V. Mitrofanova, I.G. Pozdeeva, Fluid Dynamics **50** 646 (2015)
2. O.V. Mitrofanova, I.G. Pozdeeva, A.B. Kruglov, V.B. Kruglov, Nuclear Phys. Eng. **3** (2012)
3. O.V. Mitrofanova, I.G. Pozdeeva, *Power balance at generation of acoustic fluctuations in the impact swirl flow* Materials of the International Conference "Nonlinear problems of the theory of hydrodynamic stability and turbulence" (2014)

Flat plate film cooling at the coolant supply into triangular and cylindrical craters

Artem A. Khalatov[1,2*], *Nadiia A.* Panchenko[1,2] and *Sergey D.* Severin[1]

[1]Institute of Engineering Thermophysics of NAS of Ukraine, Ukraine
[2]National Technical University of Ukraine «Igor Sikorsky Kyiv Polytechnic Institute», Ukraine

Abstract. The results are given of the film cooling numerical simulation of three different schemes including single-array of the traditional round inclined holes, as well as inclined holes arranged in the cylindrical or triangular dimples (craters). The results of simulation showed that at the medium and high values of the blowing ratio ($m > 1.0$) the scheme with coolant supply into triangular craters improves the adiabatic film cooling efficiency by 1.5...2.7 times compared to the traditional array of inclined holes, or by 1.3…1.8 times compared to the scheme with coolant supply into cylindrical craters. The greater film cooling efficiency with the coolant supply into triangular craters is explained by decrease in the intensity of secondary vortex structures ("kidney" vortex). This is due to the partial destruction and transformation of the coolant jets structure interacting with front wall of the crater. Simultaneously, the film cooling uniformity is increased in the span-wise direction.

1 Introduction

The film cooling is the most efficient technique for protection of turbine blades in advanced high-performance gas turbines. The traditional film cooling technique (Fig.1 a) has a number of disadvantages; the primary of them is the low film cooling efficiency at the high value of blowing ratio ($m > 1.0$). It is caused by the appearance of specific vortex structures in the flow contributing to the detachment of coolant jets from the cooled surface and the hot gas flowing under the coolant jets. The analysis of published papers [1-4] showed the search for the alternative film cooling techniques is one of the main scientific and technical trends of the modern gas turbine engineering towards the greater film cooling efficiency, lower coolant consumption, and relatively simpler production technology.

In paper [5] the new configuration towards the film cooling efficiency growth with coolant supply into the craters of cylindrical shape (Fig.1 b) was proposed. The positive effect of such indentation shape was achieved via the partial flow structure destruction, and breaking of cooling jets while their interaction with front walls of craters, leading to a reduction in the power of "negative" vortex structures.

[*] Corresponding author: artem.khalatov@mail.ru

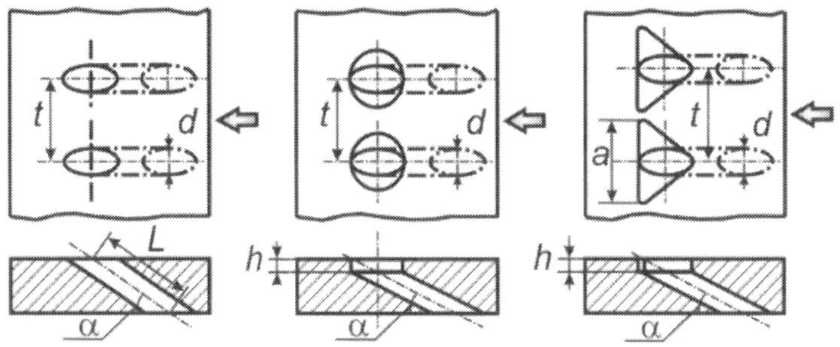

a b c

Fig. 1. The investigated film cooling schemes: a – traditional array of inclined round holes; b, c – single array of inclined round holes with coolant supply into cylindrical and triangular craters, respectively.

In the Institute of Engineering Thermophysics of the National Academy of Sciences of Ukraine the new film cooling configuration was proposed with a coolant supply through the single array of inclined round holes [6] arranged in the triangular-shaped craters (Fig.1 c). This leads to the transformation of separate coolant jets into the semi-bounded two-dimensional jet, reducing the coolant discreteness when supplying onto protected surface.

2 Numerical simulation of film cooling

The numerical simulation of three film cooling schemes was performed using ANSYS CFX 14 commercial code. All investigated schemes have the identical relative pitch and angle of hole inclination ($t/d = 3$, $\alpha = 30^o$); the depth of craters h is 0.5 d; the diameter of cylindrical craters D is 2.0 d; and the base width of triangular craters a is 2.25 d. The diameter of film cooling holes d is 0.8 mm. The non-dimensional length of the film cooling area x/d is 40.

The boundary conditions were taken corresponding to the actual vane of power gas turbine. The blowing ratio m is 0.5…2.5; the jet to main flow density factor DR is 1.8…2.1; the flow turbulence Tu is 1%; the main flow speed is 400 m/s; the cooling jets and main flow temperature is 500°C and 1100°C, respectively.

The investigation was made using the RANS SST turbulence model, which demonstrated acceptable results still obtained in earlier film cooling studies of the authors [7, 8]. In calculations the unstructured combined computational grids were used, consisting of 1,1...1,5 million cells. The computational grid inflation near the solid surface included 20 grid cells. The y^+ parameter (non-dimensional normal coordinate) was about the unity order, which satisfies conditions of the SST turbulence model correct application at the film cooling computer numerical modelling.

3 Results and discussion

The results of the numerical studies (Fig.2) have shown that at the medium and high value of the blowing ratio ($m > 1.0$) the scheme with coolant supply into triangular craters increases the average (throughout the cooled surface) adiabatic film cooling efficiency by 1.5...2.7 times compared to the traditional scheme of a single array of inclined round holes, and by 1.3…1.8 times compared to the scheme with coolant supply into cylindrical craters.

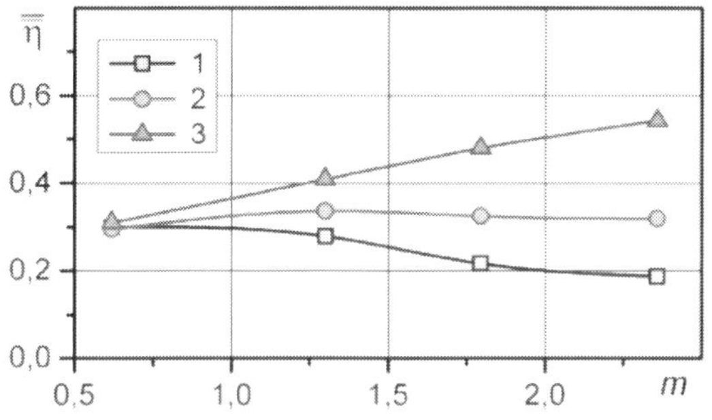

Fig. 2. The average adiabatic film cooling efficiency versus the blowing ratio:
1 – traditional array of inclined holes; 2, 3 – single array of inclined round holes with a coolant supply into cylindrical and triangular craters, respectively.

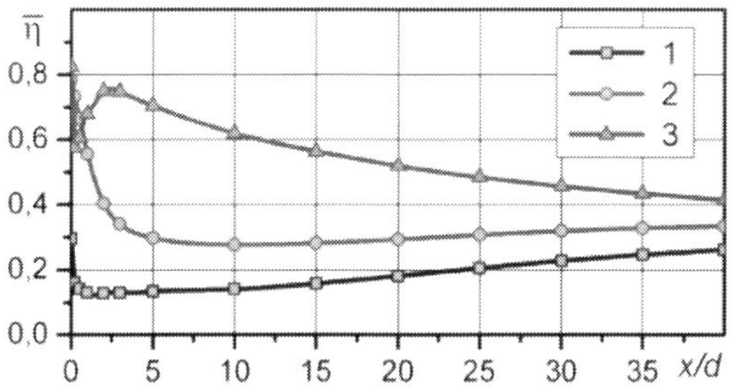

Fig. 3. The average adiabatic film cooling efficiency for m = 1.8: 1 – traditional array of inclined holes; 2, 3 – single array of inclined round holes with coolant supply into cylindrical and triangular craters, respectively.

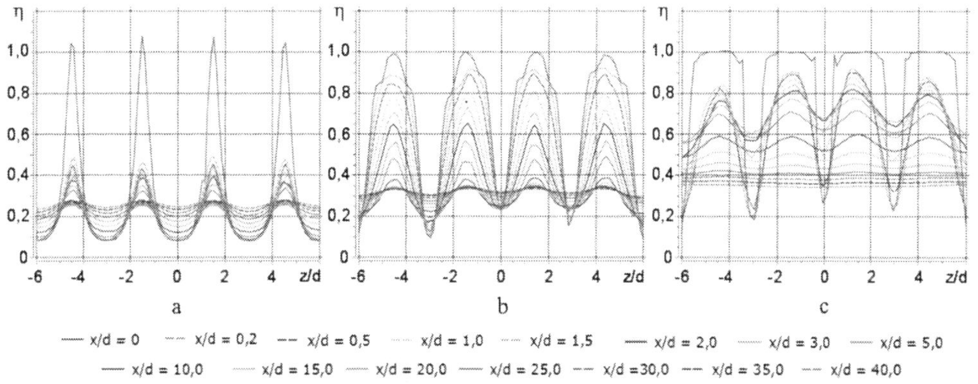

Fig. 4. The local laterally adiabatic film cooling efficiency for different values of x/d for three investigated schemes, m = 1,8: 1 – traditional array of inclined holes; 2, 3 – coolant supply into cylindrical and triangular craters, respectively.

As follows from Fig. 2, within the studied range of blowing ratio the average film cooling efficiency for the traditional array of holes is decreased with the blowing ratio growth. It is approximately constant (weak maximum at $m = 1.30$) at the coolant supply into cylindrical craters, while increases at the coolant supply into triangular-shaped craters.

The comparison of average adiabatic film cooling efficiency at $m = 1.8$ is given in Fig. 3 for three investigated film cooling schemes. As seen, significant difference in the film cooling efficiency is observed for considered cooling schemes both in the initial and primary axial distance area ($x/d = 0\ldots20$), especially for the scheme with a coolant supply into triangular craters.

In addition, for the scheme with a coolant supply into triangular craters, the significant increase in the span-wise film cooling uniformity is observed (Fig. 4). This is due to specific feature of the coolant jets interaction with rectangular-shaped front edge of the triangular craters. In this case there is a transformation of coolant round jets into the semi-bound two-dimensional jet over the cooled surface. This provides a more uniform coverage of the cooled surface by means of the coolant and leads to significant decrease in the intensity of "negative" vortex structures, contributing to cooling jets detachment from the cooled surface and the hot gas supply from the main flow to the cooled surface at the high blowing ratio.

Conclusions

The new method of the film cooling with coolant supply through round holes arranged in the indentations of the triangular shape significantly improves the average adiabatic film cooling efficiency both at the medium and high values of the blowing ratio ($m \geq 1.0$). This increase is by 1.5...2.7 times compared to the traditional array of inclined round holes, and by 1.3…1.8 times compared to the scheme with coolant supply into cylindrical craters. The greater film cooling efficiency with a coolant supply into triangular craters is explained by decrease in the secondary vortex structures ("kidney" vortices) and transformation of cylindrical cooling jets into two-dimensional semi-bounded jet due to interaction with front walls of triangular craters. At the high blowing ratio, the lifting force contributing to the cooling jets separation from the cooled surface is reduced and the film cooling uniformity in the span-wise direction is substantially improved due to decrease in the cooling jets discreteness over the cooled surface.

References

1. R.J. Goldstein, E.R.G. Eckert, Int. J. Heat Mass Transfer **17,** 595 (1974).
2. Y. Lu, A. Dhungel, S.V. Ekkad, R.S. Bunker, J. Turbomach. **125**, 57 (2003)
3. Y. Lu, A. Dhungel, S.V. Ekkad, R.S. Bunker, J. Turbomach. **131**, Paper 011003 (2009)
4. A.A. Khalatov, I.I. Borisov, N.A. Panchenko, Yu.J. Dashevsky, A.S. Kovalenko, Thermophys. Aeromech. **21,** 545 (2014)
5. J.R. Dorrington, D.G. Bogard, R.S. Bunker, Proc. of ASME Turbo Expo 2007, Paper GT2007–27992, 749 (2007)
6. A.A. Khalatov, S.D. Severin, M.V. Bezludna, I.V. Novohatska, State Register of Patents of Ukraine, Kiev, UA, Pat. No. 113452, (2017)
7. A.A. Khalatov, M.V. Bezliudna, Yu.J. Dashevsky, Aerospace Technic Technology **7,** 77, (2014)
8. A.A. Khalatov, I.I. Borisov, S.D. Severin, Eastern-European J. Enterprise Technologies **3,** 4 (2012)

Heat and mass transfer in the mushroom-shaped head of mantle plume

Anatoly Kirdyashkin[1], *Alexey* Kirdyashkin[1,2,*], and *Vladimir* Gurov[1]

[1]V.S. Sobolev Institute of Geology and Mineralogy, Siberian Branch of the Russian Academy of Sciences, pr. Akademika Koptyuga 3, Novosibirsk, 630090, Russia
[2]Novosibirsk State University, ul. Pirogova 2, Novosibirsk, 630090, Russia

Abstract. The results of experimental and theoretical modeling of free-convection flows in the melt of the plume conduit and in the mushroom-shaped head are presented. It was shown that the plumes with the mushroom-shaped heads can be responsible for the batholith formation. The main parameters of such plumes are estimated.

1 Introduction

The mantle plumes manifest themselves as localized conduits of heat transfer from the core-mantle boundary to the Earth's surface and represent magma melt conduits surrounded by the mantle [1]. In the model [1], a thermochemical mantle plume forms at the core–mantle boundary at the locations where local chemical doping has decreased the melting point to the value of $T_{mc} < T_1$, where T_1 is the temperature of the core-mantle boundary. In this case, the melting of the mantle matter proceeds and the plume forms. The outer core is a heat source. The chemical dope is formed by the reactions of lower mantle minerals with hydrogen and/or methane brought to the lower mantle from the outer core [1].

2 Structure and main parameters of plumes with mushroom-shaped heads

Let us consider ***a structure of a thermochemical plume.*** According to the laboratory modeling, in its vertical section, the plume conduit is a system of convection cells (Fig. 1A). In narrowed zones at the boundaries of cells, the ascending convection flow transits to the opposite side of the conduit. The boundary layer of the descending convection flow is located along the sidewall of the conduit. The conduit boundary changes with time. Melting occurs along the ascending flow domain, whereas crystallization (at the conduit boundary and in the zones of conduit narrowing) along the descending one. The conduit surface is a "traveling wave;" nonstationary conductive heat transfer takes place between the conduit boundary and the surrounding mantle. The diameter $d = 2R$ (R is the conduit radius) of the plume conduit is comparable to the diameter of the plume base. The diameter of narrowing is $d_n = 2R_n = 0.5d$ (R_n is the radius of narrowed zone). According to [2], a mantle plume

* Corresponding author: aak@igm.nsc.ru

originates and ascends from the core-mantle boundary as a thermochemical plume to the level $x_{tc} = 148 - 185$ km. For the formation and the ascent of the thermochemical plume to the level $x = x_{tc}$ the diffusion process is a controlling factor. For $x > x_{tc}$ thermal processes represent the controlling factor rather than diffusion processes. The plume lifetime is 10 – 70 Ma. We adopt, as the thermal power scale, the thermal power transferred from the plume conduit to the surrounding mantle under steady-state thermal conduction conditions. This thermal power is defined by the relation: $N1 = 0.5\pi\lambda_{sm}\Delta TH$, where λ_{sm} is the thermal

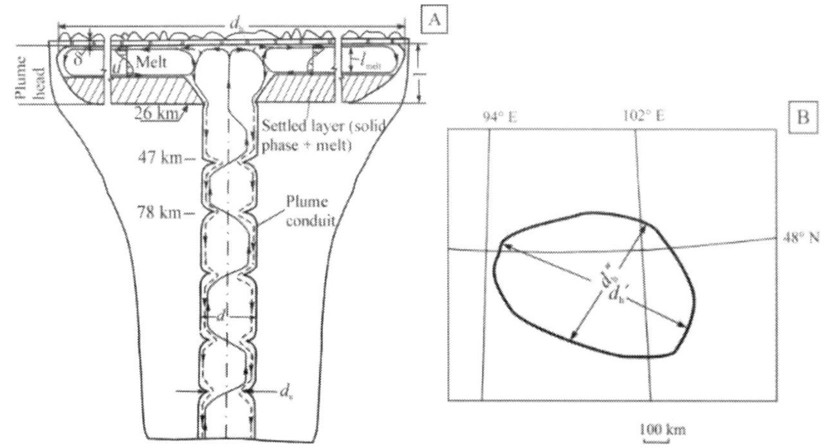

Fig. 1. Plume with the mushroom-shaped head. (A) Schematic of the thermochemical plume and free-convection flows in the plume conduit and the plume head on the data of our laboratory modeling. The scale is with reference to the diameter of the Khangai plume conduit (d = 32.5 km). (B) Contour of projection of the Khangai plume head on the Earth's surface constructed using data [3].

conductivity of the surrounding mantle averaged over the height of the plume conduit, $\Delta T = T_m - T_0$ is the temperature difference averaged over the height of the plume conduit, T_m is the temperature of the plume conduit boundary, T_0 is the temperature of the surrounding mantle, H is the height of the melted plume conduit. Let us consider the relative thermal power of the plume $\mathrm{Ka} = N/N_1$, where N is the thermal power of the plume source, i.e. the thermal power transferred from the plume base to the conduit. Thus, the Ka number refers to the ratio of the thermal power at the plume base to that released to the mantle at steady-state thermal conduction.

Further, we obtain ***relations for the thermal power and plume diameters***. The thermal power transferred from the plume base to the conduit is $N = \Delta GC/\beta$, where ΔG is the mass flow rate of magmatic melt erupted on the surface which is determined using geological data, β is the thermal expansion coefficient, and C is the mantle specific heat. We obtain the diameter of the plume conduit d using the equality [2]: $d = [N(a\nu)^{1/3}/0.045\pi\lambda\Delta T_s^{4/3}(\beta g)^{1/3}]^{1/2}$, where a is the thermal diffusivity of melt in the plume conduit, ν is the kinematic viscosity of the melt, λ is the thermal conductivity of the melt, $\Delta T_s = (T_l - T_{mc})/1.57$ is the temperature difference across the boundary layer at the plume base, and g is the gravitational acceleration.

In the case of plumes having the relative thermal power $1.9 < \mathrm{Ka} < 10$, after the melt erupts on the surface through the eruption conduit, the surrounding crustal block melts and a mushroom-shaped plume head of thickness l forms (Fig. 1A). Plumes with mushroom-shaped heads can be responsible for the formation of large intrusive bodies (batholiths). In particular, these plumes can be responsible for the formation of the largest granitic batholiths such as the Khangai (Fig. 1B) and Khentei ones. The Khangai and Khentei batholiths are located respectively in Central and North-Eastern Mongolia.

After melt erupted to the surface, the thermal power transferred to the plume roof is $\Delta N = N - N_{liml}$, where $N = \mathrm{Ka}N_l$ is the thermal power transferred from the base of the plume to its conduit, $N_{liml} = \mathrm{Ka}_{liml}N_l$ is the limiting thermal power transferred from the plume conduit to the surrounding mantle by unsteady-state thermal conduction. The value of Ka_{liml} is set equal to 1.15 in accordance with the results of laboratory modeling [2]. The limiting thermal power is $N_{liml} = 1.6 \cdot 10^{10}$ W. The thermal power transferred from the plume roof to the atmosphere through the block of thickness δ above the plume roof is much less than the thermal power ΔN. Therefore melting occurs along the base of the block located in the crust above the plume roof, resulting in the formation of a mushroom-shaped head of the plume, i.e., a root batholith (Fig. 1A).

Consider the case when the mushroom-shaped head of the plume reaches its maximum diameter $d_{h,max}$ at the instant $t = t_{melt}$. This occurs when thermal power $\Delta N = N - N_{liml}$ becomes equal to the thermal power transferred to the atmosphere through the block of thickness δ. The heat flux through the block is $q_{cb} = \lambda_\delta(\Delta T_\delta/\delta)$, where $\lambda_\delta = 3.5$ W/(m · °C) is the thermal conductivity of the block, $\Delta T_\delta = T_r - T_{sf} = 1200$ °C, T_r is the temperature of the plume roof, and T_{sf} is the temperature of the ground surface. The amount of heat transferred through the block corresponds to the thermal power $\Delta N = (\mathrm{Ka} - \mathrm{Ka}_{liml})N_l$. The surface area of the mushroom-shaped plume head $S_h = \pi d_h^2/4 = \Delta N/q_{cb}$ (d_h is the plume head diameter). Then, using the equalities for q_{cb} and ΔN, we obtain $S_h = N_l\delta(\mathrm{Ka} - \mathrm{Ka}_{liml})/\lambda_\delta\Delta T_\delta$, $d_h = [4N_l\delta(\mathrm{Ka} - \mathrm{Ka}_{liml})/\pi\lambda_\delta\Delta T_\delta]^{1/2}$.

The main parameters of plumes with mushroom-shaped heads have been estimated. Melt pressure under the plume roof exceeds the lithostatic pressure for the value ΔP [1]. In view of the necking of the plume conduit, the superlithostatic pressure is $\Delta P = \rho_0\beta g(H - x_{cr})(T_m - T_0)(R_n/R)^2$, where ρ_0 is the density of the surrounding mantle, x_{cr} is the height of the eruption conduit. The thickness of the crustal block above the mushroom-shaped plume head δ is defined by relation: $\delta = (2F/c\rho_{cb}g)^{2/5}[\lambda_\delta\Delta T_\delta/4\pi N_l(\mathrm{Ka} - \mathrm{Ka}_{liml})]^{1/5}$, where F is the total pressure force on the top of the mushroom-shaped plume head, $c = 0.28$ for a wide range of minerals [1], ρ_{cb} is the average density of the crustal block. For Khangai plume we obtain δ = 5.1 km, for Khentei plume δ = 5.9 km. For a constant thermal power of the plume ($N = \mathrm{Ka}N_l$ = const), the volume ΔV of magmatic melt intruded into the block per unit of time under the effect of the superlithostatic pressure ΔP is determined using the relation $\Delta V = \beta N/\rho C$, where ρ is the density of melt. Probably, a large mass of magma intrudes into more ancient rocks in the weakest places. It pushes those rocks apart when moving upwards. Thus, magma intrudes into fault zones in the crustal block above the plume head under the effect of superlithostatic pressure ΔP. In these zones, localized surface manifestations of batholiths are most probable (Fig. 1A). With the closure of a fault, another fault forms in the block, and magma intrudes into this new fault, etc.

For Khangai plume d = 32.5 km, d_h = 349 km, for Khentei plume d = 29 km, d_h = 329 km. The distribution of the Khangai (Fig. 1B) and Khentei batholiths is asymmetrical. The maximum size of these batholiths d_h' is observed in longitudinal direction, while the minimum size d_h'' is observed in latitudinal direction [3]. For example, for the Khangai batholith $d_h' \approx 740$ km, $d_h'' \approx 317$ km (Fig. 1B), and $d_h'/d_h'' \approx 2$. For the Khentei batholith $d_h' \approx 700$ km, $d_h'' \approx 224$ km, and $d_h'/d_h'' \approx 3$. Probably, the elongation of the batholiths in the longitudinal direction is due to the effect of the Coriolis force, which manifests itself at a relatively low kinematic viscosity of melt $\nu = 1 - 10$ m^2/s.

Within the accuracy of evaluation, the thickness of the molten crustal layer (plume head) l is comparable to the plume conduit diameter d. The solid phase fraction for the molten layer can be determined knowing the normative composition of melt. The scheme of free-convection flows in the plume conduit and in the plume head is shown on Fig. 1A. The experimental modeling shows that the ascending free-convection flow occurs along the axis of the plume conduit. The horizontal flows are due to the horizontal temperature gradient in the layer.

To a first approximation, we consider the plume head as a flat layer of thickness l, length d_h' and width d_h''. The plume head consists of two symmetric horizontal layers of length $0.5d_h'$. The thermal power transmitted to the axial plane of the layers is equal to ΔN, and consequently, the thermal power transmitted to each layer is equal to $0.5\Delta N$. Heat is removed from the plume roof having a temperature T_r = const. Experimental and theoretical investigations of free convection flows in the horizontal liquid layer heated from the butt-end are presented in [4]. Consider the maximum temperature difference $\Delta T_{max} = T_{melt} - T_r$, where T_{melt} is the temperature of melt in plume conduit. The difference ΔT_{max} is defined using the relation: $\Delta T_{max} = 8(Q/\lambda)\mathrm{Ra}_Q^{-1/3}$, where $Q = \Delta N/2d_h''$ is the thermal power delivered per meter run, $\mathrm{Ra}_Q = \beta g Q l^3/a\nu\lambda$ is the Rayleigh number. The maximum horizontal velocity of free-convection flow is constant along the layer. It is determined using the relation $u_{max} = 0.707(a/l)\mathrm{Ra}_Q^{1/3}$. The mean flow velocity is $u = 0.42u_{max}$. For the Khentei plume $N = 7.6 \cdot 10^{10}$W and $\Delta N = 6 \cdot 10^{10}$W. Then $Q = 1.3 \cdot 10^5$ W/m for d_h'' = 224 km. For $l = d = 2.9 \cdot 10^4$ m, λ = 3.5 W/(m · °C), ν = 1 – 10 m^2/s, $\beta = 10^{-5}$ °C^{-1}and $a = 10^{-6}$ m^2/s we obtain $\mathrm{Ra}_Q = 9.2 \cdot 10^{18} - 9.2 \cdot 10^{19}$. During the melting of the plume head in the crustal layer the fraction of unmelted crystallites comprises 1/3 from the overall composition of this layer. In that case, the effective kinematic viscosity of melt ranges from 3 to 30 m^2/s. The corresponding Rayleigh numbers are $\mathrm{Ra}_Q = 3.1 \cdot 10^{18} - 3.1 \cdot 10^{19}$. Then, during the melting of the plume head we have ΔT_{max} = 0.10 – 0.21 °C, $u_{max} = 3.5 \cdot 10^{-5} - 7.6 \cdot 10^{-5}$ m/s. The mean flow velocity in the upper and lower halves of the layer is $u = 1.5 \cdot 10^{-5} - 3.2 \cdot 10^{-5}$ m/s. As a consequence of the small superadiabatic temperature differences in the melt of the plume head ΔT_{max}, the melt temperature T_{melt} appears to be approximately equal to the melting temperature of the plume head material T_m. At constant quantity of heat transferred to the plume head (ΔN = const) favorable conditions for the settling of solid phase exist in the plume head within the time of batholith formation (t_b = 15 – 35 Ma). The time t_{melt} = 4.8 Ma. During the process of melting of the plume head the settling of suspended solids (refractory minerals) also proceeds. The settled particles make up the settled layer of the plume head. The residual melt forms the upper layer of thickness l_{melt} (Fig. 1A).

3 Summary

The plume conduit has the cellular structure. There are narrowed zones at the convection cell boundaries. For the relative thermal power 1.9 < Ka < 10 the flattened upper cell is mushroom-shaped. The main parameters of the plume with the mushroom-shaped head have been evaluated. These plumes can be responsible for the batholith formation.

The work was supported by state assignment project No. 0330-2016-0016.

References

1. N.L. Dobretsov, A.A. Kirdyashkin, A.G. Kirdyashkin, V.A. Vernikovsky, Lithos **100**, 66 (2008)
2. I.N. Gladkov, V.E. Distanov, A.A. Kirdyashkin, A.G. Kirdyashkin, Fluid Dynamics **47**, 433 (2012)
3. M.I. Kuzmin, V.V. Yarmolyuk, V.A. Kravchinsky, Earth Sci. Rev. **102**, 29 (2010)
4. A.A. Kirdyashkin, A.G. Kirdyashkin, N.V. Surkov, Rus. Geology Geophys. **47**, 73 (2006)

Crystallization differentiation of melt in the mushroom-shaped plume head

Anatoly Kirdyashkin[1], and *Alexey* Kirdyashkin[1,2,*]

[1]V.S. Sobolev Institute of Geology and Mineralogy, Siberian Branch of the Russian Academy of Sciences, pr. Akademika Koptyuga 3, Novosibirsk, 630090, Russia
[2]Novosibirsk State University, ul. Pirogova 2, Novosibirsk, 630090, Russia

Abstract. The influence of crystallization differentiation on the compositional changes in melt of the mushroom-shaped plume head has been clarified. According to our calculations, crystallization differentiation produces residual melt composition approaching the composition of the normal granites.

1 Introduction

The compositional changes in melt of the plume head is determined by stages: 1) after settling of refractory minerals; 2) after settling of plagioclase in the melt subsequent to the first stage. Crystallization differentiation proceeds under thermal and hydrodynamic conditions described in our previous paper [1]. The compositional change is analyzed using the estimates on the bulk composition of continental crust proposed by A. Ronov and A. Yaroshevsky [2]. The normative mineral composition is calculated according to the CIPW method using the MS Excel worksheet (author: G. Stern, Orlean, France).

2 Calculations of melt composition

Let us determine ***the melt composition after the settling of unmelted (refractory) minerals.*** Consider the melt drawn from the plume conduit to the plume roof. At the melt temperature T_{melt} = 1410°C the refractory minerals (RM) remain in the form of suspended solids in the melt volume. The following minerals will remain unmelted (Table 1, column 1): diopside (Di), hyperstene (Hyp), magnetite (Mgt), ilmenite (Ilm), apatite (Ap). The melt consists of orthoclase (Or), albite (Ab) and anorthite (An) components. On the average, the density of RM is approximately 25% above the melt density. Thus, the solid particles (crystallites) of RM will settle to the base of the plume head (Fig. 1A). The settling time of RM is 10 – 1600 years. Normative quartz Qtz (tridymite, 13.6 %) (Table 1, column 2) also remains as solid phase. Its density is lower than the melt density. Thus tridymite will float to the plume roof and will be suspended in the free-convection flows of the plume head.

Next the percent composition of the oxides comprising the solid phase (RM) is to be determined. We shall exemplify determination of the percent composition of oxides in

* Corresponding author: aak@igm.nsc.ru

diopside (Di = 8.9 %). The chemical formula of diopside can be represented in the form of oxides: CaO · MgO · $2SiO_2$. Molecular weight of diopside is M_{Di} = 216.6. The percent composition of the oxides for Di is given in Table 2.

Table 1. The composition of melt in the plume head for T_{melt} = 1410°C.

In the form of oxides					
1	**2**	**3**	**4**	**5**	**6**
Oxide	**Bulk composition by [2] [%]**	**After settling of RM [%]**	**The melt layer (l_{melt}) [%]**	**After settling of plagioclase [%]**	**The melt layer (l'_{melt}) [%]**
SiO_2	59.3	43.9	65.3	41.9	76.1
TiO_2	0.7	0.0	0.0	0.0	0.0
Al_2O_3	15.0	13.9	20.6	7.5	13.7
Fe_2O_3	2.4	0.0	0.0	0.0	0.0
FeO	5.6	0.0	0.0	0.0	0.0
MnO	0.1	0.1	0.2	0.1	0.2
MgO	4.9	0.6	0.9	0.0	0.0
CaO	7.2	4.5	6.8	1.4	2.5
Na_2O	2.5	2.3	3.4	1.6	2.9
K_2O	2.1	1.9	2.8	2.5	4.6
P_2O_5	0.2	0.0	0.0	0.0	0.0
Σ		67.2		55.1	
Normative composition					
Component	**wt. %**		**wt. %**		**wt. %**
Qtz	13.6		18.7		36.4
Or	12.4		16.8		27.2
Ab	21.2		29.1		24.5
An	23.5		32.4		10.7
Di	8.9		0.9		-
Hyp	15.2		-		-
Mgt	3.5		-		-
Ilm	1.3		-		-
Ap	0.5		-		-

Table 2. Oxide content for diopside.

Oxide	M_{ox}	M_{ox}/M_{Di}	$(M_{ox}/M_{Di}) \times 8.9$ %
CaO	56.1	0.26	2.3
MgO	40.3	0.19	1.7
$2SiO_2$	120.2	0.55	4.9

The oxide content of hyperstene (Hyp = 15.2 %) is determined using the same procedure as for Di. Hyperstene has the formula $(Mg, Fe)_2Si_2O_6$ (Fe = 50 %), and in the form of oxides MgO · FeO · $2SiO_2$ (M_{Hyp} = 232.3). Similarly, the oxide content is determined for Ilm (FeO · TiO_2, 1.3 %) and for Mgt (FeO · Fe_2O_3, 3.5 %). The total content for each oxide is the following: SiO_2 = 12.1 %, FeO = 6.4 %, Fe_2O_3 = 2.4 %, CaO = 2.3 %, MgO = 4.3 %, TiO_2 = 0.7 %. The settled RM, which form the lower layer of the plume head (Fig. 1A) comprise in aggregate 29.4% (Table 1, column 2). For melt remaining after the settling of RM we have normative composition given in Table 1, column 2. For a system of spheres, approximately 26% of melt remains in the intercrystalline space of the layer composed of crystallites (as a first approximation, spherical particles) of RM. The indicated weight content is 0.26 × 29.4 = 7.6 %. Then the weight percentage of each melt component is multiplied by 0.076. For example, Qtz (SiO_2) = 0.076 × 13.6 % = 1.0 %. For the other components we have: orthoclase Or (K_2O · Al_2O_3 · $6SiO_2$) = 1.0 %, albite Ab (Na_2O · Al_2O_3 · $6SiO_2$) = 1.6 %, and anorthite An (CaO · Al_2O_3 · $2SiO_2$) = 1.8 %.

 DOI: 10.1051/matecconf/201711509005

The total oxide content in the intercrystalline space can be determined using the above procedure. For oxides we obtain: SiO_2 = 3.3 %, Al_2O_3 = 1.1 %, CaO = 0.4 %, Na_2O = 0.2 %, K_2O = 0.2 %. We found the sum of the contents of each oxide for settled minerals and melt in the intercrystalline space: SiO_2 = 15.4 %, TiO_2 = 0.7 %, Al_2O_3 = 1.1 %, Fe_2O_3 = 2.4 %, FeO = 6.4 %, MgO = 4.3 %, CaO = 2.7%, Na_2O = 0.2 %, K_2O = 0.2 %, P_2O_5 = 0.2 %. Then, in terms of oxides, the overall content of the solid phase and the melt in the intercrystalline space comprises 33.6%. Hence, one can evaluate the thickness of the melt layer l_{melt} overlain the settled RM: $l_{melt} = l(1 - 0.336) = 0.664l$.

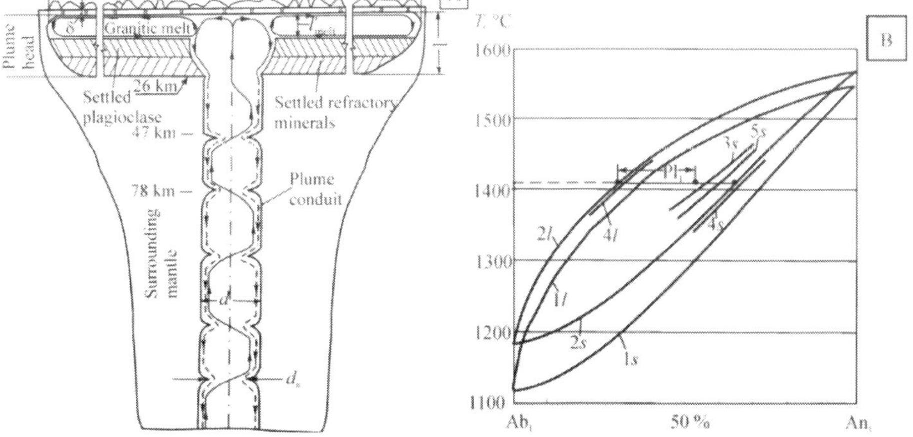

Fig. 1. The influence of crystallization differentiation on the plume head composition and structure. (A) Schematic of the plume with the mushroom-shaped head. The plume conduit diameter is d = 29 km (Khentei plume, North-Eastern Mongolia). (B) Phase diagram of plagioclase feldspars for different pressures. Solidus and liquidus lines are denoted by s and l , respectively. The numerals relate to the pressures: 1 to P = 1 bar [3], 2 to 6.3 kbar, 3 to 12 kbar, 4 to 5 kbar, 5 to 10 kbar.

For the plume head of thickness $l = d$ = 29 km (Khentei plume, Fig. 1A), l_{melt} = 19.3 km. We subtract the percent composition of the corresponding deposited oxide (with regard to the melt in the intercrystalline space) from the bulk composition of continental crust. Then we obtain the percent composition of residual melt (Table 1, column 3). Let us determine the oxide content of the melt layer of thickness l_{melt}. In Table 1 we determine the sum of column 3 ($\sum$ = 67.2%) and further multiply it by the ratio 100/67.2. The composition of the melt layer (in the oxide form) is presented in Table 1 (column 4). The corresponding normative composition is also presented in column 4. The normative quartz (Qtz, 18.7 %) occurs in the melt as the suspended solid.

Let us determine ***the composition of the melt of thickness*** l'_{melt}, ***which remains after settling of plagioclase crystallites*** (Fig. 1A). Plagioclase (Pl) consists of a solid solution between the albite and anorthite end-members. The content of the plagioclase component in the residual melt is evaluated using the phase diagram of the system Ab_1-An_1, where $Ab_1 = Ab/(Ab + An)$, $An_1 = An/(Ab+An)$, Ab is the content of albite, An is that of anorthite of the crustal layer subjected to melting (Fig. 1B). As a first approximation, it is assumed that the behavior of the phase diagram Ab_1-An_1 is unaffected by the presence of orthoclase component (16.8 %) in the melt. As indicated in our previous paper [1], the plume head is modeled by the flat liquid layer. The temperature of melt in the plume head is constant accurate to 0.2 °C. Thus, at the melt temperature of 1410°C the crystallized plagioclase occurs in relatively isothermal conditions. However, the crystallized plagioclase occurs in unstable state in the gravity field because of the density difference between the melt and the crystallized plagioclase. For example, considering the density of albite (2.6 g/sm^3) and anorthite (2.75 g/sm^3) for plagioclase N 30 containing 30 % An_1 (Fig. 1B, the solidus line)

$\rho_1 = 2.65$ g/sm^3, whereas for plagioclase N 70 containing 70 % An_1 (Fig. 1B, the liquidus line) $\rho_2 = 2.71$ g/sm^3. The density difference is $\Delta\rho = \rho_2 - \rho_1 = 0.06$ g/sm^3, that is, the density contrast is 2.2 %. The settling of solid particles of plagioclase proceeds under the conditions of horizontal free-convection flows [1]. Since heat is removed from the base of the plume head, with increasing pressure, that is, going toward the base of the plume head the melt becomes enriched with the crystallized plagioclase. The quantity of the crystallized plagioclase is determined under the following conditions: $P_b = 6.3$ kbar and $T_{melt} = 1410$ °C. The plagioclase composition corresponds to N 65, while the melt composition corresponds to N 30 (Fig. 1B). The percent composition of the crystallized plagioclase is the following: $Ab_{cryst} = Ab_{1cryst}(Ab + An)Pl_{1cryst} = 35\ \% \times 0.615 \times 0.65 = 14\ \%$, $An_{cryst} = An_{1cryst}(Ab + An)Pl_{1cryst} = 65\ \% \times 0.615 \times 0.65 = 26\ \%$. For the settled plagioclase we have $Pl_{1cryst}(Ab_{cryst} + An_{cryst}) = 0.65 \times 61.5\ \% = 40\ \%$. The quantity of melt which remains with the settled plagioclase (in the intercrystalline space) is $0.26 \times 40\ \% = 10.4\ \%$. Then the weight percentage of each melt component is multiplied by 0.104. For example, the melt in the intercrystalline space contains Qtz = $0.104 \times 18.72 = 2.0$ %. Similarly, Or = 1.8 %, Ab = 1.6 %, and An = 0.7 %.

The oxide content of the settled plagioclase is determined using the procedure described above at the first stage of calculations. Then the total oxide content of the settled plagioclase and melt in the intercrystalline space is determined. The fraction of the solid phase and melt in the intercrystalline space comprises 0.44. The thickness of the residual melt layer l'_{melt} (Fig. 1A) is equal to $l'_{melt} = (1 - 0.44)l_{melt} = 0.56l_{melt}$. For $l_{melt} = 19.3$ km, $l'_{melt} = 10.8$ km. The oxide content of the residual melt (Table 1, column 6) is calculated using the procedure identical to that for the melt composition after the settling of RM. The SiO_2 content is 76.1 %. The SiO_2 content of normal granites is 70–73 % [4]. The normative composition of the melt is also presented in the Table 1 (column 6). The normative contents of albite and anorthite are Ab = 24.5 %, and An = 10.7 %. The normative calculations for the normal granites [4] show that the albite content is Ab = 15–19 % and the anorthite content is An = 1–9 %. These values agree with our calculations for the melt components.

3 Summary

Our calculations for the components of the residual melt in the plume head agree with the normative calculations for the normal granites. The SiO_2 content of the residual melt is 76.1%. It is in agreement with the SiO_2 content of normal granites. Thus, relying on the proposed model of the plume with the mushroom-shaped head the normative composition of the melt complying with the composition of normal granites has been obtained.

The work was supported by state assignment project No. 0330-2016-0016.

References

1. A.G. Kirdyashkin, A.A. Kirdyashkin, V.V. Gurov, MATEC Web of Conferences (this issue)
2. G.V. Voitkevich, A.V. Kokin, A.E. Miroshnikov, V.G. Prokhorov, *Handbook on Geochemistry* (Nedra, Moscow, 1990) [in Russian]
3. N.L. Bowen, Amer. J. Sci., **35**, 577 (1913)
4. G.M. Saranchina, N.F. Shinkarev, *Petrography of Magmatic and Metamorphic Rocks* (Nedra, Leningrad, 1967) [in Russian]

Features of flame stabilization in near wall flows

Vladimir Lukashov*, and *Vladimir* Terekhov

Institute of Thermophysics SB RAS, 630090, Novosibirsk, 1 Lavrent'iev Ave., Russia

Abstract. The results of the experimental and numerical study of the influence of obstacle geometry on the conditions of flame stabilisation in the turbulent boundary layer with fuel injection near a porous surface are presented. The features of the detachment zones formation in the turbulent reacting flow behind the obstacle are studied. The obtained results are compared with the data on flame-off conditions when flowing over a flat surface.

1 Introduction

An active or passive impact on flow in the flame front edge can effectively control the combustion in the boundary layer. In particular, as it was shown in the works of E.P. Volchkov [1] when burning near a flat surface, through which the gaseous fuel was injected, the features of the flow formation played an important role. In the same paper, the regularities of the flame-off conditions were experimentally established, and the model was proposed. It was based on the diffusion combustion approximation, the boundary layer theory and the assumption that extinction occurs when the flame front reaches the wall. The range of conditions in which the combustion is stable near a flat surface can be significantly expanded by using the stabilizing devices. Traditionally, ribs or back steps of various geometries were used to form a recirculation zone with an extended residence time of initial reactants and reaction products mixture. The heat release and transverse momentum flow caused by injection of the fuel mixture can substantially change the regularities of this class of flows [2]. A very limited amount of works were devoted to the investigation of conditions of flame detachment in a separated flow near a flat surface with injection. In [3-5], based on visual observations, the diagrams of combustion and flame-off modes were obtained for some hydrocarbon fuels and the geometry of the barriers forming the detached flow zones. There are other ways to stabilize the flame. In particularly the detached rib may be one of the promising areas. The structure of the flow and heat transfer over a detached rib for non-reactive flows was numerically studied in [6]. It was shown that an increase in the gap between the rib and the surface leads to a change in the structure of the recirculation region. The regimes in which there was an increase in heat exchange with the wall were observed in comparison with the case of a rib placed on the surface.

In this paper, we have determined the conditions of flame-off at the detached rib located above the wall, while the main attention was paid to the peculiarities of the flow in the vicinity of the obstacle.

*Corresponding author: luka@itp.nsc.ru

DOI: 10.1051/matecconf/201711509006

2 Experimental Installation

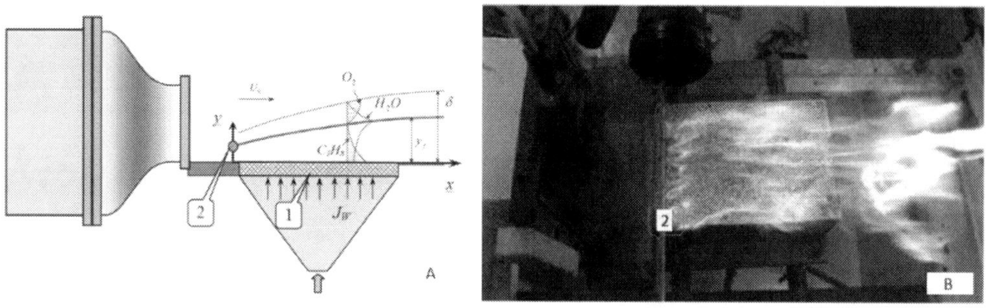

Fig. 1. The scheme of the working section of the unit for investigating the wall flow with injection and gas fuel combustion in the air stream (A). Photo (B) of the wall stream with propane injection and burning, top view.

The experimental studies were carried out in a subsonic wind tunnel with a channel section at the entrance to the working part 108x108 mm. The speed of air flow over the plate varied in the range 1...20 m/s. The diagram of the working part is shown in Figure 1. The fuel was uniformly blown into the boundary layer through the lower horizontal porous plate (1) with a plan dimension S = 95x145 mm. The injected fuel was propane (or methane) diluted with CO2. To prepare the fuel mixture, digital gas flow regulators were used. Immediately before the beginning of the porous section, a rib (2), being a ceramic cylinder 5 mm in diameter, was installed. The rib was located directly on the surface, or was fixed at a height of up to 3 mm above the wall (detached rib).

To measure the velocity distribution in the reacting flow the PIV method was used. The plane of the light sheet was oriented perpendicular to the permeable surface, which allowed obtaining data on a two-dimensional flow field (longitudinal and vertical velocity components). Because of the peculiarities of the method used to diagnose the dynamics of the wall flow, the measurements were carried out sequentially and statistically independently in overlapping regions of the space of ~ 40x40 mm in size. When using the optical method in the wall flow above the permeable wall, it is required to ensure a uniform dusting of the flow by light scattering particles. For this purpose, separate, independent dusting of TiO2 particles with both the main stream and the fuel mixture was used.

3 Theoretical model

Under the conditions of the experimental studies the numerical simulation was carried out. This allowed obtaining a more complete picture of the processes in the turbulent reactive separated flow in the boundary layer behind the barrier. In particular, the simulation allowed obtaining detailed information on the distribution of concentrations of all the mixture components, refining the data on the parameters in the immediate vicinity of the walls and identifying some other factors that were not directly measured. Numerical simulation in this paper was based on the URANS approach. We solved non-stationary averaged equations of Navier-Stokes, energy and concentrations. As a model of turbulence, the SST k-model was used, thermal and diffusion turbulent flows were determined according to the gradient model. The interaction of turbulent pulsations and combustion was modeled according to the concept of the EDC vortex dissipation.

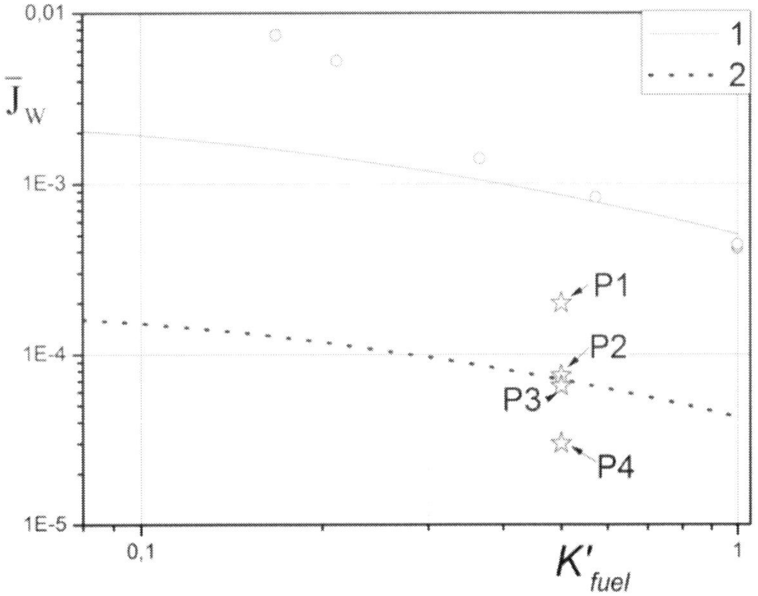

Fig. 2. Range of parameters of stable diffusion combustion in the boundary layer with the injection of the hydrocarbon mixture with CO_2 (above the line). Lines: 1– CH_4, 2– C_3H_8. Figures - the experimental data of the authors for the CH_4/CO_2 fuel mixture, points P1...P4 – simulation results for propane.

4 Results and discussion

In the diffusion combustion approximation in the boundary layer with injection of a hydrocarbon fuel mixture diluted with an inert diluent, there is a unique relationship [1] for a relative intensity of the injection $\bar{J}_W$; and K^{I}_{fuel} is the mass fraction of the fuel component in the fuel mixture. The estimates for a number of hydrocarbons are shown in Figure 2. It is seen that at $K^{I}_{fuel} > 0.4$, these boundaries adequately describe the experimental data. In order to expand the combustion area, it is necessary to use a flame stabilizer.

First of all one can note that in the boundary layer without a barrier only one stable combustion mode is observed [1] when the flame develops starting from a certain distance from the start of the field of injection. In the case of the detached flow in such regimes, a few attachment points were observed in addition to the previously described combustion regime in the boundary layer. At high blowing rates and high velocities of the incoming flow the flame can be stabilized in the separation zone behind the obstacle, and the front of the flame develops downstream the obstacle (P1 point on Figure 2). The second mode, in which the combustion was divided into two regions: directly behind the barrier in the recirculation zone and substantially lower in the flow in the boundary layer, was observed as the intensity of the injection was reduced (P2 point). With further reduction in fuel injection through the wall, a third regime was possible, in which combustion was observed only in the recirculation region behind the obstacle (P3 point). In the conditions indicated as P4 the combustion became impossible. The detachment of the rib from the surface further reduces the minimum injection intensity sufficient to ensure combustion. This is due to the peculiarities of forming the detached wall flow, which significantly change if the rib is placed above the section of porous blowing.

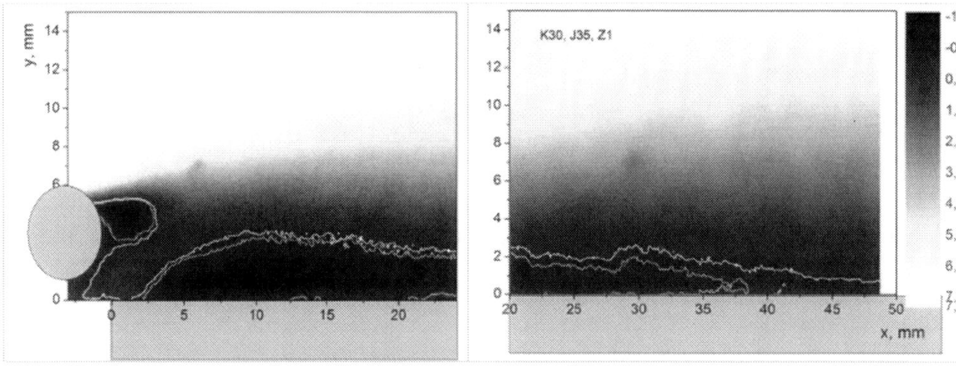

Fig.3. The measured longitudinal velocity component. The fuel mixture C_3H_8/CO_2 with propane mass fraction K^1_{C3H8}=0.3, injection $\bar{J}_w$ =0.35%, air flow velocity U_0=7.5 m/s.

Figure 3 shows the results of measurements of the field of the longitudinal velocity component when the propane-air flame is stabilized behind the detached rib. It is evident that two regions of the recurrent flow are formed by isotachs ±0.1 m/s (light lines in the picture).

The visible flame leading edge in the section of the PIV system laser sheet plane was at 6-8 mm downstream from the surface of the detached rib. The formation of the region of the recirculating near wall flow is apparently determined by two factors: the heat release in the flame front and the transverse flow of matter specified by injection of the fuel mixture.

5 Conclusion

The results of the experimental and numerical study of the turbulent reacting flow behind a detached rib near a porous surface with fuel injected through the wall have been analyzed. In a separated flow, unlike a boundary layer on a flat surface, several combustion regimes may be observed. The diagram was obtained for the flame-off condition both for the case of a plane surface and for a surface with an obstacle above it, and a comparison was made with the model proposed by E. Volchkov [1].

This work was supported by the RFBR grant № 15-08-05662a.

References

1. E.P. Volchkov, V.V. Lukashov, V.V. Terekhov, K. Hanjalic, Comb. Flame **160** 1999 (2013)
2. V.V. Terekhov, V.I. Terekhov, J. Appl. Mech. Tech. Phys. **58**, 254 (2017)
3. A. Ramachandra, B.N. Raghunandan, Comb. Sci.Techn. **36** 109 (1983)
4. B.N. Raghunandan, G.P. Yogesh, XXII Int. Symp. on Combust. 1501 (1988)
5. T.A. Rohmat, H. Katoh, T. Obara, T. Yoshihashi, S. Ohyagi, AIAA J. **36**, 1945 (1998)
6. T.V. Bogatko, V.I. Terekhov, J. Appl. Mech. and Tech. Physics **56**, 57 (2015)

AUTHOR INDEX

AUTHOR INDEX

AUTHOR INDEX